Instructor's

Elementary Linear Algebra
A Matrix Approach

Second Edition

Spence   Insel   Friedberg

PEARSON
Prentice Hall

Upper Saddle River, NJ 07458

Editorial Director, Computer Science, Engineering, and Advanced Mathematics: *Marcia J. Horton*
Senior Editor: *Holly Stark*
Editorial Assistant: *Jennifer Lonschein*
Senior Managing Editor: *Scott Disanno*
Supplement Cover Designer: *Daniel Sandin*
Manufacturing Buyer: *Lisa McDowell*

PEARSON
Prentice Hall

© 2008 by Pearson Education, Inc.
Pearson Prentice Hall
Pearson Education, Inc.
Upper Saddle River, NJ 07458

All rights reserved. No part of this book may be reproduced in any form or by any means, without permission in writing from the publisher.

The author and publisher of this book have used their best efforts in preparing this book. These efforts include the development, research, and testing of the theories and programs to determine their effectiveness. The author and publisher make no warranty of any kind, expressed or implied, with regard to these programs or the documentation contained in this book. The author and publisher shall not be liable in any event for incidental or consequential damages in connection with, or arising out of, the furnishing, performance, or use of these programs.

Pearson Prentice Hall™ is a trademark of Pearson Education, Inc.

This work is protected by United States copyright laws and is provided solely for the use of instructors in teaching their courses and assessing student learning. Dissemination or sale of any part of this work (including on the World Wide Web) will destroy the integrity of the work and is not permitted. The work and materials from it should never be made available to students except by instructors using the accompanying text in their classes. All recipients of this work are expected to abide by these restrictions and to honor the intended pedagogical purposes and the needs of other instructors who rely on these materials.

Printed in the United States of America
10 9 8 7 6 5 4 3 2 1

ISBN 0-13-230752-9
       978-0-13-230752-9

Pearson Education Ltd., *London*
Pearson Education Australia Pty. Ltd., *Sydney*
Pearson Education Singapore, Pte. Ltd.
Pearson Education North Asia Ltd., *Hong Kong*
Pearson Education Canada, Inc., *Toronto*
Pearson Educación de Mexico, S.A. de C.V.
Pearson Education—Japan, *Tokyo*
Pearson Education Malaysia, Pte. Ltd.
Pearson Education, Inc., *Upper Saddle River, New Jersey*

Contents

Preface   iv

1 Matrices, Vectors, and Systems of Linear Equations   1

- 1.1 Matrices and Vectors 1
- 1.2 Linear Combinations, Matrix-Vector Products, and Special Matrices 3
- 1.3 Systems of Linear Equations 6
- 1.4 Gaussian Elimination 9
- 1.5 Applications of Systems of Linear Equations 12
- 1.6 The Span of a Set of Vectors 15
- 1.7 Linear Dependence and Linear Independence 18
- Chapter 1 Review Exercises 22
- Chapter 1 MATLAB Exercises 25

2 Matrices and Linear Transformations   27

- 2.1 Matrix Multiplication 27
- 2.2 Applications of Matrix Multiplication 29
- 2.3 Invertibility and Elementary Matrices 33
- 2.4 The Inverse of a Matrix 39
- 2.5 Partitioned Matrices and Block Multiplication 44
- 2.6 The LU Decomposition of a Matrix 47
- 2.7 Linear Transformations and Matrices 52
- 2.8 Composition and Invertibility of Linear Transformations 57
- Chapter 2 Review Exercises 61
- Chapter 2 MATLAB Exercises 63

3 Determinants   68

- 3.1 Cofactor Expansion 68
- 3.2 Properties of Determinants 70
- Chapter 3 Review Exercises 74
- Chapter 3 MATLAB Exercises 76

4 Subspaces and Their Properties — 77

- 4.1 Subspaces 77
- 4.2 Basis and Dimension 81
- 4.3 The Dimension of Subspaces Associated with a Matrix 85
- 4.4 Coordinate Systems 90
- 4.5 Matrix Representations of Linear Operators 94
- Chapter 4 Review Exercises 100
- Chapter 4 MATLAB Exercises 103

5 Eigenvalues, Eigenvectors, and Diagonalization — 106

- 5.1 Eigenvalues and Eigenvectors 106
- 5.2 The Characteristic Polynomial 108
- 5.3 Diagonalization of Matrices 113
- 5.4 Diagonalization of Linear Operators 117
- 5.5 Applications of Eigenvalues 122
- Chapter 5 Review Exercises 130
- Chapter 5 MATLAB Exercises 132

6 Orthogonality — 135

- 6.1 The Geometry of Vectors 135
- 6.2 Orthogonal Vectors 141
- 6.3 Orthogonal Projections 145
- 6.4 Least-Squares Approximations and Orthogonal Projection Matrices 153
- 6.5 Orthogonal Matrices and Operators 155
- 6.6 Symmetric Matrices 161
- 6.7 Singular Value Decomposition 168
- 6.8 Principal Component Analysis 176
- 6.9 Rotations of $\mathcal{R}^3$ and Computer Graphics 179
- Chapter 6 Review Exercises 185
- Chapter 6 MATLAB Exercises 189

7 Vector Spaces — 195

- 7.1 Vector Spaces and Their Subspaces 195
- 7.2 Linear Transformations 203
- 7.3 Basis and Dimension 209
- 7.4 Matrix Representations of Linear Operators 217
- 7.5 Inner Product Spaces 223
- Chapter 7 Review Exercises 228
- Chapter 7 MATLAB Exercises 234

Preface

This Instructor's Manual contains solutions to the exercises in the second edition of *Elementary Linear Algebra: A Matrix Approach*. It is intended for the use of instructors rather than students, and so many solutions are written more succinctly than those in the Student Solutions Manual (ISBN 0-13-239734-X). In a cluster of similar exercises (such as Exercises 27–34 in Section 1.4), we usually work only one or two in detail and provide answers to the others. The Student Solutions Manual, which is available for student purchase, contains detailed solutions to selected odd-numbered exercises.

Additional materials for use with our book are available at

www.math.ilstu.edu/matrix

On this site, you will find data files for the technology exercises in our book that can be used with MATLAB or Texas Instrument calculators. There is also an appendix on mathematical proof, written by the authors, for use in a linear algebra course in which mathematical proof is an emphasis.

Other resources for an instructor are available on the publisher's website, whose address is

www.prenhall.com/spence

Planning Your Course

The chart below lists the sections of the text, categorized as essential material and supplementary material/applications. The 26 sections listed as essential material contain the material described in the Linear Algebra Curriculum Study Group's core syllabus as well as a thorough introduction to linear transformations. Some of these sections contain optional subsections (for example, Sections 3.1, 3.2, and 5.2) that can be included or excluded at the discretion of the instructor. The sections listed as supplementary material/applications may also be omitted depending on the nature and objectives of your course. In a semester course of 3 or 4 hours, there should be time to include some of the supplementary material or applications. We believe that a first course in linear algebra is strengthened significantly by the inclusion of applications and therefore recommend that, whenever possible, at least one application from each of Sections 1.5, 2.2, and 5.5 be included.

Essential Material	Supplementary Material and Applications
1.1 Matrices and Vectors	
1.2 Linear Combinations, Matrix-Vector Products, and Special Matrices	
1.3 Systems of Linear Equations	
1.4 Gaussian Elimination	
	1.5 Applications of Systems of Linear Equations
1.6 The Span of a Set of Vectors	
1.7 Linear Dependence and Linear Independence	
2.1 Matrix Multiplication	
	2.2 Applications of Matrix Multiplication
2.3 Invertibility and Elementary Matrices	
2.4 The Inverse of a Matrix	
	2.5 Partitioned Matrices and Block Multiplication
	2.6 The *LU* Decomposition of a Matrix
2.7 Linear Transformations and Matrices	
2.8 Composition and Invertibility of Linear Transformations	
3.1 Cofactor Expansion	
3.2 Properties of Determinants	
4.1 Subspaces	
4.2 Basis and Dimension	
4.3 The Dimensions of Subspaces Associated with a Matrix	
4.4 Coordinate Systems	
	4.5 Matrix Representations of Linear Operators
5.1 Eigenvalues and Eigenvectors	
5.2 The Characteristic Polynomial	
5.3 Diagonalization of Matrices	
	5.4 Diagonalization of Linear Operators
	5.5 Applications of Eigenvalues

Essential Material	**Supplementary Material and Applications**

Essential Material

- **6.1** The Geometry of Vectors
- **6.2** Orthogonal Vectors
- **6.3** Orthogonal Projections
- **6.4** Least-Squares Approximations and Orthogonal Projection Matrices
- **6.5** Orthogonal Matrices and Operators
- **6.6** Symmetric Matrices

Supplementary Material and Applications

- **6.7** Singular Value Decomposition
- **6.8** Principal Component Analysis
- **6.9** Rotations of $\mathcal{R}^3$ and Computer Graphics

- **7.1** Vector Spaces and Their Subspaces
- **7.2** Linear Transformations
- **7.3** Basis and Dimension
- **7.4** Matrix Representations of Linear Operators
- **7.5** Inner Product Spaces

- **Appendix A** Sets
- **Appendix B** Functions
- **Appendix C** Complex Numbers
- **Appendix D** MATLAB
- **Appendix E** The Uniqueness of the Reduced Row Echelon Form

Chapter 1

Matrices, Vectors, and Systems of Linear Equations

1.1 MATRICES AND VECTORS

1. $\begin{bmatrix} 8 & -4 & 20 \\ 12 & 16 & 4 \end{bmatrix}$

2. $\begin{bmatrix} -2 & 1 & -5 \\ -3 & -4 & -1 \end{bmatrix}$

3. $\begin{bmatrix} 6 & -4 & 24 \\ 8 & 10 & -4 \end{bmatrix}$

4. $\begin{bmatrix} 8 & -3 & 11 \\ 13 & 18 & 11 \end{bmatrix}$

5. $\begin{bmatrix} 2 & 4 \\ 0 & 6 \\ -4 & 8 \end{bmatrix}$

6. $\begin{bmatrix} 4 & 7 \\ -1 & 10 \\ 1 & 9 \end{bmatrix}$

7. $\begin{bmatrix} 3 & -1 & 3 \\ 5 & 7 & 5 \end{bmatrix}$

8. $\begin{bmatrix} 4 & 7 \\ -1 & 10 \\ 1 & 9 \end{bmatrix}$

9. $\begin{bmatrix} 2 & 3 \\ -1 & 4 \\ 5 & 1 \end{bmatrix}$

10. $\begin{bmatrix} 1 & -1 & 7 \\ 1 & 1 & -3 \end{bmatrix}$

11. $\begin{bmatrix} -1 & -2 \\ 0 & -3 \\ 2 & -4 \end{bmatrix}$

12. $\begin{bmatrix} -1 & -2 \\ 0 & -3 \\ 2 & -4 \end{bmatrix}$

13. $\begin{bmatrix} -3 & 1 & -2 & -4 \\ -1 & -5 & 6 & 2 \end{bmatrix}$

14. $\begin{bmatrix} -12 & 0 \\ 6 & 15 \\ -3 & -9 \\ 0 & 6 \end{bmatrix}$

15. $\begin{bmatrix} -6 & 2 & -4 & -8 \\ -2 & -10 & 12 & 4 \end{bmatrix}$

16. $\begin{bmatrix} -8 & 4 & -2 & -0 \\ 0 & 10 & -6 & 4 \end{bmatrix}$

17. not possible

18. $\begin{bmatrix} 7 & -3 & 3 & 4 \\ 1 & 0 & -3 & -4 \end{bmatrix}$

19. $\begin{bmatrix} 7 & 1 \\ -3 & 0 \\ 3 & -3 \\ 4 & -4 \end{bmatrix}$

20. $\begin{bmatrix} 1 & 1 & 4 & 12 \\ 3 & 25 & -24 & -2 \end{bmatrix}$

21. not possible

22. $\begin{bmatrix} 12 & 4 \\ -4 & 20 \\ 8 & -24 \\ 16 & -8 \end{bmatrix}$

23. $\begin{bmatrix} -7 & -1 \\ 3 & 0 \\ -3 & 3 \\ -4 & 4 \end{bmatrix}$

24. $\begin{bmatrix} -7 & -1 \\ 3 & 0 \\ -3 & 3 \\ -4 & 4 \end{bmatrix}$

25. -2

26. 0

27. $\begin{bmatrix} 3 \\ 0 \\ 2\pi \end{bmatrix}$

28. $\begin{bmatrix} -2 \\ 1.6 \\ 5 \end{bmatrix}$

29. $\begin{bmatrix} 2 \\ 2e \end{bmatrix}$

30. $\begin{bmatrix} 0.4 \\ 0 \end{bmatrix}$

31. $[2 \ -3 \ 0.4]$

32. $[2e \ 12 \ 0]$

33. $\begin{bmatrix} 150 \\ 150\sqrt{3} \\ 10 \end{bmatrix}$ mph

34. (a) The swimmer's velocity is $\mathbf{u} = \begin{bmatrix} \sqrt{2} \\ \sqrt{2} \end{bmatrix}$ mph.

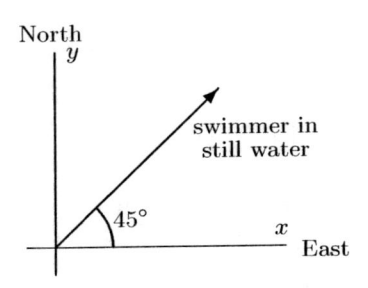

Figure for Exercise 34(a)

(b) The water's velocity is $\mathbf{v} = \begin{bmatrix} 0 \\ 1 \end{bmatrix}$ mph. So the new velocity of the swimmer is $\mathbf{u} + \mathbf{v} = \begin{bmatrix} \sqrt{2} \\ \sqrt{2}+1 \end{bmatrix}$ mph. The corresponding speed is $\sqrt{5 + 2\sqrt{2}} \approx 2.798$ mph.

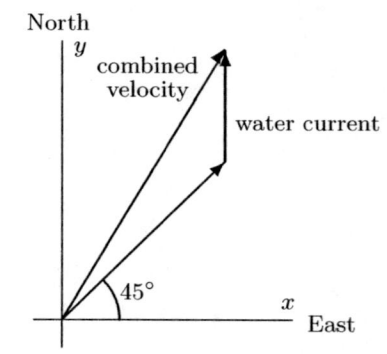

Figure for Exercise 34(b)

35. (a) $\begin{bmatrix} 150\sqrt{2} + 50 \\ 150\sqrt{2} \end{bmatrix}$ mph

    (b) $50\sqrt{37 + 6\sqrt{2}} \approx 337.21$ mph

36. The three components of the vector represent, respectively, the average blood pressure, average pulse rate, and the average cholesterol reading of the 20 people.

37. True   38. True   39. True

40. False, a scalar multiple of the zero matrix is the zero matrix.

41. False, the transpose of an $m \times n$ matrix is an $n \times m$ matrix.

42. True

43. False, the rows of B are 1×4 vectors.

44. False, the $(3,4)$-entry of a matrix lies in row 3 and column 4.

45. True

46. False, an $m \times n$ matrix has mn entries.

47. True   48. True   49. True

50. False, matrices must have the same size to be equal.

51. True   52. True   53. True

54. True   55. True   56. True

57. Suppose that A and B are $m \times n$ matrices.
    (a) The jth column of $A + B$ and $\mathbf{a}_j + \mathbf{b}_j$ are $m \times 1$ vectors. The ith component of the jth column of $A + B$ is the (i,j)-entry of $A+B$, which is $a_{ij}+b_{ij}$. By definition, the ith components of $\mathbf{a}_j$ and $\mathbf{b}_j$ are a_{ij} and b_{ij}, respectively. So the ith component of $\mathbf{a}_j + \mathbf{b}_j$ is also $a_{ij} + b_{ij}$. Thus the jth columns of $A + B$ and $\mathbf{a}_j + \mathbf{b}_j$ are equal.

    (b) The proof is similar to the proof of (a).

58. Since A is an $m \times n$ matrix, $0A$ is also an $m \times n$ matrix. Because the (i,j)-entry of $0A$ is $0a_{ij} = 0$, we see that $0A$ equals the $m \times n$ zero matrix.

59. Since A is an $m \times n$ matrix, $1A$ is also an $m \times n$ matrix. Because the (i,j)-entry of $1A$ is $1a_{ij} = a_{ij}$, we see that $1A$ equals A.

60. Because both A and B are $m \times n$ matrices, both $A+B$ and $B+A$ are $m \times n$ matrices. The (i,j)-entry of $A+B$ is $a_{ij}+b_{ij}$, and the (i,j)-entry of $B+A$ is $b_{ij}+a_{ij}$. Since $a_{ij}+b_{ij} = b_{ij}+a_{ij}$ by the commutative property of addition of real numbers, the (i,j)-entries of $A+B$ and $B+A$ are equal for all i and j. Thus, since the matrices $A+B$ and $B+A$ have the same size and all pairs of corresponding entries are equal, $A+B = B+A$.

61. If O is the $m \times n$ zero matrix, then both A and $A + O$ are $m \times n$ matrices; so we need only show they have equal corresponding entries. The (i,j)-entry of $A + O$ is $a_{ij} + 0 = a_{ij}$, which is the (i,j)-entry of A.

62. The proof is similar to the proof of Exercise 61.

63. The matrices $(st)A$, tA, and $s(tA)$ are all $m \times n$ matrices; so we need only show that the corresponding entries of $(st)A$ and $s(tA)$ are equal. The (i,j)-entry of $s(tA)$ is s times the (i,j)-entry of tA, and so it equals $s(ta_{ij}) = st(a_{ij})$, which is the (i,j)-entry of $(st)A$. Therefore $(st)A = s(tA)$.

64. The matrices $(s+t)A$, sA, and tA are $m \times n$ matrices. Hence the matrices $(s+t)A$ and $sA+tA$ are $m \times n$ matrices; so we need only show they have equal corresponding entries. The (i,j)-entry of $sA + tA$ is the sum of the (i,j)-entries of sA and tA, that is, $sa_{ij}+ta_{ij}$. And the (i,j)-entry of $(s+t)A$ is $(s+t)a_{ij} = sa_{ij}+ta_{ij}$.

65. The matrices $(sA)^T$ and sA^T are $n \times m$ matrices; so we need only show they have equal corresponding entries. The (i,j)-entry of $(sA)^T$ is the (j,i)-entry of sA, which is sa_{ji}. The (i,j)-entry of sA^T is the product of s and the (i,j)-entry of A^T, which is also sa_{ji}.

66. The matrix A^T is an $n \times m$ matrix; so the matrix $(A^T)^T$ is an $m \times n$ matrix. Thus we need only show that $(A^T)^T$ and A have equal corresponding entries. The (i,j)-entry of $(A^T)^T$ is

the (j,i)-entry of A^T, which in turn is the (i,j)-entry of A.

67. If $i \neq j$, then the (i,j)-entry of a square zero matrix is 0. Because such a matrix is square, it is a diagonal matrix.

68. If B is a diagonal matrix, then B is square. Hence cB is square, and the (i,j)-entry of cB is $cb_{ij} = c \cdot 0 = 0$ if $i \neq j$. Thus cB is a diagonal matrix.

69. If B is a diagonal matrix, then B is square. Since B^T is the same size as B in this case, B^T is square. If $i \neq j$, then the (i,j)-entry of B^T is $b_{ji} = 0$. So B^T is a diagonal matrix.

70. Suppose that B and C are $n \times n$ diagonal matrices. Then $B + C$ is also an $n \times n$ matrix. Moreover, if $i \neq j$, the (i,j)-entry of $B + C$ is $b_{ij} + c_{ij} = 0 + 0 = 0$. So $B + C$ is a diagonal matrix.

71. $\begin{bmatrix} 2 & 5 \\ 5 & 8 \end{bmatrix}$ and $\begin{bmatrix} 2 & 5 & 6 \\ 5 & 7 & 8 \\ 6 & 8 & 4 \end{bmatrix}$

72. Let A be a symmetric matrix. Then $A = A^T$. So the (i,j)-entry of A equals the (i,j)-entry of A^T, which is the (j,i)-entry of A.

73. Let O be a square zero matrix. The (i,j)-entry of O is zero, whereas the (i,j)-entry of O^T is the (j,i)-entry of O, which is also zero. So $O = O^T$, and hence O is a symmetric matrix.

74. By Theorem 1.2(b), $(cB)^T = cB^T = cB$.

75. By Theorem 1.1(a) and Theorem 1.2(a) and (c), we have
$$(B + B^T)^T = B^T + (B^T)^T = B^T + B = B + B^T.$$

76. By Theorem 1.2(a), $(B + C)^T = B^T + C^T = B + C$.

77. No. Consider $\begin{bmatrix} 2 & 5 & 6 \\ 5 & 7 & 8 \\ 6 & 8 & 4 \end{bmatrix}$ and $\begin{bmatrix} 2 & 6 \\ 5 & 8 \end{bmatrix}$.

78. Let A be a diagonal matrix. If $i \neq j$, then $a_{ij} = 0$ and $a_{ji} = 0$ by definition. Also, $a_{ij} = a_{ji}$ if $i = j$. So every entry of A equals the corresponding entry of A^T. Therefore $A = A^T$.

79. The (i,i)-entries must all equal zero. By equating the (i,i)-entries of A^T and $-A$, we obtain $a_{ii} = -a_{ii}$, and so $a_{ii} = 0$.

80. Take $B = \begin{bmatrix} 0 & 1 \\ -1 & 0 \end{bmatrix}$. If C is any 2×2 skew-symmetric matrix, then $C^T = -C$. Therefore $c_{12} = -c_{21}$. By Exercise 79, $c_{11} = c_{22} = 0$. So
$$C = \begin{bmatrix} 0 & -c_{21} \\ c_{21} & 0 \end{bmatrix} = -c_{21}\begin{bmatrix} 0 & 1 \\ -1 & 0 \end{bmatrix} = -c_{21}B.$$

81. Let $A_1 = \frac{1}{2}(A + A^T)$ and $A_2 = \frac{1}{2}(A - A^T)$. It is easy to show that $A = A_1 + A_2$. By Exercises 75 and 74, A_1 is symmetric. Also, by Theorem 1.2(b), (a), and (c), we have
$$A_2^T = \frac{1}{2}(A - A^T)^T = \frac{1}{2}[A^T - (A^T)^T]$$
$$= \frac{1}{2}(A^T - A) = -\frac{1}{2}(A - A^T) = -A_2.$$

82. (a) Because the (i,i)-entry of $A + B$ is $a_{ii} + b_{ii}$, we have
$$\text{trace}(A + B)$$
$$= (a_{11} + b_{11}) + \cdots + (a_{nn} + b_{nn})$$
$$= (a_{11} + \cdots + a_{nn}) + (b_{11} + \cdots + b_{nn})$$
$$= \text{trace}(A) + \text{trace}(B).$$

(b) The proof is similar to the proof of (a).

(c) The proof is similar to the proof of (a).

83. The ith component of $a\mathbf{p} + b\mathbf{q}$ is $ap_i + bq_i$, which is nonnegative. Also, the sum of the components of $a\mathbf{p} + b\mathbf{q}$ is
$$(ap_1 + bq_1) + \cdots + (ap_n + bq_n)$$
$$= a(p_1 + \cdots + p_n) + b(q_1 + \cdots + q_n)$$
$$= a(1) + b(1) = a + b = 1.$$

84. (a) $\begin{bmatrix} 6.5 & -0.5 & -1.9 & -2.8 \\ 9.6 & -2.9 & 1.5 & -3.0 \\ 17.4 & 0.4 & -15.5 & 5.2 \\ -1.0 & -3.7 & -7.3 & 17.5 \\ 5.2 & 1.4 & 3.5 & 16.8 \end{bmatrix}$

(b) $\begin{bmatrix} -1.3 & 3.4 & -4.0 & 10.4 \\ 3.0 & 4.9 & -2.4 & 6.6 \\ -3.9 & -4.1 & 9.4 & -8.6 \\ 1.7 & -0.1 & -14.5 & -0.2 \\ -4.7 & 4.1 & -0.7 & -1.8 \end{bmatrix}$

(c) $\begin{bmatrix} 3.9 & 7.4 & 10.3 & -0.1 & 1.9 \\ 0.8 & -0.3 & -1.1 & -2.5 & 2.3 \\ -2.6 & 0.2 & -7.2 & -9.7 & 2.1 \\ 1.6 & 0.2 & 0.6 & 11.6 & 10.6 \end{bmatrix}$

1.2 LINEAR COMBINATIONS, MATRIX-VECTOR PRODUCTS, AND SPECIAL MATRICES

1. $\begin{bmatrix} 12 \\ 14 \end{bmatrix}$ 2. $\begin{bmatrix} -5 \\ 4 \\ 7 \end{bmatrix}$ 3. $\begin{bmatrix} 9 \\ 0 \\ 10 \end{bmatrix}$ 4. $\begin{bmatrix} 22 \\ 32 \end{bmatrix}$

5. $\begin{bmatrix} a \\ b \end{bmatrix}$ 6. $[18]$ 7. $\begin{bmatrix} 22 \\ 5 \end{bmatrix}$ 8. $\begin{bmatrix} a \\ b \\ c \end{bmatrix}$

9. $\begin{bmatrix} sa \\ tb \\ uc \end{bmatrix}$ 10. $[6]$ 11. $\begin{bmatrix} 2 \\ -6 \\ 10 \end{bmatrix}$ 12. $\begin{bmatrix} -3 \\ 4 \\ 2 \end{bmatrix}$

13. $\begin{bmatrix} -1 \\ 6 \end{bmatrix}$ 14. $\begin{bmatrix} 3 \\ -1 \\ 2 \end{bmatrix}$ 15. $\begin{bmatrix} 21 \\ 13 \end{bmatrix}$ 16. $\begin{bmatrix} 26 \\ 9 \end{bmatrix}$

17. $\frac{1}{2}\begin{bmatrix} \sqrt{2} & -\sqrt{2} \\ \sqrt{2} & \sqrt{2} \end{bmatrix}, \frac{1}{2}\begin{bmatrix} -\sqrt{2} \\ \sqrt{2} \end{bmatrix}$

18. $\begin{bmatrix} 1 & 0 \\ 0 & 1 \end{bmatrix}, \mathbf{e}_1$

19. $\frac{1}{2}\begin{bmatrix} 1 & -\sqrt{3} \\ \sqrt{3} & 1 \end{bmatrix}, \frac{1}{2}\begin{bmatrix} 3 - \sqrt{3} \\ 3\sqrt{3} + 1 \end{bmatrix}$

20. $\frac{1}{2}\begin{bmatrix} \sqrt{3} & -1 \\ 1 & \sqrt{3} \end{bmatrix}, \frac{1}{2}\begin{bmatrix} \sqrt{3} - 2 \\ 1 + 2\sqrt{3} \end{bmatrix}$

21. $\frac{1}{2}\begin{bmatrix} -\sqrt{3} & 1 \\ -1 & -\sqrt{3} \end{bmatrix}, \frac{1}{2}\begin{bmatrix} \sqrt{3} - 3 \\ 3\sqrt{3} + 1 \end{bmatrix}$

22. $\frac{-1}{\sqrt{2}}\begin{bmatrix} 1 & 1 \\ -1 & 1 \end{bmatrix}, \frac{-1}{\sqrt{2}}\begin{bmatrix} 1 \\ -3 \end{bmatrix}$ 23. $\begin{bmatrix} 3 \\ 2 \end{bmatrix}$

24. $\frac{1}{2}\begin{bmatrix} 4\sqrt{3} + 1 \\ \sqrt{3} - 4 \end{bmatrix}$ 25. $\frac{1}{2}\begin{bmatrix} 3 - \sqrt{3} \\ 3\sqrt{3} + 1 \end{bmatrix}$

26. $\frac{1}{2}\begin{bmatrix} 2 - 5\sqrt{3} \\ 2\sqrt{3} + 5 \end{bmatrix}$ 27. $\frac{1}{2}\begin{bmatrix} 3 \\ -3\sqrt{3} \end{bmatrix}$

28. $\begin{bmatrix} \sqrt{3} \\ 1 \end{bmatrix}$ 29. $\begin{bmatrix} 1 \\ 1 \end{bmatrix} = (1)\begin{bmatrix} 1 \\ 0 \end{bmatrix} + (1)\begin{bmatrix} 0 \\ 1 \end{bmatrix}$

30. $\begin{bmatrix} 1 \\ -1 \end{bmatrix} = \frac{1}{4}\begin{bmatrix} 4 \\ -4 \end{bmatrix}$ 31. not possible

32. $\begin{bmatrix} 1 \\ 1 \end{bmatrix} = (1)\begin{bmatrix} 1 \\ 0 \end{bmatrix} + (1)\begin{bmatrix} 0 \\ 1 \end{bmatrix}$ 33. not possible

34. $\begin{bmatrix} 1 \\ 1 \end{bmatrix} = (1)\begin{bmatrix} 1 \\ 0 \end{bmatrix} + (-1)\begin{bmatrix} 0 \\ -1 \end{bmatrix} + 0\begin{bmatrix} 0 \\ 0 \end{bmatrix}$

35. $\begin{bmatrix} -1 \\ 11 \end{bmatrix} = 3\begin{bmatrix} 1 \\ 3 \end{bmatrix} + (-2)\begin{bmatrix} 2 \\ -1 \end{bmatrix}$

36. $\begin{bmatrix} 1 \\ 1 \end{bmatrix} = 0\begin{bmatrix} 1 \\ 0 \end{bmatrix} + 0\begin{bmatrix} 0 \\ -1 \end{bmatrix} + (1)\begin{bmatrix} 1 \\ 1 \end{bmatrix}$

37. $\begin{bmatrix} 3 \\ 8 \end{bmatrix} = 7\begin{bmatrix} 1 \\ 2 \end{bmatrix} + (-2)\begin{bmatrix} 2 \\ 3 \end{bmatrix} + 0\begin{bmatrix} -2 \\ -5 \end{bmatrix}$

38. $\begin{bmatrix} a \\ b \end{bmatrix} = \left(\frac{a + 2b}{3}\right)\begin{bmatrix} 1 \\ 1 \end{bmatrix} + \left(\frac{a - b}{3}\right)\begin{bmatrix} 2 \\ -1 \end{bmatrix}$

39. not possible 40. $\mathbf{u} = 4\begin{bmatrix} 0 \\ 1 \\ 2 \end{bmatrix} + (-2)\begin{bmatrix} -1 \\ 3 \\ 0 \end{bmatrix}$

41. $\mathbf{u} = 0\begin{bmatrix} 2 \\ -1 \\ 2 \end{bmatrix} + 1\begin{bmatrix} 3 \\ -2 \\ 1 \end{bmatrix} + 0\begin{bmatrix} -4 \\ 1 \\ 3 \end{bmatrix}$

42. $\mathbf{u} = 5\begin{bmatrix} 1 \\ 0 \\ 0 \end{bmatrix} + 6\begin{bmatrix} 0 \\ 1 \\ 0 \end{bmatrix} + 7\begin{bmatrix} 0 \\ 0 \\ 1 \end{bmatrix}$

43. $\mathbf{u} = (-4)\begin{bmatrix} 1 \\ 0 \\ 0 \end{bmatrix} + (-5)\begin{bmatrix} 0 \\ 1 \\ 0 \end{bmatrix} + (-6)\begin{bmatrix} 0 \\ 0 \\ 1 \end{bmatrix}$

44. $\mathbf{u} = 0\begin{bmatrix} 1 \\ -1 \\ 1 \end{bmatrix} + 0\begin{bmatrix} 0 \\ -2 \\ 3 \end{bmatrix} + 1\begin{bmatrix} -1 \\ 3 \\ 2 \end{bmatrix}$

45. True

46. False. If the coefficients of the linear combination $3\begin{bmatrix} 2 \\ 2 \end{bmatrix} + (-6)\begin{bmatrix} 1 \\ 1 \end{bmatrix} = \begin{bmatrix} 0 \\ 0 \end{bmatrix}$ were positive, the sum could not equal the zero vector.

47. True   48. True   49. True

50. False, the matrix-vector product of a 2×3 matrix and a 3×1 vector is a 2×1 vector.

51. False, the matrix-vector product is a linear combination of the *columns* of the matrix.

52. False, the product of a matrix and a standard vector is a column of the matrix.

53. True

54. False, the matrix-vector product of an $m \times n$ matrix and a vector in $\mathcal{R}^n$ yields a vector in $\mathcal{R}^m$.

55. False, every vector in $\mathcal{R}^2$ is a linear combination of two *nonparallel* vectors.

56. True

57. False, a standard vector is a vector with a single component equal to 1 and the others equal to 0.

58. True

59. False, consider $A = \begin{bmatrix} 1 & -1 \\ -1 & 1 \end{bmatrix}$ and $\mathbf{u} = \begin{bmatrix} 1 \\ 1 \end{bmatrix}$.

60. True

61. False, $A_\theta \mathbf{u}$ is the vector obtained by rotating $\mathbf{u}$ by a *counterclockwise* rotation of the angle θ.

62. False, consider $A = \begin{bmatrix} 1 & -1 \\ -1 & 1 \end{bmatrix}$, $\mathbf{u} = \begin{bmatrix} 1 \\ 1 \end{bmatrix}$, and $\mathbf{v} = \begin{bmatrix} 2 \\ 2 \end{bmatrix}$.

63. True   64. True

65. If $\theta = 0$, then $A_\theta = I_2$. So $A_\theta \mathbf{v} = I_2 \mathbf{v} = \mathbf{v}$ by Theorem 1.3(h).

66. We have $A_{180°}\mathbf{v} = \begin{bmatrix} -1 & 0 \\ 0 & -1 \end{bmatrix}\mathbf{v} = -I_2\mathbf{v} = -\mathbf{v}$.

67. Let $\mathbf{v} = \begin{bmatrix} a \\ b \end{bmatrix}$. Then $A_\theta(A_\beta \mathbf{v})$

$= \begin{bmatrix} \cos\theta & -\sin\theta \\ \sin\theta & \cos\theta \end{bmatrix} \left(\begin{bmatrix} \cos\beta & -\sin\beta \\ \sin\beta & \cos\beta \end{bmatrix} \begin{bmatrix} a \\ b \end{bmatrix} \right)$

$= \begin{bmatrix} \cos\theta & -\sin\theta \\ \sin\theta & \cos\theta \end{bmatrix} \begin{bmatrix} a\cos\beta - b\sin\beta \\ a\sin\beta + b\cos\beta \end{bmatrix}$

$= \begin{bmatrix} a\cos\theta\cos\beta - b\cos\theta\sin\beta \\ a\sin\theta\cos\beta - b\sin\theta\sin\beta \end{bmatrix}$
$\quad + \begin{bmatrix} -a\sin\theta\sin\beta - b\sin\theta\cos\beta \\ a\cos\theta\sin\beta + b\cos\theta\cos\beta \end{bmatrix}$

$= \begin{bmatrix} a\cos(\theta+\beta) - b\sin(\theta+\beta) \\ a\sin(\theta+\beta) + b\cos(\theta+\beta) \end{bmatrix}$

$= A_{\theta+\beta} \mathbf{v}.$

68. Let $\mathbf{u} = \begin{bmatrix} a \\ b \end{bmatrix}$. Then
$A_\theta^T(A_\theta \mathbf{u})$

$= \begin{bmatrix} \cos\theta & \sin\theta \\ -\sin\theta & \cos\theta \end{bmatrix} \left(\begin{bmatrix} \cos\theta & -\sin\theta \\ \sin\theta & \cos\theta \end{bmatrix} \begin{bmatrix} a \\ b \end{bmatrix} \right)$

$= \begin{bmatrix} \cos\theta & \sin\theta \\ -\sin\theta & \cos\theta \end{bmatrix} \begin{bmatrix} a\cos\theta - b\sin\theta \\ a\sin\theta + b\cos\theta \end{bmatrix}$

$= \begin{bmatrix} a\cos^2\theta - b\sin\theta\cos\theta \\ -a\sin\theta\cos\theta + b\sin^2\theta \end{bmatrix}$
$\quad + \begin{bmatrix} a\sin^2\theta + b\sin\theta\cos\theta \\ a\sin\theta\cos\theta + b\cos^2\theta \end{bmatrix}$

$= \begin{bmatrix} a(\sin^2\theta + \cos^2\theta) \\ b(\sin^2\theta + \cos^2\theta) \end{bmatrix} = \begin{bmatrix} a \\ b \end{bmatrix} = \mathbf{u}.$

Similarly, $A_\theta(A_\theta^T \mathbf{u}) = \mathbf{u}$.

69. (a) As in Example 3, the populations are given by the entries of $A \begin{bmatrix} 400 \\ 300 \end{bmatrix} = \begin{bmatrix} 349 \\ 351 \end{bmatrix}$; so there will be 349,000 people in the city and 351,000 in the suburbs.

(b) Computing $A \begin{bmatrix} 349 \\ 351 \end{bmatrix} = \begin{bmatrix} 307.180 \\ 392.820 \end{bmatrix}$, we see that there will be 307,180 people in the city and 392,820 in the suburbs.

70. $A\mathbf{u} = a \begin{bmatrix} 1 \\ 4 \\ 7 \end{bmatrix} + b \begin{bmatrix} 2 \\ 5 \\ 8 \end{bmatrix} + c \begin{bmatrix} 3 \\ 6 \\ 9 \end{bmatrix}$

71. $A\mathbf{u} = \begin{bmatrix} -1 & 0 \\ 0 & 1 \end{bmatrix} \begin{bmatrix} a \\ b \end{bmatrix} = \begin{bmatrix} -a \\ b \end{bmatrix}$, the reflection of $\mathbf{u}$ about the y-axis

72. We have

$A(A\mathbf{u}) = A\left(\begin{bmatrix} -1 & 0 \\ 0 & 1 \end{bmatrix} \begin{bmatrix} a \\ b \end{bmatrix} \right)$

$= \begin{bmatrix} -1 & 0 \\ 0 & 1 \end{bmatrix} \begin{bmatrix} -a \\ b \end{bmatrix} = \begin{bmatrix} a \\ b \end{bmatrix} = \mathbf{u}.$

73. $B = \begin{bmatrix} 1 & 0 \\ 0 & -1 \end{bmatrix}$

74. (a) $C = A_{180°} = \begin{bmatrix} -1 & 0 \\ 0 & -1 \end{bmatrix}$

(b) We have

$A(C\mathbf{u}) = \begin{bmatrix} -1 & 0 \\ 0 & 1 \end{bmatrix} \left(\begin{bmatrix} -1 & 0 \\ 0 & -1 \end{bmatrix} \begin{bmatrix} a \\ b \end{bmatrix} \right)$

$= \begin{bmatrix} -1 & 0 \\ 0 & 1 \end{bmatrix} \begin{bmatrix} -a \\ -b \end{bmatrix} = \begin{bmatrix} a \\ -b \end{bmatrix}.$

In a similar fashion, we have $C(A\mathbf{u}) = \begin{bmatrix} a \\ -b \end{bmatrix} = B\mathbf{u}$ and $B(C\mathbf{u}) = C(B\mathbf{u}) = A\mathbf{u}$.

(c) The first equation shows that reflecting about the x-axis can be accomplished by either first rotating by 180° and then reflecting about the y-axis, or first reflecting about the y-axis and then rotating by 180°.

The second equation shows that reflecting about the y-axis may be accomplished either by first rotating by 180° and then reflecting about the x-axis, or first reflecting about the x-axis and then rotating by 180°.

75. $A\mathbf{u} = \begin{bmatrix} a \\ 0 \end{bmatrix}$, the projection of $\mathbf{u}$ on the x-axis

76. This exercise is similar to Exercise 72.

77. If $\mathbf{v} = \begin{bmatrix} a \\ 0 \end{bmatrix}$, then $A\mathbf{v} = \begin{bmatrix} 1 & 0 \\ 0 & 0 \end{bmatrix} \begin{bmatrix} a \\ 0 \end{bmatrix} = \begin{bmatrix} a \\ 0 \end{bmatrix} = \mathbf{v}.$

78. $B = \begin{bmatrix} 0 & 0 \\ 0 & 1 \end{bmatrix}$

79. (a) We have

$A(C\mathbf{u}) = \begin{bmatrix} 1 & 0 \\ 0 & 0 \end{bmatrix} \left(\begin{bmatrix} -1 & 0 \\ 0 & -1 \end{bmatrix} \begin{bmatrix} a \\ b \end{bmatrix} \right)$

$= \begin{bmatrix} 1 & 0 \\ 0 & 0 \end{bmatrix} \begin{bmatrix} -a \\ -b \end{bmatrix} = \begin{bmatrix} -a \\ 0 \end{bmatrix},$

and

$C(A\mathbf{u}) = \begin{bmatrix} -1 & 0 \\ 0 & -1 \end{bmatrix} \left(\begin{bmatrix} 1 & 0 \\ 0 & 0 \end{bmatrix} \begin{bmatrix} a \\ b \end{bmatrix} \right)$

$$= \begin{bmatrix} -1 & 0 \\ 0 & -1 \end{bmatrix} \begin{bmatrix} a \\ 0 \end{bmatrix} = \begin{bmatrix} -a \\ 0 \end{bmatrix}.$$

(b) Rotating a vector by 180° and then projecting the result on the x-axis is equivalent to projecting a vector on the x-axis and then rotating the result by 180°.

80. The sum of the two linear combinations
$$a\mathbf{u}_1 + b\mathbf{u}_2 \quad \text{and} \quad c\mathbf{u}_1 + d\mathbf{u}_2$$
is
$$(a\mathbf{u}_1 + b\mathbf{u}_2) + (c\mathbf{u}_1 + d\mathbf{u}_2) = (a+c)\mathbf{u}_1 + (b+d)\mathbf{u}_2,$$
which is also a linear combination of $\mathbf{u}_1$ and $\mathbf{u}_2$.

81. Write $\mathbf{v} = a_1\mathbf{u}_1 + a_2\mathbf{u}_2$ and $\mathbf{w} = b_1\mathbf{u}_1 + b_2\mathbf{u}_2$, where a_1, a_2, b_1, and b_2 are scalars. A linear combination of $\mathbf{v}$ and $\mathbf{w}$ has the form
$$c\mathbf{v} + d\mathbf{w} = c(a_1\mathbf{u}_1 + a_2\mathbf{u}_2) + d(b_1\mathbf{u}_1 + b_2\mathbf{u}_2)$$
$$= (ca_1 + db_1)\mathbf{u}_1 + (ca_2 + db_2)\mathbf{u}_2,$$
which is also a linear combination of $\mathbf{u}_1$ and $\mathbf{u}_2$.

82. The proof is similar to that of Exercise 81.

83. We have
$$A(c\mathbf{u}) = (cu_1)\mathbf{a}_1 + (cu_2)\mathbf{a}_2 + \cdots + (cu_n)\mathbf{a}_n$$
$$= c(u_1\mathbf{a}_1 + u_2\mathbf{a}_2 + \cdots + u_n\mathbf{a}_n) = c(A\mathbf{u}).$$
Similarly, $(cA)\mathbf{u} = c(A\mathbf{u})$.

84. We have
$$(A + B)\mathbf{u} = u_1(\mathbf{a}_1 + \mathbf{b}_1) + \cdots + u_n(\mathbf{a}_n + \mathbf{b}_n)$$
$$= u_1\mathbf{a}_1 + u_1\mathbf{b}_1 + \cdots + u_n\mathbf{a}_n + u_n\mathbf{b}_n$$
$$= (u_1\mathbf{a}_1 + \cdots + u_n\mathbf{a}_n)$$
$$\quad + (u_1\mathbf{b}_1 + \cdots + u_n\mathbf{b}_n)$$
$$= A\mathbf{u} + B\mathbf{u}.$$

85. We have $A\mathbf{e}_j =$
$0\mathbf{a}_1 + \cdots + 0\mathbf{a}_{j-1} + 1\mathbf{a}_j + 0\mathbf{a}_{j+1} + \cdots + 0\mathbf{a}_n = \mathbf{a}_j$.

86. Suppose $B\mathbf{w} = A\mathbf{w}$ for all $\mathbf{w}$. Let $\mathbf{w} = \mathbf{e}_j$. Then $B\mathbf{e}_j = A\mathbf{e}_j$. From Theorem 1.3(e), it follows that $\mathbf{b}_j = \mathbf{a}_j$ for all j. So $B = A$.

87. The vector $A\mathbf{0}$ is an $m \times 1$ vector. By definition
$$A\mathbf{0} = 0\mathbf{a}_1 + 0\mathbf{a}_2 + \cdots + 0\mathbf{a}_n = \mathbf{0}.$$

88. Every column of O is the $m \times 1$ zero vector. So
$$O\mathbf{v} = v_1\mathbf{0} + v_2\mathbf{0} + \cdots + v_n\mathbf{0} = \mathbf{0}.$$

89. The jth column of I_n is $\mathbf{e}_j$. So
$$I_n\mathbf{v} = v_1\mathbf{e}_1 + v_2\mathbf{e}_2 + \cdots + v_n\mathbf{e}_n = \mathbf{v}.$$

90. Using $\mathbf{p} = \begin{bmatrix} 400 \\ 300 \end{bmatrix}$, we compute $A\mathbf{p}, A(A\mathbf{p}), \ldots$ until we have ten vectors. From the final vector, we see that there will be 155,610 people living in the city and 544,389 people living in the suburbs after ten years.

91. (a) $\begin{bmatrix} 24.6 \\ 45.0 \\ 26.0 \\ -41.4 \end{bmatrix}$ (b) $\begin{bmatrix} 134.1 \\ 44.4 \\ 7.6 \\ 104.8 \end{bmatrix}$ (c) $\begin{bmatrix} 128.4 \\ 80.6 \\ 63.5 \\ 25.8 \end{bmatrix}$

(d) $\begin{bmatrix} 653.09 \\ 399.77 \\ 528.23 \\ -394.52 \end{bmatrix}$

1.3 SYSTEMS OF LINEAR EQUATIONS

1. (a) $\begin{bmatrix} 0 & -1 & 2 \\ 1 & 3 & 0 \end{bmatrix}$ (b) $\begin{bmatrix} 0 & -1 & 2 & 0 \\ 1 & 3 & 0 & -1 \end{bmatrix}$

2. (a) $\begin{bmatrix} 2 & -1 & 3 \end{bmatrix}$ (b) $\begin{bmatrix} 2 & -1 & 3 & 4 \end{bmatrix}$

3. (a) $\begin{bmatrix} 1 & 2 \\ -1 & 3 \\ -3 & 4 \end{bmatrix}$ (b) $\begin{bmatrix} 1 & 2 & 3 \\ -1 & 3 & 2 \\ -3 & 4 & 1 \end{bmatrix}$

4. (a) $\begin{bmatrix} 1 & 0 & 2 & -1 \\ 2 & -1 & 0 & 1 \end{bmatrix}$

(b) $\begin{bmatrix} 1 & 0 & 2 & -1 & 3 \\ 2 & -1 & 0 & 1 & 0 \end{bmatrix}$

5. (a) $\begin{bmatrix} 0 & 2 & -3 \\ -1 & 1 & 2 \\ 2 & 0 & 1 \end{bmatrix}$ (b) $\begin{bmatrix} 0 & 2 & -3 & 4 \\ -1 & 1 & 2 & -6 \\ 2 & 0 & 1 & 0 \end{bmatrix}$

6. (a) $\begin{bmatrix} 1 & -2 & 1 & 7 \\ 1 & -2 & 0 & 10 \\ 2 & -4 & 4 & 8 \end{bmatrix}$

(b) $\begin{bmatrix} 1 & -2 & 1 & 7 & 5 \\ 1 & -2 & 0 & 10 & 3 \\ 2 & -4 & 4 & 8 & 7 \end{bmatrix}$

7. $\begin{bmatrix} 0 & 2 & -4 & 4 & 2 \\ -2 & 6 & 3 & -1 & 1 \\ 1 & -1 & 0 & 2 & -3 \end{bmatrix}$

8. $\begin{bmatrix} -3 & 3 & 0 & -6 & 9 \\ -2 & 6 & 3 & -1 & 1 \\ 0 & 2 & -4 & 4 & 2 \end{bmatrix}$

9. $\begin{bmatrix} 1 & -1 & 0 & 2 & -3 \\ 0 & 4 & 3 & 3 & -5 \\ 0 & 2 & -4 & 4 & 2 \end{bmatrix}$

10. $\begin{bmatrix} -2 & 6 & 3 & -1 & 1 \\ 1 & -1 & 0 & 2 & -3 \\ 0 & 2 & -4 & 4 & 2 \end{bmatrix}$

11. $\begin{bmatrix} 1 & -1 & 0 & 2 & -3 \\ -2 & 6 & 3 & -1 & 1 \\ 0 & 1 & -2 & 2 & 1 \end{bmatrix}$

12. $\begin{bmatrix} 1 & -1 & 0 & 2 & -3 \\ -2 & 0 & 15 & -13 & -5 \\ 0 & 2 & -4 & 4 & 2 \end{bmatrix}$

13. $\begin{bmatrix} 1 & -1 & 0 & 2 & -3 \\ -2 & 6 & 3 & -1 & 1 \\ -8 & 26 & 8 & 0 & 6 \end{bmatrix}$

14. $\begin{bmatrix} 1 & -1 & 0 & 2 & -3 \\ -2 & 6 & 3 & -1 & 1 \\ 2 & 0 & -4 & 8 & -4 \end{bmatrix}$

15. $\begin{bmatrix} -2 & 4 & 0 \\ -1 & 1 & -1 \\ 2 & -4 & 6 \\ -3 & 2 & 1 \end{bmatrix}$

16. $\begin{bmatrix} 1 & -2 & 0 \\ -\frac{1}{2} & \frac{1}{2} & -\frac{1}{2} \\ 2 & -4 & 6 \\ -3 & 2 & 1 \end{bmatrix}$

17. $\begin{bmatrix} 1 & -2 & 0 \\ -1 & 1 & -1 \\ 0 & 0 & 6 \\ -3 & 2 & 1 \end{bmatrix}$

18. $\begin{bmatrix} 1 & -2 & 0 \\ -1 & 1 & -1 \\ 2 & -4 & 6 \\ 0 & -4 & 1 \end{bmatrix}$

19. $\begin{bmatrix} 1 & -2 & 0 \\ 2 & -4 & 6 \\ -1 & 1 & -1 \\ -3 & 2 & 1 \end{bmatrix}$

20. $\begin{bmatrix} 1 & -2 & 0 \\ -3 & 2 & 1 \\ 2 & -4 & 6 \\ -1 & 1 & -1 \end{bmatrix}$

21. $\begin{bmatrix} 1 & -2 & 0 \\ -1 & 1 & -1 \\ 2 & -4 & 6 \\ -1 & 0 & 3 \end{bmatrix}$

22. $\begin{bmatrix} -1 & 0 & -2 \\ -1 & 1 & -1 \\ 2 & -4 & 6 \\ -3 & 2 & 1 \end{bmatrix}$

23. Yes, because $1(1) - 4(-2) + 3(-1) = 6$ and $1(-5) - 2(-1) = -3$. Alternatively,

$$\begin{bmatrix} 1 & -4 & 0 & 3 \\ 0 & 0 & 1 & -2 \end{bmatrix} \begin{bmatrix} 1 \\ -2 \\ -5 \\ -1 \end{bmatrix} = \begin{bmatrix} 6 \\ -3 \end{bmatrix}.$$

24. No, because $1(2) - 4(0) + 3(1) = 5 \neq 6$. Alternatively, if A is the coefficient matrix, and the given vector is $\mathbf{v}$, then $A\mathbf{v} = \begin{bmatrix} 5 \\ -3 \end{bmatrix} \neq \begin{bmatrix} 6 \\ -3 \end{bmatrix}$.

25. No, because the left side of the second equation yields $1(2) - 2(1) = 0 \neq -3$. Alternatively,

$$\begin{bmatrix} 1 & -4 & 0 & 3 \\ 0 & 0 & 1 & -2 \end{bmatrix} \begin{bmatrix} 3 \\ 0 \\ 2 \\ 1 \end{bmatrix} = \begin{bmatrix} 6 \\ 0 \end{bmatrix} \neq \begin{bmatrix} 6 \\ -3 \end{bmatrix}.$$

26. Yes, the components of the vector satisfy both equations. Alternatively, if the given vector is $\mathbf{v}$, then $A\mathbf{v} = \begin{bmatrix} 6 \\ -3 \end{bmatrix}$.

27. no   28. yes   29. yes   30. yes

31. yes   32. no   33. yes   34. yes

35. no   36. yes   37. no   38. no

39. $\begin{array}{l} x_1 = 2 + x_2 \\ x_2 \text{ free} \end{array}$ 40. $\begin{array}{l} x_1 = -4 \\ x_2 = 5 \end{array}$

41. $\begin{array}{l} x_1 = 6 + 2x_2 \\ x_2 \text{ free} \end{array}$ 42. $\begin{array}{l} x_1 = 5 + 4x_2 \\ x_2 \text{ free} \end{array}$

43. not consistent   44. $\begin{array}{l} x_1 = -6 \\ x_2 = 3 \end{array}$

45. $\begin{array}{l} x_1 = 4 + 2x_2 \\ x_2 \text{ free} \\ x_3 = 3 \end{array}$ 46. not consistent

47. $\begin{array}{l} x_1 = 3x_4 \\ x_2 = 4x_4 \\ x_3 = -5x_4 \\ x_4 \text{ free} \end{array}$ and $\begin{bmatrix} x_1 \\ x_2 \\ x_3 \\ x_4 \end{bmatrix} = x_4 \begin{bmatrix} 3 \\ 4 \\ -5 \\ 1 \end{bmatrix}$

48. $\begin{array}{l} x_1 = 9 + x_3 - 3x_4 \\ x_2 = 8 - 2x_3 + 5x_4 \\ x_3 \text{ free} \\ x_4 \text{ free} \end{array}$ and

$\begin{bmatrix} x_1 \\ x_2 \\ x_3 \\ x_4 \end{bmatrix} = x_3 \begin{bmatrix} 1 \\ -2 \\ 1 \\ 0 \end{bmatrix} + x_4 \begin{bmatrix} -3 \\ 5 \\ 0 \\ 1 \end{bmatrix} + \begin{bmatrix} 9 \\ 8 \\ 0 \\ 0 \end{bmatrix}$

49. $\begin{array}{l} x_1 \text{ free} \\ x_2 = -3 \\ x_3 = -4 \\ x_4 = 5 \end{array}$ and $\begin{bmatrix} x_1 \\ x_2 \\ x_3 \\ x_4 \end{bmatrix} = x_1 \begin{bmatrix} 1 \\ 0 \\ 0 \\ 0 \end{bmatrix} + \begin{bmatrix} 0 \\ -3 \\ -4 \\ 5 \end{bmatrix}$

50. $\begin{array}{l} x_1 = -3 + 2x_2 \\ x_2 \text{ free} \\ x_3 = -4 \\ x_4 = 5 \end{array}$ and

$\begin{bmatrix} x_1 \\ x_2 \\ x_3 \\ x_4 \end{bmatrix} = x_2 \begin{bmatrix} 2 \\ 1 \\ 0 \\ 0 \end{bmatrix} + \begin{bmatrix} -3 \\ 0 \\ -4 \\ 5 \end{bmatrix}$

51. $\begin{array}{l} x_1 = 6 - 3x_2 + 2x_4 \\ x_2 \text{ free} \\ x_3 = 7 - 4x_4 \\ x_4 \text{ free} \end{array}$ and

$\begin{bmatrix} x_1 \\ x_2 \\ x_3 \\ x_4 \end{bmatrix} = x_2 \begin{bmatrix} -3 \\ 1 \\ 0 \\ 0 \end{bmatrix} + x_4 \begin{bmatrix} 2 \\ 0 \\ -4 \\ 1 \end{bmatrix} + \begin{bmatrix} 6 \\ 0 \\ 7 \\ 0 \end{bmatrix}$

52. x_1 free
$x_2 = -4 - 3x_4$
$x_3 = 9 - 2x_4$
x_4 free
and
$$\begin{bmatrix} x_1 \\ x_2 \\ x_3 \\ x_4 \end{bmatrix} = x_1 \begin{bmatrix} 1 \\ 0 \\ 0 \\ 0 \end{bmatrix} + x_4 \begin{bmatrix} 0 \\ -3 \\ -2 \\ 1 \end{bmatrix} + \begin{bmatrix} 0 \\ -4 \\ 9 \\ 0 \end{bmatrix}$$

53. not consistent

54. x_1 free
x_2 free
$x_3 = 3x_4 - 2x_6$
x_4 free
$x_5 = x_6$
x_6 free
and
$$\begin{bmatrix} x_1 \\ x_2 \\ x_3 \\ x_4 \\ x_5 \\ x_6 \end{bmatrix} = x_1 \begin{bmatrix} 1 \\ 0 \\ 0 \\ 0 \\ 0 \\ 0 \end{bmatrix} + x_2 \begin{bmatrix} 0 \\ 1 \\ 0 \\ 0 \\ 0 \\ 0 \end{bmatrix} + x_4 \begin{bmatrix} 0 \\ 0 \\ 3 \\ 1 \\ 0 \\ 0 \end{bmatrix} + x_6 \begin{bmatrix} 0 \\ 0 \\ -2 \\ 0 \\ 1 \\ 1 \end{bmatrix}$$

55. All variables are either free or basic, so if there are k free variables, there must be $n - k$ basic variables.

56. Because R is in reduced row echelon form, the leading entry must equal 1, and every other entry in the column must be 0. So this column equals $\mathbf{e}_4$.

57. False, the system $0x_1 + 0x_2 = 1$ has no solutions.

58. False, a system of linear equations has 0, 1, or infinitely many solutions.

59. True

60. False, the matrix $\begin{bmatrix} 2 & 0 \\ 0 & 0 \end{bmatrix}$ is in row echelon form.

61. True   62. True

63. False, the matrices $\begin{bmatrix} 2 & 0 \\ 0 & 0 \end{bmatrix}$ and $\begin{bmatrix} 1 & 0 \\ 0 & 0 \end{bmatrix}$ are both row echelon forms for $\begin{bmatrix} 2 & 0 \\ 0 & 0 \end{bmatrix}$.

64. True   65. True

66. False, the system
$$0x_1 + 0x_2 = 1$$
$$0x_1 + 0x_2 = 0$$
is inconsistent, but its augmented matrix is
$$\begin{bmatrix} 0 & 0 & 1 \\ 0 & 0 & 0 \end{bmatrix}.$$

67. True   68. True

69. False, the coefficient matrix of a system of m linear equations in n variables is an $m \times n$ matrix.

70. True   71. True   72. True

73. False, multiplying every entry of some row of a matrix by a *nonzero* scalar is an elementary row operation.

74. True

75. False, the system may be inconsistent; consider $0x_1 + 0x_2 = 1$.

76. True

77. If $[R \ \mathbf{c}]$ is in reduced row echelon form, then so is R. If we apply the same row operations to A that were applied to $[A \ \mathbf{b}]$ to produce $[R \ \mathbf{c}]$, we obtain the matrix R. So R is the reduced row echelon form of A.

78. The row operations that reduce A to R may be applied to $[A \ \mathbf{0}]$ and do not affect its last column. The resulting matrix is $[R \ \mathbf{0}]$, which is in reduced row echelon form.

79. If we let $\mathbf{0}_n$ be the $n \times 1$ zero vector, then, by Theorem 1.2(f), $A\mathbf{0}_n = \mathbf{0}$. So $\mathbf{0}_n$ is a solution of $A\mathbf{x} = \mathbf{0}$, and hence $A\mathbf{x} = \mathbf{0}$ is consistent.

80. Let R be the reduced row echelon form of A. Then by Exercise 77, $[R \ \mathbf{c}]$ is the reduced row echelon form of $[A \ \mathbf{b}]$ for some vector $\mathbf{c}$. By hypothesis, $[R \ \mathbf{c}]$ contains no row whose only nonzero entry lies in the last column. So the system $A\mathbf{x} = \mathbf{b}$ is consistent.

81. The ranks of the possible reduced row echelon forms are 0, 1, and 2. Considering each of these ranks, we see that there are 7 possible reduced row echelon forms:
$$\begin{bmatrix} 0 & 0 & 0 \\ 0 & 0 & 0 \end{bmatrix}, \begin{bmatrix} 1 & * & * \\ 0 & 0 & 0 \end{bmatrix}, \begin{bmatrix} 0 & 1 & * \\ 0 & 0 & 0 \end{bmatrix}, \begin{bmatrix} 0 & 0 & 1 \\ 0 & 0 & 0 \end{bmatrix},$$
$$\begin{bmatrix} 1 & 0 & * \\ 0 & 1 & * \end{bmatrix}, \begin{bmatrix} 1 & * & 0 \\ 0 & 0 & 1 \end{bmatrix}, \text{ and } \begin{bmatrix} 0 & 1 & 0 \\ 0 & 0 & 1 \end{bmatrix}.$$

82. As in the solution to Exercise 81, there are 11 possible reduced row echelon forms:
$$\begin{bmatrix} 0 & 0 & 0 & 0 \\ 0 & 0 & 0 & 0 \end{bmatrix}, \begin{bmatrix} 1 & * & * & * \\ 0 & 0 & 0 & 0 \end{bmatrix}, \begin{bmatrix} 0 & 1 & * & * \\ 0 & 0 & 0 & 0 \end{bmatrix},$$
$$\begin{bmatrix} 0 & 0 & 1 & * \\ 0 & 0 & 0 & 0 \end{bmatrix}, \begin{bmatrix} 0 & 0 & 0 & 1 \\ 0 & 0 & 0 & 0 \end{bmatrix}, \begin{bmatrix} 1 & 0 & * & * \\ 0 & 1 & * & * \end{bmatrix},$$
$$\begin{bmatrix} 1 & * & 0 & 0 \\ 0 & 0 & 1 & * \end{bmatrix}, \begin{bmatrix} 1 & * & * & 0 \\ 0 & 0 & 0 & 1 \end{bmatrix}, \begin{bmatrix} 0 & 1 & 0 & * \\ 0 & 0 & 1 & * \end{bmatrix},$$
$$\begin{bmatrix} 0 & 1 & * & 0 \\ 0 & 0 & 0 & 1 \end{bmatrix}, \text{ and } \begin{bmatrix} 0 & 0 & 1 & 0 \\ 0 & 0 & 0 & 1 \end{bmatrix}.$$

83. There are three cases. If the operation interchanges rows i and j of A, then interchanging rows i and j of B produces A. If the operation multiplies row i of A by the nonzero scalar c, then multiplying row i of B by $\frac{1}{c}$ produces A. Finally, if the operation adds k times row i to row j of A, then adding $-k$ times row i to row j of B produces A.

84. The system $x_1 = 1$ has only the solution 1, but the system $0x_1 = 0 \cdot 1$ has infinitely many solutions.

85. Multiplying the second equation by c produces a system whose augmented matrix is obtained from the augmented matrix of the original system by the elementary row operation of multiplying the second row by c. From the statement on page 33, the two systems are equivalent.

86. The solution is similar to that of Exercise 85.

1.4 GAUSSIAN ELIMINATION

1. $x_1 = -2 - 3x_2$
   x_2 free

2. $x_1 = 3 + x_2$
   x_2 free

3. $x_1 = 4$
   $x_2 = 5$

4. $x_1 = 1 + 2x_3$
   $x_2 = -2 - x_3$
   x_3 free

5. not consistent

6. $x_1 = 3 + 2x_2 + x_3$
   x_2 free
   x_3 free

7. $x_1 = -1 + 2x_2$
   x_2 free
   $x_3 = 2$

8. $x_1 = -1 - 4x_4$
   $x_2 = 3x_4$
   $x_3 = 1 - 2x_4$
   x_4 free

9. $x_1 = 1 + 2x_3$
   $x_2 = -2 - x_3$
   x_3 free
   $x_4 = -3$

10. not consistent

11. $x_1 = -4 - 3x_2 + x_4$
    x_2 free
    $x_3 = 3 - 2x_4$
    x_4 free

12. $x_1 = 3 + 2x_3$
    $x_2 = -4 - 3x_3$
    x_3 free

13. not consistent

14. not consistent

15. $x_1 = -2 + x_5$
    x_2 free
    $x_3 = 3 - 3x_5$
    $x_4 = -1 - 2x_5$
    x_5 free

16. $x_1 = -3 + x_2 + 2x_5$
    x_2 free
    x_3 free
    $x_4 = -1 - 2x_5$
    x_5 free

17. The augmented matrix can be transformed to $\begin{bmatrix} -1 & 4 & 3 \\ 0 & r+12 & 11 \end{bmatrix}$ using an elementary row operation. Therefore the system is inconsistent if $r + 12 = 0$, that is, $r = -12$.

18. The augmented matrix can be transformed to $\begin{bmatrix} -1 & 4 & 6 \\ 0 & r+12 & 16 \end{bmatrix}$ using two elementary row operations. So the system is inconsistent if $r + 12 = 0$, that is, $r = -12$.

19. The augmented matrix can be transformed to $\begin{bmatrix} 1 & -2 & 0 \\ 0 & 0 & r \end{bmatrix}$. So the system is inconsistent if $r \neq 0$.

20. The augmented matrix can be transformed to $\begin{bmatrix} 1 & 0 & -3 \\ 0 & r & 0 \end{bmatrix}$. So the system is inconsistent for no value of r.

21. The augmented matrix can be transformed to $\begin{bmatrix} 1 & -3 & -2 \\ 0 & r+6 & 0 \end{bmatrix}$. So the system is inconsistent for no value of r.

22. The augmented matrix is $\begin{bmatrix} -2 & 1 & 5 \\ r & 4 & 3 \end{bmatrix}$. Add $\frac{r}{2}$ times the first row to the second row to obtain $\begin{bmatrix} -2 & 1 & 5 \\ 0 & 4+\frac{r}{2} & 3+\frac{5}{2}r \end{bmatrix}$. The system is inconsistent if $4 + \frac{r}{2} = 0$ and $3 + \frac{5}{2}r \neq 0$. So $r = -8$.

23. The augmented matrix can be transformed to $\begin{bmatrix} -1 & r & 2 \\ 0 & r^2-9 & 2r+6 \end{bmatrix}$. For the system to be inconsistent, we need $r^2 - 9 = 0$ and $2r + 6 \neq 0$. So $r = \pm 3$ and $r \neq -3$. Therefore $r = 3$.

24. The argument is similar to that of Exercise 23. The system is inconsistent if $r = -4$.

25. The augmented matrix can be transformed to $\begin{bmatrix} 1 & -1 & 2 & 4 \\ 0 & r+3 & -7 & -10 \end{bmatrix}$. Because this matrix does not contain a row whose only nonzero entry lies in the last column, the system is never inconsistent.

26. The augmented matrix can be transformed to $\begin{bmatrix} 1 & 2 & -4 & 1 \\ 0 & 0 & r-8 & 5 \end{bmatrix}$. If $r = 8$, then this matrix contains a row whose only nonzero entry lies in

the last column, and so the system is inconsistent if $r = 8$.

27. The augmented matrix can be transformed to
$$\begin{bmatrix} 1 & r & 5 \\ 0 & 6-3r & s-15 \end{bmatrix}.$$
    (a) We need $6 - 3r = 0$ and $s - 15 \neq 0$. So $r = 2$ and $s \neq 15$.
    (b) We need $6 - 3r \neq 0$, that is, $r \neq 2$.
    (c) We need $6 - 3r = 0$ and $s - 15 = 0$. So $r = 2$ and $s = 15$.

28. The augmented matrix can be transformed to
$$\begin{bmatrix} -1 & 4 & s \\ 0 & r+8 & 6+2s \end{bmatrix}.$$
    (a) We need $r + 8 = 0$ and $6 + 2s \neq 0$. So $r = -8$ and $s \neq -3$.
    (b) We need $r + 8 \neq 0$, that is, $r \neq -8$.
    (c) We need $r + 8 = 0$ and $6 + 2s = 0$. So $r = -8$ and $s = -3$.

29. (a) $r = -8$, $s \neq -2$ (b) $r \neq -8$
    (c) $r = -8$, $s = -2$

30. (a) $r = -12$, $s \neq 2$ (b) $r \neq -12$
    (c) $r = -12$, $s = 2$

31. (a) $r = \frac{5}{2}$, $s \neq -6$ (b) $r \neq \frac{5}{2}$
    (c) $r = \frac{5}{2}$, $s = -6$

32. (a) $r = -2$, $s \neq -15$ (b) $r \neq -2$
    (c) $r = -2$, $s = -15$

33. (a) $r = 3$, $s \neq \frac{2}{3}$ (b) $r \neq 3$
    (c) $r = 3$, $s = \frac{2}{3}$

34. (a) $r = -2$, $s \neq 6$ (b) $r \neq -2$
    (c) $r = -2$, $s = 6$

35. The reduced row echelon form of the matrix is
$$R = \begin{bmatrix} 1 & 0 & 0 & -2 \\ 0 & 1 & 0 & 1 \\ 0 & 0 & 1 & -3 \\ 0 & 0 & 0 & 0 \\ 0 & 0 & 0 & 0 \end{bmatrix}.$$
The rank of the given matrix equals the number of nonzero rows in R, which is 3. The nullity of the given matrix equals its number of columns minus its rank, which is $4 - 3 = 1$.

36. The rank is 2, and the nullity is 2.
37. The rank is 2, and the nullity is 3.
38. The rank is 4, and the nullity is 2.
39. The rank is 3, and the nullity is 1.
40. The rank is 3, and the nullity is 2.
41. The rank is 2, and the nullity is 3.
42. The rank is 3, and the nullity is 3.
43. Let x_1, x_2, and x_3 be the number of days that mines 1, 2, and 3, respectively, must operate to supply the desired amounts.
    (a) The requirements may be written as the matrix equation
$$\begin{bmatrix} 1 & 1 & 2 \\ 1 & 2 & 2 \\ 2 & 1 & 0 \end{bmatrix} \begin{bmatrix} x_1 \\ x_2 \\ x_3 \end{bmatrix} = \begin{bmatrix} 80 \\ 100 \\ 40 \end{bmatrix}.$$
    The reduced row echelon form of the augmented matrix of this system is
$$\begin{bmatrix} 1 & 0 & 0 & 10 \\ 0 & 1 & 0 & 20 \\ 0 & 0 & 1 & 25 \end{bmatrix}.$$
    So $x_1 = 10$, $x_2 = 20$, $x_3 = 25$.
    (b) A system of equations similar to that in (a) yields the reduced row echelon form
$$\begin{bmatrix} 1 & 0 & 0 & 10 \\ 0 & 1 & 0 & 60 \\ 0 & 0 & 1 & -15 \end{bmatrix}.$$
    Because $x_3 = -15$ is impossible for this problem, these amounts cannot be supplied.

44. Let x_1, x_2, and x_3 denote the number of pounds of the three types of fertilizer, respectively, needed to satisfy the requirements.
    (a) The given requirements yield the system
$$\begin{aligned} x_1 + x_2 + x_3 &= 600 \\ .10x_1 + .08x_2 + .06x_3 &= .075(600) \\ .03x_1 + .06x_2 + .01x_3 &= .05(600). \end{aligned}$$
    This system has the solution $x_1 = -18.75$, $x_2 = 487.5$, and $x_3 = 131.25$. So this mixture is impossible.
    (b) A similar approach yields the solution $x_1 = 375$, $x_2 = 150$, and $x_3 = 75$.

45. Let x_1, x_2, and x_3 be the amounts of the three supplements, respectively, that must be used.
    (a) The given requirements yield the system
$$\begin{aligned} 10x_1 + 15x_2 + 36x_3 &= 660 \\ 10x_1 + 20x_2 + 44x_3 &= 820 \\ 15x_1 + 15x_2 + 42x_3 &= 750, \end{aligned}$$
    which has the solution
$$\begin{aligned} x_1 &= 18 - 1.2x_3 \\ x_2 &= 32 - 1.6x_3 \\ x_3 &\text{ free.} \end{aligned}$$
    Because the solution must be nonnegative, we need $x_3 \leq 15$ and $x_3 \leq 20$. This yields a maximum value of $x_3 = 15$.

(b) No. A similar approach yields an inconsistent system.

46. Let x_1, x_2, and x_3 be the amounts of A, B, and C, respectively, that must be blended.

    (a) The given requirements yield the system
    $$\begin{aligned} x_1 + x_2 + x_3 &= 100 \\ 40x_1 + 32x_2 + 24x_3 &= 35(100) \\ 30x_1 + 62x_2 + 94x_3 &= 50(100), \end{aligned}$$
    which has the solution
    $$\begin{aligned} x_1 &= 37.5 + x_3 \\ x_2 &= 62.5 - 2x_3 \\ x_3 &\text{ free.} \end{aligned}$$
    Letting $x_3 = 0$, we obtain $x_1 = 37.5$ and $x_2 = 62.5$.

    (b) In order that x_1 and x_2 be nonnegative, we need $x_3 \geq 0$ and $2x_3 \leq 62.5$. So we take $x_3 = 31.25$ for a maximum value of x_3.

47. We need $f(-1) = 14$, $f(1) = 4$, and $f(3) = 10$. These conditions produce the system
    $$\begin{aligned} a - b + c &= 14 \\ a + b + c &= 4 \\ 9a + 3b + c &= 10. \end{aligned}$$
    This system has the solution $a = 2$, $b = -5$, $c = 7$. So $f(x) = 2x^2 - 5x + 7$.

48. $f(x) = -3x^2 + 8x - 5$

49. $f(x) = 4x^2 - 7x + 2$

50. $f(x) = -x^3 + 6x^2 + 4x - 12$.

51. Column j is $\mathbf{e}_3$. Each pivot column has exactly one nonzero entry, which is 1, and hence it is a standard vector. Also because of the definition of the reduced row echelon form, the sequence of pivot columns must be $\mathbf{e}_1, \mathbf{e}_2, \ldots$. Hence the third pivot column must be $\mathbf{e}_3$.

52. As noted in the solution to Exercise 51, column j equals $\mathbf{e}_4$, and because $\mathbf{e}_1, \mathbf{e}_2$, and $\mathbf{e}_3$ are among the previous columns, it follows that $j \geq 4$. Because the fourth component of column j is 1, the only nonzero entry, it follows that $i = 4$.

53. True

54. False. For example, the matrix $\begin{bmatrix} 0 & 1 \\ 2 & 0 \end{bmatrix}$ can be reduced to I_2 by interchanging its rows and then multiplying the first row by $\frac{1}{2}$, or by multiplying the second row by $\frac{1}{2}$ and then interchanging rows.

55. True   56. True   57. True   58. True

59. False. By definition, rank A + nullity A equals the number of columns of A. So, for a 5×8 matrix, we have $3 + 2 \neq 8$.

60. False, we need only repeat one equation to produce an equivalent system with a different number of equations.

61. True   62. True   63. True

64. False, the augmented matrix $\begin{bmatrix} 1 & 0 & 2 \\ 0 & 1 & 3 \\ 0 & 0 & 0 \end{bmatrix}$ contains a zero row, but the corresponding system has the unique solution $x_1 = 2$, $x_3 = 3$.

65. False, the augmented matrix $\begin{bmatrix} 0 & 0 & 1 \\ 0 & 0 & 0 \end{bmatrix}$ contains a zero row, but the system is inconsistent.

66. True   67. True

68. False, the sum of the rank and nullity of a matrix equals the number of *columns* in the matrix.

69. True   70. True

71. False, the third pivot position in a matrix may be in any column to the right of column 2.

72. True

73. If the rank of a matrix is 0, then its reduced row echelon form has only zero rows, which means that the original matrix must have only zero rows, and hence must be the zero matrix.

74. The 4×7 zero matrix has rank 0, and the rank of any matrix must be nonnegative. Hence the smallest possible rank is 0.

75. The largest possible rank is 4. The reduced row echelon form is a 4×7 matrix and hence has at most 4 nonzero rows. So the rank must be less than or equal to 4. On the other hand, the 4×7 matrix whose first four columns are $\mathbf{e}_1, \mathbf{e}_2, \mathbf{e}_3$, and $\mathbf{e}_4$ has rank 4.

76. The largest possible rank is 4. By the first boxed result on page 48, the rank of a matrix equals the number of its pivot columns. Clearly a 7×4 matrix can have at most 4 pivot columns.

77. The smallest possible nullity is 3. Note that if the rank of a 4×7 matrix A equals 4, then its nullity is $7 - \text{rank } A = 7 - 4 = 3$. On the other hand, from the solution to Exercise 75, we see that rank $A \leq 4$. So
    $$\text{nullity } A = 7 - \text{rank } A \geq 7 - 4 = 3.$$

78. The smallest possible nullity is 0. The solution is similar to that of Exercise 77.

12 Chapter 1 Matrices, Vectors, and Systems of Linear Equations

79. The largest possible rank is the minimum of m and n. If $m \leq n$, the solution is similar to that of Exercise 75. If $n \leq m$, the solution is similar to that of Exercise 76.

80. The smallest possible nullity is $n - m$ if $m \leq n$ and 0 if $m > n$. By Exercise 79, the rank of a matrix A equals the minimum p of m and n. So nullity $A = n - $ rank $A = n - p$. If $m \leq n$, then $p = m$, so nullity $A = n - m$. If $n < m$, then $p = n$; so nullity $A = 0$.

81. No. Let R be the reduced row echelon form of A. By Exercise 79, rank $A \leq 3$; so R has a zero row. Thus we can choose c so that $[R \;\; c]$ has a row equal to $[0 \; 0 \; 0 \; 1]$. By appropriate elementary row operations, we can transform $[R \;\; c]$ into a matrix of the form $[A \;\; b]$. So, by Theorem 1.5, the system $A\mathbf{x} = \mathbf{b}$ is not consistent.

82. For the solution to be unique, the solution must have no free variables; so nullity $A = 0$. Therefore rank $A = n - $ nullity $A = n$.

83. There are either no solutions or infinitely many solutions. Let the system be $A\mathbf{x} = \mathbf{b}$, and let R be the reduced row echelon form of A. Each nonzero row of R corresponds to a basic variable. Since there are fewer equations than variables, if the system is consistent, there must be free variables. Therefore the system is either inconsistent or has infinitely many solutions.

84. (a) $\begin{array}{l} x_1 + x_2 = 2 \\ x_1 + x_2 = 3 \\ x_1 + x_2 = 4 \end{array}$ (b) $\begin{array}{l} x_1 + x_2 = 3 \\ 2x_1 + x_2 = 4 \\ 3x_1 + x_2 = 5 \end{array}$

(c) $\begin{array}{l} x_1 + x_2 = 3 \\ 2x_1 + 2x_2 = 6 \\ 3x_1 + 3x_2 = 9 \end{array}$

85. Let $[R \;\; \mathbf{c}]$ denote the reduced row echelon form of $[A \;\; \mathbf{b}]$. Then R is the reduced row echelon form of A. If rank $A = m$, then R contains no nonzero rows. Hence $[R \;\; \mathbf{c}]$ contains no row in which the only nonzero entry lies in the last column. So $A\mathbf{x} = \mathbf{b}$ is consistent for every $\mathbf{b}$ by Theorem 1.5.

86. Let $[R \;\; \mathbf{c}]$ denote the reduced row echelon form of $[A \;\; \mathbf{b}]$. Then R is the reduced row echelon form of A. If $A\mathbf{x} = \mathbf{b}$ is inconsistent, then $[R \;\; \mathbf{c}]$ contains the row $[0 \; 0 \; \ldots \; 0 \; 1]$. The corresponding row of R is a zero row, and every other nonzero row of $[R \;\; \mathbf{c}]$ corresponds to a nonzero row of R. Thus rank $[A \;\; \mathbf{b}] = 1 + $ rank A; so the ranks of $[A \;\; \mathbf{b}]$ and A are not equal. Conversely, the reduced row echelon form of A equals the reduced row echelon form of $[A \;\; \mathbf{b}]$ with its last column removed. Thus if the ranks of these matrices are not equal, we must have rank $[A \;\; \mathbf{b}] = 1 + $ rank A. This can happen only if $[R \;\; \mathbf{c}]$ contains the row $[0 \; 0 \; \ldots \; 0 \; 1]$. So the matrix equation $A\mathbf{x} = \mathbf{b}$ is inconsistent.

87. Yes, $A(c\mathbf{u}) = c(A\mathbf{u}) = c \cdot \mathbf{0} = \mathbf{0}$; so $c\mathbf{u}$ is a solution of $A\mathbf{x} = \mathbf{0}$.

88. Yes, $A(\mathbf{u} + \mathbf{v}) = A\mathbf{u} + A\mathbf{v} = \mathbf{0} + \mathbf{0} = \mathbf{0}$; so $\mathbf{u} + \mathbf{v}$ is a solution of $A\mathbf{x} = \mathbf{0}$.

89. We have $A(\mathbf{u} - \mathbf{v}) = A\mathbf{u} - A\mathbf{v} = \mathbf{b} - \mathbf{b} = \mathbf{0}$; so $\mathbf{u} - \mathbf{v}$ is a solution of $A\mathbf{x} = \mathbf{0}$.

90. We have $A(\mathbf{u} + \mathbf{v}) = A\mathbf{u} + A\mathbf{v} = \mathbf{b} + \mathbf{0} = \mathbf{b}$; so $\mathbf{u} + \mathbf{v}$ is a solution of $A\mathbf{x} = \mathbf{b}$.

91. If $A\mathbf{x} = \mathbf{b}$ is consistent, then there exists a vector $\mathbf{u}$ such that $A\mathbf{u} = \mathbf{b}$. So $A(c\mathbf{u}) = c(A\mathbf{u}) = c\mathbf{b}$. Hence $c\mathbf{u}$ is a solution of $A\mathbf{x} = c\mathbf{b}$, and therefore $A\mathbf{x} = c\mathbf{b}$ is consistent.

92. As in Exercise 87, there exist vectors $\mathbf{u}_1$ and $\mathbf{u}_2$ such that $A\mathbf{u}_1 = \mathbf{b}_1$ and $A\mathbf{u}_2 = \mathbf{b}_2$. Therefore $A(\mathbf{u}_1 + \mathbf{u}_2) = A\mathbf{u}_1 + A\mathbf{u}_2 = \mathbf{b}_1 + \mathbf{b}_2$. Hence $A\mathbf{x} = \mathbf{b}_1 + \mathbf{b}_2$ is consistent.

93. No. If $\mathbf{u} + \mathbf{v}$ were a solution of $A\mathbf{x} = \mathbf{b}$, then
$$\mathbf{b} = A(\mathbf{u} + \mathbf{v}) = A\mathbf{u} + A\mathbf{v} = \mathbf{b} + \mathbf{b} = 2\mathbf{b};$$
so $\mathbf{b} = \mathbf{0}$. Therefore the result is not true if $\mathbf{b} \neq \mathbf{0}$.

94. $\begin{array}{l} x_1 = 4.9927 + 1.1805x_4 + 8.5341x_5 \\ x_2 = 7.1567 + 3.0513x_4 + 15.3103x_5 \\ x_3 = -2.5738 + 5.2366x_4 + 15.1360x_5 \\ x_4 \quad \text{free} \\ x_5 \quad \text{free} \end{array}$

95. $\begin{array}{l} x_1 = 2.32 + 0.32x_5 \\ x_2 = -6.44 + 0.56x_5 \\ x_3 = 0.72 - 0.28x_5 \\ x_4 = 5.92 + 0.92x_5 \\ x_5 \quad \text{free} \end{array}$

96. The system is not consistent.

97. 3, 2   **98.** 5, 0   **99.** 4, 1

1.5 APPLICATIONS OF SYSTEMS OF LINEAR EQUATIONS

1. True   **2.** True

3. False, the net production vector is $\mathbf{x} - C\mathbf{x}$. The vector $C\mathbf{x}$ is the total output of the economy that is consumed during the production process.

4. False, see Kirchoff's voltage law.

5. True   **6.** True

7. $50(.22) = \$11$ million

8. $100(.20) = $20 million

9. The smallest entry in the row labeled *services* lies in the column labeled *services*; so the services sector is least dependent on services.

10. The largest entry in the row labeled *services* lies in the column labeled *entertainment*; so the services sector is most dependent on entertainment.

11. The smallest entry in the column labeled *agriculture* lies in the row labeled *entertainment*; so the agriculture sector is least dependent on entertainment.

12. The largest entry in the column labeled *agriculture* lies in the row labeled *manufacturing*; so the agriculture sector is most dependent on manufacturing.

13. Let $\mathbf{x} = \begin{bmatrix} 30 \\ 40 \\ 30 \\ 20 \end{bmatrix}$. The totals of the inputs from each sector of the economy that are consumed during the production process are the components of $C\mathbf{x}$, which are $16.1 million of agriculture, $17.8 million of manufacturing, $18 million of services, and $10.1 million of entertainment.

14. As in Exercise 13, the totals of the inputs from each sector of the economy that are consumed during the production process are $10.5 million of agriculture, $11.9 million of manufacturing, $11.8 million of services, and $6.8 million of entertainment.

15. The net production for each sector is given by the components of $\mathbf{x} - C\mathbf{x}$, where $\mathbf{x}$ is the vector in the solution to Exercise 13. These values are $13.9 million of agriculture, $22.2 million of manufacturing, $12 million of services, and $9.9 million of entertainment.

16. As in Exercise 15, the net production for each sector is $9.5 million of agriculture, $18.1 million of manufacturing, $8.2 million of services, and $3.2 million of entertainment.

17. (a) The gross production vector is $\mathbf{x} = \begin{bmatrix} 40 \\ 30 \\ 35 \end{bmatrix}$.

    If C is the input-output matrix, then the net production vector is $\mathbf{x} - C\mathbf{x} =$

    $\begin{bmatrix} 40 \\ 30 \\ 35 \end{bmatrix} - \begin{bmatrix} .2 & .20 & .3 \\ .4 & .30 & .1 \\ .2 & .25 & .3 \end{bmatrix} \begin{bmatrix} 40 \\ 30 \\ 35 \end{bmatrix} = \begin{bmatrix} 15.5 \\ 1.5 \\ 9.0 \end{bmatrix}$.

    So the net productions are $15.5 million of transportation, $1.5 million of food, and $9 million of oil.

    (b) Let $\mathbf{d} = \begin{bmatrix} 32 \\ 48 \\ 24 \end{bmatrix}$. In order to obtain the gross production, we must solve the system $(I_3 - C)\mathbf{x} = \mathbf{d}$, which yields the solution $\mathbf{x} = \begin{bmatrix} 128 \\ 160 \\ 128 \end{bmatrix}$. So the gross productions required are $128 million of transportation, $160 million of food, and $128 million of oil.

18. (a) As in Exercise 17(a), the net productions required are $24 million of metals, $8 million of nonmetals, and $14 million of services.

    (b) As in Exercise 17(b), the gross productions required are $370 million of metals, $680 million of nonmetals, and $400 million of services.

19. (a) $\begin{bmatrix} .1 & .4 \\ .3 & .2 \end{bmatrix}$

    (b) As in Exercise 17(a), the net productions required are $34 million of electricity and $22 million of oil.

    (c) As in Exercise 17(b), the gross productions required are $128 million of electricity and $138 million of oil.

20. (a) $\begin{bmatrix} .1 & .2 \\ .1 & .7 \end{bmatrix}$

    (b) As in Exercise 17(a), the net productions required are $12 million of nongovernment and $7 million of government.

    (c) As in Exercise 17(b), the gross productions required are $94 million of nongovernment and $198 million of government.

21. (a) As in Exercise 17(a), the net productions required are $49 million of finance, $10 million of goods, and $18 million of services.

    (b) As in Exercise 17(b), the gross productions required are $75 million of finance, $125 million of goods, and $100 million of services.

    (c) As in Exercise 17(b), the gross productions required are $75 million of finance, $104 million of goods, and $114 million of services.

14 Chapter 1 Matrices, Vectors, and Systems of Linear Equations

22. (a) As in Exercise 17(a), the net productions required are $20 million of agriculture, $21 million of manufacturing, and $10 million of services.

(b) As in Exercise 17(b), the gross productions required are $194 million of agriculture, $226.5 million of manufacturing, and $196.5 million of services.

23. We are given that $\mathbf{v} = \mathbf{p} - C^T\mathbf{p}$, and we know that $\mathbf{x} - C\mathbf{x} = \mathbf{d}$. So we have

$$\mathbf{p} = \mathbf{v} + C^T\mathbf{p} \quad \text{and} \quad \mathbf{x} = C\mathbf{x} + \mathbf{d}.$$

Therefore

$$\mathbf{p}^T\mathbf{x} = (\mathbf{v} + C^T\mathbf{p})^T\mathbf{x} = [\mathbf{v}^T + (C^T\mathbf{p})^T]\mathbf{x}$$
$$= [\mathbf{v}^T + \mathbf{p}^T C]\mathbf{x} = \mathbf{v}^T\mathbf{x} + \mathbf{p}^T C\mathbf{x}.$$

Also

$$\mathbf{p}^T\mathbf{x} = \mathbf{p}^T(C\mathbf{x} + \mathbf{d}) = \mathbf{p}^T C\mathbf{x} + \mathbf{p}^T\mathbf{d}.$$

Hence

$$\mathbf{v}^T\mathbf{x} + \mathbf{p}^T C\mathbf{x} = \mathbf{p}^T C\mathbf{x} + \mathbf{p}^T\mathbf{d}.$$

So $\mathbf{v}^T\mathbf{x} = \mathbf{p}^T\mathbf{d}$.

24. (a) Observe that

$$C^T\mathbf{p} = \begin{bmatrix} .1 & .2 & .3 \\ .2 & .4 & .1 \\ .1 & .2 & .1 \end{bmatrix} \begin{bmatrix} p_1 \\ p_2 \\ p_3 \end{bmatrix}$$
$$= \begin{bmatrix} .1p_1 + .2p_2 + .3p_3 \\ .2p_1 + .4p_2 + .1p_3 \\ .1p_1 + .2p_2 + .1p_3 \end{bmatrix}.$$

It takes .1 ton of agricultural input to produce one ton of agricultural output. So $.1p_1$ is the value of the agricultural input needed to produce 1 ton of agricultural output. Likewise, $.2p_2$ is the value of the agricultural input needed to produce 1 ton of manufacturing output, and $.3p_3$ is the value of the agricultural input needed to produce one ton of services output. So the first component of $C^T\mathbf{p}$ gives the total value of the inputs from each of the three sectors needed to produce a ton of agricultural output. A similar interpretation applies to the other components of $C^T\mathbf{p}$.

(b) The vector $\mathbf{p} - C^T\mathbf{p}$ gives the difference between the selling price of a ton of each sector's output and the value of the inputs needed to produce the output.

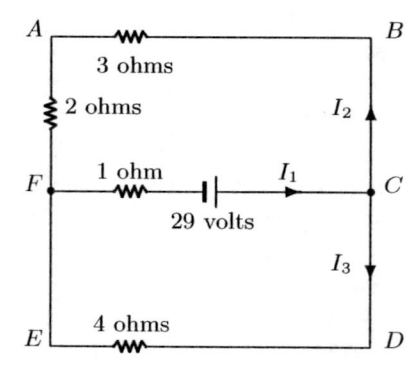

25.

Figure for Exercise 25

Applying Kirchoff's voltage law to the closed path $FCBAF$ in the network above, we obtain the equation

$$3I_2 + 2I_2 + 1I_1 = 29.$$

Similarly, from the closed path $FCDEF$, we obtain

$$1I_1 + 4I_3 = 29.$$

At the junction C, Kirchoff's current law yields the equation

$$I_1 = I_2 + I_3.$$

Thus the currents I_1, I_2, and I_3 satisfy the system

$$\begin{aligned} I_1 + 5I_2 &= 29 \\ I_1 + 4I_3 &= 29 \\ I_1 - I_2 - I_3 &= 0. \end{aligned}$$

Since the reduced row echelon form of the augmented matrix of this system is

$$\begin{bmatrix} 1 & 0 & 0 & 9 \\ 0 & 1 & 0 & 4 \\ 0 & 0 & 1 & 5 \end{bmatrix},$$

this system has $I_1 = 9$, $I_2 = 4$, and $I_3 = 5$ as its unique solution.

26. As in Exercise 25, the currents I_1, I_2, and I_3 satisfy the system

$$\begin{aligned} I_1 + 2I_2 &= 35 \\ 2I_2 - 6I_3 &= 30 \\ I_1 - I_2 - I_3 &= 0. \end{aligned}$$

So $I_1 = 11$, $I_2 = 12$, and $I_3 = -1$.

27. As in Exercise 25, the currents I_1, I_2, and I_3 satisfy the system

$$\begin{aligned} 2I_1 + I_2 &= 60 \\ -I_2 + 6I_3 &= 0 \\ I_1 - I_2 - I_3 &= 0. \end{aligned}$$

So $I_1 = 21$, $I_2 = 18$, and $I_3 = 3$.

28. As in Exercise 25, the currents I_1, I_2, and I_3 satisfy the system

$$\begin{aligned} 2I_1 + 3I_2 &= 30 \\ 3I_2 + 6I_3 &= 54 \\ I_1 - I_2 + I_3 &= 0. \end{aligned}$$

So $I_1 = 3$, $I_2 = 8$, and $I_3 = 5$.

29. As in Exercise 25, the currents I_1, I_2, and I_3 satisfy the system

$$\begin{aligned} 4I_1 + 2I_2 &= 60 \\ I_2 - 2I_3 - I_5 + I_6 &= 0 \\ 2I_4 + I_5 &= 30 \\ I_1 - I_2 - I_3 &= 0 \\ I_2 - I_4 + I_5 &= 0 \\ I_4 - I_5 - I_6 &= 0. \end{aligned}$$

Therefore $I_1 = I_4 = 12.5$, $I_2 = I_6 = 7.5$, and $I_3 = I_5 = 5$.

30.

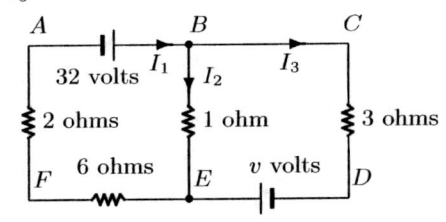

Figure for Exercise 30

Applying Kirchoff's voltage law to the closed path $ABEFA$ in the network above, we obtain the equation

$$1I_2 + 6I_1 + 2I_1 = 32.$$

Similarly, from the closed path $BCDEB$, we obtain

$$1(-I_2) + 3I_3 = v.$$

At junction B, Kirchoff's current law yields the equation

$$I_1 = I_2 + I_3.$$

Thus the currents I_1, I_2, and I_3 satisfy the system

$$\begin{aligned} 8I_1 + I_2 &= 32 \\ -I_2 + 3I_3 &= v \\ I_1 - I_2 - I_3 &= 0. \end{aligned}$$

The reduced row echelon form of the augmented matrix of this system is

$$\begin{bmatrix} 1 & 0 & 0 & \frac{v+128}{35} \\ 0 & 1 & 0 & \frac{-8v+96}{35} \\ 0 & 0 & 1 & \frac{9v+32}{35} \end{bmatrix}.$$

Hence for $I_2 = 0$, we need $-8v + 96 = 0$, that is, $v = 12$.

31. Solving the system $(I_6 - C)\mathbf{x} = \mathbf{d}$, we obtain the gross production vector, which is

$$\mathbf{x} = \begin{bmatrix} 454.0403 \\ 508.6149 \\ 618.7457 \\ 623.7464 \\ 810.8047 \\ 640.3667 \end{bmatrix},$$

where each entry is rounded to 4 places after the decimal point.

1.6 THE SPAN OF A SET OF VECTORS

1. Yes. By the boxed result on page 68, the given vector is in the span if and only if there exists a solution of $A\mathbf{x} = \mathbf{b}$, where

$$A = \begin{bmatrix} 1 & -1 & 1 \\ 0 & 1 & 1 \\ 1 & 1 & 3 \end{bmatrix} \quad \text{and} \quad \mathbf{b} = \begin{bmatrix} -1 \\ 4 \\ 7 \end{bmatrix}.$$

The reduced row echelon form of the augmented matrix associated with this system is

$$\begin{bmatrix} 1 & 0 & 2 & 3 \\ 0 & 1 & 1 & 4 \\ 0 & 0 & 0 & 0 \end{bmatrix}.$$

Thus, by Theorem 1.5, $A\mathbf{x} = \mathbf{b}$ is consistent, and hence the given vector is in the span of the given set.

2. No. The given vector is in the span if and only if there exists a solution of $A\mathbf{x} = \mathbf{b}$, where

$$A = \begin{bmatrix} 1 & -1 & 1 \\ 0 & 1 & 1 \\ 1 & 1 & 3 \end{bmatrix} \quad \text{and} \quad \mathbf{b} = \begin{bmatrix} 0 \\ 0 \\ 1 \end{bmatrix}.$$

The reduced row echelon form of the augmented matrix associated with this system is

$$\begin{bmatrix} 1 & 0 & 2 & 0 \\ 0 & 1 & 1 & 0 \\ 0 & 0 & 0 & 1 \end{bmatrix}.$$

Because of the form of the last row, it follows that the system is inconsistent, and hence the given vector is not in the span of the given set.

3. no
4. no
5. yes
6. yes
7. no
8. yes
9. no
10. yes
11. yes
12. no
13. yes
14. no
15. no
16. yes

16   Chapter 1   Matrices, Vectors, and Systems of Linear Equations

17. As in the solution to Exercise 1, $\mathbf{v}$ is in the span of $\mathcal{S}$ if and only if the system $A\mathbf{x} = \mathbf{v}$ is consistent, where
$$A = \begin{bmatrix} 1 & -1 \\ 0 & 3 \\ -1 & 2 \end{bmatrix}.$$
With elementary row operations, its augmented matrix $\begin{bmatrix} 1 & -1 & 2 \\ 0 & 3 & r \\ -1 & 2 & -1 \end{bmatrix}$ can be transformed to
$$\begin{bmatrix} 1 & -1 & 2 \\ 0 & 1 & 1 \\ 0 & 0 & r-3 \end{bmatrix}.$$
This system is consistent if and only if $r - 3 = 0$, that is, $r = 3$. Therefore $\mathbf{v}$ is in the span of $\mathcal{S}$ if and only if $r = 3$.

18. $r = 2$ 19. $r = -6$ 20. $r = -3$

21. No. Let $A = \begin{bmatrix} 1 & -2 \\ -1 & 2 \end{bmatrix}$. The reduced row echelon form of A is $\begin{bmatrix} 1 & -2 \\ 0 & 0 \end{bmatrix}$, which has rank 1. By Theorem 1.6(c), the set is not a generating set for $\mathcal{R}^2$.

22. yes   23. yes   24. no

25. Yes. Let
$$A = \begin{bmatrix} 1 & -1 & 1 \\ 0 & 1 & 2 \\ -2 & 4 & -2 \end{bmatrix}.$$
The reduced row echelon form of A is
$$\begin{bmatrix} 1 & 0 & 0 \\ 0 & 1 & 0 \\ 0 & 0 & 1 \end{bmatrix},$$
which has rank 3. By Theorem 1.6(c), the set is a generating set for $\mathcal{R}^3$.

26. no   27. no   28. yes

29. Yes. The reduced row echelon form of A is I_2, which has rank 2. By Theorem 1.6(b), the system $A\mathbf{x} = \mathbf{b}$ is consistent for every $\mathbf{b}$ in $\mathcal{R}^2$.

30. no   31. no   32. yes

33. The reduced row echelon form of A is $[\mathbf{e}_1 \ \mathbf{e}_2]$, which has rank 2. By Theorem 1.6(b), the system is inconsistent for some $\mathbf{b}$ in $\mathcal{R}^3$.

34. no   35. yes   36. yes

37. The desired set is $\left\{ \begin{bmatrix} 1 \\ 3 \end{bmatrix}, \begin{bmatrix} 0 \\ 1 \end{bmatrix} \right\}$. If we delete either vector, then the span of $\mathcal{S}$ consists of all multiples of the other vector. Because neither vector in $\mathcal{S}$ is a multiple of the other, neither can be deleted.

38. One possible set is $\left\{ \begin{bmatrix} -1 \\ 1 \end{bmatrix}, \begin{bmatrix} 1 \\ 0 \end{bmatrix} \right\}$. The second vector from $\mathcal{S}$ may be deleted because it is a multiple of the first. Neither of the two remaining vectors can be deleted for the same reason given in Exercise 37.

39. One possible set is $\left\{ \begin{bmatrix} 1 \\ 0 \\ -1 \end{bmatrix}, \begin{bmatrix} 0 \\ 1 \\ 0 \end{bmatrix} \right\}$. The reason is the same as that given in Exercise 38.

40. One possible set is $\left\{ \begin{bmatrix} 1 \\ -1 \\ 2 \end{bmatrix}, \begin{bmatrix} 2 \\ -3 \\ 0 \end{bmatrix} \right\}$. The third vector in $\mathcal{S}$ is the zero vector, and so can be deleted. Neither of the two remaining vectors can be deleted for the same reason given in Exercise 37.

41. One possible set is $\left\{ \begin{bmatrix} 1 \\ -2 \\ 1 \end{bmatrix} \right\}$. The last two vectors in $\mathcal{S}$ are multiples of the first, and so can be deleted.

42. The desired set is $\mathcal{S}$ itself. The rank of the matrix whose columns are the three vectors in $\mathcal{S}$ is 3; so by Theorem 1.6, $\mathcal{S}$ is a generating set for $\mathcal{R}^3$. On the other hand, if a vector is deleted from $\mathcal{S}$, then the rank of the matrix whose columns are the remaining vectors is less than 3, and hence the smaller set is not a generating set for $\mathcal{R}^3$, again by Theorem 1.6.

43. One possible set is $\left\{ \begin{bmatrix} -1 \\ 0 \\ 1 \end{bmatrix}, \begin{bmatrix} 0 \\ 1 \\ 2 \end{bmatrix} \right\}$. Since the reduced row echelon form of the matrix whose columns are the given vectors is $\begin{bmatrix} 1 & 0 & -1 \\ 0 & 1 & 2 \\ 0 & 0 & 0 \end{bmatrix}$, the equation $\begin{bmatrix} -1 & 0 \\ 0 & 1 \\ 1 & 2 \end{bmatrix} \mathbf{x} = \begin{bmatrix} 1 \\ 2 \\ 3 \end{bmatrix}$ has a solution. So the third vector in $\mathcal{S}$ is a linear combination of the first two and hence can be deleted. Neither of the two remaining vectors in $\mathcal{S}$ can be deleted for the reason given in Exercise 37.

44. One possible set is $\left\{\begin{bmatrix}1\\0\\0\end{bmatrix}, \begin{bmatrix}1\\1\\0\end{bmatrix}, \begin{bmatrix}1\\1\\1\end{bmatrix}\right\}$. The reduced row echelon form of the matrix whose columns are the given vectors is
$$\begin{bmatrix} 1 & 0 & 0 & 0 \\ 0 & 1 & 0 & -1 \\ 0 & 0 & 1 & 1 \end{bmatrix}.$$
Using the reasoning in Exercises 43 and 42, we see that the fourth vector in $\mathcal{S}$ can be deleted, but none of the three remaining vectors can be deleted.

45. True 46. True 47. True
48. False, by Theorem 1.6(c), we need rank $A = m$ for $A\mathbf{x} = \mathbf{b}$ to be consistent for every $\mathbf{b}$.
49. True 50. True 51. True
52. False, the sets $\mathcal{S}_1 = \{\mathbf{e}_1\}$ and $\mathcal{S}_2 = \{2\mathbf{e}_1\}$ have the same spans, but are not equal.
53. False, the sets $\mathcal{S}_1 = \{\mathbf{e}_1\}$ and $\mathcal{S}_2 = \{\mathbf{e}_1, 2\mathbf{e}_1\}$ have equal spans, but do not contain the same number of vectors.
54. False, $\mathcal{S} = \{\mathbf{e}_1\}$ and $\mathcal{S} \cup \{2\mathbf{e}_1\}$ have equal spans, but $2\mathbf{e}_1$ is not in $\mathcal{S}$.
55. True 56. True 57. True 58. True
59. True 60. True 61. True 62. True
63. True 64. True
65. (a) 2
    (b) Every choice of the scalars a and b yields a different vector of the form $a\mathbf{u}_1 + b\mathbf{u}_2$. Thus there are infinitely many vectors in the span.
66. $\left\{\begin{bmatrix}1\\0\\0\end{bmatrix}, \begin{bmatrix}0\\1\\0\end{bmatrix}\right\}, \left\{\begin{bmatrix}1\\1\\0\end{bmatrix}, \begin{bmatrix}2\\3\\0\end{bmatrix}\right\}, \left\{\begin{bmatrix}4\\5\\0\end{bmatrix}, \begin{bmatrix}5\\6\\0\end{bmatrix}\right\}$
67. By Exercise 79 in Section 1.4, rank A is less than m. So, by Theorem 1.6, there is some $\mathbf{b}$ in $\mathcal{R}^m$ such that $A\mathbf{x} = \mathbf{b}$ is inconsistent.
68. There must be at least m vectors in the set. Let k be the number of vectors in the set, and let A be the matrix whose columns are these vectors. So A is an $m \times k$ matrix. By Theorem 1.6, rank $A = m$. Thus $k \geq m$ by the solution to Exercise 79 in Section 1.4.
69. Suppose that $\mathcal{S}_1 = \{\mathbf{u}_1, \mathbf{u}_2, \ldots, \mathbf{u}_k\}$ is a generating set for $\mathcal{R}^n$ and $\mathcal{S}_2 = \{\mathbf{u}_1, \mathbf{u}_2, \ldots, \mathbf{u}_r\}$, where $r \geq k$. Let $\mathbf{v}$ be in $\mathcal{R}^n$. Then, for some scalars $a_1, a_2, \ldots, a_k$, we have
$$\mathbf{v} = a_1\mathbf{u}_1 + a_2\mathbf{u}_2 + \cdots + a_k\mathbf{u}_k$$
$$= a_1\mathbf{u}_1 + \cdots + a_k\mathbf{u}_k + 0\mathbf{u}_{k+1} + \cdots + 0\mathbf{u}_r.$$
So $\mathcal{S}_2$ is also a generating set for $\mathcal{R}^n$.

70. If $\mathbf{w}$ is in Span $\{\mathbf{u} + \mathbf{v}, \mathbf{u} - \mathbf{v}\}$, then for some scalars a and b we may write
$$\mathbf{w} = a(\mathbf{u} + \mathbf{v}) + b(\mathbf{u} - \mathbf{v}) = (a+b)\mathbf{u} + (a-b)\mathbf{v}.$$
So $\mathbf{w}$ is also in Span $\{\mathbf{u}, \mathbf{v}\}$.
On the other hand, if $\mathbf{z}$ is in Span $\{\mathbf{u}, \mathbf{v}\}$, then for some scalars c and d, we may write $\mathbf{z} = c\mathbf{u} + d\mathbf{v}$. It follows that
$$\mathbf{z} = \frac{c+d}{2}(\mathbf{u} + \mathbf{v}) + \frac{c-d}{2}(\mathbf{u} - \mathbf{v}).$$
Thus $\mathbf{z}$ is in Span $\{\mathbf{u} + \mathbf{v}, \mathbf{u} - \mathbf{v}\}$. So Span $\{\mathbf{u}, \mathbf{v}\}$ and Span $\{\mathbf{u} + \mathbf{v}, \mathbf{u} - \mathbf{v}\}$ are each contained in one another and hence are equal.

71. Use the reasoning in Exercise 70 and the relationship
$$a_1\mathbf{u}_1 + a_2\mathbf{u}_2 + \cdots + a_k\mathbf{u}_k$$
$$= \frac{a_1}{c_1}(c_1\mathbf{u}_1) + \frac{a_2}{c_2}(c_2\mathbf{u}_2) + \cdots + \frac{a_k}{c_k}(c_k\mathbf{u}_k),$$
which holds for all scalars $a_1, a_2, \ldots, a_k$.

72. Use the reasoning in Exercise 70 and the relationship
$$a_1\mathbf{u}_1 + a_2\mathbf{u}_2 + \cdots + a_k\mathbf{u}_k$$
$$= a_1(\mathbf{u}_1 + c\mathbf{u}_2) + (a_2 - ca_1)\mathbf{u}_2$$
$$+ a_3\mathbf{u}_3 + \cdots + a_k\mathbf{u}_k,$$
which holds for all scalars $a_1, a_2, \ldots, a_k$.

73. No, let $A = \begin{bmatrix} 1 & 0 \\ 1 & 0 \end{bmatrix}$. Then $R = \begin{bmatrix} 1 & 0 \\ 0 & 0 \end{bmatrix}$. The span of the columns of A equals all multiples of $\begin{bmatrix} 1 \\ 1 \end{bmatrix}$, whereas, the span of the columns of R equals all multiples of $\mathbf{e}_1$.

74. Let $\mathbf{v}$ be in the span of $\mathcal{S}_1 = \{\mathbf{u}_1, \mathbf{u}_2, \ldots, \mathbf{u}_k\}$, and let $\mathcal{S}_2 = \{\mathbf{u}_1, \mathbf{u}_2, \ldots, \mathbf{u}_m\}$, where $k \leq m$. We can write $\mathbf{v} = a_1\mathbf{u}_1 + a_2\mathbf{u}_2 + \cdots + a_k\mathbf{u}_k$ for some scalars $a_1, a_2, \ldots, a_k$. Since
$$\mathbf{v} = a_1\mathbf{u}_1 + a_2\mathbf{u}_2 + \cdots + a_k\mathbf{u}_k + 0\mathbf{u}_{k+1} + \cdots + 0\mathbf{u}_m,$$
we see that $\mathbf{v}$ is also in Span $\mathcal{S}_2$.

75. Let $\mathcal{S} = \{\mathbf{u}_1, \mathbf{u}_2, \ldots, \mathbf{u}_k\}$. If $\mathbf{u}$ and $\mathbf{v}$ are in Span $\mathcal{S}$, then $\mathbf{u} = a_1\mathbf{u}_1 + a_2\mathbf{u}_2 + \cdots + a_k\mathbf{u}_k$ and $\mathbf{v} = b_1\mathbf{u}_1 + b_2\mathbf{u}_2 + \cdots + b_k\mathbf{u}_k$ for some scalars $a_1, a_2, \ldots, a_k$ and $b_1, b_2, \ldots, b_k$. For any scalar c, we have
$$\mathbf{u} + c\mathbf{v} = (a_1\mathbf{u}_1 + \cdots + a_k\mathbf{u}_k)$$

18   Chapter 1   Matrices, Vectors, and Systems of Linear Equations

$$+ c(b_1\mathbf{u}_1 + \cdots + b_k\mathbf{u}_k)$$
$$= (a_1 + cb_1)\mathbf{u}_1 + \cdots + (a_k + cb_k)\mathbf{u}_k.$$

So $\mathbf{u} + c\mathbf{v}$ is in Span S.

76. Let $V = \text{Span }\mathcal{S}$. If $\mathcal{S}$ contains only the zero vector, then $\text{Span }\mathcal{S} = \{\mathbf{0}\}$. If $\mathcal{S}$ contains a nonzero vector $\mathbf{u}$, then $\text{Span }\mathcal{S}$ contains all multiples of $\mathbf{u}$, and hence is infinite.

77. Let B be obtained by performing a single elementary row operation on a matrix A. If B is obtained by interchanging two rows of A, then the set of rows of A is the same as that of B. So the spans of their rows must be equal.

    If B is obtained by multiplying row i of A by a nonzero scalar c, then the result follows from Exercise 71 by taking $c_i = c$ and $c_j = 1$ if $j \ne i$. Finally, if B is obtained by multiplying row 2 of A by the scalar c and adding it to row 1, the result follows from Exercise 72. The general case follows by a similar argument.

78. Let R be the reduced row echelon form of A. By applying appropriate elementary row operations to A, we obtain a sequence $A, B_1, B_2, \ldots, B_k, R$, where each matrix after the first is formed from the previous matrix by an elementary row operation. By Exercise 77, the span of the rows of any of these matrices equals the span of the rows of the previous matrix. Thus the span of the rows of A equals the span of the rows of R.

79. Yes. As in the analysis in Exercise 1, a vector $\mathbf{v}$ is in the span of $\{\mathbf{u}_1, \mathbf{u}_2, \mathbf{u}_3, \mathbf{u}_4\}$ if and only if the system $[\mathbf{u}_1\ \mathbf{u}_2\ \mathbf{u}_3\ \mathbf{u}_4]\mathbf{x} = \mathbf{v}$ is consistent. Since the reduced row echelon form of the associated augmented matrix does not contain the row $[0\ 0\ 0\ 0\ 1]$, $\mathbf{v}$ is in the span of $\{\mathbf{u}_1, \mathbf{u}_2, \mathbf{u}_3, \mathbf{u}_4\}$.

80. no   81. no   82. yes

1.7 LINEAR DEPENDENCE AND LINEAR INDEPENDENCE

1. Yes, the second vector is a multiple of the first.
2. No, neither vector is a multiple of the other.
3. Yes, the second vector is a multiple of the first.
4. Yes, the set contains the zero vector.
5. No, the first two vectors are linearly independent because neither is a multiple of the other. The third vector is not a linear combination of the first two because its first component is not zero. So, by Theorem 1.9, the set of 3 vectors is not linearly dependent.
6. yes, by property 4 of linearly dependent and linearly independent sets
7. yes   8. yes   9. no
10. yes   11. yes   12. yes
13. $\left\{ \begin{bmatrix} 1 \\ -2 \\ 3 \end{bmatrix} \right\}$, because the second vector is a multiple of the first
14. $\left\{ \begin{bmatrix} 1 \\ 0 \\ 2 \end{bmatrix}, \begin{bmatrix} 3 \\ -1 \\ 1 \end{bmatrix} \right\}$, because neither vector is a multiple of the other
15. $\left\{ \begin{bmatrix} -3 \\ 2 \\ 0 \end{bmatrix}, \begin{bmatrix} 1 \\ 6 \\ 0 \end{bmatrix} \right\}$, because the third vector is zero, and neither of the first two vectors are multiples of the other
16. $\left\{ \begin{bmatrix} 0 \\ 0 \\ 1 \end{bmatrix}, \begin{bmatrix} 0 \\ 1 \\ 2 \end{bmatrix}, \begin{bmatrix} 1 \\ 2 \\ 3 \end{bmatrix} \right\}$ because the fourth vector is the sum of the first three, and none of the first three can be deleted for a reason similar to that in the solution to Exercise 5
17. $\left\{ \begin{bmatrix} 2 \\ -3 \\ 5 \end{bmatrix}, \begin{bmatrix} 1 \\ 0 \\ 2 \end{bmatrix} \right\}$ 18. $\left\{ \begin{bmatrix} 1 \\ 0 \\ -1 \end{bmatrix} \right\}$
19. $\left\{ \begin{bmatrix} 4 \\ 3 \end{bmatrix}, \begin{bmatrix} -2 \\ 5 \end{bmatrix} \right\}$ 20. $\left\{ \begin{bmatrix} 1 \\ 2 \\ -3 \end{bmatrix}, \begin{bmatrix} 4 \\ -6 \\ 2 \end{bmatrix} \right\}$
21. $\left\{ \begin{bmatrix} -2 \\ 0 \\ 3 \end{bmatrix}, \begin{bmatrix} 0 \\ 4 \\ 0 \end{bmatrix} \right\}$ 22. $\left\{ \begin{bmatrix} 2 \\ 1 \\ 0 \end{bmatrix}, \begin{bmatrix} 3 \\ 2 \\ 1 \end{bmatrix} \right\}$
23. No, let
$$A = \begin{bmatrix} 1 & -1 & 1 \\ -1 & 0 & 2 \\ -2 & 1 & 1 \end{bmatrix},$$
which has reduced row echelon form
$$\begin{bmatrix} 1 & 0 & -2 \\ 0 & 1 & -3 \\ 0 & 0 & 0 \end{bmatrix}.$$
So rank $A = 2$. By Theorem 1.8, the set is linearly dependent.
24. yes
25. Yes, let
$$A = \begin{bmatrix} 1 & 1 & 1 \\ 2 & -3 & 2 \\ 0 & 1 & -2 \\ -1 & -2 & 3 \end{bmatrix},$$

1.7 Linear Dependence and Linear Independence

which has reduced row echelon form
$$\begin{bmatrix} 1 & 0 & 0 \\ 0 & 1 & 0 \\ 0 & 0 & 1 \\ 0 & 0 & 0 \end{bmatrix}.$$

So rank $A = 3$. By Theorem 1.8, the set is linearly independent.

26. no **27.** yes **28.** no **29.** no

30. yes

31. $-3\begin{bmatrix} -1 \\ 1 \\ 2 \end{bmatrix} + 0\begin{bmatrix} 0 \\ 1 \\ 2 \end{bmatrix} = \begin{bmatrix} 3 \\ -3 \\ -6 \end{bmatrix}$

32. $\begin{bmatrix} 0 \\ 0 \\ 0 \end{bmatrix} = 0\begin{bmatrix} -2 \\ 3 \\ -4 \end{bmatrix} + 0\begin{bmatrix} 4 \\ -3 \\ 2 \end{bmatrix}$

33. $5\begin{bmatrix} 0 \\ 1 \\ 1 \end{bmatrix} + 4\begin{bmatrix} 1 \\ 0 \\ -1 \end{bmatrix} = \begin{bmatrix} 4 \\ 5 \\ 1 \end{bmatrix}$

34. $\begin{bmatrix} 4 \\ 6 \\ -2 \end{bmatrix} = 6\begin{bmatrix} 1 \\ 2 \\ -1 \end{bmatrix} + 2\begin{bmatrix} -1 \\ -3 \\ 2 \end{bmatrix}$

35. $1\begin{bmatrix} 1 \\ -1 \end{bmatrix} + 5\begin{bmatrix} 0 \\ 1 \end{bmatrix} + 0\begin{bmatrix} 3 \\ -2 \end{bmatrix} = \begin{bmatrix} 1 \\ 4 \end{bmatrix}$

36. $-3\begin{bmatrix} 1 \\ 0 \\ 3 \end{bmatrix} + 4\begin{bmatrix} 2 \\ -1 \\ 1 \end{bmatrix} = \begin{bmatrix} 5 \\ -4 \\ -5 \end{bmatrix}$

37. $5\begin{bmatrix} 1 \\ 2 \\ -1 \end{bmatrix} - 3\begin{bmatrix} 0 \\ 1 \\ -1 \end{bmatrix} + 3\begin{bmatrix} -1 \\ -2 \\ 0 \end{bmatrix} = \begin{bmatrix} 2 \\ 1 \\ -2 \end{bmatrix}$

38. $5\begin{bmatrix} 1 \\ 0 \\ -1 \\ -1 \\ 1 \end{bmatrix} + 2\begin{bmatrix} -1 \\ 1 \\ 1 \\ 0 \\ 1 \end{bmatrix} + 3\begin{bmatrix} -1 \\ -1 \\ 2 \\ 1 \\ 0 \end{bmatrix} = \begin{bmatrix} 0 \\ -1 \\ 3 \\ -2 \\ 7 \end{bmatrix}$

39. all real numbers, because the second vector is a multiple of the first

40. no value of r, because the second components of the first two vectors are both 0

41. When $r = -2$, the sum of the first two vectors equals the third.

42. Apply elementary row operations to the matrix whose columns are the given vectors to obtain
$$\begin{bmatrix} 1 & 0 & -1 & -1 \\ 0 & -1 & 1 & 9 \\ 0 & 0 & 2 & r+17 \\ 0 & 0 & 0 & -r-9 \end{bmatrix}.$$

The rank of this matrix is less than 4 if and only if $r = -9$, so by Theorem 1.8 the vectors are linearly dependent if and only if $r = -9$.

43. Every vector in $\mathcal{R}^2$ is a linear combination of two nonparallel vectors in $\mathcal{R}^2$. Hence the given set is linearly dependent for every value of r.

44. $r = 0$ **45.** every r **46.** no r **47.** 4

48. 7 **49.** no r **50.** every r

51. The solution of the system is
$$\begin{aligned} x_1 &= 4x_2 - 2x_3 \\ x_2 &\quad \text{free} \\ x_3 &\quad \text{free.} \end{aligned}$$

So
$$\begin{bmatrix} x_1 \\ x_2 \\ x_3 \end{bmatrix} = \begin{bmatrix} 4x_2 - 2x_3 \\ x_2 \\ x_3 \end{bmatrix} = x_2\begin{bmatrix} 4 \\ 1 \\ 0 \end{bmatrix} + x_3\begin{bmatrix} -2 \\ 0 \\ 1 \end{bmatrix}.$$

52. $\begin{bmatrix} x_1 \\ x_2 \\ x_3 \end{bmatrix} = x_3\begin{bmatrix} -5 \\ 3 \\ 1 \end{bmatrix}$

53. $\begin{bmatrix} x_1 \\ x_2 \\ x_3 \\ x_4 \end{bmatrix} = x_2\begin{bmatrix} -3 \\ 1 \\ 0 \\ 0 \end{bmatrix} + x_4\begin{bmatrix} -2 \\ 0 \\ 6 \\ 1 \end{bmatrix}$

54. $\begin{bmatrix} x_1 \\ x_2 \\ x_3 \\ x_4 \end{bmatrix} = x_3\begin{bmatrix} 0 \\ 0 \\ 1 \\ 0 \end{bmatrix} + x_4\begin{bmatrix} -4 \\ 2 \\ 0 \\ 1 \end{bmatrix}$

55. $\begin{bmatrix} x_1 \\ x_2 \\ x_3 \\ x_4 \end{bmatrix} = x_3\begin{bmatrix} -4 \\ 3 \\ 1 \\ 0 \end{bmatrix} + x_4\begin{bmatrix} 2 \\ -5 \\ 0 \\ 1 \end{bmatrix}$

56. $\begin{bmatrix} x_1 \\ x_2 \\ x_3 \\ x_4 \end{bmatrix} = x_2\begin{bmatrix} 2 \\ 1 \\ 0 \\ 0 \end{bmatrix} + x_4\begin{bmatrix} 1 \\ 0 \\ -3 \\ 1 \end{bmatrix}$

57. $\begin{bmatrix} x_1 \\ x_2 \\ x_3 \\ x_4 \\ x_5 \\ x_6 \end{bmatrix} = x_2\begin{bmatrix} 0 \\ 1 \\ 0 \\ 0 \\ 0 \\ 0 \end{bmatrix} + x_4\begin{bmatrix} -1 \\ 0 \\ 2 \\ 1 \\ 0 \\ 0 \end{bmatrix} + x_6\begin{bmatrix} -3 \\ 0 \\ -1 \\ 0 \\ 0 \\ 1 \end{bmatrix}$

58. $\begin{bmatrix} x_1 \\ x_2 \\ x_3 \\ x_4 \\ x_5 \end{bmatrix} = x_5\begin{bmatrix} 0.0 \\ -8.5 \\ -1.5 \\ -2.0 \\ 1.0 \end{bmatrix}$

59. $\begin{bmatrix} x_1 \\ x_2 \\ x_3 \\ x_4 \end{bmatrix} = x_3\begin{bmatrix} 0 \\ 0 \\ 1 \\ 0 \end{bmatrix} + x_4\begin{bmatrix} 2 \\ -3 \\ 0 \\ 1 \end{bmatrix}$

60. $\begin{bmatrix} x_1 \\ x_2 \\ x_3 \\ x_4 \\ x_5 \end{bmatrix} = x_2 \begin{bmatrix} 2 \\ 1 \\ 0 \\ 0 \\ 0 \end{bmatrix} + x_4 \begin{bmatrix} -2 \\ 0 \\ 1 \\ 1 \\ 0 \end{bmatrix} + x_5 \begin{bmatrix} -4 \\ 0 \\ -3 \\ 0 \\ 1 \end{bmatrix}$

61. $\begin{bmatrix} x_1 \\ x_2 \\ x_3 \\ x_4 \\ x_5 \\ x_6 \end{bmatrix} = x_2 \begin{bmatrix} -2 \\ 1 \\ 0 \\ 0 \\ 0 \\ 0 \end{bmatrix} + x_3 \begin{bmatrix} 1 \\ 0 \\ 1 \\ 0 \\ 0 \\ 0 \end{bmatrix} + x_5 \begin{bmatrix} -2 \\ 0 \\ 0 \\ -4 \\ 1 \\ 0 \end{bmatrix} + x_6 \begin{bmatrix} 1 \\ 0 \\ 0 \\ -3 \\ 0 \\ 1 \end{bmatrix}$

62. $\begin{bmatrix} x_1 \\ x_2 \\ x_3 \\ x_4 \\ x_5 \\ x_6 \end{bmatrix} = x_2 \begin{bmatrix} 1 \\ 1 \\ 0 \\ 0 \\ 0 \\ 0 \end{bmatrix} + x_4 \begin{bmatrix} 2 \\ 0 \\ -3 \\ 1 \\ 0 \\ 0 \end{bmatrix} + x_6 \begin{bmatrix} -2 \\ 0 \\ -1 \\ 0 \\ 2 \\ 1 \end{bmatrix}$

63. True

64. False, by Theorem 1.8 the *columns* are linearly independent. Consider the matrix $\begin{bmatrix} 1 & 0 \\ 0 & 1 \\ 0 & 0 \end{bmatrix}$.

65. False, by Theorem 1.8, the columns are linearly independent. See the matrix in the solution to Exercise 64.

66. True 67. True 68. True

69. False, consider the equation $I_2 \mathbf{x} = \mathbf{0}$.

70. True

71. False, if $\mathbf{v} \ne \mathbf{0}$.

72. False, consider the set $\left\{ \begin{bmatrix} 1 \\ 0 \end{bmatrix}, \begin{bmatrix} 0 \\ 1 \end{bmatrix}, \begin{bmatrix} 1 \\ 1 \end{bmatrix} \right\}$.

73. False, consider $n = 3$ and the set $\left\{ \begin{bmatrix} 1 \\ 0 \\ 0 \end{bmatrix}, \begin{bmatrix} 2 \\ 0 \\ 0 \end{bmatrix} \right\}$.

74. True 75. True 76. True

77. False, consider the example in Exercise 64.

78. True

79. False, $\{\mathbf{u}_1, \mathbf{u}_2, \ldots, \mathbf{u}_k\}$ is linearly independent *if the only scalars* such that
$$c_1 \mathbf{u}_1 + c_2 \mathbf{u}_2 + \cdots + c_k \mathbf{u}_k = \mathbf{0}$$
are $c_1 = c_2 = \cdots = c_k = 0$.

80. True 81. True 82. True

83. $\begin{bmatrix} 1 & 0 \\ 0 & 1 \end{bmatrix}$ 84. $\begin{bmatrix} 1 & 0 \\ 0 & 0 \end{bmatrix}$

85. In $\mathcal{R}^3$, let $\mathbf{u}_1 = \mathbf{e}_1$, $\mathbf{u}_2 = \mathbf{e}_2$, and $\mathbf{v} = \mathbf{e}_1 + \mathbf{e}_2$. The sets $\{\mathbf{u}_1, \mathbf{u}_2\}$, and $\{\mathbf{v}\}$ are both linearly independent, but the set $\{\mathbf{u}_1, \mathbf{u}_2, \mathbf{v}\}$ is linearly dependent because $\mathbf{v} = \mathbf{u}_1 + \mathbf{u}_2$.

86. Let $a_1, a_2, \ldots, a_k$ be scalars such that
$$a_1 \mathbf{v} + a_2 \mathbf{u}_2 + \cdots + a_k \mathbf{u}_k = \mathbf{0}.$$
Substituting for $\mathbf{v}$ in this equation produces
$$\begin{aligned} \mathbf{0} &= a_1 \mathbf{v} + a_2 \mathbf{u}_2 + \cdots + a_k \mathbf{u}_k \\ &= a_1(c_1 \mathbf{u}_1 + c_2 \mathbf{u}_2 + \cdots + c_k \mathbf{u}_k) \\ &\quad + a_2 \mathbf{u}_2 + \cdots + a_k \mathbf{u}_k \\ &= a_1 c_1 \mathbf{u}_1 + (a_1 c_2 + a_2) \mathbf{u}_2 + \cdots \\ &\quad + (a_1 c_k + a_k) \mathbf{u}_k. \end{aligned}$$
Because $\{\mathbf{u}_1, \mathbf{u}_2, \ldots, \mathbf{u}_k\}$ is linearly independent, all the coefficients of this linear combination are 0, that is,
$$a_1 c_1 = a_1 c_2 + a_2 = \ldots = a_1 c_k + a_k = 0.$$
But if $a_1 c_1 = 0$ and $c_1 \ne 0$, then $a_1 = 0$. Thus these equalities reduce to $a_1 = a_2 = \ldots = a_k = 0$. This proves that $\{\mathbf{v}, \mathbf{u}_2, \ldots, \mathbf{u}_k\}$ is linearly independent.

87. Let $\{\mathbf{u}, \mathbf{v}\}$ be linearly independent, and assume that
$$a(\mathbf{u} + \mathbf{v}) + b(\mathbf{u} - \mathbf{v}) = \mathbf{0}$$
for some scalars a and b. Then
$$(a + b)\mathbf{u} + (a - b)\mathbf{v} = \mathbf{0}.$$
Because $\{\mathbf{u}, \mathbf{v}\}$ is linearly independent, we have $a + b = 0$ and $a - b = 0$. So $a = b = 0$. Thus $\{\mathbf{u} + \mathbf{v}, \mathbf{u} - \mathbf{v}\}$ is linearly independent.

Now let $\{\mathbf{u} + \mathbf{v}, \mathbf{u} - \mathbf{v}\}$ be linearly independent, and assume that $c\mathbf{u} + d\mathbf{v} = \mathbf{0}$ for some scalars c and d. We may write
$$c\mathbf{u} + d\mathbf{v} = \frac{c+d}{2}(\mathbf{u}+\mathbf{v}) + \frac{c-d}{2}(\mathbf{u}-\mathbf{v}).$$
Because $\{\mathbf{u} + \mathbf{v}, \mathbf{u} - \mathbf{v}\}$ is linearly independent, we have
$$\frac{c+d}{2} = 0 = \frac{c-d}{2},$$
from which it follows that $c = d = 0$. So $\{\mathbf{u}, \mathbf{v}\}$ is linearly independent.

88. The proof is similar to that of Exercise 87.

89. Suppose $a_1, a_2, \ldots, a_k$ are scalars such that
$$a_1(c_1 \mathbf{u}_1) + a_2(c_2 \mathbf{u}_2) + \cdots + a_k(c_k \mathbf{u}_k) = \mathbf{0},$$

that is,
$$(a_1 c_1)\mathbf{u}_1 + (a_2 c_2)\mathbf{u}_2 + \cdots + (a_k c_k)\mathbf{u}_k = \mathbf{0}.$$

Because $\{\mathbf{u}_1, \mathbf{u}_2, \ldots, \mathbf{u}_k\}$ is linearly independent, we have $a_1 c_1 = a_2 c_2 = \cdots = a_k c_k = 0$. Thus, since $c_1, c_2, \ldots, c_k$ are nonzero, it follows that $a_1 = a_2 = \cdots = a_k = 0$.

90. If $\mathbf{u}_1 = \mathbf{0}$, then write
$$1\mathbf{u}_1 + 0\mathbf{u}_2 + \cdots + 0\mathbf{u}_k = \mathbf{0},$$
from which it follows that $\{\mathbf{u}_1, \mathbf{u}_2, \ldots, \mathbf{u}_k\}$ is linearly dependent. If $\mathbf{u}_i$ is in the span of $\{\mathbf{u}_1, \mathbf{u}_2, \ldots, \mathbf{u}_{i-1}\}$, then
$$\mathbf{u}_i = c_1 \mathbf{u}_1 + c_2 \mathbf{u}_2 + \cdots + c_{i-1} \mathbf{u}_{i-1}$$
for some scalars $c_1, c_2, \ldots, c_{i-1}$. Because we can write
$$\mathbf{0} = c_1 \mathbf{u}_1 + c_2 \mathbf{u}_2 + \cdots + c_{i-1} \mathbf{u}_{i-1}$$
$$+ (-1)\mathbf{u}_i + 0\mathbf{u}_{i+1} + \cdots + 0\mathbf{u}_k,$$
it follows that $\{\mathbf{u}_1, \mathbf{u}_2, \ldots, \mathbf{u}_k\}$ is linearly dependent.

91. Let $\mathcal{S}' = \{\mathbf{u}_1, \mathbf{u}_2, \ldots, \mathbf{u}_r\}$ be a nonempty subset of the linearly independent set
$$\mathcal{S} = \{\mathbf{u}_1, \mathbf{u}_2, \ldots, \mathbf{u}_r, \mathbf{u}_{r+1}, \ldots, \mathbf{u}_k\}.$$
For some scalars $c_1, c_2, \ldots, c_r$, suppose that
$$c_1 \mathbf{u}_1 + c_2 \mathbf{u}_2 + \cdots + c_r \mathbf{u}_r = \mathbf{0}.$$
Then
$$c_1 \mathbf{u}_1 + c_2 \mathbf{u}_2 + \cdots + c_r \mathbf{u}_r + 0\mathbf{u}_{r+1} + \cdots + 0\mathbf{u}_k = \mathbf{0}.$$
Because $\mathcal{S}$ is linearly independent, it follows that $c_1 = c_2 = \cdots = c_r = 0$. So $\mathcal{S}'$ is linearly independent.

92. This exercise states the contrapositive of the result in Exercise 91.

93. Suppose that $\mathbf{v}$ is in the span of $\mathcal{S}$ and that
$$\mathbf{v} = c_1 \mathbf{u}_1 + c_2 \mathbf{u}_2 + \cdots + c_k \mathbf{u}_k$$
and
$$\mathbf{v} = d_1 \mathbf{u}_1 + d_2 \mathbf{u}_2 + \cdots + d_k \mathbf{u}_k$$
for scalars $c_1, c_2, \ldots, c_k, d_1, d_2, \ldots, d_k$. Subtracting the second equation from the first yields
$$\mathbf{0} = (c_1 - d_1)\mathbf{u}_1 + (c_2 - d_2)\mathbf{u}_2 + \cdots + (c_k - d_k)\mathbf{u}_k.$$
Because $\mathcal{S}$ is linearly independent, it follows that $0 = c_1 - d_1 = c_2 - d_2 = \cdots = c_k - d_k$. So $c_1 = d_1, c_2 = d_2, \cdots, c_k = d_k$.

94. Let $\mathcal{S} = \{\mathbf{u}_1, \mathbf{u}_2, \ldots, \mathbf{u}_k\}$ be a nonempty subset of $\mathcal{R}^n$. If every vector in the span of $\mathcal{S}$ can be written as a unique linear combination of $\mathbf{u}_1, \mathbf{u}_2, \ldots, \mathbf{u}_k$, then $\mathcal{S}$ is linearly independent. To show that $\mathcal{S}$ is linearly independent, suppose that
$$c_1 \mathbf{u}_1 + c_2 \mathbf{u}_2 + \cdots + c_k \mathbf{u}_k = \mathbf{0}$$
for some scalars $c_1, c_2, \ldots, c_k$. But also
$$0\mathbf{u}_1 + 0\mathbf{u}_2 + \cdots + 0\mathbf{u}_k = \mathbf{0}.$$
By uniqueness $c_1 = 0, c_2 = 0, \cdots, c_k = 0$. So $\mathcal{S}$ is linearly independent.

95. Because $\mathcal{S}$ is linearly dependent, there are scalars $c_1, c_2, \ldots, c_k$, not all zero, such that
$$c_1 \mathbf{u}_1 + c_2 \mathbf{u}_2 + \cdots + c_k \mathbf{u}_k = \mathbf{0}.$$
So
$$A(c_1 \mathbf{u}_1 + c_2 \mathbf{u}_2 + \cdots + c_k \mathbf{u}_k) = A\mathbf{0}.$$
Hence
$$c_1 A\mathbf{u}_1 + c_2 A\mathbf{u}_2 + \cdots + c_k A\mathbf{u}_k = \mathbf{0}.$$
Therefore $\{A\mathbf{u}_1, A\mathbf{u}_2, \ldots, A\mathbf{u}_k\}$ is linearly dependent.

96. In $\mathcal{R}^2$, let $\mathcal{S} = \{\mathbf{e}_1, \mathbf{e}_2\}$ and $A = \begin{bmatrix} 1 & 2 \\ 1 & 2 \end{bmatrix}$. Then $\mathcal{S}$ is linearly independent, but
$$\mathcal{S}' = \{A\mathbf{e}_1, A\mathbf{e}_2\} = \left\{ \begin{bmatrix} 1 \\ 1 \end{bmatrix}, \begin{bmatrix} 2 \\ 2 \end{bmatrix} \right\}$$
is linearly dependent.

97. Suppose
$$c_1 A\mathbf{u}_1 + c_2 A\mathbf{u}_2 + \cdots + c_k A\mathbf{u}_k = \mathbf{0}$$
for some scalars $c_1, c_2, \ldots, c_k$. Then
$$A(c_1 \mathbf{u}_1 + c_2 \mathbf{u}_2 + \cdots + c_k \mathbf{u}_k) = \mathbf{0}.$$
By Theorem 1.8, it follows that
$$c_1 \mathbf{u}_1 + c_2 \mathbf{u}_2 + \cdots + c_k \mathbf{u}_k = \mathbf{0}.$$
Because $\mathcal{S}$ is linearly independent, it follows that $c_1 = c_2 = \cdots = c_k = 0$.

98. If B is obtained from A by a row interchange, then the sets of rows of A and B are the same; so the result is immediate. If B is obtained from A by a scaling operation, then the result follows from Exercise 89. So we need to prove the

result only for row addition operations. To simplify the notation, we consider only the operation of adding k times the first row of A to the second row. Let the rows of A be denoted by $\mathbf{u}_1, \mathbf{u}_2, \ldots, \mathbf{u}_m$. Then the rows of B are given by $\mathbf{u}_1, \mathbf{u}_2 + k\mathbf{u}_1, \mathbf{u}_3, \ldots, \mathbf{u}_m$. Suppose for scalars $c_1, c_2, \ldots, c_m$ we have

$$c_1\mathbf{u}_1 + c_2(\mathbf{u}_2 + k\mathbf{u}_1) + c_3\mathbf{u}_3 + \cdots + c_m\mathbf{u}_m = \mathbf{0}.$$

Then

$$(c_1 + c_2 k)\mathbf{u}_1 + c_2\mathbf{u}_2 + c_3\mathbf{u}_3 + \cdots + c_m\mathbf{u}_m = \mathbf{0}.$$

Because the rows of A are linearly independent, we have $c_1 + c_2 k = 0, c_2 = 0, c_3 = 0, \ldots, c_m = 0$. It follows that $c_1 = c_2 = \cdots = c_m = 0$, and so the rows of B are linearly independent.

99. Let $\mathbf{u}_1, \mathbf{u}_2, \ldots, \mathbf{u}_k$ be the nonzero rows of a matrix A in reduced row echelon form, and suppose

$$a_1\mathbf{u}_1 + a_2\mathbf{u}_2 + \cdots + a_k\mathbf{u}_k = \mathbf{0}$$

for some scalars $a_1, a_2, \ldots, a_k$. Each $\mathbf{u}_i$ has a component equal to 1 in a position where all the other vectors have components of 0. By equating these components on both sides of the equation, we obtain $a_1 = 0, a_2 = 0, \cdots, a_k = 0$. So $\mathbf{u}_1, \mathbf{u}_2, \ldots, \mathbf{u}_k$ are linearly independent.

100. Let R be the reduced row echelon form of A. Suppose A has linearly independent rows. By repeatedly applying the result in Exercise 98, we see that R has linearly independent rows. So R has no zero rows, and therefore rank $A = m$. Conversely, if rank $A = m$, then R has no zero rows, and so, by Exercise 99, the rows of R are linearly independent. There is a sequence of elementary row operations that transforms R into A. Beginning with R, apply Exercise 98 repeatedly to deduce that the rows of A are linearly independent.

101. Let $\mathbf{v}_j$ denote the jth vector in the set. The set is linearly dependent and $\mathbf{v}_5 = 2\mathbf{v}_1 - \mathbf{v}_3 + \mathbf{v}_4$.

102. The set is linearly independent.

103. The set is linearly independent.

104. The set is linearly independent.

CHAPTER 1 REVIEW EXERCISES

1. False, the columns are 3×1 vectors.
2. True   3. True   4. True
5. True   6. True   7. True
8. False, the nonzero entry has to be the last entry.
9. False. For the matrix $\begin{bmatrix} 1 & 0 & 2 \\ 0 & 1 & 3 \\ 0 & 0 & 0 \end{bmatrix}$, which is in reduced row echelon form, the associated system has the unique solution $x_1 = 2, x_2 = 3$.
10. True   11. True   12. True
13. False, in $A = \begin{bmatrix} 1 & 2 \end{bmatrix}$, the columns are linearly dependent, but rank $A = 1$, the number of rows in A.
14. True   15. True
16. False, the subset $\left\{ \begin{bmatrix} 1 \\ 2 \\ 3 \end{bmatrix}, \begin{bmatrix} 2 \\ 4 \\ 6 \end{bmatrix} \right\}$ of $\mathcal{R}^3$ is linearly dependent.
17. False, consider the example in Exercise 16.
18. (a) Misuse, "inconsistent" applies to systems of linear equations.
    (b) Misuse, "solution" applies to systems of linear equations.
    (c) Misuse, "equivalent" applies to systems of linear equations.
    (d) Misuse, "nullity" applies to matrices.
    (e) Misuse, "span" applies to finite subsets of $\mathcal{R}^n$.
    (f) Misuse, "generating set" applies to finite subsets of $\mathcal{R}^n$.
    (g) Misuse, "homogeneous" applies to systems of linear equations.
    (h) Misuse, "linearly independent" applies to finite subsets of $\mathcal{R}^n$.
19. (a) There is at most one solution. (See the table at the end of Section 1.7.)
    (b) There is at least one solution. (See the table at the end of Section 1.7.)
20. undefined because A and B^T don't have the same number of rows
21. $\begin{bmatrix} 3 & 2 \\ -2 & 7 \\ 4 & 3 \end{bmatrix}$ 22. $\begin{bmatrix} -3 \\ 15 \\ 9 \end{bmatrix}$
23. undefined because A has 2 columns and D^T has 3 rows
24. $\begin{bmatrix} -4 & 9 \\ -4 & -1 \\ -12 & 1 \end{bmatrix}$ 25. $\begin{bmatrix} 3 \\ 3 \end{bmatrix}$
26. undefined because A^T and B don't have the same number of rows
27. undefined because C^T and D don't have the same number of columns

28. The boat has a velocity of
$$\mathbf{v} = \frac{\sqrt{2}}{2}\begin{bmatrix}-10\\-10\end{bmatrix} = -\sqrt{2}\begin{bmatrix}5\\5\end{bmatrix} \text{ mph},$$
and the passenger has a velocity of
$$\mathbf{w} = \frac{\sqrt{2}}{2}\begin{bmatrix}-2\\2\end{bmatrix} = \sqrt{2}\begin{bmatrix}-1\\1\end{bmatrix} \text{ mph}$$
with respect to the boat. So with respect to the riverbank, the passenger has a velocity of $\mathbf{v} + \mathbf{w} = \begin{bmatrix}-6\sqrt{2}\\-4\sqrt{2}\end{bmatrix}$ mph and a speed of
$$\sqrt{(-6\sqrt{2})^2 + (-4\sqrt{2})^2} = \sqrt{104} \approx 10.20 \text{ mph}.$$

29. The components are the average values of sales at all stores during January of last year for produce, meats, dairy, and processed foods, respectively.

30. $\begin{bmatrix}13\\-1\\6\end{bmatrix}$ 31. $\begin{bmatrix}0\\-4\\3\\-2\end{bmatrix}$ 32. $\frac{1}{\sqrt{2}}\begin{bmatrix}3\\1\end{bmatrix}$

33. $\frac{1}{2}\begin{bmatrix}2\sqrt{3}-1\\-2-\sqrt{3}\end{bmatrix}$ 34. $\begin{bmatrix}2 & -1\\1 & 3\\3 & 6\end{bmatrix}\begin{bmatrix}3\\-4\end{bmatrix}$

35. Vector $\mathbf{v}$ is in the span of $\mathcal{S}$ if and only if $\mathbf{v}$ is a linear combination of the vectors in $\mathcal{S}$, which is true if and only if the system $A\mathbf{x} = \mathbf{v}$ is consistent, where
$$A = \begin{bmatrix}-1 & 1 & 1\\5 & 3 & -1\\2 & 4 & 1\end{bmatrix}.$$
Any solution of the system gives the coefficients of a particular linear combination. The augmented matrix for this system has reduced row echelon form
$$\begin{bmatrix}1 & 0 & -\frac{1}{2} & -\frac{3}{2}\\0 & 1 & \frac{1}{2} & \frac{7}{2}\\0 & 0 & 0 & 0\end{bmatrix}.$$
Hence the solution is
$$x_1 = -\frac{3}{2} + \frac{1}{2}x_3$$
$$x_2 = \frac{7}{2} - \frac{1}{2}x_3$$
$$x_3 \quad \text{free}.$$
So taking $x_3 = 1$, we obtain $x_1 = -1$, $x_2 = 3$. Thus one possibility is
$$\mathbf{v} = \begin{bmatrix}5\\3\\11\end{bmatrix} = (-1)\begin{bmatrix}-1\\5\\2\end{bmatrix} + 3\begin{bmatrix}1\\3\\4\end{bmatrix} + 1\begin{bmatrix}1\\-1\\1\end{bmatrix}.$$

36. $\mathbf{v}$ is not in the span of $\mathcal{S}$.
37. $\mathbf{v}$ is not in the span of $\mathcal{S}$.
38. $\mathbf{v}$ is not in the span of $\mathcal{S}$.

39. $x_1 = 1 - 2x_2 + x_3$
    x_2 free
    x_3 free

40. $x_1 = 25/24$
    $x_2 = 31/12$
    $x_3 = -5/8$

41. not consistent 42. not consistent

43. $x_1 = 7 - 5x_3 - 4x_4$
    $x_2 = -5 + 3x_3 + 3x_4$
    x_3 free
    x_4 free

44. $x_1 = 15 - 7x_4$
    $x_2 = -9 + 4x_4$
    $x_3 = -5 + 2x_4$
    x_4 free

45. The given matrix is in reduced row echelon form. Hence its rank is 1, and its nullity is $5 - 1 = 4$.

46. The given matrix is in reduced row echelon form. Hence its rank is 1, and its nullity is $5 - 1 = 4$.

47. The reduced row echelon form of the given matrix is
$$\begin{bmatrix}1 & 0 & 0 & 1 & 2.125\\0 & 1 & 0 & -1 & -1.250\\0 & 0 & 1 & 0 & 2.375\end{bmatrix}.$$
Thus its rank is 3 and its nullity is $5 - 3 = 2$.

48. The reduced row echelon form of the given matrix is $[\mathbf{e}_1 \ \mathbf{e}_2 \ \mathbf{e}_3]$. Thus its rank is 3 and its nullity is $3 - 3 = 0$.

49. Let x_1, x_2, x_3, respectively, be the appropriate numbers of the three fruit packs. We must solve
$$10x_1 + 10x_2 + 5x_3 = 500$$
$$10x_1 + 15x_2 + 10x_3 = 750$$
$$10x_2 + 5x_3 = 300.$$
The unique solution of this system is $x_1 = 20$, $x_2 = 10$, $x_3 = 40$.

50. The matrix whose columns are the vectors in the set is a 3×4 matrix. Since its rank equals the number of its pivot columns, its rank cannot equal 4. Thus, by Theorem 1.6, the set is not a generating set for $\mathcal{R}^4$.

51. The reduced row echelon form of the matrix whose columns are the vectors in the set is I_3. Thus its rank is 3, and so the set is a generating set for $\mathcal{R}^3$.

52. yes 53. no

54. For an $m \times n$ matrix A, the system $A\mathbf{x} = \mathbf{b}$ is consistent for every vector $\mathbf{b}$ in $\mathcal{R}^m$ if and only if the rank of A equals m. Since the rank of $\begin{bmatrix}1 & 2\\3 & 6\end{bmatrix}$ is 1, $A\mathbf{x} = \mathbf{b}$ is not consistent for every vector $\mathbf{b}$ in $\mathcal{R}^2$.

55. yes **56.** yes **57.** yes
58. no **59.** no
60. The set is linearly dependent. By Theorem 1.8, the set is linearly independent if and only if the rank of the matrix whose columns are the given vectors equals the number of columns of the matrix. The reduced row echelon form of this matrix is
$$\begin{bmatrix} 1 & 0 & 1 \\ 0 & 1 & 2 \\ 0 & 0 & 0 \end{bmatrix}.$$
Therefore its rank is 2, not 3.

61. linearly independent

62. linearly dependent

63. In this exercise, we have a set of 4 vectors from $\mathcal{R}^3$. Since, by property 4 on page 81, no subset of $\mathcal{R}^n$ containing more than n vectors is linearly independent, this set is linearly dependent.

64. By inspection, we see that the third vector is twice the second, so
$$\begin{bmatrix} 2 \\ 4 \\ 2 \end{bmatrix} = 0 \begin{bmatrix} 1 \\ -1 \\ 3 \end{bmatrix} + 2 \begin{bmatrix} 1 \\ 2 \\ 1 \end{bmatrix}.$$

65. $\begin{bmatrix} 3 \\ 3 \\ 8 \end{bmatrix} = 2 \begin{bmatrix} 1 \\ 2 \\ 3 \end{bmatrix} + 1 \begin{bmatrix} 1 \\ -1 \\ 2 \end{bmatrix}$

66. $\begin{bmatrix} 3 \\ 3 \\ 2 \\ 1 \end{bmatrix} = 3 \begin{bmatrix} 3 \\ 1 \\ 4 \\ 1 \end{bmatrix} - 2 \begin{bmatrix} 3 \\ 0 \\ 5 \\ 1 \end{bmatrix}$

67. $\begin{bmatrix} 1 \\ -1 \\ 1 \\ -1 \end{bmatrix} = 0 \begin{bmatrix} 1 \\ -1 \\ 1 \\ 2 \end{bmatrix} + 2 \begin{bmatrix} 1 \\ 0 \\ 1 \\ 0 \end{bmatrix} + (-1) \begin{bmatrix} 1 \\ 1 \\ 1 \\ 1 \end{bmatrix}$

68. The solution of the system is
$$x_1 = -2x_2 + x_3 - x_4$$
$$x_2 \text{ free}$$
$$x_3 \text{ free}$$
$$x_4 \text{ free}.$$
Therefore
$$\begin{bmatrix} x_1 \\ x_2 \\ x_3 \\ x_4 \end{bmatrix} = \begin{bmatrix} -2x_2 + x_3 - x_4 \\ x_2 \\ x_3 \\ x_4 \end{bmatrix}$$
$$= x_2 \begin{bmatrix} -2 \\ 1 \\ 0 \\ 0 \end{bmatrix} + x_3 \begin{bmatrix} 1 \\ 0 \\ 1 \\ 0 \end{bmatrix} + x_4 \begin{bmatrix} -1 \\ 0 \\ 0 \\ 1 \end{bmatrix}.$$

69. $\begin{bmatrix} x_1 \\ x_2 \\ x_3 \end{bmatrix} = x_3 \begin{bmatrix} -3 \\ 2 \\ 1 \end{bmatrix}$

70. $\begin{bmatrix} x_1 \\ x_2 \\ x_3 \\ x_4 \end{bmatrix} = x_3 \begin{bmatrix} 13 \\ -5 \\ 1 \\ 0 \end{bmatrix} + x_4 \begin{bmatrix} -8 \\ 3 \\ 0 \\ 1 \end{bmatrix}$

71. $\begin{bmatrix} x_1 \\ x_2 \\ x_3 \\ x_4 \end{bmatrix} = x_4 \begin{bmatrix} -2 \\ 5 \\ 0 \\ 1 \end{bmatrix}$

72. (a) The vector $\mathbf{v} + \mathbf{w}$ is a solution of $A\mathbf{x} = \mathbf{b}$ because $A(\mathbf{v}+\mathbf{w}) = A\mathbf{v}+A\mathbf{w} = \mathbf{b}+\mathbf{0} = \mathbf{b}$.
(b) Let $\mathbf{u}$ be a solution of $A\mathbf{x} = \mathbf{b}$, and define $\mathbf{w} = \mathbf{u} - \mathbf{v}$. Then
$$A\mathbf{w} = A(\mathbf{u} - \mathbf{v}) = A\mathbf{u} - A\mathbf{v} = \mathbf{b} - \mathbf{b} = \mathbf{0}.$$
Therefore $\mathbf{w}$ is a solution of $A\mathbf{x} = \mathbf{0}$, and $\mathbf{u} = \mathbf{v} + \mathbf{w}$.

73. We prove the equivalent result: Suppose that $\mathbf{w}_1$ and $\mathbf{w}_2$ are linear combinations of vectors $\mathbf{v}_1$ and $\mathbf{v}_2$. If $\mathbf{v}_1$ and $\mathbf{v}_2$ are linearly dependent, then $\mathbf{w}_1$ and $\mathbf{w}_2$ are linearly dependent.

By assumption, one of $\mathbf{v}_1$ or $\mathbf{v}_2$ is a multiple of the other, say $\mathbf{v}_1 = k\mathbf{v}_2$ for some scalar k. Thus, for some scalars a_1, a_2, b_1, b_2, we have
$$\mathbf{w}_1 = a_1\mathbf{v}_1 + a_2\mathbf{v}_2 = a_1 k\mathbf{v}_2 + a_2\mathbf{v}_2$$
$$= (a_1 k + a_2)\mathbf{v}_2$$
$$\mathbf{w}_2 = b_1\mathbf{v}_1 + b_2\mathbf{v}_2 = b_1 k\mathbf{v}_2 + b_2\mathbf{v}_2$$
$$= (b_1 k + b_2)\mathbf{v}_2.$$
Let $c_1 = a_1 k + a_2$ and $c_2 = b_1 k + b_2$. Then $\mathbf{w}_1 = c_1\mathbf{v}_2$ and $\mathbf{w}_2 = c_2\mathbf{v}_2$. If $\mathbf{w}_1 = \mathbf{0}$ or $\mathbf{w}_2 = \mathbf{0}$, then $\mathbf{w}_1$ and $\mathbf{w}_2$ are linearly dependent. Otherwise, $c_2 \neq 0$, and
$$\mathbf{w}_1 = c_1\mathbf{v}_2 = c_1 \left(\frac{1}{c_2} \mathbf{w}_2 \right) = \frac{c_1}{c_2}\mathbf{w}_2,$$
proving that $\mathbf{w}_1$ and $\mathbf{w}_2$ are linearly dependent.

74. (a) Because elementary row operations performed on the first n columns of $[A \ \mathbf{0}]$ do not change the last column, the reduced row echelon form of $[A \ \mathbf{0}]$ is $[R \ \mathbf{0}]$.
(b) Because elementary row operations performed on the first k columns of A do not affect the last $n - k$ columns, the reduced row echelon form of $[\mathbf{a}_1 \ \mathbf{a}_2 \ \cdots \ \mathbf{a}_k]$ is $[\mathbf{r}_1 \ \mathbf{r}_2 \ \cdots \ \mathbf{r}_k]$. In other words, the reduced row echelon form of the matrix obtained by deleting the last $n - k$ columns of A is obtained by deleting the last $n - k$ columns of R.

(c) First, observe that cA can be transformed into A by means of m elementary row operations, namely, multiplying each row of A by $1/c$. Since A can be transformed into R by means of elementary row operations, it follows that cA can be transformed into R by means of elementary row operations. Thus the reduced row echelon form of cA is R.

(d) Because $[I_m \ A]$ is in reduced row echelon form, its reduced row echelon form is $[I_m \ A]$.

(e) The same elementary row operations performed on A to obtain R will transform $[A \ cA]$ into $[R \ cR]$, which is in reduced row echelon form. Thus the reduced row echelon form of $[A \ cA]$ is $[R \ cR]$.

CHAPTER 1 MATLAB EXERCISES

1. (a) $A \begin{bmatrix} 1.5 \\ -2.2 \\ 2.7 \\ 4 \end{bmatrix} = \begin{bmatrix} 3.38 \\ 8.86 \\ 16.11 \\ 32.32 \\ 15.13 \end{bmatrix}$

   (b) $\begin{bmatrix} 13.45 \\ -4.30 \\ -1.89 \\ 7.78 \\ 10.69 \end{bmatrix}$ (c) $\begin{bmatrix} 20.18 \\ -11.79 \\ 7.71 \\ 8.52 \\ 0.28 \end{bmatrix}$

2. (a) $\begin{bmatrix} -0.3 & 8.5 & -12.3 & 3.9 \\ 27.5 & -9.0 & -22.3 & -2.7 \\ -11.6 & 4.9 & 16.2 & -2.1 \\ 8.0 & 12.7 & 34.2 & -24.7 \end{bmatrix}$

   (b) $\begin{bmatrix} -7.1 & 20.5 & -13.3 & 6.9 \\ 10.5 & -30.0 & -22.1 & -14.3 \\ -7.0 & -31.7 & 16.4 & 27.3 \\ -14.6 & 19.3 & -9.6 & -23.9 \end{bmatrix}$

   (c) $\begin{bmatrix} 1.30 & 4.1 & -2.75 & 3.15 \\ 4.10 & 2.4 & 1.90 & 1.50 \\ -2.75 & 1.9 & 3.20 & 4.65 \\ 3.15 & 1.5 & 4.65 & -5.10 \end{bmatrix}$

   (d) $\begin{bmatrix} 0.00 & -2.00 & -.55 & .95 \\ 2.00 & 0.00 & -3.20 & -4.60 \\ .55 & 3.20 & 0.00 & -2.55 \\ -.95 & 4.60 & 2.55 & 0.00 \end{bmatrix}$

   (e) $P^T = P$, $Q^T = -Q$, $P + Q = A$

   (f) $\begin{bmatrix} 17.67 \\ -15.87 \\ -9.83 \\ -44.27 \end{bmatrix}$ (g) $\begin{bmatrix} -143.166 \\ -154.174 \\ -191.844 \\ -202.945 \end{bmatrix}$

   (h) $\begin{bmatrix} -64.634 \\ 93.927 \\ -356.424 \\ -240.642 \end{bmatrix}$

   (i) $A \begin{bmatrix} 3.5 \\ -1.2 \\ 4.1 \\ 2.0 \end{bmatrix} = \begin{bmatrix} -3.30 \\ 6.94 \\ 3.50 \\ 19.70 \end{bmatrix}$, $\mathbf{w} = \begin{bmatrix} 3.5 \\ -1.2 \\ 4.1 \\ 2.0 \end{bmatrix}$

   (j) $M\mathbf{u} = B(A\mathbf{u})$ for every $\mathbf{u}$ in $\mathcal{R}^4$

3. (a) $\begin{bmatrix} -0.0864 \\ 3.1611 \end{bmatrix}$ (b) $\begin{bmatrix} -1.6553 \\ 2.6944 \end{bmatrix}$

   (c) $\begin{bmatrix} -1.6553 \\ 2.6944 \end{bmatrix}$ (d) $\begin{bmatrix} 1.0000 \\ 3.0000 \end{bmatrix}$

4. (a) The following sequence of elementary row operations transforms A into reduced row echelon form: $-\frac{3.1}{1.1}\mathbf{r}_1 + \mathbf{r}_2 \to \mathbf{r}_2$,
   $-\frac{7.1}{1.1}\mathbf{r}_1 + \mathbf{r}_3 \to \mathbf{r}_3$, $-2\mathbf{r}_1 + \mathbf{r}_4 \to \mathbf{r}_4$,
   $-\frac{21.4091}{7.1363}\mathbf{r}_2 + \mathbf{r}_3 \to \mathbf{r}_3$, $\mathbf{r}_3 \leftrightarrow \mathbf{r}_4$,
   $-\frac{2.7}{1.1}\mathbf{r}_3 + \mathbf{r}_1 \to \mathbf{r}_1$, $-\frac{0.6909}{1.1}\mathbf{r}_3 + \mathbf{r}_2 \to \mathbf{r}_2$,
   $\frac{2}{7.1364}\mathbf{r}_2 + \mathbf{r}_1 \to \mathbf{r}_1$, $-\frac{1}{7.1364}\mathbf{r}_2 \to \mathbf{r}_2$,
   $\frac{1}{1.1}\mathbf{r}_3 \to \mathbf{r}_3$, $\frac{1}{1.1}\mathbf{r}_1 \to \mathbf{r}_1$.
   (Other sequences are possible.)

   (b) $\begin{bmatrix} 1 & 0 & 2.0000 & 0 & .1569 & 9.2140 \\ 0 & 1 & 1.0000 & 0 & .8819 & -.5997 \\ 0 & 0 & 0 & 1 & -.2727 & -3.2730 \\ 0 & 0 & 0 & 0 & 0 & 0 \end{bmatrix}$

5. Answers are given correct to 4 places after the decimal point.

   (a) $\begin{bmatrix} x_1 \\ x_2 \\ x_3 \\ x_4 \\ x_5 \\ x_6 \end{bmatrix} = \begin{bmatrix} -8.2142 \\ -0.4003 \\ 0.0000 \\ 3.2727 \\ 0.0000 \\ 0.0000 \end{bmatrix} + x_3 \begin{bmatrix} -2.0000 \\ -1.0000 \\ 1.0000 \\ 0.0000 \\ 0.0000 \\ 0.0000 \end{bmatrix}$
   $+ x_5 \begin{bmatrix} -0.1569 \\ -0.8819 \\ 0.0000 \\ 0.2727 \\ 1.0000 \\ 0.0000 \end{bmatrix} + x_6 \begin{bmatrix} -9.2142 \\ 0.5997 \\ 0.0000 \\ 3.2727 \\ 0.0000 \\ 1.0000 \end{bmatrix}$

   (b) inconsistent

(c) $\begin{bmatrix} x_1 \\ x_2 \\ x_3 \\ x_4 \\ x_5 \\ x_6 \end{bmatrix} = \begin{bmatrix} -9.0573 \\ 1.4815 \\ 0.0000 \\ 4.0000 \\ 0.0000 \\ 0.0000 \end{bmatrix} + x_3 \begin{bmatrix} -2.0000 \\ -1.0000 \\ 1.0000 \\ 0.0000 \\ 0.0000 \\ 0.0000 \end{bmatrix}$

$+ x_5 \begin{bmatrix} -0.1569 \\ -0.8819 \\ 0.0000 \\ 0.2727 \\ 1.0000 \\ 0.0000 \end{bmatrix} + x_6 \begin{bmatrix} -9.2142 \\ 0.5997 \\ 0.0000 \\ 3.2727 \\ 0.0000 \\ 1.0000 \end{bmatrix}$

(d) inconsistent

6. By solving the equation

$$(I_5 - C)\mathbf{x} = \mathbf{d},$$

we see that the gross production for each of the respective sectors is \$264.2745 billion, \$265.7580 billion, \$327.9525 billion, \$226.1281 billion, and \$260.6357 billion.

7. (a) linearly dependent

$\begin{bmatrix} 0 \\ 1 \\ 1 \\ 2 \\ 2 \\ 1 \end{bmatrix} = \begin{bmatrix} 1 \\ 2 \\ -1 \\ 3 \\ 2 \\ 1 \end{bmatrix} + \begin{bmatrix} 1 \\ 0 \\ 1 \\ 1 \\ 0 \\ 1 \end{bmatrix} - \begin{bmatrix} 2 \\ 1 \\ -1 \\ 2 \\ 0 \\ 1 \end{bmatrix}$

(b) linearly independent

8. Let $\mathbf{v}_1, \mathbf{v}_2, \ldots, \mathbf{v}_5$ denote the vectors in the set S_1 in the order listed in Exercise 7(a), and let $A = \begin{bmatrix} \mathbf{v}_1 & \mathbf{v}_2 & \ldots & \mathbf{v}_5 \end{bmatrix}$.

(a) For the vector $\mathbf{u}$ given in (a), the equation $A\mathbf{x} = \mathbf{u}$ is not consistent; so $\mathbf{u}$ is not in the span of S_1.

(b) For the vector $\mathbf{u}$ given in (b), the equation $A\mathbf{x} = \mathbf{u}$ is consistent, and

$$\mathbf{u} = 2\mathbf{v}_1 - \mathbf{v}_2 + 0\mathbf{v}_3 + \mathbf{v}_4 + 0\mathbf{v}_5.$$

(c) yes, $2\mathbf{v}_1 - \mathbf{v}_2 + 0\mathbf{v}_3 + \mathbf{v}_4 + 0\mathbf{v}_5$

(d) no

Chapter 2

Matrices and Linear Transformations

2.1 MATRIX MULTIPLICATION

1. AB is defined and has size 2×2.
2. AB is defined and has size 2×6.
3. AB is undefined because A has 3 columns and B has 2 rows.
4. Let A be any 3×4 matrix and B be any 2×3 matrix.
5. $C\mathbf{y} = \begin{bmatrix} 22 \\ -18 \end{bmatrix}$ 6. $\begin{bmatrix} 28 \\ 8 \end{bmatrix}$
7. $\mathbf{xz} = \begin{bmatrix} 14 & -2 \\ 21 & -3 \end{bmatrix}$
8. $B\mathbf{y}$ is undefined because B is a 2×2 matrix, $\mathbf{y}$ is a 3×1 matrix, and $2 \neq 3$.
9. $AC\mathbf{x}$ is undefined because $AC\mathbf{x} = A(C\mathbf{x})$ and $C\mathbf{x}$ is undefined because C is a 2×3 matrix, $\mathbf{x}$ is a 2×1 matrix, and $3 \neq 2$.
10. $\begin{bmatrix} 9 \\ 17 \end{bmatrix}$ 11. $\begin{bmatrix} 5 & 0 \\ 25 & 20 \end{bmatrix}$
12. $\begin{bmatrix} -1 & 8 & -7 \\ 17 & 24 & 19 \end{bmatrix}$
13. $\begin{bmatrix} 29 & 56 & 23 \\ 7 & 8 & 9 \end{bmatrix}$ 14. $\begin{bmatrix} 19 & 2 \\ 7 & 6 \end{bmatrix}$
15. CB^T is undefined because C is a 2×3 matrix, B^T is a 2×2 matrix, and $3 \neq 2$.
16. CB is undefined because C is a 2×3 matrix, B is a 2×2 matrix, and $3 \neq 2$.
17. $A^3 = \begin{bmatrix} -35 & -30 \\ 45 & 10 \end{bmatrix}$ 18. $\begin{bmatrix} -5 & -10 \\ 15 & 10 \end{bmatrix}$
19. C^2 is undefined because C is a 2×3 matrix and the number of columns of C does not equal the number of rows of C.
20. $\begin{bmatrix} 53 & 36 \\ 9 & 8 \end{bmatrix}$

22. $(AB)C = A(BC) = \begin{bmatrix} 15 & 40 & 5 \\ 115 & 200 & 105 \end{bmatrix}$
23. $(AB)^T = B^T A^T = \begin{bmatrix} 5 & 25 \\ 0 & 20 \end{bmatrix}$
24. $\mathbf{z}(AC) = (\mathbf{z}A)C = \begin{bmatrix} -24 & 32 & -68 \end{bmatrix}$
25. -2 26. 12 27. 24 28. -1
29. $\begin{bmatrix} -4 \\ -9 \\ -2 \end{bmatrix}$ 30. $\begin{bmatrix} 1 \\ -6 \\ 1 \end{bmatrix}$
31. $\begin{bmatrix} 7 \\ 16 \end{bmatrix}$ 32. $\begin{bmatrix} 3 \\ 7 \end{bmatrix}$
33. False, the product is not defined unless $n = m$.
34. False, if A is a 2×3 matrix and B is a 3×4 matrix.
35. False, see Example 5.
36. True
37. False, if A is a 2×3 matrix and B is a 3×2 matrix.
38. False, $(AB)^T = B^T A^T$.
39. True 40. True
41. False, see the box titled "Row-Column Rule for the (i,j)-Entry of a Matrix Product."
42. False, it is the sum of the products of corresponding entries from the ith row of A and the jth column of B.
43. True
44. False, $(A+B)C = AC + BC$.
45. False. If $A = B = \begin{bmatrix} 0 & 1 \\ 1 & 0 \end{bmatrix}$, then $AB = \begin{bmatrix} 1 & 0 \\ 0 & 1 \end{bmatrix}$.
46. True
47. False, let $A = \begin{bmatrix} 1 & 0 \\ 0 & 0 \end{bmatrix}$ and $B = \begin{bmatrix} 0 & 0 \\ 1 & 0 \end{bmatrix}$.

48. True **49.** True **50.** True

51. (a) The number of people living in single-unit houses is $.70v_1 + .95v_2$. The number of people living in multiple-unit housing is $.30v_1 + .05v_2$. These results may be expressed as the matrix equation

$$\begin{bmatrix} .70 & .95 \\ .30 & .05 \end{bmatrix} \begin{bmatrix} v_1 \\ v_2 \end{bmatrix} = \begin{bmatrix} u_1 \\ u_2 \end{bmatrix}.$$

So take
$$B = \begin{bmatrix} .70 & .95 \\ .30 & .05 \end{bmatrix}.$$

(b) Because $A \begin{bmatrix} v_1 \\ v_2 \end{bmatrix}$ represents the number of people living in the city and suburbs after one year, it follows from (a) that $BA \begin{bmatrix} v_1 \\ v_2 \end{bmatrix}$ gives the number of single-unit and multiple-unit dwellers after one year.

52. (a) $B = \begin{bmatrix} .6 & .3 \\ .3 & .5 \\ .1 & .2 \end{bmatrix}$ (b) $A = \begin{bmatrix} .6 & .4 & .5 \\ .4 & .6 & .5 \end{bmatrix}$

(c) $C = \begin{bmatrix} .53 & .48 \\ .47 & .52 \end{bmatrix}$

53. (a)
$$\begin{array}{c} \text{Today} \\ \text{Hot} \quad \text{Bag} \\ \text{Lunch} \quad \text{Lunch} \end{array}$$
$$\begin{array}{c} \text{Next Hot Lunch} \\ \text{Day Bag Lunch} \end{array} A = \begin{bmatrix} .3 & .4 \\ .7 & .6 \end{bmatrix}$$

(b) $A^3 \begin{bmatrix} u_1 \\ u_2 \end{bmatrix} = \begin{bmatrix} 109.1 \\ 190.9 \end{bmatrix}$. Approximately 109 students will buy hot lunches and 191 students will bring bag lunches 3 school days from today.

(c) $A^{100} \begin{bmatrix} u_1 \\ u_2 \end{bmatrix} = \begin{bmatrix} 109.0909 \\ 190.9091 \end{bmatrix}$ (rounded to four places after the decimal)

54. We prove that $s(AC) = A(sC)$. The remaining part is proved similarly. Note that both $s(AC)$ and $A(sC)$ are $k \times n$ matrices. The jth column of $s(AC)$ is $s(A\mathbf{c}_j)$. On the other hand, the jth column of sC is $s\mathbf{c}_j$. So the jth column of $A(sC)$ is $A(s\mathbf{c}_j)$, which equals $s(A\mathbf{c}_j)$ by Theorem 1.3(c).

55. We prove that $C(P + Q) = CP + CQ$. Note that $P + Q$ is an $n \times p$ matrix, and so $C(P+Q)$ is an $m \times p$ matrix. Also CP and CQ are both $m \times p$ matrices; so $CP + CQ$ is an $m \times p$ matrix. Hence the matrices on both sides of the equation have the same size. The jth column of $P + Q$ is $\mathbf{p}_j + \mathbf{q}_j$; so the jth column of $C(P+Q)$ is $C(\mathbf{p}_j + \mathbf{q}_j)$, which equals $C\mathbf{p}_j + C\mathbf{q}_j$ by Theorem 1.3(c). On the other hand, the jth columns of CP and CQ are $C\mathbf{p}_j$ and $C\mathbf{q}_j$, respectively. So the jth column of $CP + CQ$ is $C\mathbf{p}_j + C\mathbf{q}_j$. Thus $C(P + Q)$ and $CP + CQ$ have the same corresponding columns, and hence are equal.

56. The proof is similar to that of Exercise 55.

57. The proof is similar to that of Exercise 55.

58. The matrix $BA = \begin{bmatrix} -1 & 0 \\ 0 & 1 \end{bmatrix}$ reflects $\mathcal{R}^2$ about the y-axis.

59. By the row-column rule, the (i,j)-entry of AB is
$$a_{i1}b_{1j} + a_{i2}b_{2j} + \cdots + a_{in}b_{nj}.$$
Suppose $i < j$. A typical term above has the form $a_{ik}b_{kj}$ for $k = 1, 2, \ldots, n$. If $k < j$, then $b_{kj} = 0$ because B is lower triangular. If $k \geq j$, then $k > i$; so $a_{ik} = 0$ because A is lower triangular. Thus every term is 0, and therefore AB is lower triangular.

60. (a) The ith component of $\mathbf{a}_j$ is a_{ij}. If A is a diagonal matrix, then $a_{ij} = 0$ if $i \neq j$, and hence every component of $\mathbf{a}_j$, except perhaps the jth component, is zero. Therefore the jth column of A is $\mathbf{a}_j = a_{jj}\mathbf{e}_j$. The converse is proved similarly.

(b) From (a), Theorem 1.3(c), and Theorem 1.3(a), the jth column of AB is
$$A\mathbf{b}_j = A(b_{jj}\mathbf{e}_j) = b_{jj}(A\mathbf{e}_j)$$
$$= b_{jj}\mathbf{a}_j = b_{jj}(a_{jj}\mathbf{e}_j) = (b_{jj}a_{jj})\mathbf{e}_j$$
$$= a_{jj}b_{jj}\mathbf{e}_j.$$
Thus AB is a diagonal matrix by (a).

61. By the row-column rule, the (i,j)-entry of AB is
$$a_{i1}b_{1j} + a_{i2}b_{2j} + \cdots + a_{in}b_{nj}.$$
Suppose $i > j$. A typical term above has the form $a_{ik}b_{kj}$ for $k = 1, 2, \ldots, n$. If $k < i$, then $a_{ik} = 0$ because A is upper triangular. If $k \geq i$, then $k > j$; so $b_{kj} = 0$ because B is upper triangular. Thus every term is 0, and therefore AB is upper triangular.

62. $B = \begin{bmatrix} 1 & 2 \\ 3 & 1 \\ 1 & 0 \\ 0 & 1 \end{bmatrix}$

63. Let $A = \begin{bmatrix} 1 & 0 \\ 0 & 0 \end{bmatrix}$ and $B = \begin{bmatrix} 0 & 0 \\ 1 & 0 \end{bmatrix}$. Then $AB = O$, but $BA = \begin{bmatrix} 0 & 0 \\ 1 & 0 \end{bmatrix} \neq O$.

64. For the matrices in the solution to Exercise 47, we have rank $AB = 0$ and rank $BA = 1$.

65. The trace of AB is the sum of the diagonal entries of AB, that is,
$$[a_{11}b_{11} + a_{12}b_{21} + \cdots + a_{1n}b_{n1}]$$
$$+ [a_{21}b_{12} + a_{22}b_{22} + \cdots + a_{2n}b_{n2}] + \cdots$$
$$+ [a_{m1}b_{1m} + a_{m2}b_{2m} + \cdots + a_{mn}b_{nm}].$$
By adding the first terms within each bracket, then the second terms, etc., we obtain the trace of BA.

66. Row r of EB is the same as row s of B, and the other rows of EB are zero.

67. Using (b) and (g) of Theorem 2.1, we have
$$(ABC)^T = ((AB)C)^T = C^T(AB)^T$$
$$= C^T(B^T A^T) = C^T B^T A^T.$$

68. (a) By Theorem 2.1(g), AB is symmetric if and only if $AB = (AB)^T$, which is true if and only if $AB = B^T A^T = BA$.
   (b) Let $A = \begin{bmatrix} 1 & 0 \\ 0 & 0 \end{bmatrix}$ and $B = \begin{bmatrix} 0 & 1 \\ 1 & 0 \end{bmatrix}$. Then
   $AB = \begin{bmatrix} 0 & 1 \\ 0 & 0 \end{bmatrix}$ and $BA = \begin{bmatrix} 0 & 0 \\ 1 & 0 \end{bmatrix}$.

69. (a), (b), (c), and (d) have the same answer, namely, $\begin{bmatrix} -1 & 0 \\ 0 & -1 \end{bmatrix}$.

   (e)

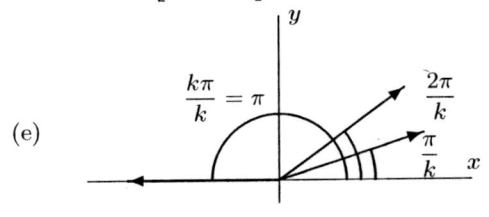

   Figure for Exercise 69(e)

70. (b) not valid in general
   (c) We have
   $$(A + B)^2 = (A + B)(A + B)$$
   $$= A(A + B) + B(A + B)$$
   $$= AA + AB + BA + BB$$
   $$= A^2 + AB + BA + B^2.$$
   (d) The equation in (c) holds if and only if $AB = BA$.
   (e) If $AB = BA$, then from (c) we have
   $$(A + B)^2 = A^2 + AB + AB + B^2$$
   $$= A^2 + 2AB + B^2.$$

71. (b) The population of the city is 205,688. The population of the suburbs is 994,332.
   (c) The population of the city is 200,015. The population of the suburbs is 999,985.
   (d) Eventually the population in the city will be 200,000, and the population in the suburbs will be 1,000,000.

2.2 APPLICATIONS OF MATRIX MULTIPLICATION

1. False, the population may be decreasing.
2. False, the population may continue to grow without bound.
3. False, this is only the case for $i = 1$.
4. True   5. True
6. False, $\mathbf{z} = BA\mathbf{x}$.   7. True
8. False, a $(0,1)$-matrix need not be square, see Example 2.
9. False, a $(0,1)$-matrix need not be symmetric, see Example 2.
10. (a) $A = \begin{bmatrix} 3.0 & 2.0 & 0 \\ 0.6 & 0.0 & 0 \\ 0.0 & 0.2 & 0 \end{bmatrix}$
   (b) $\begin{bmatrix} 420 \\ 60 \\ 12 \end{bmatrix}$, $\begin{bmatrix} 15588 \\ 2786 \\ 166 \end{bmatrix}$
   (c) The only solution of $(A - I_3)\mathbf{z} = \mathbf{0}$ is $\mathbf{0}$.
11. (a) all of them
    (b) 0 from the first and 1 from the second
    (c) $\begin{bmatrix} a \\ b \end{bmatrix}$ in even-numbered years, and $\begin{bmatrix} b \\ a \end{bmatrix}$ in odd-numbered years
12. (a) $A = \begin{bmatrix} 0 & 1.2 & 1 \\ .5 & 0 & 0 \\ 0 & q & 0 \end{bmatrix}$
    (b) The population approaches $\mathbf{0}$.
    (c) The population increases with an (approximate) distribution given by the vector $\begin{bmatrix} 3400 \\ 1700 \\ 1500 \end{bmatrix}$.
    (d) $q = 0.8$. The stable distribution is (approximately) given by the vector $\begin{bmatrix} 1233 \\ 617 \\ 493 \end{bmatrix}$.

30 Chapter 2 Matrices and Linear Transformations

(e) Over time, the population distribution approaches $\begin{bmatrix} 500 \\ 250 \\ 200 \end{bmatrix}$.

(f) $q = 0.8$

(g) The general solution of $(A - I_3)\mathbf{x}$ is

$$x_1 = 2.50x_3$$
$$x_2 = 1.25x_3$$
$$x_3 \text{ free,}$$

and its vector form is

$$\begin{bmatrix} x_1 \\ x_2 \\ x_3 \end{bmatrix} = x_3 \begin{bmatrix} 2.50 \\ 1.25 \\ 1.00 \end{bmatrix}.$$

For $x_3 = 200$, we obtain (e).

13. (a) Using the notation on page 108, we have $p_1 = q$ and $p_2 = .5$. Also, $b_1 = 0$ because females under age 1 do not give birth. Likewise, $b_2 = 2$ and $b_3 = 1$. So the Leslie matrix is

$$A = \begin{bmatrix} b_1 & b_2 & b_3 \\ p_1 & 0 & 0 \\ 0 & p_2 & 0 \end{bmatrix} = \begin{bmatrix} 0 & 2 & 1 \\ q & 0 & 0 \\ 0 & .5 & 0 \end{bmatrix}.$$

(b) If $q = .8$, then

$$A = \begin{bmatrix} 0 & 2 & 1 \\ .8 & 0 & 0 \\ 0 & .5 & 0 \end{bmatrix} \text{ and } \mathbf{x}_0 = \begin{bmatrix} 300 \\ 1180 \\ 130 \end{bmatrix}.$$

The population in 50 years is given by

$$A^{50}\mathbf{x}_0 \approx 10^9 \begin{bmatrix} 9.28 \\ 5.40 \\ 1.96 \end{bmatrix}.$$

So the population appears to grow without bound.

(c) If $q = .2$, then

$$A = \begin{bmatrix} 0 & 2 & 1 \\ .2 & 0 & 0 \\ 0 & .5 & 0 \end{bmatrix}.$$

As in (b), we compute

$$A^{50}\mathbf{x}_0 \approx 10^{-3} \begin{bmatrix} .3697 \\ 1.009 \\ .0689 \end{bmatrix}$$

and conclude that the population appears to approach zero.

(d) $q = .4$. The stable distribution is $\begin{bmatrix} 400 \\ 160 \\ 80 \end{bmatrix}$.

(e) For $q = .4$ and $\mathbf{x}_0 = \begin{bmatrix} 210 \\ 240 \\ 180 \end{bmatrix}$, the respective vectors $A^5\mathbf{x}_0$, $A^{10}\mathbf{x}_0$, and $A^{30}\mathbf{x}_0$ equal

$$\begin{bmatrix} 513.60 \\ 144.96 \\ 114.00 \end{bmatrix}, \begin{bmatrix} 437.36 \\ 189.99 \\ 85.17 \end{bmatrix}, \text{ and } \begin{bmatrix} 499.98 \\ 180.01 \\ 89.99 \end{bmatrix}.$$

It appears that the populations approach

$$\begin{bmatrix} 450 \\ 180 \\ 90 \end{bmatrix}.$$

(f) By Theorem 1.8, we need $\text{rank}(A - I_3) < 3$. We can use elementary row operations to transform the matrix

$$A - I_3 = \begin{bmatrix} 0 & 2 & 1 \\ q & 0 & 0 \\ 0 & .5 & 0 \end{bmatrix}$$

to

$$\begin{bmatrix} 1 & -2 & -1 \\ 0 & 1 & -2 \\ 0 & 0 & -2(1 - 2q) + q \end{bmatrix}.$$

For rank $(A - I_3) < 3$, we need

$$-2(1 - 2q) + q = 0,$$

that is, $q = .4$. This is the value obtained in (d).

(g) For $q = .4$, the solution of $(A - I_3)\mathbf{x} = \mathbf{0}$ is

$$\begin{bmatrix} x_1 \\ x_2 \\ x_3 \end{bmatrix} = x_3 \begin{bmatrix} 5 \\ 2 \\ 1 \end{bmatrix}.$$

For $x_3 = 90$, we obtain the solution in (e).

14. (a) $A = \begin{bmatrix} 0 & b & 10 \\ .1 & 0 & 0 \\ 0 & .2 & 0 \end{bmatrix}$

(b) The population increases with an (approximate) distribution given by the vector

$$\begin{bmatrix} 503000 \\ 4700 \\ 8500 \end{bmatrix}.$$

(c) The population approaches **0**.

(d) $b = 8$. The population stabilizes to the (approximate) distribution $\begin{bmatrix} 6796 \\ 680 \\ 136 \end{bmatrix}$.

(e) The population stabilizes to the (approximate) distribution $\begin{bmatrix} 7296 \\ 730 \\ 146 \end{bmatrix}$.

(f) $b = 8$

(g) Let $A = \begin{bmatrix} 0 & 8 & 10 \\ .1 & 0 & 0 \\ 0 & .2 & 0 \end{bmatrix}$. Then $A^k \mathbf{p}$ approaches $\mathbf{q}$ as k grows large. It is instructive to compute A^k for a large value of k, treating the matrix entries as rational numbers. Using MATLAB, this can be done using the command `format rat` (see the subsection "Displaying Data" in Appendix D). In this case,

$$B = A^{100} = \begin{bmatrix} \frac{5}{11} & \frac{10}{11} & \frac{50}{11} \\ \frac{1}{22} & \frac{5}{11} & \frac{5}{11} \\ \frac{1}{110} & \frac{1}{11} & \frac{1}{11} \end{bmatrix}.$$

Observe that $AB = B$, and hence A^{100} yields a limiting value of A^k (although a smaller value of k may also be adequate). Thus

$$\mathbf{q} = \begin{bmatrix} \frac{5}{11} & \frac{10}{11} & \frac{50}{11} \\ \frac{1}{22} & \frac{5}{11} & \frac{5}{11} \\ \frac{1}{110} & \frac{1}{11} & \frac{1}{11} \end{bmatrix} \begin{bmatrix} p_1 \\ p_2 \\ p_3 \end{bmatrix} = c \begin{bmatrix} 50 \\ 5 \\ 1 \end{bmatrix},$$

where $c = \dfrac{1}{110}(p_1 + 10p_2 + 10p_3)$.

15. (a) $\begin{bmatrix} 0 & 2 & b \\ .2 & 0 & 0 \\ 0 & .5 & 0 \end{bmatrix}$

(b) The population approaches $\mathbf{0}$.

(c) The population grows without bound.

(d) $b = 6$, $\begin{bmatrix} 1600 \\ 320 \\ 160 \end{bmatrix}$

(e) Over time, it approaches $\begin{bmatrix} 1500 \\ 300 \\ 150 \end{bmatrix}$.

(f) $b = 6$

(g) Let $A = \begin{bmatrix} 0 & 2 & 6 \\ .2 & 0 & 0 \\ 0 & .5 & 0 \end{bmatrix}$. As in 14(g), computing A^{100} in rational format, we obtain

$$B = A^{100} = \begin{bmatrix} \frac{5}{13} & \frac{25}{13} & \frac{30}{13} \\ \frac{1}{13} & \frac{5}{13} & \frac{6}{13} \\ \frac{1}{26} & \frac{5}{26} & \frac{3}{13} \end{bmatrix}.$$

Observe that $AB = B$, and hence A^{100} yields a limiting value of A^k. So

$$\mathbf{q} = \begin{bmatrix} \frac{5}{13} & \frac{25}{13} & \frac{30}{13} \\ \frac{1}{13} & \frac{5}{13} & \frac{6}{13} \\ \frac{1}{26} & \frac{5}{26} & \frac{3}{13} \end{bmatrix} \begin{bmatrix} p_1 \\ p_2 \\ p_3 \end{bmatrix} = c \begin{bmatrix} 10 \\ 2 \\ 1 \end{bmatrix},$$

where $c = \dfrac{1}{26}(p_1 + 5p_2 + 6p_3)$.

16. (a) $A = \begin{bmatrix} 1.0 & 1.0 & 0 \\ 0.5 & 0.0 & 0 \\ 0.0 & 0.3 & 0 \end{bmatrix}$ (b) $\begin{bmatrix} 80 \\ 30 \\ 6 \end{bmatrix}$ and $\begin{bmatrix} 110 \\ 40 \\ 9 \end{bmatrix}$

17. Let p and q be the amounts of donations and interest received by the foundation, and let n and a be the net income and fund raising costs, respectively. Then

$$n = .7p + .9q$$
$$a = .3p + .1q,$$

and hence

$$\begin{bmatrix} n \\ a \end{bmatrix} = \begin{bmatrix} .7 & .9 \\ .3 & .1 \end{bmatrix} \begin{bmatrix} p \\ q \end{bmatrix}.$$

Next, let r and c be the amounts of net income used for research and clinic maintenance, respectively. Then

$$r = .4n$$
$$c = .6n$$
$$a = a,$$

and hence

$$\begin{bmatrix} r \\ c \\ a \end{bmatrix} = \begin{bmatrix} .4 & 0 \\ .6 & 0 \\ 0 & 1 \end{bmatrix} \begin{bmatrix} n \\ a \end{bmatrix}.$$

Finally, let m and f be the material and personnel costs of the foundation, respectively. Then

$$m = .8r + .5c + .7a$$
$$f = .2r + .5c + .3a,$$

and hence

$$\begin{bmatrix} m \\ f \end{bmatrix} = \begin{bmatrix} .8 & .5 & .7 \\ .2 & .5 & .4 \end{bmatrix} \begin{bmatrix} r \\ c \\ a \end{bmatrix}.$$

Combining these matrix equations, we have

$$\begin{bmatrix} m \\ f \end{bmatrix} = \begin{bmatrix} .8 & .5 & .7 \\ .2 & .5 & .4 \end{bmatrix} \begin{bmatrix} .4 & 0 \\ .6 & 0 \\ 0 & 1 \end{bmatrix} \begin{bmatrix} n \\ a \end{bmatrix}$$

$$= \begin{bmatrix} .8 & .5 & .7 \\ .2 & .5 & .4 \end{bmatrix} \begin{bmatrix} .4 & 0 \\ .6 & 0 \\ 0 & 1 \end{bmatrix} \begin{bmatrix} .7 & .9 \\ .3 & .1 \end{bmatrix} \begin{bmatrix} p \\ q \end{bmatrix}$$

$$= \begin{bmatrix} 0.644 & 0.628 \\ 0.356 & 0.372 \end{bmatrix} \begin{bmatrix} p \\ q \end{bmatrix}.$$

18. $M = \begin{bmatrix} 0.48 & 0.00 \\ 0.52 & 0.65 \\ 0.00 & 0.35 \end{bmatrix}$

19. (a) There are no nonstop flights from any of the cities 1, 2, and 3 to either of the cities 4 and 5, and vice versa.

(b) $A^2 = \begin{bmatrix} B^2 & O_1 \\ O_2 & C^2 \end{bmatrix}$, $A^3 = \begin{bmatrix} B^3 & O_1 \\ O_2 & C^2 \end{bmatrix}$,

and $A^k = \begin{bmatrix} B^k & O_1 \\ O_2 & C^2 \end{bmatrix}$

(c) It is impossible to fly between any of the cities 1, 2, and 3 and either city 4 or city 5.

20. (a) $(1,3), (1,4), (2,4), (2,5), (3,5)$
(b) none
(c) There are 3 ways in which countries 1 and 4 are linked by three intermediate countries.

21. (a) We need only find entries a_{ij} such that $a_{ij} = a_{ji} = 1$. The friends are 1 and 2, 1 and 4, 2 and 3, and 3 and 4.

(b) The (i,j)-entry of A^2 is

$$a_{i1}a_{1j} + a_{i2}a_{2j} + a_{i3}a_{3j} + a_{i4}a_{4j}.$$

The kth term equals 1 if and only if $a_{ik} = 1$ and $a_{kj} = 1$, that is, person i likes person k and person k likes person j. Otherwise, the term is 0. So the (i,j)-entry of A^2 equals the number of people who like person j and are liked by person i.

(c) We have

$$B = \begin{bmatrix} 0 & 1 & 0 & 1 \\ 1 & 0 & 1 & 0 \\ 0 & 1 & 0 & 1 \\ 1 & 0 & 1 & 0 \end{bmatrix},$$

which is symmetric because $B = B^T$.

(d) Because $B^3 = B^2B$, the (i,i)-entry of B^3 equals a sum of terms of the form $c_{ik}b_{ki}$, where c_{ik} equals a sum of terms of the form $b_{ij}b_{jk}$. Therefore the (i,i)-entry of B^3 consists of terms of the form $b_{ij}b_{jk}b_{ki}$. The (i,i)-entry of B^3 is positive if and only if some term $b_{ij}b_{jk}b_{ki}$ is positive. This occurs if and only if $b_{ij} = 1 = b_{jk} = b_{ki}$, that is, there are friends k and j who are also friends of person i, that is, person i is in a clique.

(e) We have

$$B^3 = \begin{bmatrix} 0 & 4 & 0 & 4 \\ 4 & 0 & 4 & 0 \\ 0 & 4 & 0 & 4 \\ 4 & 0 & 4 & 0 \end{bmatrix};$$

so the (i,i)-entry is 0 for every i. Therefore there are no cliques.

22. (a) $(1,5)$ (b) $(1,2)$ and $(2,4)$

(c) $\begin{bmatrix} 5 & 0 & 2 & 1 & 5 \\ 0 & 3 & 1 & 0 & 2 \\ 2 & 1 & 5 & 2 & 3 \\ 1 & 0 & 2 & 3 & 1 \\ 5 & 2 & 3 & 1 & 7 \end{bmatrix}$

23. (a) Let $B = A^T$. The (i,j)-entry of AA^T equals

$$a_{i1}b_{1j} + a_{i2}b_{2j} + \cdots + a_{i5}b_{5j}$$
$$= a_{i1}a_{j1} + a_{i2}a_{j2} + \cdots + a_{i5}a_{j5}.$$

A typical term in this sum has the form $a_{ik}a_{jk}$, which equals 1 if and only if both students i and j want to take course k. So this entry equals the number of courses students i and j both want to take.

(b) Use the row-column rule to perform each calculation.

(c) Students 1 and 2 want to take one of the same courses, and students 9 and 1 want to take three of the same courses.

(d) For each i, the ith diagonal entry of AA^T represents the number of courses that student i wants to take.

24. (a) We have

$$(A+B)^2 = (A+B)(A+B)$$
$$= (A+B)A + (A+B)B$$
$$= AA + BA + AB + BB$$
$$= A^2 + AB + AB + B^2$$
$$= A^2 + 2AB + B^2.$$

The proof for $k = 3$ is similar.

(b) The proof for matrices is identical to the proof for scalars.

(c) Because $B^k = O$ for $k \geq 3$, we have

$$A^k = (I_4 + B)^k = I_4^k + kI_4^{k-1}B$$
$$+ \frac{k!}{2!(k-2)!}I_4^{k-2}B^2$$
$$+ \frac{k!}{3!(k-3)!}I_4^{k-3}B^3$$

$$+ O + \cdots + O$$
$$= I_4 + kB + \frac{k!}{2(k-2)!}B^2$$
$$= I_4 + kB + \frac{k(k-1)}{2}B^2.$$

25. (a)

k	Sun	Noble	Honored	MMQ
1	100	300	500	7700
2	100	400	800	7300
3	100	500	1200	6800

(b)

k	Sun	Noble	Honored	MMQ
9	100	1100	5700	1700
10	100	1200	6800	500
11	100	1300	8000	-800

(c) The tribe will cease to exist because every member is required to marry a member of the MMQ, and the number of members of the MMQ decreases to zero.

(d) We must find k such that
$$s_k + n_k + h_k > m_k,$$
that is, there are enough members of the MMQ for the other classes to marry. From equation (6), this inequality is equivalent to
$$s_0 + (n_0 + ks_0) + \left(h_0 + kn_0 + \frac{k(k-1)}{2}s_0\right)$$
$$> m_0 - kn_0 - \frac{k(k+1)}{2}s_0.$$

If we let $s_0 = 100$, $n_0 = 200$, $k_0 = 300$, and $m_0 = 8000$ and simplify the inequality above, we obtain
$$k^2 + 5k - 74 > 0.$$
The smallest value of k that satisfies this inequality is $k = 7$.

26. (a) By definition, each entry of A equals 0 or 1.
(b) Person 3 can send a message to person 1 using person 2 as an intermediary.
(c) The $(3,1)$-entry of A^2 equals
$$a_{31}a_{11} + a_{32}a_{21} + \cdots + a_{36}a_{61}.$$
Each term in this sum is 0 or 1, and as in (b), $a_{3i}a_{i1} = 1$ only when person 3 can send a message to person 1 using person i as an intermediary. So the $(3,1)$-entry of A^2 equals the number of people to whom person 3 can send a message and who in turn can send a message to person 1.

(d) The (i,j)-entry of A^2 equals the number of ways that person i can send a message to person j in two stages.
(e) The (i,j)-entry of A^m equals the number of ways that person i can send a message to person j in m stages.
(f) Yes, person 5
(g) 1, 0, 3, 1
(h) 6

2.3 INVERTIBILITY AND ELEMENTARY MATRICES

1. No, we must have $AB = BA = I_n$. In this case $AB \neq I_n$.
2. yes 3. yes 4. no 5. yes
6. no 7. no 8. no

9. $(A^T)^{-1} = (A^{-1})^T = \begin{bmatrix} 1 & 2 & 1 \\ 2 & 0 & 1 \\ 3 & 1 & -1 \end{bmatrix}$

10. $(B^T)^{-1} = \begin{bmatrix} 2 & 0 & 3 \\ -1 & 0 & -2 \\ 3 & 4 & 1 \end{bmatrix}$

11. $(AB)^{-1} = B^{-1}A^{-1} = \begin{bmatrix} 3 & 7 & 2 \\ 4 & 4 & -4 \\ 0 & 7 & 6 \end{bmatrix}$

12. $(BA)^{-1} = \begin{bmatrix} 11 & -7 & 14 \\ 7 & -4 & 1 \\ -1 & 1 & 6 \end{bmatrix}$

13. $(AB^T)^{-1} = \begin{bmatrix} 5 & 7 & 3 \\ -3 & -4 & -1 \\ 12 & 7 & 12 \end{bmatrix}$

14. $(A^TB^T)^{-1} = \begin{bmatrix} 11 & 7 & -1 \\ -7 & -4 & 1 \\ 14 & 7 & 6 \end{bmatrix}$

15. $\begin{bmatrix} 1 & 0 \\ -1 & 1 \end{bmatrix}$ 16. $\begin{bmatrix} 1 & 3 \\ 0 & 1 \end{bmatrix}$

17. $\begin{bmatrix} 1 & 0 & 0 \\ 2 & 1 & 0 \\ 0 & 0 & 1 \end{bmatrix}$ 18. $\begin{bmatrix} 0 & 0 & 1 \\ 0 & 1 & 0 \\ 1 & 0 & 0 \end{bmatrix}$

19. $\begin{bmatrix} 1 & 0 & 0 & 0 \\ 0 & .25 & 0 & 0 \\ 0 & 0 & 1 & 0 \\ 0 & 0 & 0 & 1 \end{bmatrix}$ 20. $\begin{bmatrix} 1 & 0 & 0 \\ 0 & 1 & 0 \\ 0 & 0 & \frac{1}{4} \end{bmatrix}$

21. $\begin{bmatrix} 1 & 0 & 0 & 0 \\ 0 & 0 & 0 & 1 \\ 0 & 0 & 1 & 0 \\ 0 & 1 & 0 & 0 \end{bmatrix}$ 22. $\begin{bmatrix} 1 & 0 & 0 & 0 \\ 0 & 1 & 0 & -2 \\ 0 & 0 & 1 & 0 \\ 0 & 0 & 0 & 1 \end{bmatrix}$

23. $\begin{bmatrix} -1 & 0 \\ 0 & 1 \end{bmatrix}$ 24. $\begin{bmatrix} 1 & 0 \\ 2 & 1 \end{bmatrix}$ 25. $\begin{bmatrix} 0 & 1 \\ 1 & 0 \end{bmatrix}$

26. $\begin{bmatrix} 1 & 0 \\ 0 & -1 \end{bmatrix}$ 27. $\begin{bmatrix} 0 & 1 \\ 1 & 0 \end{bmatrix}$ 28. $\begin{bmatrix} 1 & 2 \\ 0 & 1 \end{bmatrix}$

29. $\begin{bmatrix} 1 & 0 & 0 \\ 0 & 1 & 0 \\ 0 & -5 & 1 \end{bmatrix}$ 30. $\begin{bmatrix} 1 & 0 & 0 \\ 0 & 1 & 0 \\ 1 & 0 & 1 \end{bmatrix}$

31. $\begin{bmatrix} 1 & 0 & 0 \\ 0 & 0 & 1 \\ 0 & 1 & 0 \end{bmatrix}$ 32. $\begin{bmatrix} 1 & 0 & 0 \\ 0 & 1 & 0 \\ -2 & 0 & 1 \end{bmatrix}$

33. False, the $n \times n$ zero matrix is not invertible.

34. True   35. True

36. False, let

$$A = \begin{bmatrix} 1 & 0 & 0 \\ 0 & 1 & 0 \end{bmatrix} \quad \text{and} \quad B = \begin{bmatrix} 1 & 0 \\ 0 & 1 \\ 0 & 0 \end{bmatrix}.$$

Then $AB = I_2$, but neither A nor B is square; so neither is invertible.

37. True   38. True

39. False, see the comment below the definition of *inverse*.

40. True   41. True   42. True

43. False, $(AB)^{-1} = B^{-1}A^{-1}$.

44. False, an elementary matrix is a matrix that can be obtained by performing one elementary row operation on an identity matrix.

45. True

46. False, $\begin{bmatrix} 2 & 0 \\ 0 & 1 \end{bmatrix}$ and $\begin{bmatrix} 1 & 0 \\ 0 & 3 \end{bmatrix}$ are elementary matrices, but $\begin{bmatrix} 2 & 0 \\ 0 & 1 \end{bmatrix}\begin{bmatrix} 1 & 0 \\ 0 & 3 \end{bmatrix} = \begin{bmatrix} 2 & 0 \\ 0 & 3 \end{bmatrix}$, which is not an elementary matrix.

47. True   48. True   49. True   50. True

51. False, see Theorem 2.4(a).

52. True

53. By the row-column rule, the $(1,1)$-entry of

$$(A_\alpha)^T A_\alpha = \begin{bmatrix} \cos\alpha & \sin\alpha \\ -\sin\alpha & \cos\alpha \end{bmatrix}\begin{bmatrix} \cos\alpha & -\sin\alpha \\ \sin\alpha & \cos\alpha \end{bmatrix}$$

is

$$\cos^2\alpha + \sin^2\alpha = \sin^2\alpha + \cos^2\alpha = 1.$$

Likewise, its $(2,2)$-entry is

$$(-\sin\alpha)^2 + \cos^2\alpha = \sin^2\alpha + \cos^2\alpha = 1.$$

Its $(1,2)$-entry is

$$\cos\alpha(-\sin\alpha) + \sin\alpha\cos\alpha = 0,$$

and its $(2,1)$-entry is

$$-\sin\alpha\cos\alpha + \cos\alpha\sin\alpha = 0.$$

Therefore $(A_\alpha)^T A_\alpha = \begin{bmatrix} 1 & 0 \\ 0 & 1 \end{bmatrix} = I_2$. Similarly, $A_\alpha(A_\alpha)^T = I_2$. Thus, by the definition of *inverse*, we have $(A_\alpha)^T = (A_\alpha)^{-1}$.

54. (a) We have

$$BA = \left(\frac{1}{ad-bc}\begin{bmatrix} d & -b \\ -c & a \end{bmatrix}\right)\begin{bmatrix} a & b \\ c & d \end{bmatrix}$$

$$= \frac{1}{ad-bc}\begin{bmatrix} ad-bc & db-bd \\ -ac+ac & -bc+ad \end{bmatrix}$$

$$= I_2.$$

Similarly, $AB = I_2$, so A is invertible and

$$A^{-1} = \frac{1}{ad-bc}\begin{bmatrix} d & -b \\ -c & a \end{bmatrix}.$$

(b) Assume A is invertible and $ad-bc = 0$. Let $C = \begin{bmatrix} d & -b \\ -c & a \end{bmatrix}$. Then

$$AC = \begin{bmatrix} a & b \\ c & d \end{bmatrix}\begin{bmatrix} d & -b \\ -c & a \end{bmatrix}$$

$$= \begin{bmatrix} ad-bc & -ab+ba \\ cd-dc & -cb+da \end{bmatrix}$$

$$= O.$$

So

$$O = A^{-1}O = A^{-1}(AC)$$
$$= (A^{-1}A)C = I_2 C = C.$$

Since $C = O$, we have $a = b = c = d = 0$. Consequently, $A = O$. But this is a contradiction because the zero matrix is not invertible.

55. By the boxed result on page 127, every elementary matrix is invertible. By Theorem 2.2(b), the product of invertible matrices is invertible. Hence the product of elementary matrices is invertible.

56. We prove the contrapositive. Suppose $A\mathbf{u} = A\mathbf{v}$. Then

$$\mathbf{u} = I_n\mathbf{u} = (A^{-1}A)\mathbf{u} = A^{-1}(A\mathbf{u})$$
$$= A^{-1}(A\mathbf{v}) = (A^{-1}A)\mathbf{v} = I_n\mathbf{v} = \mathbf{v}.$$

Thus $\mathbf{u} = \mathbf{v}$.

57. Since Q is invertible, $Q\mathbf{u} = \mathbf{0}$ if and only if $\mathbf{u} = \mathbf{0}$, and hence
$$a_1\mathbf{u}_1 + a_2\mathbf{u}_2 + \cdots + a_k\mathbf{u}_k = \mathbf{0}$$
if and only if
$$a_1 Q\mathbf{u}_1 + a_2 Q\mathbf{u}_2 + \cdots + a_k Q\mathbf{u}_k = \mathbf{0}$$
for any scalars $a_1, a_2, \ldots, a_k$. Thus the set $\{\mathbf{u}_1, \mathbf{u}_2, \ldots, \mathbf{u}_k\}$ is linearly independent if and only if the set $\{Q\mathbf{u}_1, Q\mathbf{u}_2, \ldots, Q\mathbf{u}_k\}$ is linearly independent.

58. We must prove that the inverse of A^{-1} is A. We have
$$A^{-1}A = I_n = AA^{-1}.$$
So the result follows by definition.

59. Using Theorem 2.1(b) and Theorem 2.2(b), we have
$$(ABC)^{-1} = [(AB)C]^{-1} = C^{-1}(AB)^{-1}$$
$$= C^{-1}(B^{-1}A^{-1}) = C^{-1}B^{-1}A^{-1}.$$

60. Suppose that A and AB are invertible. Then A^{-1} is invertible and
$$A^{-1}(AB) = (A^{-1}A)B = I_n B = B.$$
Hence B is the product of invertible matrices. Therefore B is invertible by Theorem 2.2(b).

61. By Theorem 1.6, we need only show that $A\mathbf{x} = \mathbf{b}$ is consistent for every $\mathbf{b}$ in $\mathcal{R}^n$. For any $\mathbf{b}$ in $\mathcal{R}^n$, let $\mathbf{u} = B\mathbf{b}$. Then
$$A\mathbf{u} = A(B\mathbf{b}) = (AB)\mathbf{b} = I_n \mathbf{b} = \mathbf{b}.$$
So $A\mathbf{x} = \mathbf{b}$ is consistent.

62. We first establish the hint. If the first column of B is not a pivot column, then $\mathbf{b}_1 = \mathbf{0}$. Hence $A\mathbf{b}_1 = A\mathbf{0} = \mathbf{0}$, which implies that the first column of AB is not a pivot column. Now suppose that column $\mathbf{b}_k$ of B is not a pivot column, where $k > 1$. Then $\mathbf{b}_k$ is a linear combination of the preceding columns of B. So there are scalars $c_1, c_2, \ldots, c_{k-1}$ such that
$$\mathbf{b}_k = c_1 \mathbf{b}_1 + c_2 \mathbf{b}_2 + \cdots + c_{k-1}\mathbf{b}_{k-1}.$$
Now multiply both sides of this equation on the left by A to obtain
$$A\mathbf{b}_k = c_1 A\mathbf{b}_1 + c_2 A\mathbf{b}_2 + \cdots + c_{k-1} A\mathbf{b}_{k-1},$$
which implies that the kth column of AB is a linear combination of the preceding columns, and hence the kth column of AB is not a pivot column.

We are now ready to prove the result. Let s and t be the number of nonpivot columns of AB and B, respectively. The preceding paragraph proves that $t \le s$, and so $p - s \le p - t$. But $p - s$ and $p - t$ are the number of pivot columns of AB and B, respectively, and the number of pivot columns of a matrix equals its rank. Therefore
$$\operatorname{rank} AB = p - s \le p - t \le \operatorname{rank} B.$$

63. By Theorem 1.6, $B\mathbf{x} = \mathbf{b}$ has a solution for every $\mathbf{b}$ in $\mathcal{R}^n$. So, for every standard vector $\mathbf{e}_i$, there is a vector $\mathbf{u}_i$ that satisfies $B\mathbf{u}_i = \mathbf{e}_i$. Let
$$C = [\mathbf{u}_1 \ \mathbf{u}_2 \ \cdots \ \mathbf{u}_n].$$
Then
$$BC = [B\mathbf{u}_1 \ B\mathbf{u}_2 \ \cdots \ B\mathbf{u}_n]$$
$$= [\mathbf{e}_1 \ \mathbf{e}_2 \ \cdots \ \mathbf{e}_n]$$
$$= I_n.$$

64. By Exercise 62, we have
$$n = \operatorname{rank} I_n = \operatorname{rank} AB \le \operatorname{rank} B \le n.$$
So $\operatorname{rank} B = n$, and hence, by Exercise 63, there exists an $n \times n$ matrix C such that $BC = I_n$. If we show $BA = I_n$, then we are done. Now
$$BA = BAI_n = BA(BC) = B(AB)C$$
$$= BI_n C = BC = I_n.$$

65. Suppose $\operatorname{rank} B = n$. Then, by Exercise 63, there is an $n \times n$ matrix C such that $BC = I_n$. By Exercise 64, C is invertible and $B = C^{-1}$. So B is invertible.

66. Suppose M is invertible. Then there is a matrix $\begin{bmatrix} C & D \\ E & F \end{bmatrix}$ such that
$$\begin{bmatrix} A & O_1 \\ O_2 & B \end{bmatrix} \begin{bmatrix} C & D \\ E & F \end{bmatrix} = \begin{bmatrix} C & D \\ E & F \end{bmatrix} \begin{bmatrix} A & O_1 \\ O_2 & B \end{bmatrix}$$
$$= \begin{bmatrix} I_n & O_1 \\ O_2 & I_m \end{bmatrix}.$$
This equation implies $AC = CA = I_n$ and $FB = BF = I_m$; so A and B are invertible. Likewise, if A and B are invertible, it is easy to show that $\begin{bmatrix} A^{-1} & O_1 \\ O_2 & B^{-1} \end{bmatrix}$ is the inverse of M.

36 Chapter 2 Matrices and Linear Transformations

67. From the column correspondence property, it follows that the third column of A equals
$$\mathbf{a}_1 + 2\mathbf{a}_2 = \begin{bmatrix} 3 \\ -1 \end{bmatrix} + 2\begin{bmatrix} 2 \\ 5 \end{bmatrix} = \begin{bmatrix} 7 \\ 9 \end{bmatrix}.$$
Therefore $A = \begin{bmatrix} 3 & 2 & 7 \\ -1 & 5 & 9 \end{bmatrix}$.

68. Using an approach similar to that of Exercise 67, we obtain
$$A = \begin{bmatrix} 2 & 4 & 1 & -4 & 2 & 10 \\ 0 & 0 & -1 & -2 & 3 & 7 \\ -1 & -2 & 2 & 7 & 0 & 3 \\ 1 & 2 & 0 & -3 & 1 & 4 \end{bmatrix}.$$

69. Using an approach similar to that of Exercise 67, we obtain
$$A = \begin{bmatrix} -1 & 1 & 1 & 4 & 13 \\ 2 & -2 & -1 & 1 & 3 \\ -1 & 1 & 0 & 3 & 8 \end{bmatrix}.$$

70. Using an approach similar to that of Exercise 67, we obtain
$$A = \begin{bmatrix} 1 & 2 & 0 & 1 & 1 & 2 \\ 2 & 4 & -1 & 3 & -4 & -1 \\ 3 & 6 & 4 & -1 & -6 & -7 \\ -1 & -2 & -2 & 1 & 1 & 2 \end{bmatrix}.$$

71. $\begin{bmatrix} 1 & 2 & 3 & 13 \\ 2 & 4 & 5 & 23 \end{bmatrix}$

72. $\begin{bmatrix} 1 & -1 & 0 & -2 & -3 & 2 \\ 1 & -1 & 2 & 4 & 5 & -6 \\ 1 & -1 & 1 & 1 & 1 & -2 \end{bmatrix}$

73. $\begin{bmatrix} 1 & -1 & 1 & 1 & 1 & -1 \\ 0 & 0 & 2 & 6 & 1 & -7 \\ 1 & -1 & 0 & 2 & 1 & 3 \end{bmatrix}$

74. $\begin{bmatrix} 1 & 0 & -1 & -3 & 1 & 4 \\ 2 & -1 & -1 & -8 & 3 & 9 \\ -1 & 1 & 1 & 5 & -2 & -6 \\ 0 & 1 & 1 & 2 & -1 & -3 \end{bmatrix}$

75. $\mathbf{a}_2 = (-2)\mathbf{a}_1 + 0\mathbf{a}_3$
76. $\mathbf{a}_3 = 0\mathbf{a}_1 + 1\mathbf{a}_3$
77. $\mathbf{a}_4 = 2\mathbf{a}_1 + (-3)\mathbf{a}_3$
78. $\mathbf{a}_5 = 3\mathbf{a}_1 + (-5)\mathbf{a}_3$
79. $\mathbf{b}_3 = 1\mathbf{b}_1 + (-1)\mathbf{b}_2 + 0\mathbf{b}_5$
80. $\mathbf{b}_4 = (-3)\mathbf{b}_1 + 2\mathbf{b}_2 + 0\mathbf{b}_5$
81. $\mathbf{b}_5 = 0\mathbf{b}_1 + 0\mathbf{b}_2 + 1\mathbf{b}_5$
82. $\mathbf{b}_6 = 3\mathbf{b}_1 + (-2)\mathbf{b}_2 + (-1)\mathbf{b}_5$

83. Let R be the reduced row echelon form of A. Because $\mathbf{u}$ and $\mathbf{v}$ are linearly independent, $\mathbf{a}_1 = \mathbf{u} \neq \mathbf{0}$, and hence $\mathbf{a}_1$ is a pivot column. Thus $\mathbf{r}_1 = \mathbf{e}_1$. Since $\mathbf{a}_2 = 2\mathbf{u} = 2\mathbf{a}_1$, it follows that $\mathbf{r}_2 = 2\mathbf{r}_1 = 2\mathbf{e}_1$ by the column correspondence property. Since $\mathbf{u}$ and $\mathbf{v}$ are linearly independent, it is easy to show that $\mathbf{u}$ and $\mathbf{u} + \mathbf{v}$ are linearly independent, and hence $\mathbf{a}_3$ is not a linear combination of $\mathbf{a}_1$ and $\mathbf{a}_2$. Thus $\mathbf{a}_3$ is a pivot column, and so $\mathbf{r}_3 = \mathbf{e}_2$. Finally,
$$\mathbf{a}_4 = \mathbf{a}_3 - \mathbf{u} = \mathbf{a}_3 - \mathbf{a}_1,$$
and hence $\mathbf{r}_4 = \mathbf{r}_3 - \mathbf{r}_1$ by the column correspondence property. Therefore
$$R = \begin{bmatrix} 1 & 2 & 0 & -1 \\ 0 & 0 & 1 & 1 \\ 0 & 0 & 0 & 0 \end{bmatrix}.$$

84. (a) A vector $\mathbf{u}$ in $\mathcal{R}^n$ is a solution of $A\mathbf{x} = \mathbf{e}_j$ if and only if $A\mathbf{u} = \mathbf{e}_j$, that is, $\mathbf{u} = A^{-1}\mathbf{e}_j$. But $A^{-1}\mathbf{e}_j$ is the jth column of A^{-1} by Theorem 1.3(d).
(b) From the proof in (a), the jth column of A^{-1} is completely determined since it is the unique solution of $A\mathbf{x} = \mathbf{e}_j$.
(c) By Theorem 1.6, we need only show that $A\mathbf{x} = \mathbf{b}$ has a solution for every $\mathbf{b}$ in $\mathcal{R}^n$. Let $\mathbf{u}_j$ be the jth column of A^{-1}. Then from (a), $A\mathbf{u}_j = \mathbf{e}_j$. Let $\mathbf{b}$ be in $\mathcal{R}^n$, and let
$$\mathbf{w} = b_1\mathbf{u}_1 + b_2\mathbf{u}_2 + \cdots + b_n\mathbf{u}_n.$$
Then
$$\begin{aligned} A\mathbf{w} &= A(b_1\mathbf{u}_1 + b_2\mathbf{u}_2 + \cdots + b_n\mathbf{u}_n) \\ &= b_1 A\mathbf{u}_1 + b_2 A\mathbf{u}_2 + \cdots + b_n A\mathbf{u}_n \\ &= b_1\mathbf{e}_1 + b_2\mathbf{e}_2 + \cdots + b_n\mathbf{e}_n \\ &= \mathbf{b}. \end{aligned}$$

85. (a) Suppose that $\mathbf{a}_j = \mathbf{0}$. Then $\mathbf{a}_j = 0\mathbf{a}_j$, and hence, by the column correspondence property, $\mathbf{r}_j = 0\mathbf{r}_j$. Thus $\mathbf{r}_j = \mathbf{0}$. Similarly, if $\mathbf{r}_j = \mathbf{0}$, then $\mathbf{a}_j = \mathbf{0}$.
(b) We prove the equivalent statement: A set of columns of A is linearly dependent if and only if the corresponding set of columns of R is linearly dependent.
To simplify the notation, suppose without loss of generality that the first k columns of A form a linearly dependent set. Then there are scalars $c_1, c_2, \ldots, c_k$, not all zero, such that
$$c_1\mathbf{a}_1 + c_2\mathbf{a}_2 + \cdots + c_k\mathbf{a}_k = \mathbf{0}.$$

By the column correspondence property, it follows that
$$c_1 \mathbf{r}_1 + c_2 \mathbf{r}_2 + \cdots + c_k \mathbf{r}_k = \mathbf{0},$$
and hence the corresponding columns of R are linearly dependent. Similarly, if a set of columns of R is linearly dependent, then the corresponding set of columns of A is linearly dependent.

86. Let $1 \le n_1 < n_2 < \cdots < n_k \le n$ be the column numbers that correspond to the pivot columns of R. Then the ith column of R^T is column n_i of $R^T R$.

    To justify this, observe that $\mathbf{r}_{n_i} = \mathbf{e}_i$, and hence if $C = R^T R$, then
    $$\mathbf{c}_{n_i} = R^T \mathbf{r}_{n_i} = R^T \mathbf{e}_i,$$
    which is the ith column of R^T.

87. (a) Note that, in the form described for R^T, the first r rows are the transposes of the standard vectors of $\mathcal{R}^m$, and the remaining rows are zero rows. As we learned on page 48, the standard vectors $\mathbf{e}'_1, \mathbf{e}'_2, \ldots, \mathbf{e}'_r$ of $\mathcal{R}^m$ must appear among the columns of R. Thus their transposes occur among the rows of R^T. By Theorem 2.4(b), every nonpivot column of R is a linear combination of $\mathbf{e}'_1, \mathbf{e}'_2, \ldots, \mathbf{e}'_r$. Thus, by appropriate row addition operations, the rows of R^T that correspond to the nonpivot columns of R can be changed to zero rows. Finally, by appropriate row interchanges, the first r rows of R^T can be changed to the transposes of $\mathbf{e}'_1, \mathbf{e}'_2, \ldots, \mathbf{e}'_r$. This is the form described for R^T.

    (b) The reduced row echelon form of R^T given in (a) has r nonzero rows. Hence
    $$\operatorname{rank} R^T = r = \operatorname{rank} R.$$

88. PAQ^T is the $m \times n$ matrix whose (i,i)-entry equals 1 for $1 \le i \le \operatorname{rank} A$, and whose other entries equal 0.

    To justify this, let S be the reduced row echelon form of R^T. Then $QR^T = S$, and hence $S^T = (QR^T)^T = R^{TT}Q^T = RQ^T$. So $PAQ^T = RQ^T = S^T$. Now apply Exercise 86, which describes S, to obtain the result.

89. Condition (a) implies there exist invertible matrices P_1 and P_2 such that $P_1 A = R = P_2 B$, where R is the common reduced row echelon form of A and B. So $P_2^{-1} P_1 A = B$. Choose $P = P_2^{-1} P_1$, and we are done.

Now assume condition (b) holds. Consider any vector $\mathbf{u}$ in $\mathcal{R}^n$. If $A\mathbf{u} = \mathbf{0}$, then
$$B\mathbf{u} = (PA)\mathbf{u} = P(A\mathbf{u}) = P\mathbf{0} = \mathbf{0},$$
and conversely, if $B\mathbf{u} = \mathbf{0}$, then
$$A\mathbf{u} = (P^{-1}B)\mathbf{u} = P^{-1}(B\mathbf{u}) = P^{-1}\mathbf{0} = \mathbf{0}.$$
Thus $A\mathbf{u} = \mathbf{0}$ if and only if $B\mathbf{u} = \mathbf{0}$. It follows, as in the discussion in this section, that there is a column correspondence property between the columns of A and the columns of B. Since both A and B have column correspondence properties with their respective reduced row echelon forms, and since these reduced row echelon forms are completely determined by these column correspondence properties (see the proof of the uniqueness of the reduced row echelon form of a matrix), it follows that A and B have the same reduced row echelon form.

90. The condition is that A is invertible. Suppose A is invertible and $AB = AC$. Then $A^{-1}(AB) = A^{-1}(AC)$; so $B = C$. Clearly, if $B = C$, then $AB = AC$.

    Now suppose $AB = AC$ if and only if $B = C$. We claim A is invertible. By Exercise 65, A is invertible if $\operatorname{rank} A = n$. By Theorem 1.8, we need to show nullity $A = 0$. Suppose $A\mathbf{u} = \mathbf{0}$. Because $A\mathbf{0} = \mathbf{0}$, we have $A\mathbf{u} = A\mathbf{0}$; so $\mathbf{u} = \mathbf{0}$. Therefore nullity $A = 0$.

91. Let
$$A = \begin{bmatrix} a & b & c \\ p & q & r \end{bmatrix}.$$
We first prove the result for the operation of interchanging rows 1 and 2 of A. In this case, performing this operation on I_2 yields
$$E = \begin{bmatrix} 0 & 1 \\ 1 & 0 \end{bmatrix},$$
and
$$EA = \begin{bmatrix} 0 & 1 \\ 1 & 0 \end{bmatrix} \begin{bmatrix} a & b & c \\ p & q & r \end{bmatrix} = \begin{bmatrix} p & q & r \\ a & b & c \end{bmatrix},$$
which is the result of interchanging rows 1 and 2 of A.

Next, we prove the result for the operation of multiplying row 1 of A by the nonzero scalar k. In this case, performing this operation on I_2 yields
$$E = \begin{bmatrix} k & 0 \\ 0 & 1 \end{bmatrix},$$

and
$$EA = \begin{bmatrix} k & 0 \\ 0 & 1 \end{bmatrix} \begin{bmatrix} a & b & c \\ p & q & r \end{bmatrix} = \begin{bmatrix} ka & kb & kc \\ p & q & r \end{bmatrix},$$
which is the result of multiplying row 1 of A by the k. The proof for the operation of multiplying row 2 of A by k is similar.

Finally, we prove the result for the operation of adding k times the second row of A to the first. In this case,
$$E = \begin{bmatrix} 1 & k \\ 0 & 1 \end{bmatrix}.$$
Then
$$EA = \begin{bmatrix} 1 & k \\ 0 & 1 \end{bmatrix} \begin{bmatrix} a & b & c \\ p & q & r \end{bmatrix}$$
$$= \begin{bmatrix} a+kp & b+kq & c+kr \\ p & q & r \end{bmatrix},$$
which is the result of adding k times the second row of A to the first. The proof for the operation of adding k times the first row to the second is similar.

92. Suppose (a) is true, that is, A and B have a common reduced row echelon form R. Then there exist invertible matrices P and Q such that $PA = R = QB$. So if $A\mathbf{x} = \mathbf{0}$, we have $Q^{-1}PA\mathbf{x} = Q^{-1}\mathbf{0} = \mathbf{0}$, or $B\mathbf{x} = \mathbf{0}$, and conversely.

 Suppose (b) is true, and let R and S be the reduced row echelon forms of A and B, respectively. Since $A\mathbf{x}$ and $B\mathbf{x}$ are each linear combinations of the columns of A and B, respectively, where the coefficients are the components of $\mathbf{x}$, it follows that a linear combination of the columns of A equals $\mathbf{0}$ if and only if the corresponding linear combination of the columns of B equals $\mathbf{0}$. Thus by the column correspondence property between a matrix and its reduced row echelon form, a linear combination of the columns of R equals $\mathbf{0}$ if and only if the corresponding linear combination of the columns of S equals $\mathbf{0}$. Since these reduced row echelon forms are entirely determined by these column correspondence properties (see the proof of the uniqueness of the reduced row echelon form of a matrix), it follows that $R = S$.

93. We illustrate the result for $n = 3$. Suppose E is the elementary matrix that corresponds to adding k times the first row to the third row. Then
$$E = \begin{bmatrix} 1 & 0 & 0 \\ 0 & 1 & 0 \\ k & 0 & 1 \end{bmatrix}.$$

We see that E can be also obtained from I_3 by adding k times the third column to the first. The converse is proved similarly.

94. We assume $n = 3$ and use the matrix E from Exercise 93. Let
$$A = \begin{bmatrix} a & b & c \\ d & e & f \\ g & h & i \end{bmatrix}.$$
Then
$$AE = \begin{bmatrix} a & b & c \\ d & e & f \\ g & h & i \end{bmatrix} \begin{bmatrix} 1 & 0 & 0 \\ 0 & 1 & 0 \\ k & 0 & 1 \end{bmatrix}$$
$$= \begin{bmatrix} a+kc & b & c \\ d+kf & e & f \\ g+ki & h & i \end{bmatrix}.$$

The other operations are handled similarly.

More generally, suppose that an $m \times n$ matrix A is transformed into a matrix B by means of an elementary column operation. Then A^T is transformed into B^T by the corresponding elementary row operation. Thus there exists an elementary matrix E such that $B^T = EA^T$. Hence
$$B = (EA^T)^T = AE^T.$$
Now observe that E^T is the elementary matrix obtained from I_n by performing the corresponding elementary column operation.

95. (a) $A^{-1} = \begin{bmatrix} -7 & 2 & 3 & -2 \\ 5 & -1 & -2 & 1 \\ 1 & 0 & 0 & 1 \\ -3 & 1 & 1 & -1 \end{bmatrix}$

 (b) $B^{-1} = \begin{bmatrix} 3 & 2 & -7 & -2 \\ -2 & -1 & 5 & 1 \\ 0 & 0 & 1 & 1 \\ 1 & 1 & -3 & -1 \end{bmatrix}$

 and
$$C^{-1} = \begin{bmatrix} -7 & -2 & 3 & 2 \\ 5 & 1 & -2 & -1 \\ 1 & 1 & 0 & 0 \\ -3 & -1 & 1 & 1 \end{bmatrix}$$

 (c) B^{-1} can be obtained by interchanging columns 1 and 3 of A^{-1}, and C^{-1} can be obtained by interchanging columns 2 and 4 of A^{-1}.

 (d) B^{-1} can be obtained by interchanging columns i and j of A^{-1}.

(e) Let E be the elementary matrix that corresponds to interchanging rows i and j. Then $B = EA$. So by Theorem 2.2(b), we have $B^{-1} = (EA)^{-1} = A^{-1}E^{-1}$. It follows from Exercise 94 that B^{-1} is obtained from A^{-1} by performing the elementary column operation on A^{-1} that is associated with E^{-1}. But because E is associated with a row interchange, we have $E^2 = I_n$, and hence $E^{-1} = E$.

96. (a) Use a calculator or computer to show that $A^T - A = O$.

(b) First compute A^{-1} to obtain
$$\begin{bmatrix} 10 & -5 & 11 & 6 \\ -5 & 10 & -26 & 7 \\ 11 & -26 & 69 & -18 \\ 6 & 7 & -18 & 6 \end{bmatrix}.$$
Now show that A^{-1} is symmetric as in (a).

(c) By Theorem 2.2(c), the transpose of A is invertible and $(A^{-1})^T = (A^T)^{-1} = A^{-1}$. So A^{-1} is symmetric.

97. (a) Use a calculator or computer to show that the reduced row echelon forms of A and A^2 have no zero rows, and hence the ranks of both matrices equal 4. Now use Exercise 65.

(b) Use a calculator or computer to obtain
$$(A^2)^{-1} = (A^{-1})^2$$
$$= \begin{bmatrix} 113 & -22 & -10 & -13 \\ -62 & 13 & 6 & 6 \\ -22 & 4 & 3 & 2 \\ 7 & -2 & -1 & 0 \end{bmatrix}.$$

(c) A^3 is invertible and $(A^3)^{-1} = (A^{-1})^3$. Using Theorem 2.2(b) and (b) of this exercise, we have that $A^3 = A^2 A$ is invertible, and
$$(A^3)^{-1} = (A^2 A)^{-1} = A^{-1}(A^2)^{-1}$$
$$= A^{-1}(A^{-1})^2 = (A^{-1})^3.$$

(d) For any invertible $k \times k$ matrix A and any positive integer n, A^n is invertible and $(A^n)^{-1} = (A^{-1})^n$. To prove this assertion, we use the fact that $AA^{-1} = A^{-1}A$ to obtain that
$$A^n(A^{-1})^n = AA \cdots AA^{-1}A^{-1} \cdots A^{-1}$$
$$= (A \cdots A)(AA^{-1})(A^{-1} \cdots A^{-1})$$
$$= I_k.$$

Similarly, $(A^{-1})^n A^n = I_k$. Therefore A^n is invertible and $(A^n)^{-1} = (A^{-1})^n$.

98. $A^{-1} = \begin{bmatrix} 12 & 3 & -8 & 2 \\ -2 & -1 & 2 & -1 \\ -2 & 0 & 1 & 0 \\ -1 & 0 & 0 & 1 \end{bmatrix}$, $A^{-1}\mathbf{b} = \begin{bmatrix} 51 \\ -9 \\ -9 \\ -2 \end{bmatrix}$

99. Observe that
$$A^{-1} = \begin{bmatrix} 10 & -2 & -1 & -1 \\ -6 & 1 & -1 & 2 \\ -2 & 0 & 1 & 0 \\ 1 & 0 & 1 & -1 \end{bmatrix}.$$

(a) The solution of $A\mathbf{x} = \mathbf{e}_1$ is $\begin{bmatrix} 10 \\ -6 \\ -2 \\ 1 \end{bmatrix}$.

(b) The solutions of $A\mathbf{x} = \mathbf{e}_i$ are the corresponding columns of A^{-1}.

2.4 THE INVERSE OF A MATRIX

1. We use the algorithm for matrix inversion to determine if A is invertible. First, form the matrix
$$\begin{bmatrix} 1 & 3 & | & 1 & 0 \\ 1 & 2 & | & 0 & 1 \end{bmatrix}.$$
Its reduced row echelon form is
$$\begin{bmatrix} 1 & 0 & | & -2 & 3 \\ 0 & 1 & | & 1 & -1 \end{bmatrix}.$$
Thus the algorithm implies that A is invertible and
$$A^{-1} = \begin{bmatrix} -2 & 3 \\ 1 & -1 \end{bmatrix}.$$

2. By Theorem 2.6(b), an $n \times n$ matrix is invertible if and only if its reduced row echelon form is I_n. The reduced row echelon form of the given matrix is
$$\begin{bmatrix} 1 & 2 \\ 0 & 0 \end{bmatrix} \neq I_2.$$
So this matrix is not invertible.

3. not invertible

4. not invertible

5. $\begin{bmatrix} 5 & -3 \\ -3 & 2 \end{bmatrix}$

6. not invertible

7. not invertible

8. $\begin{bmatrix} -10 & 3 & 5 \\ 3 & -1 & -1 \\ 1 & 0 & -1 \end{bmatrix}$

9. $\dfrac{1}{3}\begin{bmatrix} -7 & 2 & 3 \\ -6 & 0 & 3 \\ 8 & -1 & -3 \end{bmatrix}$

10. not invertible

11. not invertible

12. $\begin{bmatrix} 4 & 1 & -2 \\ 2 & 0 & -1 \\ -1 & -1 & 1 \end{bmatrix}$

13. $\begin{bmatrix} -1 & -5 & 3 \\ 1 & 2 & -1 \\ 1 & 4 & -2 \end{bmatrix}$

14. not invertible

15. not invertible

16. $\begin{bmatrix} -16 & 1 & -3 & 9 \\ 5 & 0 & 1 & -3 \\ 5 & -1 & 1 & -2 \\ -2 & 0 & 0 & 1 \end{bmatrix}$.

17. $\dfrac{1}{3}\begin{bmatrix} 1 & 1 & 1 & -2 \\ 1 & 1 & -2 & 1 \\ 1 & -2 & 1 & 1 \\ -2 & 1 & 1 & 1 \end{bmatrix}$

18. $\dfrac{1}{70}\begin{bmatrix} -25 & 7 & -12 & 14 \\ -20 & 21 & -11 & 7 \\ -35 & -14 & -21 & 7 \\ -20 & 0 & 10 & 0 \end{bmatrix}$

19. $A^{-1}B = \begin{bmatrix} -1 & 3 & -4 \\ 1 & -2 & 3 \end{bmatrix}$

20. $A^{-1}B = \begin{bmatrix} 14 & 1 \\ 9 & 0 \end{bmatrix}$

21. $A^{-1}B = \begin{bmatrix} -1 & -4 & 7 & -7 \\ 2 & 6 & -6 & 10 \end{bmatrix}$. Using the algorithm for computing $A^{-1}B$, we form the matrix

$$\begin{bmatrix} 2 & 2 & | & 2 & 4 & 2 & 6 \\ 2 & 1 & | & 0 & -2 & 8 & -4 \end{bmatrix}$$

and compute its reduced row echelon form

$$\begin{bmatrix} 1 & 0 & | & -1 & 4 & 7 & -7 \\ 0 & 1 & | & 2 & 6 & -6 & 10 \end{bmatrix} = [I_2 \ A^{-1}B].$$

22. $A^{-1}B = \begin{bmatrix} 4 & -17 \\ -1 & -9 \\ -2 & 6 \end{bmatrix}$.

23. $A^{-1}B = \begin{bmatrix} 1.0 & -0.5 & 1.5 & 1.0 \\ 6.0 & 12.5 & -11.5 & 12.0 \\ -2.0 & -5.5 & 5.5 & -5.0 \end{bmatrix}$.

24. $A^{-1}B = \begin{bmatrix} 1 & -1 & 0 & -2 & -3 \\ 1 & -1 & -2 & -8 & -11 \\ -1 & 1 & 1 & 5 & 7 \end{bmatrix}$

25. $A^{-1}B = \begin{bmatrix} -5 & -1 & -6 \\ -1 & 1 & 0 \\ 4 & 1 & 3 \\ 3 & 1 & 2 \end{bmatrix}$.

26. $\begin{bmatrix} 15 & -6 & -11 & -57 & 21 & 68 \\ 2 & -1 & -1 & -8 & 3 & 9 \\ -13 & 5 & 9 & 49 & -18 & -58 \\ 0 & 1 & 1 & 2 & -1 & -3 \end{bmatrix}$

27. $R = \begin{bmatrix} 1 & 0 & -1 \\ 0 & 1 & -3 \end{bmatrix}$ and $P = \begin{bmatrix} -1 & -1 \\ -2 & -1 \end{bmatrix}$. By the discussion on page 136, the reduced row echelon form of $[A \ I_n]$ is $[R \ P]$. In this exercise,

$$[R \ P] = \begin{bmatrix} 1 & 0 & -1 & | & -1 & -1 \\ 0 & 1 & -3 & | & -2 & -1 \end{bmatrix}.$$

28. $R = \begin{bmatrix} 1 & 0 & 0 \\ 0 & 1 & 0 \\ 0 & 0 & 1 \end{bmatrix}$ and $P = \begin{bmatrix} 1 & 1 & -1 \\ -1 & -2 & 3 \\ -1 & -1 & 2 \end{bmatrix}$.

29. $R = \begin{bmatrix} 1 & 0 & -2 & -1 \\ 0 & 1 & 1 & -1 \\ 0 & 0 & 0 & 0 \end{bmatrix}$ and one possibility for P is $P = \begin{bmatrix} -1 & 0 & 0 \\ 0 & 1 & 0 \\ 2 & -3 & 1 \end{bmatrix}$.

30. $R = \begin{bmatrix} 1 & -2 & 0 & 2 & 3 \\ 0 & 0 & 1 & -3 & -5 \end{bmatrix}$ and $P = \begin{bmatrix} -1 & 1 \\ 2 & -1 \end{bmatrix}$.

31. $R = \begin{bmatrix} 1 & 0 & 0 & 0 \\ 0 & 1 & 0 & 0 \\ 0 & 0 & 1 & 0 \\ 0 & 0 & 0 & 1 \end{bmatrix}$ and $P = \begin{bmatrix} -4 & -15 & -8 & 1 \\ 1 & 4 & 2 & 0 \\ 1 & 3 & 2 & 0 \\ -4 & -13 & -7 & 1 \end{bmatrix}$.

32. $R = \begin{bmatrix} 1 & -1 & 0 & -1 & 2 \\ 0 & 0 & 1 & -3 & 3 \\ 0 & 0 & 0 & 0 & 0 \end{bmatrix}$ and $P = \begin{bmatrix} 1 & 0 & 0 \\ 1 & 1 & 0 \\ -2 & 3 & 1 \end{bmatrix}$.

33. $R = \begin{bmatrix} 1 & 0 & 0 & 5 & 2.5 \\ 0 & 1 & 0 & -4 & -1.5 \\ 0 & 0 & 1 & -3 & -1.5 \\ 0 & 0 & 0 & 0 & 0 \end{bmatrix}$ and $P = \dfrac{1}{6}\begin{bmatrix} 0 & -5 & 4 & 1 \\ 0 & 7 & -2 & 1 \\ 0 & 1 & -2 & 1 \\ 6 & 4 & -2 & -2 \end{bmatrix}$.

34. $R = \begin{bmatrix} 1 & 0 & 0 & -3 & 1 & 3 \\ 0 & 1 & 0 & 2 & -1 & -2 \\ 0 & 0 & 1 & 0 & 0 & -1 \\ 0 & 0 & 0 & 0 & 0 & 0 \end{bmatrix}$ and $P = \begin{bmatrix} 0 & 0 & -1 & 1 \\ 1 & 0 & 1 & 0 \\ -1 & 0 & -1 & 1 \\ 0 & 1 & 2 & -1 \end{bmatrix}$

35. True
36. False, let
$$A = \begin{bmatrix} 1 & 0 & 0 \\ 0 & 1 & 0 \end{bmatrix} \quad \text{and} \quad B = \begin{bmatrix} 1 & 0 \\ 0 & 1 \\ 0 & 0 \end{bmatrix}.$$
Then $AB = I_2$, but A is not square; so it is not invertible.
37. True 38. True 39. True 40. True
41. True 42. True 43. True 44. True
45. True 46. True 47. True
48. False, if $A = I_2$ and $B = -I_2$, then $A + B = O$, which is not invertible.
49. True
50. False, $C = A^{-1}B$.
51. False, if $A = O$, then A^{-1} does not exist.
52. True 53. True 54. True
55. Let A be an $n \times n$ invertible matrix.
To prove that (a) implies (e), consider the system $A\mathbf{x} = \mathbf{b}$, where $\mathbf{b}$ is in $\mathcal{R}^n$. If we let $\mathbf{u} = A^{-1}\mathbf{b}$, then
$$A\mathbf{u} = A(A^{-1}\mathbf{b}) = (AA^{-1})\mathbf{b} = I_n\mathbf{b} = \mathbf{b},$$
and so $\mathbf{u}$ is a solution, that is, the system is consistent.

To prove that (a) implies (h), suppose that $\mathbf{u}$ is a solution of $A\mathbf{x} = \mathbf{0}$, that is, $A\mathbf{u} = \mathbf{0}$. Then
$$\mathbf{u} = (A^{-1}A)\mathbf{u} = A^{-1}(A\mathbf{u}) = A^{-1}\mathbf{0} = \mathbf{0}.$$

56. (a) $\begin{bmatrix} 1 & 2 \\ 2 & 3 \end{bmatrix} \begin{bmatrix} x_1 \\ x_2 \end{bmatrix} = \begin{bmatrix} 9 \\ 3 \end{bmatrix}$

(b) $A^{-1} = \begin{bmatrix} -3 & 2 \\ 2 & -1 \end{bmatrix}$

(c) The solution is $A^{-1}\mathbf{b} = \begin{bmatrix} -21 \\ 15 \end{bmatrix}$.

57. (a) $\begin{bmatrix} -1 & -3 \\ 2 & 5 \end{bmatrix} \begin{bmatrix} x_1 \\ x_2 \end{bmatrix} = \begin{bmatrix} -6 \\ 4 \end{bmatrix}$

(b) The reduced row echelon form of
$$\begin{bmatrix} -1 & -3 & | & 1 & 0 \\ 2 & 5 & | & 0 & 1 \end{bmatrix}$$
is
$$\begin{bmatrix} 1 & 0 & | & 5 & 3 \\ 0 & 1 & | & -2 & -1 \end{bmatrix}.$$
So $A^{-1} = \begin{bmatrix} 5 & 3 \\ -2 & -1 \end{bmatrix}$.

(c) We have
$$\begin{bmatrix} x_1 \\ x_2 \end{bmatrix} = A^{-1}\mathbf{b} = \begin{bmatrix} 5 & 3 \\ -2 & -1 \end{bmatrix} \begin{bmatrix} -6 \\ 4 \end{bmatrix}$$
$$= \begin{bmatrix} -18 \\ 8 \end{bmatrix}.$$

58. (a) $\begin{bmatrix} 1 & 1 & 1 \\ 2 & 1 & 4 \\ 3 & 2 & 6 \end{bmatrix} \begin{bmatrix} x_1 \\ x_2 \\ x_3 \end{bmatrix} = \begin{bmatrix} 4 \\ 7 \\ -1 \end{bmatrix}$

(b) $A^{-1} = \begin{bmatrix} 2 & 4 & -3 \\ 0 & -3 & 2 \\ -1 & -1 & 1 \end{bmatrix}$

(c) $\begin{bmatrix} 39 \\ -23 \\ -12 \end{bmatrix}$

59. (a) $\begin{bmatrix} -1 & 0 & 1 \\ 1 & 2 & -2 \\ 2 & -1 & 1 \end{bmatrix} \begin{bmatrix} x_1 \\ x_2 \\ x_3 \end{bmatrix} = \begin{bmatrix} -4 \\ 3 \\ 1 \end{bmatrix}$

(b) $A^{-1} = \dfrac{1}{5} \begin{bmatrix} 0 & 1 & 2 \\ 5 & 3 & 1 \\ 5 & 1 & 2 \end{bmatrix}$

(c) $\begin{bmatrix} 1 \\ -2 \\ -3 \end{bmatrix}$

60. (a) $\begin{bmatrix} 1 & 1 & 1 \\ 2 & 1 & 1 \\ 3 & 0 & 1 \end{bmatrix} \begin{bmatrix} x_1 \\ x_2 \\ x_3 \end{bmatrix} = \begin{bmatrix} -5 \\ -3 \\ 2 \end{bmatrix}$

(b) $A^{-1} = \begin{bmatrix} -1 & 1 & 0 \\ -1 & 2 & -1 \\ 3 & -3 & 1 \end{bmatrix}$

(c) $\begin{bmatrix} 2 \\ -3 \\ -4 \end{bmatrix}$

61. (a) $\begin{bmatrix} 2 & 3 & -4 \\ -1 & -1 & 2 \\ 0 & -1 & 1 \end{bmatrix} \begin{bmatrix} x_1 \\ x_2 \\ x_3 \end{bmatrix} = \begin{bmatrix} -6 \\ 5 \\ 3 \end{bmatrix}$

(b) $A^{-1} = \begin{bmatrix} 1 & 1 & 2 \\ 1 & 2 & 0 \\ 1 & 2 & 1 \end{bmatrix}$

(c) $\begin{bmatrix} 5 \\ 4 \\ 7 \end{bmatrix}$

62. (a) $\begin{bmatrix} 1 & 0 & -1 & 1 \\ 2 & -1 & -1 & 0 \\ -1 & 1 & 1 & 1 \\ 0 & 1 & 1 & 1 \end{bmatrix} \begin{bmatrix} x_1 \\ x_2 \\ x_3 \\ x_4 \end{bmatrix} = \begin{bmatrix} 3 \\ -2 \\ 4 \\ -1 \end{bmatrix}$

(b) $A^{-1} = \begin{bmatrix} 0 & 0 & -1 & 1 \\ 1 & -2 & -3 & 2 \\ -1 & 1 & 1 & 0 \\ 0 & 1 & 2 & -1 \end{bmatrix}$

(c) $\begin{bmatrix} -5 \\ -7 \\ -1 \\ 7 \end{bmatrix}$

63. (a) $\begin{bmatrix} 1 & -2 & -1 & 1 \\ 1 & 1 & 0 & -1 \\ -1 & -1 & 1 & 1 \\ -3 & 1 & 2 & 0 \end{bmatrix} \begin{bmatrix} x_1 \\ x_2 \\ x_3 \\ x_4 \end{bmatrix} = \begin{bmatrix} 4 \\ -2 \\ 1 \\ -1 \end{bmatrix}$

(b) $A^{-1} = \begin{bmatrix} -1 & 0 & 1 & -1 \\ -3 & -2 & 1 & -2 \\ 0 & 1 & 1 & 0 \\ -4 & -3 & 2 & -3 \end{bmatrix}$

(c) $\begin{bmatrix} -2 \\ -5 \\ -1 \\ -5 \end{bmatrix}$

64. (b) We may rewrite the equation in (a) as
$$I_2 = -A^2 + 3A = A(3I_2 - A).$$
Thus A is invertible by Theorem 2.6(j). Furthermore,
$$A^{-1} = A^{-1}I_2$$
$$= A^{-1}A(3I_2 - A) = 3I_2 - A.$$

65. (b) We may rewrite the equation in (a) as
$$I_3 = \frac{1}{4}A(A^2 - 5A + 9I_3).$$
Thus A is invertible by Theorem 2.6(j). Furthermore,
$$A^{-1} = A^{-1}I_2 = A^{-1}\frac{1}{4}A(A^2 - 5A + 9I_3)$$
$$= \frac{1}{4}(A^2 - 5A + 9I_3).$$

66. The equation $A^2 = I_n$ may be written $AA = I_n$. So A is invertible by Theorem 2.6(j). As in Exercise 55, we have $A^{-1} = A$.

67. (a) If $k = 1$, then $A = I_n$, and the result is clear. If $k > 1$, rewrite the equation $A^k = I_n$ as $A(A^{k-1}) = I_n$. Theorem 2.6(j) shows that A is invertible.
    (b) From (a), we have $A^{-1} = A^{k-1}$.

68. We have rank $PA \le$ rank A by Exercise 62 of Section 2.3. Furthermore, by the same exercise,
$$\text{rank } A = \text{rank } P^{-1}(PA) \le \text{rank } PA.$$

Combining these two inequalities, we obtain the required equality.

69. (a) Suppose rank $R = r$. Then each of the last $m - r$ rows of R are zero, and therefore each of the last $m - r$ rows of RB are zero. Thus, rank $RB \le$ rank R.
    (b) Let R be the reduced row echelon form of A. Then rank $R =$ rank A. By Theorem 2.3, there exists an invertible matrix P such that $PA = R$. By Exercise 68 and (a), we have
$$\text{rank } AB = \text{rank } P(AB) = \text{rank}(PA)B$$
$$= \text{rank } RB \le \text{rank } R = \text{rank } A.$$

70. Applying Exercise 69 twice, we obtain
$$\text{rank } AQ \le \text{rank } A = \text{rank } A(QQ^{-1})$$
$$= \text{rank}(AQ)Q^{-1} \le \text{rank } AQ.$$
So rank $AQ =$ rank A.

71. By Theorem 2.3, there exists an invertible matrix P such that $PA = R$, where R is the reduced row echelon form of A. Then by Theorem 2.2(c), Exercise 70, and Exercise 87(b) of Section 2.3, we have
$$\text{rank } A^T = \text{rank}(QR)^T \text{ rank } R^T Q^T$$
$$= \text{rank } R^T = \text{rank } R = \text{rank } A.$$

72. Let A be the matrix whose columns are the vectors in $\mathcal{S}$. By (d) and (g) of Theorem 2.6, the span of $\mathcal{S}$ is $\mathcal{R}^n$ if and only if $\mathcal{S}$ is linearly independent.

73. Clearly conditions 2 and 3 of the definition of reduced row echelon form are satisfied for R if they are satisfied for $[R\ S]$. Let $A = [R\ S]$. To show that condition 1 is satisfied, suppose that $\mathbf{r}_i = \mathbf{0}$ but for some $j > i$, $\mathbf{r}_j \ne \mathbf{0}$. Because A is in reduced row echelon form, $\mathbf{a}_i \ne \mathbf{0}$ since $\mathbf{a}_j \ne \mathbf{0}$. So the leading nonzero entry of $\mathbf{a}_i$ is to the right of the leading nonzero entry of $\mathbf{a}_j$. This contradicts condition 2 for A.

74. (a) $\begin{array}{l} x_1 = \quad\quad x_3 \\ x_2 = 10 - 2x_3 \\ x_3 \quad \text{free} \end{array}$
    (b) A is not invertible.

75. (a) $\begin{array}{l} x_1 = -3 + x_3 \\ x_2 = \ 4 - 2x_3 \\ x_3 \quad \text{free} \end{array}$
    (b) A is not invertible.

76. (a) $x_1 = -13 + x_3$
$x_2 = 14 - 2x_3$
x_3 free

(b) A is not invertible.

77. In Exercise 19(c) of Section 1.5, we have two sectors, oil and electricity. The input-output matrix is given by
$$C = \begin{bmatrix} .1 & .4 \\ .3 & .2 \end{bmatrix}.$$
As in Example 5, we need to compute 3 times the second column of $(I_2 - C)^{-1}$. Since
$$3(I_2 - C)^{-1} = 3\begin{bmatrix} \frac{4}{3} & \frac{2}{3} \\ \frac{1}{2} & \frac{3}{2} \end{bmatrix} = \begin{bmatrix} 4.0 & 2.0 \\ 1.5 & 4.5 \end{bmatrix},$$
the amount required is \$2 million of electricity and \$4.5 million of oil.

78. The amount required is \$1.2 million of non-government input and \$0.4 million of government input. The solution is similar to that of Exercise 77.

79. The amount required is \$12.5 million of finance, \$15 million of goods, and \$65 million of services. The solution is similar to that of Exercise 77.

80. The amount required is \$10 million of agriculture, \$37.5 million of manufacturing, and \$7.5 million of services. The solution is similar to that of Exercise 77.

81. Suppose the net production of sector i must be increased by k units, where $k > 0$. The gross production vector is given by
$$(I_n - C)^{-1}\mathbf{d} + k\mathbf{p}_i,$$
where C is the input-output matrix, $\mathbf{d}$ is the original demand vector, and $\mathbf{p}_i$ is the ith column of $(I_n - C)^{-1}$. All the entries of $\mathbf{p}_i$ are positive. Hence the gross production of every sector of the economy must be increased.

82. Let $A_1 = B^T$ and $B_1 = A^T$. We want to compute $E = AB^{-1}$. Now
$$E^T = (AB^{-1})^T = (B^{-1})^T A^T$$
$$= (B^T)^{-1} A^T = A_1^{-1} B_1.$$
Thus if we apply the algorithm to compute $E^T = A_1^{-1} B_1$, the desired matrix can be obtained by taking the transpose, $(E^T)^T = E$.

83. (a) By Theorem 2.3, there exists an invertible $m \times m$ matrix P such that $PA = R$. So $A = P^{-1}R$. If the jth pivot column of A is $\mathbf{u}_j$ $(j = 1, 2, \ldots, m)$, then the corresponding column of $P^{-1}R$ is $P^{-1}\mathbf{e}_j$, which is the jth column of P^{-1}. So
$$P^{-1} = [\mathbf{u}_1 \ \mathbf{u}_2 \ \cdots \ \mathbf{u}_m],$$
and hence $P = [\mathbf{u}_1 \ \mathbf{u}_2 \ \cdots \ \mathbf{u}_m]^{-1}$.

(b) If rank $A < m$, then (at least) the last row of R is zero. Let E be the elementary $m \times m$ matrix that corresponds to multiplying the last row by 2. Then $ER = R$. By Theorem 2.3, there exists an invertible $m \times m$ matrix P such that $PA = R$. Also $PA = ER$, and so $(E^{-1}P)A = R$. But $E^{-1}P$ is an invertible matrix and does not equal P because $E^{-1} \neq I_n$.

84. (a) Write $A = I_n^{-1} A I_n$, and let $P = I_n$.

(b) Since $B = P^{-1}AP$ for some invertible matrix P, it follows that
$$A = PBP^{-1} = (P^{-1})^{-1}BP^{-1}.$$
So B is similar to A using P^{-1} as the appropriate invertible matrix.

(c) We are given that
$$B = P^{-1}AP \quad \text{and} \quad C = Q^{-1}BQ$$
for invertible matrices P and Q. Hence
$$C = Q^{-1}(P^{-1}AP)Q$$
$$= (PQ)^{-1}A(PQ).$$
So A is similar to C using PQ as the appropriate invertible matrix.

85. (a) We are given that $I_n = P^{-1}AP$ for some invertible matrix P. It follows that
$$A = PI_nP^{-1} = PP^{-1} = I_n.$$

(b) We are given that $O = P^{-1}AP$ for some invertible matrix P. It follows that $A = POP^{-1} = O$.

(c) We are given that A is similar to $B = cI_n$. It follows that
$$A = PBP^{-1} = P(cI_n)P^{-1}$$
$$= cPP^{-1} = cI_n.$$
Thus $A = B$.

86. We are given that $B = P^{-1}AP$ for some invertible matrix P. If A is invertible, then B is a product of invertible matrices; so B is invertible. Also
$$B^{-1} = (P^{-1}AP)^{-1} = P^{-1}A^{-1}(P^{-1})^{-1}$$
$$= P^{-1}A^{-1}P.$$
So A^{-1} is similar to B^{-1}.

44   Chapter 2   Matrices and Linear Transformations

87. We are given that $B = P^{-1}AP$ for some invertible matrix P. So
$$B^T = (P^{-1}AP)^T = (P^{-1})^T A^T P^T$$
$$= (P^T)^{-1} A^T P^T.$$

It follows that A^T is similar to B^T using P^T as the appropriate invertible matrix.

88. We are given that $B = P^{-1}AP$ for some invertible matrix P. By Exercises 68 and 70, we have
$$\text{rank } B = \text{rank }(P^{-1}AP) = \text{rank } P^{-1}(AP)$$
$$= \text{rank } AP = \text{rank } A.$$

89. The reduced row echelon form of A is I_4.
90. The only solution of $A\mathbf{x} = \mathbf{0}$ is $\mathbf{0}$.
91. rank $A = 4$ 92. rank $P = 4$

2.5 PARTITIONED MATRICES AND BLOCK MULTIPLICATION

1. $[-4 \mid 2]$ 2. $\left[\dfrac{-2}{7}\right]$ 3. $\begin{bmatrix} -2 \\ 7 \end{bmatrix}$

4. $\begin{bmatrix} -2 \\ 7 \end{bmatrix}$ 5. $\left[\begin{array}{cc|cc} -2 & 4 & 6 & 0 \\ -1 & 8 & 8 & 2 \\ \hline 11 & 8 & -8 & 10 \\ 3 & 6 & 1 & 4 \end{array}\right]$

6. $\left[\begin{array}{ccc|c} -2 & 4 & 6 & 0 \\ -1 & 8 & 8 & 2 \\ 11 & 8 & -8 & 10 \\ \hline 3 & 6 & 1 & 4 \end{array}\right]$

7. $\left[\begin{array}{c|ccc} -2 & 4 & 6 & 0 \\ \hline -1 & 8 & 8 & 2 \\ 11 & 8 & -8 & 10 \\ 3 & 6 & 1 & 4 \end{array}\right]$

8. $\left[\begin{array}{ccc|c} 0 & 0 & 0 & 0 \\ 0 & 0 & 0 & 0 \\ 0 & 0 & 0 & 0 \\ \hline 0 & 0 & 0 & 9 \end{array}\right]$ 9. $\left[\begin{array}{c} 3 \quad 6 \\ 9 \quad 12 \\ \hline 2 \quad 4 \\ 6 \quad 8 \end{array}\right]$

10. $\begin{bmatrix} 1 & 2 \\ 1 & 1 \end{bmatrix}$ 11. $\left[\begin{array}{cc|cc} 1 & 1 & 2 & 1 \\ \hline 1 & 0 & 1 & -1 \\ 0 & 1 & -1 & 1 \end{array}\right]$

12. $\begin{bmatrix} A_{60°} & A_{70°} \\ A_{70°} & A_{80°} \end{bmatrix}$ 13. $[16 \; -4]$

14. $[7 \; 5 \; 10]$ 15. $[16 \; 9 \; 24]$
16. $[12 \; 7 \; -6]$ 17. $[-2 \; -3 \; 1]$
18. $[8 \; 3 \; 4]$ 19. $[-12 \; -3 \; 2]$
20. $[-7 \; -4 \; -17]$

21. $AB = \begin{bmatrix} -1 & 0 \\ -2 & 0 \\ 3 & 0 \end{bmatrix} + \begin{bmatrix} 8 & 2 \\ -4 & -1 \\ -8 & -2 \end{bmatrix} + \begin{bmatrix} 9 & -6 \\ 12 & -8 \\ 0 & 0 \end{bmatrix}$

22. $BC = \begin{bmatrix} -2 & -1 & 1 \\ 8 & 4 & -4 \\ 6 & 3 & -3 \end{bmatrix} + \begin{bmatrix} 0 & 0 & 0 \\ 4 & 3 & -2 \\ -8 & -6 & 4 \end{bmatrix}$

23. $CB = \begin{bmatrix} -2 & 0 \\ -4 & 0 \end{bmatrix} + \begin{bmatrix} 4 & 1 \\ 12 & 3 \end{bmatrix} + \begin{bmatrix} -3 & 2 \\ -6 & 4 \end{bmatrix}$

24.
$$CA = \begin{bmatrix} 2 & 4 & 6 \\ 4 & 8 & 12 \end{bmatrix} + \begin{bmatrix} 2 & -1 & 4 \\ 6 & -3 & 12 \end{bmatrix}$$
$$+ \begin{bmatrix} 3 & 2 & 0 \\ 6 & 4 & 0 \end{bmatrix}$$

25.
$$B^T A = \begin{bmatrix} -1 & -2 & -3 \\ 0 & 0 & 0 \end{bmatrix} + \begin{bmatrix} 8 & -4 & 16 \\ 2 & -1 & 4 \end{bmatrix}$$
$$+ \begin{bmatrix} -9 & -6 & 0 \\ 6 & 4 & 0 \end{bmatrix}$$

26. $AC^T = \begin{bmatrix} 2 & 4 \\ 4 & 8 \\ -6 & -12 \end{bmatrix} + \begin{bmatrix} 2 & 6 \\ -1 & -3 \\ -2 & -6 \end{bmatrix} + \begin{bmatrix} -3 & -6 \\ -4 & -8 \\ 0 & 0 \end{bmatrix}$

27. $A^T B = \begin{bmatrix} -1 & 0 \\ -2 & 0 \\ -3 & 0 \end{bmatrix} + \begin{bmatrix} 8 & 2 \\ -4 & -1 \\ 16 & 4 \end{bmatrix} + \begin{bmatrix} -9 & 6 \\ -6 & 4 \\ 0 & 0 \end{bmatrix}$

28.
$$CA^T = \begin{bmatrix} 2 & 4 & -6 \\ 4 & 8 & -12 \end{bmatrix} + \begin{bmatrix} 2 & -1 & -2 \\ 6 & 3 & -6 \end{bmatrix}$$
$$+ \begin{bmatrix} -3 & -4 & 0 \\ -6 & -8 & 0 \end{bmatrix}$$

29. True   30. True
31. False, for example $\mathbf{v}$ can be in $\mathcal{R}^2$ and $\mathbf{w}$ can be in $\mathcal{R}^3$, then $\mathbf{v}\mathbf{w}^T$ is a 2×3 matrix.
32. True
33. False, if either $\mathbf{v}$ or $\mathbf{w}$ is $\mathbf{0}$, then $\mathbf{v}\mathbf{w}^T$ is $\mathbf{0}$.
34. False, if the matrices have sizes 2×1 and 1×2, then their product can be written as a sum of two matrices of rank 1.

35. $2I_n$ 36. $\begin{bmatrix} I & A^{-1} \\ A & I \end{bmatrix}$ 37. $\begin{bmatrix} O & AC \\ BD & O \end{bmatrix}$

38. $\begin{bmatrix} A & B \\ O & C \end{bmatrix}$ 39. $\begin{bmatrix} A^T A + C^T C & A^T B + C^T D \\ B^T A + D^T C & B^T B + D^T D \end{bmatrix}$

40. $\begin{bmatrix} 2A & A+B \\ A+B & 2B \end{bmatrix}$

41. We have
$$\begin{bmatrix} I_n & O \\ CA^{-1} & I_n \end{bmatrix} \begin{bmatrix} A & O \\ O & D-CA^{-1}B \end{bmatrix} \begin{bmatrix} I_n & A^{-1}B \\ O & I_n \end{bmatrix}$$
$$= \begin{bmatrix} A & O \\ C & D-CA^{-1}B \end{bmatrix} \begin{bmatrix} I_n & A^{-1}B \\ O & I_n \end{bmatrix}$$
$$= \begin{bmatrix} A & B \\ C & D \end{bmatrix}.$$

42. We have
$$\begin{bmatrix} A & O \\ O & D \end{bmatrix} \begin{bmatrix} A^{-1} & O \\ O & D^{-1} \end{bmatrix} = \begin{bmatrix} AA^{-1} & O \\ O & DD^{-1} \end{bmatrix}$$
$$= \begin{bmatrix} I_n & O \\ O & I_n \end{bmatrix} = I_{2n}.$$

43. This is similar to Exercise 42.

44. We have
$$\begin{bmatrix} A & B \\ O & D^{-1} \end{bmatrix} \begin{bmatrix} A^{-1} & -A^{-1}BD \\ O & D \end{bmatrix}$$
$$= \begin{bmatrix} AA^{-1} & -AA^{-1}BD + BD \\ O & D^{-1}D \end{bmatrix}$$
$$= \begin{bmatrix} I_n & O \\ O & I_n \end{bmatrix} = I_{2n}$$

45. This is similar to Exercise 44.

46. This is similar to Exercise 44.

47. We have
$$\begin{bmatrix} I_n & B \\ C & I_n \end{bmatrix} \begin{bmatrix} P & -PB \\ -CP & I_n + CPB \end{bmatrix}$$
$$= \begin{bmatrix} P - BCP & -PB + B(I_n + CPB) \\ CP - CP & -CPB + I_n + CPB \end{bmatrix}$$
$$= \begin{bmatrix} P(I_n - BC) & -PB + B + BCP \\ O & I_n \end{bmatrix}$$
$$= \begin{bmatrix} PP^{-1} & B - (PB - BCPB) \\ O & I_n \end{bmatrix}$$
$$= \begin{bmatrix} I_n & B - (I_n - BC)PB \\ O & I_n \end{bmatrix}$$
$$= \begin{bmatrix} I_n & B - P^{-1}PB \\ O & I_n \end{bmatrix}$$
$$= \begin{bmatrix} I_n & B - B \\ O & I_n \end{bmatrix} = \begin{bmatrix} I_n & O \\ O & I_n \end{bmatrix} = I_{2n}.$$

48. Notice that
$$\begin{bmatrix} A & O \\ O & B \end{bmatrix}^2 = \begin{bmatrix} A & O \\ O & B \end{bmatrix} \begin{bmatrix} A & O \\ O & B \end{bmatrix}$$
$$= \begin{bmatrix} AA + OO & AO + OB \\ OA + BO & OO + BB \end{bmatrix}$$
$$= \begin{bmatrix} A^2 & O \\ O & B^2 \end{bmatrix}.$$

We conjecture that
$$\begin{bmatrix} A & O \\ O & B \end{bmatrix}^k = \begin{bmatrix} A^k & O \\ O & B^k \end{bmatrix}$$

for any positive integer k. This is clear for $k = 1$ and, as we have just seen, for $k = 2$. If the result is true for $k - 1$, where $k > 2$, then
$$\begin{bmatrix} A & O \\ O & B \end{bmatrix}^k = \begin{bmatrix} A & O \\ O & B \end{bmatrix} \begin{bmatrix} A & O \\ O & B \end{bmatrix}^{k-1}$$
$$= \begin{bmatrix} A & O \\ O & B \end{bmatrix} \begin{bmatrix} A^{k-1} & O \\ O & B^{k-1} \end{bmatrix}$$
$$= \begin{bmatrix} AA^{k-1} + OO & AO + OB^{k-1} \\ OA^{k-1} + BO & OO + BB^{k-1} \end{bmatrix}$$
$$= \begin{bmatrix} A^k & O \\ O & B^k \end{bmatrix}.$$

So the result is true for k. In this manner, it follows that the result is true for every positive integer k. (For those familiar with mathematical induction, this proof can be given using induction.)

49. Notice that
$$\begin{bmatrix} A & B \\ O & O \end{bmatrix}^2 = \begin{bmatrix} A & B \\ O & O \end{bmatrix} \begin{bmatrix} A & B \\ O & O \end{bmatrix}$$
$$= \begin{bmatrix} AA + BO & AB + BO \\ OA + OO & OB + OO \end{bmatrix}$$
$$= \begin{bmatrix} A^2 & AB \\ O & O \end{bmatrix}$$

and
$$\begin{bmatrix} A & B \\ O & O \end{bmatrix}^3 = \begin{bmatrix} A & B \\ O & O \end{bmatrix} \begin{bmatrix} A & B \\ O & O \end{bmatrix}^2$$
$$= \begin{bmatrix} A & B \\ O & O \end{bmatrix} \begin{bmatrix} A^2 & AB \\ O & O \end{bmatrix}$$
$$= \begin{bmatrix} AA^2 + BO & AAB + BO \\ OA^2 + OO & OAB + OO \end{bmatrix}$$
$$= \begin{bmatrix} A^3 & A^2B \\ O & O \end{bmatrix}.$$

We conjecture that
$$\begin{bmatrix} A & B \\ O & O \end{bmatrix}^k = \begin{bmatrix} A^k & A^{k-1}B \\ O & O \end{bmatrix}$$
for any positive integer k. This is clear for $k=1$, and, as we have just seen, for $k=2$ and $k=3$. A proof for the general case is similar to the proof given in Exercise 48.

50. Suppose that $\begin{bmatrix} A & B \\ B & A \end{bmatrix}$ is invertible. Then there exist $n \times n$ matrices C, D, E, and F such that
$$\begin{bmatrix} AC+BE & AD+BF \\ BC+AE & BD+AF \end{bmatrix} = \begin{bmatrix} A & B \\ B & A \end{bmatrix}\begin{bmatrix} C & D \\ E & F \end{bmatrix}$$
$$= I_{2n} = \begin{bmatrix} I_n & O \\ O & I_n \end{bmatrix}.$$
Thus $BC + AE = O$, and hence, $E = -A^{-1}BC$. Furthermore,
$$I_n = AC + BE = AC - BA^{-1}BC$$
$$= (A - BA^{-1}B)C,$$
and hence $A - BA^{-1}B$ is invertible by Theorem 2.6(j).

Conversely, let $A - BA^{-1}B$ be invertible. Define $C = (A - BA^{-1}B)^{-1}$ and $E = -A^{-1}BC$. Then
$$AC + BE = (A - BA^{-1}B)C = I_n$$
and
$$BC + AE = BC - BC = O.$$
Now set $D = E$ and $F = C$. Then
$$\begin{bmatrix} A & B \\ B & A \end{bmatrix}\begin{bmatrix} C & D \\ E & F \end{bmatrix} = I_{2n},$$
and hence $\begin{bmatrix} A & B \\ B & A \end{bmatrix}$ is invertible.

51. In order to guess the form of the inverse if the matrix is invertible, consider the matrix $\begin{bmatrix} a & 0 \\ 1 & b \end{bmatrix}$, where a and b are nonzero scalars. It is easy to show that this matrix is invertible with inverse $\begin{bmatrix} -a^{-1} & 0 \\ -(ab)^{-1} & b^{-1} \end{bmatrix}$. So a reasonable guess for the inverse of $\begin{bmatrix} A & O \\ I_n & B \end{bmatrix}$ is $\begin{bmatrix} A^{-1} & O \\ -B^{-1}A^{-1} & B^{-1} \end{bmatrix}$.

Now we must verify that this guess is correct. In the product
$$\begin{bmatrix} A & O \\ I_n & B \end{bmatrix}\begin{bmatrix} A^{-1} & O \\ -B^{-1}A^{-1} & B^{-1} \end{bmatrix},$$

the upper left submatrix is
$$AA^{-1} + O(-B^{-1}A^{-1}) = I_n + O = I_n,$$
the upper right submatrix is
$$AO + OB^{-1} = O + O = O,$$
the lower left submatrix is
$$I_n A^{-1} - B(B^{-1}A^{-1}) = A^{-1} - A^{-1}$$
$$= O,$$
and the lower right submatrix is
$$I_n O + BB^{-1} = O + I_n = I_n.$$
Thus
$$\begin{bmatrix} A & O \\ I_n & B \end{bmatrix}\begin{bmatrix} A^{-1} & O \\ -B^{-1}A^{-1} & B^{-1} \end{bmatrix} = I_{2n},$$
and so $\begin{bmatrix} A & O \\ I_n & B \end{bmatrix}$ is invertible with inverse $\begin{bmatrix} A^{-1} & O \\ -B^{-1}A^{-1} & B^{-1} \end{bmatrix}.$

52. Choose j such that $a_j \neq 0$. Because
$$\mathbf{ab}^T = \begin{bmatrix} a_1 \mathbf{b}^T \\ a_2 \mathbf{b}^T \\ \vdots \\ a_m \mathbf{b}^T \end{bmatrix},$$
$\mathbf{ab}^T$ is a nonzero matrix all of whose rows are multiples of $\mathbf{b}^T$. Consequently, $\mathbf{ab}^T$ has rank 1.

53. (c) $A^k = \begin{bmatrix} B^k & * \\ O & D^k \end{bmatrix}$, where $*$ represents some 2×2 matrix.
   (d) We have
$$A^2 = AA = \begin{bmatrix} B & C \\ O & D \end{bmatrix}\begin{bmatrix} B & C \\ O & D \end{bmatrix}$$
$$= \begin{bmatrix} BB+CO & BC+CD \\ OB+DO & OC+DD \end{bmatrix}$$
$$= \begin{bmatrix} B^2 & * \\ O & D^2 \end{bmatrix}.$$
So
$$A^3 = A^2 A = \begin{bmatrix} B^2 & * \\ O & D^2 \end{bmatrix}\begin{bmatrix} B & C \\ O & D \end{bmatrix}$$
$$= \begin{bmatrix} B^2 B + O & * \\ OB + D^2 O & OC + D^2 D \end{bmatrix}$$
$$= \begin{bmatrix} B^3 & * \\ O & D^3 \end{bmatrix}.$$

2.6 THE LU DECOMPOSITION OF A MATRIX

1. $L = \begin{bmatrix} 1 & 0 & 0 \\ 3 & 1 & 0 \\ -1 & 1 & 1 \end{bmatrix}$ and $U = \begin{bmatrix} 2 & 3 & 4 \\ 0 & -1 & -2 \\ 0 & 0 & 3 \end{bmatrix}$

2. $L = \begin{bmatrix} 1 & 0 & 0 \\ 2 & 1 & 0 \\ -1 & 0 & 1 \end{bmatrix}$ and $U = \begin{bmatrix} 2 & -1 & 1 \\ 0 & 1 & 2 \\ 0 & 0 & 3 \end{bmatrix}$

3. $L = \begin{bmatrix} 1 & 0 & 0 \\ 2 & 1 & 0 \\ -3 & 1 & 1 \end{bmatrix}$ and $U = \begin{bmatrix} 1 & -1 & 2 & 1 \\ 0 & -1 & 1 & 2 \\ 0 & 0 & 1 & 1 \end{bmatrix}$

4. $L = \begin{bmatrix} 1 & 0 \\ 3 & 1 \end{bmatrix}$ and $U = \begin{bmatrix} 1 & -1 & 2 & 4 \\ 0 & 0 & -1 & -3 \end{bmatrix}$

5. $L = \begin{bmatrix} 1 & 0 & 0 \\ -1 & 1 & 0 \\ 2 & 1 & 1 \end{bmatrix}$ and
$U = \begin{bmatrix} 1 & -1 & 2 & 1 & 3 \\ 0 & 1 & 2 & -1 & 1 \\ 0 & 0 & 1 & -2 & -6 \end{bmatrix}$

6. $L = \begin{bmatrix} 1 & 0 & 0 & 0 \\ 2 & 1 & 0 & 0 \\ -1 & 0 & 1 & 0 \\ 1 & 2 & -1 & 1 \end{bmatrix}$ and
$U = \begin{bmatrix} 3 & 1 & -1 & 1 \\ 0 & 2 & 1 & 2 \\ 0 & 0 & 1 & 0 \\ 0 & 0 & 0 & -2 \end{bmatrix}$

7. $L = \begin{bmatrix} 1 & 0 & 0 & 0 \\ 2 & 1 & 0 & 0 \\ -1 & -1 & 1 & 0 \\ 0 & -1 & 0 & 1 \end{bmatrix}$ and
$U = \begin{bmatrix} 1 & 0 & -3 & -1 & -2 & 1 \\ 0 & -1 & -2 & 1 & -1 & -2 \\ 0 & 0 & 0 & 1 & 1 & 1 \\ 0 & 0 & 0 & 2 & 2 & 2 \end{bmatrix}$

8. We apply elementary row operations to transform the given matrix into an upper triangular matrix:

$\begin{bmatrix} -1 & 2 & 1 & -1 & 3 \\ 1 & -4 & 0 & 5 & -5 \\ -2 & 6 & -1 & -5 & 7 \\ -1 & -4 & 4 & 11 & -2 \end{bmatrix} \xrightarrow{r_1+r_2 \to r_2}$

$\begin{bmatrix} -1 & 2 & 1 & -1 & 3 \\ 0 & -2 & 1 & 4 & -2 \\ -2 & 6 & -1 & -5 & 7 \\ -1 & -4 & 4 & 11 & -2 \end{bmatrix} \xrightarrow{-2r_1+r_3 \to r_3}$

$\begin{bmatrix} -1 & 2 & 1 & -1 & 3 \\ 0 & -2 & 1 & 4 & -2 \\ 0 & 2 & -3 & -3 & 1 \\ -1 & -4 & 4 & 11 & -2 \end{bmatrix} \xrightarrow{-r_1+r_4 \to r_4}$

$\begin{bmatrix} -1 & 2 & 1 & -1 & 3 \\ 0 & -2 & 1 & 4 & -2 \\ 0 & 2 & -3 & -3 & 1 \\ 0 & -6 & 3 & 12 & -5 \end{bmatrix} \xrightarrow{r_2+r_3 \to r_3}$

$\begin{bmatrix} -1 & 2 & 1 & -1 & 3 \\ 0 & -2 & 1 & 4 & -2 \\ 0 & 0 & -2 & 1 & -1 \\ 0 & -6 & 3 & 12 & -5 \end{bmatrix} \xrightarrow{-3r_2+r_4 \to r_4}$

$\begin{bmatrix} -1 & 2 & 1 & -1 & 3 \\ 0 & -2 & 1 & 4 & -2 \\ 0 & 0 & -2 & 1 & -1 \\ 0 & 0 & 0 & 0 & 1 \end{bmatrix} = U.$

Since U consists of 4 rows, L is a 4×4 matrix. As in Example 3, the entries of L below the diagonal are the multipliers, and these can be obtained directly from the labels above the arrows describing the transformation of the given matrix into an upper triangular matrix. In particular, a label of the form $cr_i + r_j$ indicates that the (i,j)-entry of L is $-c$. Thus

$$L = \begin{bmatrix} 1 & 0 & 0 & 0 \\ -1 & 1 & 0 & 0 \\ 2 & -1 & 1 & 0 \\ 1 & 3 & 0 & 1 \end{bmatrix}.$$

9. $\begin{bmatrix} x_1 \\ x_2 \\ x_3 \end{bmatrix} = \begin{bmatrix} 2 \\ -1 \\ 0 \end{bmatrix}$ 10. $\begin{bmatrix} x_1 \\ x_2 \\ x_3 \end{bmatrix} = \begin{bmatrix} 1 \\ 2 \\ -1 \end{bmatrix}$

11. $\begin{bmatrix} x_1 \\ x_2 \\ x_3 \\ x_4 \end{bmatrix} = \begin{bmatrix} -7 \\ -4 \\ 2 \\ 0 \end{bmatrix} + x_4 \begin{bmatrix} 2 \\ 1 \\ -1 \\ 1 \end{bmatrix}$

12. $\begin{bmatrix} x_1 \\ x_2 \\ x_3 \\ x_4 \end{bmatrix} = \begin{bmatrix} 5 \\ 0 \\ -2 \\ 0 \end{bmatrix} + x_2 \begin{bmatrix} 1 \\ 1 \\ 0 \\ 0 \end{bmatrix} + x_4 \begin{bmatrix} 2 \\ 0 \\ -3 \\ 1 \end{bmatrix}$

13. $\begin{bmatrix} x_1 \\ x_2 \\ x_3 \\ x_4 \\ x_5 \end{bmatrix} = \begin{bmatrix} -3 \\ 3 \\ 1 \\ 0 \\ 0 \end{bmatrix} + x_4 \begin{bmatrix} -8 \\ -3 \\ 2 \\ 1 \\ 0 \end{bmatrix} + x_5 \begin{bmatrix} -28 \\ -13 \\ 6 \\ 0 \\ 1 \end{bmatrix}$

14. $\begin{bmatrix} x_1 \\ x_2 \\ x_3 \\ x_4 \end{bmatrix} = \begin{bmatrix} -2 \\ 3 \\ 1 \\ 4 \end{bmatrix}$

15.
$$\begin{bmatrix} x_1 \\ x_2 \\ x_3 \\ x_4 \\ x_5 \\ x_6 \end{bmatrix} = \begin{bmatrix} 3 \\ -4 \\ 0 \\ 2 \\ 0 \\ 0 \end{bmatrix} + x_3 \begin{bmatrix} 3 \\ -2 \\ 1 \\ 0 \\ 0 \\ 0 \end{bmatrix} + x_5 \begin{bmatrix} 1 \\ -2 \\ 0 \\ -1 \\ 1 \\ 0 \end{bmatrix} + x_6 \begin{bmatrix} -2 \\ -3 \\ 0 \\ -1 \\ 0 \\ 1 \end{bmatrix}$$

16. The system of equations can be written $A\mathbf{x} = \mathbf{b}$ for

$$A = \begin{bmatrix} -1 & 2 & 1 & -1 & 3 \\ 1 & -4 & 0 & 5 & -5 \\ -2 & 6 & -1 & -5 & 7 \\ -1 & -4 & 4 & 11 & -2 \end{bmatrix}, \mathbf{b} = \begin{bmatrix} 7 \\ -7 \\ 6 \\ 11 \end{bmatrix}.$$

We will use the matrices L and U in Exercise 8, which form an LU decomposition of A. We first solve the system $L\mathbf{y} = \mathbf{b}$, which is

$$\begin{aligned} y_1 &= 7 \\ -y_1 + y_2 &= -7 \\ 2y_1 - y_2 + y_3 &= 6 \\ y_1 + 3y_2 + y_4 &= 11. \end{aligned}$$

From the first equation, $y_1 = 7$. Substituting this value into the second equation, we obtain $y_2 = 0$. Continuing in this manner, we can solve for the other values to obtain

$$\mathbf{y} = \begin{bmatrix} y_1 \\ y_2 \\ y_3 \\ y_4 \end{bmatrix} = \begin{bmatrix} 7 \\ 0 \\ -8 \\ 4 \end{bmatrix}.$$

Next, we solve the system $U\mathbf{x} = \mathbf{y}$, which is

$$\begin{aligned} -x_1 + 2x_2 + x_3 - x_4 + 3x_5 &= 7 \\ -2x_2 + x_3 + 4x_4 - 2x_5 &= 0 \\ -2x_3 + x_4 - x_5 &= -8 \\ x_5 &= 4, \end{aligned}$$

using back substitution. From the fourth equation, we have $x_5 = 4$. Substituting this value in the third equation and solving for x_3, while treating x_4 as a free variable, we obtain

$$x_3 = 2 + \frac{1}{2}x_4.$$

Similarly, we substitute the values obtained in the third and fourth equations into the second equation to solve for x_2, and we substitute the values we now have into the first equation and solve for x_1. This gives

$$x_2 = -3 + \frac{9}{4}x_4 \quad \text{and} \quad x_1 = 1 + 4x_4.$$

Therefore the general solution of the system is

$$\begin{bmatrix} x_1 \\ x_2 \\ x_3 \\ x_4 \\ x_5 \end{bmatrix} = \begin{bmatrix} 1 \\ -3 \\ 2 \\ 0 \\ 4 \end{bmatrix} + x_4 \begin{bmatrix} 4 \\ \frac{9}{4} \\ \frac{1}{2} \\ 1 \\ 0 \end{bmatrix}.$$

17. $P = \begin{bmatrix} 1 & 0 & 0 \\ 0 & 0 & 1 \\ 0 & 1 & 0 \end{bmatrix}$, $L = \begin{bmatrix} 1 & 0 & 0 \\ -1 & 1 & 0 \\ 2 & 0 & 1 \end{bmatrix}$, and

$$U = \begin{bmatrix} 1 & -1 & 3 \\ 0 & 1 & 2 \\ 0 & 0 & -1 \end{bmatrix}$$

18. $P = \begin{bmatrix} 0 & 1 & 0 \\ 1 & 0 & 0 \\ 0 & 0 & 1 \end{bmatrix}$, $L = \begin{bmatrix} 1.0 & 0 & 0 \\ 0.0 & 1 & 0 \\ 0.5 & 0 & 1 \end{bmatrix}$, and

$$U = \begin{bmatrix} 2 & 6 & 0 \\ 0 & 2 & -1 \\ 0 & 0 & -1 \end{bmatrix}$$

19. $P = \begin{bmatrix} 1 & 0 & 0 \\ 0 & 0 & 1 \\ 0 & 1 & 0 \end{bmatrix}$, $L = \begin{bmatrix} 1 & 0 & 0 \\ -1 & 1 & 0 \\ 2 & 0 & 1 \end{bmatrix}$, and

$$U = \begin{bmatrix} 1 & 1 & -2 & -1 \\ 0 & -1 & -3 & 0 \\ 0 & 0 & 1 & 1 \end{bmatrix}$$

20. $P = \begin{bmatrix} 0 & 1 & 0 \\ 1 & 0 & 0 \\ 0 & 0 & 1 \end{bmatrix}$, $L = \begin{bmatrix} 1.0 & 0.0 & 0 \\ 0.0 & 1.0 & 0 \\ -0.5 & 0.5 & 1 \end{bmatrix}$, and

$$U = \begin{bmatrix} -2 & -3 & 2 & 2.0 \\ 0 & -1 & 4 & 3.0 \\ 0 & 0 & -2 & 0.5 \end{bmatrix}$$

21. $P = \begin{bmatrix} 0 & 1 & 0 & 0 \\ 1 & 0 & 0 & 0 \\ 0 & 0 & 1 & 0 \\ 0 & 0 & 0 & 1 \end{bmatrix}$, $L = \begin{bmatrix} 1 & 0 & 0 & 0 \\ 0 & 1 & 0 & 0 \\ -2 & 0 & 1 & 0 \\ -1 & -1 & -1 & 0 \end{bmatrix}$,

and $U = \begin{bmatrix} -1 & 2 & -1 \\ 0 & 1 & -2 \\ 0 & 0 & 1 \\ 0 & 0 & 0 \end{bmatrix}$

22. $P = \begin{bmatrix} 1 & 0 & 0 & 0 \\ 0 & 1 & 0 & 0 \\ 0 & 0 & 0 & 1 \\ 0 & 0 & 1 & 0 \end{bmatrix}$, $L = \begin{bmatrix} 1 & 0 & 0 & 0 \\ -1 & 1 & 0 & 0 \\ 2 & -1 & 1 & 0 \\ 1 & 1 & 0 & 1 \end{bmatrix}$,

and $U = \begin{bmatrix} 2 & 4 & -6 & 0 \\ 0 & 5 & -3 & 2 \\ 0 & 0 & 6 & 2 \\ 0 & 0 & 0 & -1 \end{bmatrix}$

2.6 The *LU* Decomposition of a Matrix 49

23. We use Examples 5 and 6 as a model, placing the multipliers in parentheses in appropriate matrix entries:

$$A = \begin{bmatrix} 1 & 2 & 1 & -1 \\ 2 & 4 & 1 & 1 \\ 3 & 2 & -1 & -2 \\ 2 & 5 & 3 & 0 \end{bmatrix} \xrightarrow{-2r_1+r_2 \to r_2}$$

$$\begin{bmatrix} 1 & 2 & 1 & -1 \\ (2) & 0 & -1 & 3 \\ 3 & 2 & -1 & -2 \\ 2 & 5 & 3 & 0 \end{bmatrix} \xrightarrow{-3r_1+r_3 \to r_3}$$

$$\begin{bmatrix} 1 & 2 & 1 & -1 \\ (2) & 0 & -1 & 3 \\ (3) & -4 & -4 & 1 \\ 2 & 5 & 3 & 0 \end{bmatrix} \xrightarrow{-2r_1+r_4 \to r_4}$$

$$\begin{bmatrix} 1 & 2 & 1 & -1 \\ (2) & 0 & -1 & 3 \\ (3) & -4 & -4 & 1 \\ (2) & 1 & 1 & 2 \end{bmatrix} \xrightarrow{r_2 \leftrightarrow r_4}$$

$$\begin{bmatrix} 1 & 2 & 1 & -1 \\ (2) & 1 & 1 & 2 \\ (3) & -4 & -4 & 1 \\ (2) & 0 & -1 & 3 \end{bmatrix} \xrightarrow{4r_2+r_3 \to r_3}$$

$$\begin{bmatrix} 1 & 2 & 1 & -1 \\ (2) & 1 & 1 & 2 \\ (3) & (-4) & 0 & 9 \\ (2) & 0 & -1 & 3 \end{bmatrix} \xrightarrow{r_3 \leftrightarrow r_4}$$

$$\begin{bmatrix} 1 & 2 & 1 & -1 \\ (2) & 1 & 1 & 2 \\ (2) & 0 & -1 & 3 \\ (3) & (-4) & 0 & 9 \end{bmatrix}.$$

The last matrix in the sequence contains the information necessary to construct the matrices L and U in an LU decomposition of A. L is the unit lower triangular matrix whose subdiagonal entries are the same as the subdiagonal entries of the final matrix, where parentheses are removed, if necessary. U is the upper triangular matrix obtained from the final matrix in the sequence by replacing all subdiagonal entries by zeros. Thus we obtain

$$L = \begin{bmatrix} 1 & 0 & 0 & 0 \\ 2 & 1 & 0 & 0 \\ 2 & 0 & 1 & 0 \\ 3 & -4 & 0 & 1 \end{bmatrix}$$

and

$$U = \begin{bmatrix} 1 & 2 & 1 & -1 \\ 0 & 1 & 1 & 2 \\ 0 & 0 & -1 & 3 \\ 0 & 0 & 0 & 9 \end{bmatrix}.$$

Finally, we obtain P by applying to I_4 the row interchanges that occur in the sequence at the left. Thus

$$I_4 \xrightarrow{r_2 \leftrightarrow r_4} \begin{bmatrix} 1 & 0 & 0 & 0 \\ 0 & 0 & 0 & 1 \\ 0 & 0 & 1 & 0 \\ 0 & 1 & 0 & 0 \end{bmatrix}$$

$$\xrightarrow{r_3 \leftrightarrow r_4} \begin{bmatrix} 1 & 0 & 0 & 0 \\ 0 & 0 & 0 & 1 \\ 0 & 1 & 0 & 0 \\ 0 & 0 & 1 & 0 \end{bmatrix} = P.$$

24. $P = \begin{bmatrix} 1 & 0 & 0 & 0 \\ 0 & 0 & 1 & 0 \\ 0 & 1 & 0 & 0 \\ 0 & 0 & 0 & 1 \end{bmatrix}$,

$L = \begin{bmatrix} 1 & 0 & 0 & 0 \\ 1 & 1 & 0 & 0 \\ 2 & 0 & 1 & 0 \\ -3 & -4 & -1 & 1 \end{bmatrix}$, and

$U = \begin{bmatrix} 1 & 2 & 2 & 2 & 1 \\ 0 & -1 & -1 & 0 & 1 \\ 0 & 0 & -2 & -3 & -2 \\ 0 & 0 & 0 & 0 & 0 \end{bmatrix}$

25. $\begin{bmatrix} x_1 \\ x_2 \\ x_3 \end{bmatrix} = \begin{bmatrix} -2 \\ 1 \\ 3 \end{bmatrix}$ **26.** $\begin{bmatrix} x_1 \\ x_2 \\ x_3 \end{bmatrix} = \begin{bmatrix} -4 \\ 1 \\ 0 \end{bmatrix}$

27. $\begin{bmatrix} x_1 \\ x_2 \\ x_3 \\ x_4 \end{bmatrix} = \begin{bmatrix} 16 \\ -9 \\ 3 \\ 0 \end{bmatrix} + x_4 \begin{bmatrix} -4 \\ 3 \\ -1 \\ 1 \end{bmatrix}$

28. $\begin{bmatrix} x_1 \\ x_2 \\ x_3 \\ x_4 \end{bmatrix} = \begin{bmatrix} 1.25 \\ -2.00 \\ -0.75 \\ 0.00 \end{bmatrix} + x_4 \begin{bmatrix} -4.75 \\ 4.00 \\ 0.25 \\ 1.00 \end{bmatrix}$

29. $\begin{bmatrix} x_1 \\ x_2 \\ x_3 \end{bmatrix} = \begin{bmatrix} 5 \\ 2 \\ 1 \end{bmatrix}$ **30.** $\begin{bmatrix} x_1 \\ x_2 \\ x_3 \\ x_4 \end{bmatrix} = \begin{bmatrix} 1 \\ 3 \\ 2 \\ 0 \end{bmatrix}$

31. Let

$$A = \begin{bmatrix} 1 & 2 & 1 & -1 \\ 2 & 4 & 1 & 1 \\ 3 & 2 & -1 & -2 \\ 2 & 5 & 3 & 0 \end{bmatrix} \text{ and } \mathbf{b} = \begin{bmatrix} 3 \\ 2 \\ -4 \\ 7 \end{bmatrix}.$$

Then the system can be written as the matrix equation $A\mathbf{x} = \mathbf{b}$. By Exercise 23, $PA = LU$, where

$$P = \begin{bmatrix} 1 & 0 & 0 & 0 \\ 0 & 0 & 0 & 1 \\ 0 & 1 & 0 & 0 \\ 0 & 0 & 1 & 0 \end{bmatrix}, L = \begin{bmatrix} 1 & 0 & 0 & 0 \\ 2 & 1 & 0 & 0 \\ 2 & 0 & 1 & 0 \\ 3 & -4 & 0 & 1 \end{bmatrix},$$

and $U = \begin{bmatrix} 1 & 2 & 1 & -1 \\ 0 & 1 & 1 & 2 \\ 0 & 0 & -1 & 3 \\ 0 & 0 & 0 & 9 \end{bmatrix}.$

Since P is invertible, the system $A\mathbf{x} = \mathbf{b}$ is equivalent to

$PA\mathbf{x} = P\mathbf{b}$

$= \begin{bmatrix} 1 & 0 & 0 & 0 \\ 0 & 0 & 0 & 1 \\ 0 & 1 & 0 & 0 \\ 0 & 0 & 1 & 0 \end{bmatrix} \begin{bmatrix} 3 \\ 2 \\ -4 \\ 7 \end{bmatrix} = \begin{bmatrix} 3 \\ 7 \\ -2 \\ 4 \end{bmatrix} = \mathbf{b}'.$

We can solve this system using the LU decomposition of PA given above. As in Example 4, set $\mathbf{y} = U\mathbf{x}$, and use forward substitution to solve the system $L\mathbf{y} = \mathbf{b}'$, which can be written

$$\begin{aligned} y_1 &= 3 \\ 2y_1 + y_2 &= 7 \\ 2y_1 + y_3 &= 2 \\ 3y_1 - 4y_2 + y_4 &= -4. \end{aligned}$$

The resulting solution is

$$\mathbf{y} = \begin{bmatrix} y_1 \\ y_2 \\ y_3 \\ y_4 \end{bmatrix} = \begin{bmatrix} 3 \\ 1 \\ -4 \\ -9 \end{bmatrix}.$$

Finally, to obtain the solution of the original system, use back substitution to solve $U\mathbf{x} = \mathbf{y}$, which can be written as

$$\begin{aligned} x_1 + 2x_2 + x_3 - x_4 &= 3 \\ 2x_2 + x_3 + 2x_4 &= 1 \\ -x_3 + 3x_4 &= -4 \\ 9x_4 &= 9. \end{aligned}$$

This solution is

$$\begin{bmatrix} x_1 \\ x_2 \\ x_3 \\ x_4 \end{bmatrix} = \begin{bmatrix} -3 \\ 2 \\ 1 \\ -1 \end{bmatrix}.$$

32. $\begin{bmatrix} x_1 \\ x_2 \\ x_3 \\ x_4 \\ x_5 \end{bmatrix} = x_4 \begin{bmatrix} -2.0 \\ 1.5 \\ -1.5 \\ 1.0 \\ 0.0 \end{bmatrix} + x_5 \begin{bmatrix} -3 \\ 2 \\ -1 \\ 0 \\ 1 \end{bmatrix}$

33. False, the matrices in Exercises 17–24 do not have LU decompositions.

34. True

35. False, the entries below and to the left of the diagonal entries are zeros.

36. False, consider the LU decomposition of the matrix in Exercise 1.

37. False, for example, if A is the $m \times n$ zero matrix and $U = A$, then $A = LU$, where U is any $m \times m$ unit lower triangular matrix.

38. True

39. False, the (i,j)-entry of L is $-c$.

40. True **41.** True

42. That AB is upper triangular is Exercise 61 of Section 2.1. The ith diagonal entry of $C = AB$ is the sum

$c_{ii} = a_{i1}b_{1i} + a_{i2}b_{2i} + \cdots + a_{ii}b_{ii} + \cdots + a_{in}b_{ni}.$

Since $a_{ij} = 0$ if $i > j$ and $b_{ij} = 0$ if $i > j$, it follows that $a_{ij}b_{ji} = 0$ if $i \neq j$. So the preceding equation reduces to $c_{ii} = a_{ii}b_{ii}$.

43. Because U is invertible, it can be transformed into I_n by means of elementary row operations. Since U is upper triangular, each elementary row operation can be chosen so that a multiple of a row is never added to a lower row. Note that the corresponding elementary matrix is upper triangular. Thus there exist upper triangular elementary matrices $E_1, E_2, \ldots, E_k$ such that

$E_k E_{k-1} \cdots E_1 U = I_n.$

Let $A = E_k E_{k-1} \cdots E_1$. Then $AU = I_n$, and hence $A = U^{-1}$. Furthermore, since A is the product of upper triangular matrices, $A = U^{-1}$ is upper triangular.

To find the ith diagonal entry of $A = U^{-1}$, observe that $AU = I_n$, and so, by Exercise 42, the ith diagonal entry of AU is $a_{ii}u_{ii} = 1$. It now follows that $a_{ii} = 1/u_{ii}$.

44. (a) This is Exercise 59 of Section 2.1. However, we give here an alternate proof. Observe that A^T and B^T are upper triangular, and hence $B^T A^T$ is upper triangular by Exercise 42. Therefore $AB = (B^T A^T)^T$ is lower triangular.

(b) Both A^T and B^T are upper triangular matrices whose diagonal entries are all equal to 1. Hence, by Exercise 42, the diagonal entries of $B^T A^T$ all equal $1 \cdot 1 = 1$. Thus the diagonal entries of $AB = (B^T A^T)^T$ all equal 1.

45. By means of elementary row operations, L can be transformed into a unit lower triangular matrix L_1 whose first column is $\mathbf{e}_1$. Additional elementary row operations can be applied to transform L_1 into a unit lower triangular matrix whose first two columns are $\mathbf{e}_1$ and $\mathbf{e}_2$. This process can be continued until L is transformed into I_n, which is in reduced row echelon form. Hence L has rank n, and so L is invertible. Thus L^T is in invertible upper triangular matrix whose diagonal entries all equal 1. So it follows from Exercise 43 that $(L^{-1})^T = (L^T)^{-1}$ is an upper triangular matrix with diagonal entries equal to $1/1 = 1$. Therefore, L^{-1} is a lower triangular matrix whose diagonal entries are all equal to 1.

46. Suppose $LU = L'U'$. Then $(L')^{-1}L = U'U^{-1}$. But $(L')^{-1}L$ is unit lower triangular by Exercises 45 and 44(b), and $U'U^{-1}$ is upper triangular by Exercises 43 and 42. It follows that $(L')^{-1}L$ is both upper triangular and lower triangular, and its diagonal entries are all equal to 1. Hence $(L')^{-1}L = I_n$, and so $L = L'$. Finally observe that $U'U^{-1} = (L')^{-1}L = I_n$, and therefore $U = U'$.

47. (a) The ith component of $C\mathbf{b}$ is
$$c_{i1}b_1 + c_{i2}b_2 + \cdots + c_{in}b_n.$$
Each term in this sum requires a multiplication for a total of n multiplications. Also, this component is a sum of n terms, which requires $n-1$ additions.

(b) Computing $C\mathbf{b}$ requires the computation of each of its n components. Hence, by (a), this requires exactly n^2 multiplications and $n(n-1)$ additions, for a total of $2n^2 - n$ operations, or an approximate flop count of $2n^2$.

48. By the boxed result on page 163, approximately $2n^3$ flops are required to compute A^{-1}. Also, by Exercise 47(b), approximately $2n^2$ flops are required to compute the product $A^{-1}\mathbf{b}$ for each constant vector $\mathbf{b}$. Hence the total flop count is approximately
$$2n^3 + n(2n^2) = 4n^3.$$

49. Each entry of AB requires $n-1$ additions and n multiplications for a total of $2n-1$ flops. Since AB has mp entries, a total of $(2n-1)mp$ flops are required to compute all the entries of AB.

50. Applying Exercise 49, we see that method (a) requires
$$mp(2n-1) + mq(2p-1)$$

flops, and method (b) requires
$$nq(2p-1) + mq(2n-1)$$

flops. Choose the method that requires the smaller number of flops.

51. $L = \begin{bmatrix} 1 & 0 & 0 & 0 & 0 \\ -1 & 1 & 0 & 0 & 0 \\ 2 & 3 & 1 & 0 & 0 \\ 3 & -3 & 2 & 1 & 0 \\ 2 & 0 & 1 & -1 & 1 \end{bmatrix}$ and

$U = \begin{bmatrix} 2 & -1 & 3 & 2 & 1 \\ 0 & 1 & 2 & 3 & 5 \\ 0 & 0 & 3 & -1 & 2 \\ 0 & 0 & 0 & 1 & 8 \\ 0 & 0 & 0 & 0 & 13 \end{bmatrix}$

52. $L = \begin{bmatrix} 1 & 0 & 0 & 0 \\ 2 & 1 & 0 & 0 \\ 5 & -1 & 1 & 0 \\ 0 & 2 & 0 & 1 \end{bmatrix}$ and

$U = \begin{bmatrix} -3 & 1 & 0 & 2 & 1 \\ 0 & -2 & 1 & -1 & 3 \\ 0 & 0 & 5 & -10 & 10 \\ 0 & 0 & 0 & -4 & 2 \end{bmatrix}$

53. $P = \begin{bmatrix} 0 & 1 & 0 & 0 & 0 \\ 1 & 0 & 0 & 0 & 0 \\ 0 & 0 & 1 & 0 & 0 \\ 0 & 0 & 0 & 1 & 0 \\ 0 & 0 & 0 & 0 & 1 \end{bmatrix}$,

$L = \begin{bmatrix} 1.0 & 0 & 0 & 0 & 0 \\ 0.0 & 1 & 0 & 0 & 0 \\ 0.5 & 2 & 1 & 0 & 0 \\ -0.5 & -1 & -3 & 1 & 0 \\ 1.5 & 7 & 9 & -9 & 1 \end{bmatrix}$, and

$U = \begin{bmatrix} 2 & -2 & -1.0 & 3.0 & 4 \\ 0 & 1 & 2.0 & -1.0 & 1 \\ 0 & 0 & -1.5 & -0.5 & -2 \\ 0 & 0 & 0.0 & -1.0 & -2 \\ 0 & 0 & 0.0 & 0.0 & -9 \end{bmatrix}$

54. $P = \begin{bmatrix} 1 & 0 & 0 & 0 & 0 \\ 0 & 0 & 1 & 0 & 0 \\ 0 & 1 & 0 & 0 & 0 \\ 0 & 0 & 0 & 1 & 0 \\ 0 & 0 & 0 & 0 & 1 \end{bmatrix}$,

$L = \begin{bmatrix} 1 & 0 & 0.00 & 0.0 & 0 \\ 2 & 1 & 0.00 & 0.0 & 0 \\ 3 & 0 & 1.00 & 0.0 & 0 \\ -1 & -4 & 2.50 & 1.0 & 0 \\ 3 & 4 & 0.25 & -2.9 & 1 \end{bmatrix}$, and

$$U = \begin{bmatrix} 1 & 2 & -3 & 1.0 & 4.0 \\ 0 & -1 & 3 & 0.0 & -7.0 \\ 0 & 0 & 4 & 1.0 & -4.0 \\ 0 & 0 & 0 & 2.5 & -12.0 \\ 0 & 0 & 0 & 0.0 & -17.8 \end{bmatrix}$$

2.7 LINEAR TRANSFORMATIONS AND MATRICES

1. The domain is $\mathcal{R}^3$ and the codomain is $\mathcal{R}^2$.
2. The domain is $\mathcal{R}^3$ and the codomain is $\mathcal{R}^3$.
3. The domain is $\mathcal{R}^2$ and the codomain is $\mathcal{R}^3$.
4. The domain is $\mathcal{R}^2$ and the codomain is $\mathcal{R}^3$.
5. The domain is $\mathcal{R}^3$ and the codomain is $\mathcal{R}^3$.
6. The domain is $\mathcal{R}^3$ and the codomain is $\mathcal{R}^2$.
7.
$$T_A\left(\begin{bmatrix} 3 \\ -1 \\ 2 \end{bmatrix}\right) = A \begin{bmatrix} 3 \\ -1 \\ 2 \end{bmatrix}$$
$$= \begin{bmatrix} 2 & -3 & 1 \\ 4 & 0 & -2 \end{bmatrix} \begin{bmatrix} 3 \\ -1 \\ 2 \end{bmatrix}$$
$$= \begin{bmatrix} 11 \\ 8 \end{bmatrix}$$

8. $\begin{bmatrix} 1 \\ 5 \\ -2 \end{bmatrix}$ 9. $\begin{bmatrix} 8 \\ -6 \\ 11 \end{bmatrix}$ 10. $\begin{bmatrix} 9 \\ 4 \end{bmatrix}$

11. $\begin{bmatrix} 6 \\ -7 \\ 6 \end{bmatrix}$ 12. $\begin{bmatrix} 7 \\ -8 \\ 0 \end{bmatrix}$ 13. $\begin{bmatrix} 5 \\ 22 \end{bmatrix}$

14. $\begin{bmatrix} 3 \\ 12 \\ -4 \end{bmatrix}$ 15. $\begin{bmatrix} -1 \\ 6 \\ 17 \end{bmatrix}$ 16. $\begin{bmatrix} 1 \\ 2 \end{bmatrix}$

17. $\begin{bmatrix} -3 \\ -9 \\ 2 \end{bmatrix}$ 18. $\begin{bmatrix} -5 \\ 4 \\ -6 \end{bmatrix}$

19.
$$T_{(A+C^T)}\left(\begin{bmatrix} 2 \\ 1 \\ 1 \end{bmatrix}\right) = T_A\left(\begin{bmatrix} 2 \\ 1 \\ 1 \end{bmatrix}\right) + T_{C^T}\left(\begin{bmatrix} 2 \\ 1 \\ 1 \end{bmatrix}\right)$$
$$= \begin{bmatrix} 8 \\ 9 \end{bmatrix}$$

20. $T_A(\mathbf{e}_1) = \begin{bmatrix} 2 & -3 & 1 \\ 4 & 0 & -2 \end{bmatrix} \begin{bmatrix} 1 \\ 0 \\ 0 \end{bmatrix} = \begin{bmatrix} 2 \\ 4 \end{bmatrix}$.

Similarly, $T_A(\mathbf{e}_3) = \begin{bmatrix} 1 \\ -2 \end{bmatrix}$.

21. $n = 3, \; m = 2$ 22. $n = 2, \; m = 3$
23. $n = 2, \; m = 4$ 24. $n = 4, \; m = 3$
25. $\begin{bmatrix} 0 & 1 \\ 1 & 1 \end{bmatrix}$ 26. $\begin{bmatrix} 2 & 3 \\ 4 & 5 \end{bmatrix}$
27. $\begin{bmatrix} 1 & 1 & 1 \\ 2 & 0 & 0 \end{bmatrix}$ 28. $\begin{bmatrix} 0 & 3 \\ 2 & -1 \\ 1 & 1 \end{bmatrix}$
29. $\begin{bmatrix} 1 & -1 \\ 2 & -3 \\ 0 & 0 \\ 0 & 1 \end{bmatrix}$ 30. $\begin{bmatrix} 1 & 0 & -2 \\ -3 & 4 & 0 \\ 0 & 0 & 0 \end{bmatrix}$
31. $\begin{bmatrix} 1 & -1 \\ 0 & 0 \\ 3 & 0 \\ 0 & 1 \end{bmatrix}$ 32. $\begin{bmatrix} 2 & -1 & 0 & 3 \\ -1 & 0 & 0 & 2 \\ 0 & 3 & -1 & 0 \end{bmatrix}$
33. $\begin{bmatrix} 1 & 0 & 0 \\ 0 & 1 & 0 \\ 0 & 0 & 1 \end{bmatrix}$ 34. $\begin{bmatrix} 0 & 0 & 0 \\ 0 & 0 & 0 \end{bmatrix}$

35. False, only a linear transformation has a standard matrix.
36. True
37. False, the function must also preserve vector addition.
38. True
39. False, the standard matrix has size 2×3.
40. True
41. False, the function must be linear.
42. True
43. False, the range of a function is the set of all images.
44. False, the range is contained in the codomain.
45. False, the function must be one-to-one.
46. True 47. True 48. True
49. True 50. True 51. True
52. False, $f : \mathcal{R} \to \mathcal{R}$ defined by $f(x) = x^2$ does not preserve scalar multiplication.
53. False, the functions must be linear.
54. True
55. They are equal.
56.
$$T\left(\begin{bmatrix} -2 \\ 1 \end{bmatrix}\right) = T\left(-\frac{1}{2}\begin{bmatrix} 4 \\ -2 \end{bmatrix}\right) = -\frac{1}{2}T\left(\begin{bmatrix} 4 \\ -2 \end{bmatrix}\right)$$
$$= -\frac{1}{2}\begin{bmatrix} -6 \\ 16 \end{bmatrix} = \begin{bmatrix} -3 \\ 8 \end{bmatrix}$$

Similarly, $T\left(\begin{bmatrix} 8 \\ -4 \end{bmatrix}\right) = \begin{bmatrix} -12 \\ 32 \end{bmatrix}$.

2.7 Linear Transformations and Matrices

57.
$$T\left(\begin{bmatrix}16\\4\end{bmatrix}\right) = T\left(2\begin{bmatrix}8\\2\end{bmatrix}\right) = 2T\left(\begin{bmatrix}8\\2\end{bmatrix}\right)$$
$$= 2\begin{bmatrix}2\\-4\\6\end{bmatrix} = \begin{bmatrix}4\\-8\\12\end{bmatrix}$$

and
$$T\left(\begin{bmatrix}-4\\-1\end{bmatrix}\right) = T\left(-\frac{1}{2}\begin{bmatrix}8\\2\end{bmatrix}\right) = -\frac{1}{2}T\left(\begin{bmatrix}8\\2\end{bmatrix}\right)$$
$$= -\frac{1}{2}\begin{bmatrix}2\\-4\\6\end{bmatrix} = \begin{bmatrix}-1\\2\\-3\end{bmatrix}$$

58. Using the approach of Exercise 56, we obtain
$$T\left(\begin{bmatrix}1\\-3\\-2\end{bmatrix}\right) = \begin{bmatrix}2\\1\end{bmatrix} \text{ and } T\left(\begin{bmatrix}-4\\12\\8\end{bmatrix}\right) = \begin{bmatrix}-8\\4\end{bmatrix}.$$

59. Using the approach of Exercise 56, we obtain
$$T\left(\begin{bmatrix}-4\\-8\\-12\end{bmatrix}\right) = \begin{bmatrix}-16\\12\\4\end{bmatrix}$$

and
$$T\left(\begin{bmatrix}5\\10\\15\end{bmatrix}\right) = \begin{bmatrix}20\\-15\\-5\end{bmatrix}.$$

60. Write $\begin{bmatrix}1\\2\end{bmatrix} = a\begin{bmatrix}2\\0\end{bmatrix} + b\begin{bmatrix}0\\3\end{bmatrix}$ and solve for a and b.

We obtain $\begin{bmatrix}1\\2\end{bmatrix} = \frac{1}{2}\begin{bmatrix}2\\0\end{bmatrix} + \frac{2}{3}\begin{bmatrix}0\\3\end{bmatrix}$. Hence,

$$T\left(\begin{bmatrix}1\\2\end{bmatrix}\right) = T\left(\frac{1}{2}\begin{bmatrix}2\\0\end{bmatrix} + \frac{2}{3}\begin{bmatrix}0\\3\end{bmatrix}\right)$$
$$= \frac{1}{2}T\left(\begin{bmatrix}2\\0\end{bmatrix}\right) + \frac{2}{3}T\left(\begin{bmatrix}0\\3\end{bmatrix}\right)$$
$$= \frac{1}{2}\begin{bmatrix}-4\\6\end{bmatrix} + \frac{2}{3}\begin{bmatrix}9\\-6\end{bmatrix} = \begin{bmatrix}4\\-1\end{bmatrix}.$$

61. Using the approach of Exercise 60, we obtain
$$T\left(\begin{bmatrix}-2\\6\end{bmatrix}\right) = \begin{bmatrix}16\\2\\0\end{bmatrix}.$$

62. Using the approach of Exercise 60, we obtain
$$T\left(\begin{bmatrix}-3\\3\end{bmatrix}\right) = \begin{bmatrix}24\\-9\\6\end{bmatrix}.$$

63.
$$T\left(\begin{bmatrix}-2\\3\end{bmatrix}\right) = T\left(\begin{bmatrix}2\\3\end{bmatrix} + \begin{bmatrix}-4\\0\end{bmatrix}\right)$$
$$= T\left(\begin{bmatrix}2\\3\end{bmatrix}\right) + T\left(\begin{bmatrix}-4\\0\end{bmatrix}\right)$$
$$= \begin{bmatrix}1\\2\end{bmatrix} + \begin{bmatrix}-5\\1\end{bmatrix} = \begin{bmatrix}-4\\3\end{bmatrix}$$

64.
$$T\left(\begin{bmatrix}5\\6\end{bmatrix}\right) = T(5\mathbf{e}_1 + 6\mathbf{e}_2)$$
$$= 5T(\mathbf{e}_1) + 6T\mathbf{e}_2)$$
$$= 5\begin{bmatrix}2\\3\end{bmatrix} + 6\begin{bmatrix}4\\1\end{bmatrix} = \begin{bmatrix}34\\21\end{bmatrix}$$

65.
$$T\left(\begin{bmatrix}x_1\\x_2\end{bmatrix}\right) = T(x_1\mathbf{e}_1 + x_2\mathbf{e}_2)$$
$$= x_1T(\mathbf{e}_1) + x_2T\mathbf{e}_2)$$
$$= x_1\begin{bmatrix}2\\3\end{bmatrix} + x_2\begin{bmatrix}4\\1\end{bmatrix}$$
$$= \begin{bmatrix}2x_1\\3x_1\end{bmatrix} + \begin{bmatrix}4x_2\\x_2\end{bmatrix} = \begin{bmatrix}2x_1 + 4x_2\\3x_1 + x_2\end{bmatrix}$$

66. Using the approach of Exercise 65, we obtain
$$T\left(\begin{bmatrix}x_1\\x_2\end{bmatrix}\right) = \begin{bmatrix}3x_1 - x_2\\-x_1 + 2x_2\end{bmatrix}.$$

67. Using the approach of Exercise 65, we obtain
$$T\left(\begin{bmatrix}x_1\\x_2\\x_3\end{bmatrix}\right) = \begin{bmatrix}-x_1 + 3x_2\\-x_2 - 3x_3\\2x_1 + 2x_3\end{bmatrix}.$$

68. Using the approach of Exercise 65, we obtain
$$T\left(\begin{bmatrix}x_1\\x_2\\x_3\end{bmatrix}\right) = \begin{bmatrix}-2x_1 + 2x_3\\x_1 - 3x_2 + 4x_3\end{bmatrix}.$$

69. We begin by finding the standard matrix of T. This requires that we express each of the standard vectors of $\mathcal{R}^2$ as a linear combination of $\begin{bmatrix}1\\-2\end{bmatrix}$ and $\begin{bmatrix}-1\\3\end{bmatrix}$. For example, we need to solve the equation
$$\begin{bmatrix}1\\0\end{bmatrix} = a\begin{bmatrix}1\\-2\end{bmatrix} + b\begin{bmatrix}-1\\3\end{bmatrix}.$$

54 Chapter 2 Matrices and Linear Transformations

We obtain
$$\mathbf{e}_1 = 3\begin{bmatrix} 1 \\ -2 \end{bmatrix} + 2\begin{bmatrix} -1 \\ 3 \end{bmatrix}.$$

Likewise
$$\mathbf{e}_2 = 1\begin{bmatrix} 1 \\ -2 \end{bmatrix} + 1\begin{bmatrix} -1 \\ 3 \end{bmatrix}.$$

So
$$T(\mathbf{e}_1) = 3T\left(\begin{bmatrix} 1 \\ -2 \end{bmatrix}\right) + 2T\left(\begin{bmatrix} -1 \\ 3 \end{bmatrix}\right)$$
$$= 3\begin{bmatrix} 2 \\ 1 \end{bmatrix} + 2\begin{bmatrix} 3 \\ 0 \end{bmatrix} = \begin{bmatrix} 12 \\ 3 \end{bmatrix}.$$

Similarly, $T(\mathbf{e}_2) = \begin{bmatrix} 5 \\ 1 \end{bmatrix}$. Therefore the standard matrix of T is
$$A = \begin{bmatrix} 12 & 5 \\ 3 & 1 \end{bmatrix}.$$

So
$$T\left(\begin{bmatrix} x_1 \\ x_2 \end{bmatrix}\right) = A\begin{bmatrix} x_1 \\ x_2 \end{bmatrix} = \begin{bmatrix} 12x_1 + 5x_2 \\ 3x_1 + x_2 \end{bmatrix}.$$

70. $T\left(\begin{bmatrix} x_1 \\ x_2 \end{bmatrix}\right) = \begin{bmatrix} 17x_1 + 10x_2 \\ -2x_1 - x_2 \\ -6x_1 - 4x_2 \end{bmatrix}$

71. Using the approach of Exercise 70, we obtain
$$T\left(\begin{bmatrix} x_1 \\ x_2 \\ x_3 \end{bmatrix}\right) = \begin{bmatrix} x_1 + 3x_2 - x_3 \\ 2x_1 + 3x_2 + x_3 \\ 2x_1 + 3x_2 + 2x_3 \end{bmatrix}.$$

72. T is not linear. The solution is similar to that of Exercise 75.

73. T is linear. Let $A = \begin{bmatrix} 0 & 0 \\ 2 & 0 \end{bmatrix}$. Then
$$A\mathbf{x} = \begin{bmatrix} 0 & 0 \\ 2 & 0 \end{bmatrix} \begin{bmatrix} x_1 \\ x_2 \end{bmatrix} = \begin{bmatrix} 0 \\ 2x_1 \end{bmatrix} = T(\mathbf{x}).$$

So $T = T_A$, and hence T is a linear transformation by Theorem 2.7.

ALTERNATE PROOF. We can use the definition of a linear transformation by proving that T preserves vector addition and scalar multiplication.

Let $\mathbf{u}$ and $\mathbf{v}$ be vectors in $\mathcal{R}^2$. Then we have
$$T(\mathbf{u} + \mathbf{v}) = T\left(\begin{bmatrix} u_1 \\ u_2 \end{bmatrix} + \begin{bmatrix} v_1 \\ v_2 \end{bmatrix}\right)$$

$$= T\left(\begin{bmatrix} u_1 + v_1 \\ u_2 + v_2 \end{bmatrix}\right)$$
$$= \begin{bmatrix} 0 \\ 2(u_1 + v_1) \end{bmatrix} = \begin{bmatrix} 0 \\ 2u_1 + 2v_1 \end{bmatrix}.$$

Also
$$T(\mathbf{u}) + T(\mathbf{v}) = T\left(\begin{bmatrix} u_1 \\ u_2 \end{bmatrix}\right) + T\left(\begin{bmatrix} v_1 \\ v_2 \end{bmatrix}\right)$$
$$= \begin{bmatrix} 0 \\ 2u_1 \end{bmatrix} + \begin{bmatrix} 0 \\ 2v_1 \end{bmatrix} = \begin{bmatrix} 0 \\ 2u_1 + 2v_1 \end{bmatrix}.$$

So $T(\mathbf{u} + \mathbf{v}) = T(\mathbf{u}) + T(\mathbf{v})$, and hence T preserves vector addition.

Now suppose c is any scalar. Then
$$T(c\mathbf{u}) = T\left(c\begin{bmatrix} u_1 \\ u_2 \end{bmatrix}\right) = T\left(\begin{bmatrix} cu_1 \\ cu_2 \end{bmatrix}\right)$$
$$= \begin{bmatrix} 0 \\ 2(cu_1) \end{bmatrix} = \begin{bmatrix} 0 \\ c(2u_1) \end{bmatrix}.$$

Also
$$cT(\mathbf{u}) = cT\left(\begin{bmatrix} u_1 \\ u_2 \end{bmatrix}\right)$$
$$= c\begin{bmatrix} 0 \\ 2u_1 \end{bmatrix} = \begin{bmatrix} 0 \\ c(2u_1) \end{bmatrix}.$$

So $T(c\mathbf{u}) = cT(\mathbf{u})$. Therefore T preserves scalar multiplication. Hence T is linear.

74. T is not linear. The solution is similar to that of Exercise 75.

75. T is not linear. We must show that either T does not preserve vector addition or T does not preserve scalar multiplication. For example, let $\mathbf{u} = \mathbf{e}_1$ and $\mathbf{v} = \mathbf{e}_2$. Then
$$T(\mathbf{u} + \mathbf{v}) = T(\mathbf{e}_1 + \mathbf{e}_2) = T\left(\begin{bmatrix} 1 \\ 1 \\ 0 \end{bmatrix}\right)$$
$$= 1 + 1 + 0 - 1 = 1.$$

On the other hand,
$$T(\mathbf{u}) + T(\mathbf{v}) = T(\mathbf{e}_1) + T(\mathbf{e}_2)$$
$$= T\left(\begin{bmatrix} 1 \\ 0 \\ 0 \end{bmatrix}\right) + T\left(\begin{bmatrix} 0 \\ 1 \\ 0 \end{bmatrix}\right)$$
$$= (1 + 0 + 0 - 1) + (0 + 1 + 0 - 1)$$
$$= 0.$$

So $T(\mathbf{u}+\mathbf{v}) \neq T(\mathbf{u})+T(\mathbf{v})$ for the given vectors. Therefore T does not preserve vector addition and hence is not linear.

ALTERNATE PROOF. Let $c = 4$ and $\mathbf{u} = \mathbf{e}_1$. Then

$$T(4\mathbf{u}) = T(4\mathbf{e}_1) = T\left(\begin{bmatrix} 4 \\ 0 \\ 0 \end{bmatrix}\right)$$
$$= 4 + 0 + 0 - 1 = 3.$$

On the other hand,

$$4T(\mathbf{u}) = 4T(\mathbf{e}_1) = 4T\left(\begin{bmatrix} 1 \\ 0 \\ 0 \end{bmatrix}\right)$$
$$= 4(1 + 0 + 0 - 1) = 0.$$

So $T(4\mathbf{u}) \neq 4T(\mathbf{u})$ and hence T does not preserve scalar multiplication. Therefore T is not linear.

COMMENT. For this example, we can also show that T is not linear by noting that

$$T(\mathbf{0}) = 0 + 0 + 0 - 1 = -1 \neq 0.$$

So T is not linear by Theorem 2.8(a).

76. linear 77. linear 78. not linear
79. not linear 80. linear
81. For any $\mathbf{v}$ in $\mathcal{R}^n$, we have $T_{I_n}(\mathbf{v}) = I_n\mathbf{v} = \mathbf{v} = I(\mathbf{v})$. So $T_{I_n} = I$.
82. For any $\mathbf{v}$ in $\mathcal{R}^n$, we have $T_O(\mathbf{v}) = O\mathbf{v} = \mathbf{0} = T_0(\mathbf{v})$. So $T_O = T_0$.
83. We must show that the transformation cT preserves vector addition and scalar multiplication. Let $\mathbf{u}$ and $\mathbf{v}$ be in $\mathcal{R}^n$. Because T is linear,

$$(cT)(\mathbf{u} + \mathbf{v}) = cT(\mathbf{u} + \mathbf{v})$$
$$= c(T(\mathbf{u}) + T(\mathbf{v}))$$
$$= cT(\mathbf{u}) + cT(\mathbf{v})$$
$$= (cT)(\mathbf{u}) + (cT)(\mathbf{v}).$$

Also

$$(cT)(\mathbf{u}) + (cT)(\mathbf{v}) = cT(\mathbf{u}) + cT(\mathbf{v}).$$

So cT preserves vector addition. Now suppose k is a scalar. Because T is linear,

$$(cT)(k\mathbf{u}) = cT(k\mathbf{u})$$
$$= c(kT(\mathbf{u})) = ckT(\mathbf{u}).$$

Also

$$k((cT)(\mathbf{u})) = k(cT(\mathbf{u}))$$
$$= kcT(\mathbf{u}) = ckT(\mathbf{u}).$$

So cT preserves scalar multiplication. Hence cT is linear.

84. The proof is similar to that of Exercise 83.
85. The jth column of the standard matrix of cT is $(cT)(\mathbf{e}_j) = cT(\mathbf{e}_j) = c\mathbf{a}_j$, which is the jth column of cA. So the standard matrix of cT is cA.
86. The proof is similar to that of Exercise 85.
87. By Theorem 2.9, there exists a unique matrix A such that $T(\mathbf{v}) = A\mathbf{v}$ for all $\mathbf{v}$ in $\mathcal{R}^2$. Let $A = \begin{bmatrix} a & b \\ c & d \end{bmatrix}$. Then

$$T\left(\begin{bmatrix} x_1 \\ x_2 \end{bmatrix}\right) = \begin{bmatrix} a & b \\ c & d \end{bmatrix}\begin{bmatrix} x_1 \\ x_2 \end{bmatrix} = \begin{bmatrix} ax_1 + bx_2 \\ cx_1 + dx_2 \end{bmatrix}.$$

88. Let $T: \mathcal{R}^n \to \mathcal{R}^m$ be a linear transformation. There exist unique scalars a_{ij}, where $1 \leq i \leq m$ and $1 \leq j \leq n$, such that the kth component of $T(\mathbf{x})$ is $a_{k1}x_1 + a_{k2}x_2 + \cdots + a_{kn}x_n$ for every k ($1 \leq k \leq m$).
89. (a) Let A be the matrix in (b) and show that $T = T_A$. Then T is a linear transformation by Theorem 2.7.
    (b) $\begin{bmatrix} 1 & 0 \\ 0 & 0 \end{bmatrix}$
    (c)
    $$T(T(\mathbf{v})) = T\left(T\left(\begin{bmatrix} v_1 \\ v_2 \end{bmatrix}\right)\right)$$
    $$= T\left(\begin{bmatrix} v_1 \\ 0 \end{bmatrix}\right) = \begin{bmatrix} v_1 \\ 0 \end{bmatrix} = T(\mathbf{v}).$$

90. (a) Let A be the matrix in (b) and show that $T = T_A$. Then T is a linear transformation by Theorem 2.7.
    (b) $\begin{bmatrix} 0 & 0 & 0 \\ 0 & 1 & 0 \\ 0 & 0 & 1 \end{bmatrix}$
    (c) The proof is similar to that of Exercise 89(c).

91. (a) Because it is given that T is linear, it follows from Theorem 2.9 that T is a matrix transformation.
    ALTERNATE PROOF. Let
    $$A = \begin{bmatrix} -1 & 0 \\ 0 & 1 \end{bmatrix}.$$

56 Chapter 2 Matrices and Linear Transformations

Then
$$T_A\left(\begin{bmatrix}x_1\\x_2\end{bmatrix}\right) = \begin{bmatrix}-1 & 0\\0 & 1\end{bmatrix}\begin{bmatrix}x_1\\x_2\end{bmatrix}$$
$$= \begin{bmatrix}-x_1\\x_2\end{bmatrix} = T\left(\begin{bmatrix}x_1\\x_2\end{bmatrix}\right),$$
and hence $T = T_A$.

(b) For any vector $\mathbf{v} = \begin{bmatrix}v_1\\v_2\end{bmatrix}$ in $\mathcal{R}^2$,
$$T\left(\begin{bmatrix}-v_1\\v_2\end{bmatrix}\right) = \begin{bmatrix}-(-v_1)\\v_2\end{bmatrix} = \begin{bmatrix}v_1\\v_2\end{bmatrix} = \mathbf{v},$$
and hence the range of T is $\mathcal{R}^2$.

92. (a) The solution is similar to that of Exercise 91.
    (b) $\mathcal{R}^3$

93. (a) Observe that $T = T_A$, where $A = kI_n$.
    (b) For a vector $\mathbf{v}$ in $\mathcal{R}^n$, we have $\mathbf{v} = T(\frac{1}{k}\mathbf{v})$. So every vector is an image, and so the range of T is $\mathcal{R}^n$.

94. (a) The solution is similar to that of Exercise 93.
    (b) $\mathcal{R}^n$

95. We have $T(\mathbf{u}) = T(\mathbf{v})$ if and only if
$$T(\mathbf{u}) - T(\mathbf{v}) = \mathbf{0}.$$
Because T is linear, the previous equation is true if and only if $T(\mathbf{u} - \mathbf{v}) = \mathbf{0}$.

96. Consider, for example,
$$f\left(\begin{bmatrix}x_1\\x_2\end{bmatrix}\right) = \begin{bmatrix}x_1 + x_2\\0\end{bmatrix}$$
and
$$g\left(\begin{bmatrix}x_1\\x_2\end{bmatrix}\right) = \begin{bmatrix}x_1^2 + x_2^2\\0\end{bmatrix}.$$
Then $f(\mathbf{e}_1) = g(\mathbf{e}_1)$ and $f(\mathbf{e}_2) = g(\mathbf{e}_2)$, but $f(2\mathbf{e}_1) \neq g(2\mathbf{e}_1)$

97.
$$T_{A^{-1}}(T_A(\mathbf{v})) = T_{A^{-1}}(A\mathbf{v})$$
$$= A^{-1}(A\mathbf{v})$$
$$= (A^{-1}A)\mathbf{v} = I_n\mathbf{v} = \mathbf{v}.$$
Similarly, $T_A(T_{A^{-1}}(\mathbf{v})) = \mathbf{v}$.

98. $T_{AB}(\mathbf{v}) = (AB)\mathbf{v} = A(B\mathbf{v}) = A(T_B(\mathbf{v})) = T_A(T_B(\mathbf{v}))$.

99. A vector $\mathbf{v}$ is in the range of T if and only if $\mathbf{v} = T(\mathbf{u}) = A\mathbf{u}$ for some $\mathbf{u}$ in $\mathcal{R}^n$, which is true if and only if $\mathbf{v}$ is in the span of the columns of A.

100. By Exercise 99, we have $\mathcal{R}^m$ is the range of T if and only if $\mathcal{R}^m$ equals the span of the columns of A. But by Theorem 1.6, this is true if and only if rank $A = m$.

101. Suppose that $a_1\mathbf{v}_1 + a_2\mathbf{v}_2 + \cdots + a_k\mathbf{v}_k = \mathbf{0}$ for scalars $a_1, a_2, \ldots, a_k$. Since T is linear,
$$T(a_1\mathbf{v}_1 + a_2\mathbf{v}_2 + \cdots + a_k\mathbf{v}_k) = T(\mathbf{0})$$
$$a_1T(\mathbf{v}_1) + a_2T(\mathbf{v}_2) + \cdots + a_kT(\mathbf{v}_k) = \mathbf{0}.$$
Because $\{T(\mathbf{v}_1), T(\mathbf{v}_2), \ldots, T(\mathbf{v}_k)\}$ is linearly independent, we have $0 = a_1 = a_2 = \cdots = a_k$. So $\mathcal{S}$ is linearly independent.

102. (a) Let A be the standard matrix of T. Then
$$A\begin{bmatrix}1 & 1 & 0 & -1\\2 & 1 & 1 & 2\\0 & 1 & 0 & -3\\-1 & -1 & 1 & 1\end{bmatrix}$$
$$= \begin{bmatrix}A\begin{bmatrix}1\\2\\0\\-1\end{bmatrix} & A\begin{bmatrix}1\\1\\1\\-1\end{bmatrix} & A\begin{bmatrix}0\\1\\0\\1\end{bmatrix} & A\begin{bmatrix}-1\\2\\-3\\1\end{bmatrix}\end{bmatrix}$$
$$= \begin{bmatrix}T\left(\begin{bmatrix}1\\2\\0\\-1\end{bmatrix}\right) & T\left(\begin{bmatrix}1\\1\\1\\-1\end{bmatrix}\right) & T\left(\begin{bmatrix}0\\1\\0\\1\end{bmatrix}\right) & T\left(\begin{bmatrix}-1\\2\\-3\\1\end{bmatrix}\right)\end{bmatrix}$$
$$= \begin{bmatrix}0 & -2 & 4 & 0\\1 & 1 & 6 & 0\\1 & 3 & 0 & 0\\0 & 2 & -3 & 0\end{bmatrix}.$$
Hence
$$A = \begin{bmatrix}0 & -2 & 4 & 0\\1 & 1 & 6 & 0\\1 & 3 & 0 & 0\\0 & 2 & -3 & 0\end{bmatrix}\begin{bmatrix}1 & 1 & 0 & -1\\2 & 1 & 1 & 2\\0 & 1 & 0 & -3\\-1 & -1 & 1 & 1\end{bmatrix}^{-1}$$
$$= \begin{bmatrix}22 & -6 & -8 & 10\\4 & 1 & 1 & 5\\-20 & 7 & 9 & -7\\-21 & 6 & 8 & -9\end{bmatrix}.$$
Therefore
$$T\left(\begin{bmatrix}x_1\\x_2\\x_3\\x_4\end{bmatrix}\right) = A\begin{bmatrix}x_1\\x_2\\x_3\\x_4\end{bmatrix}$$
$$= \begin{bmatrix}22x_1 - 6x_2 - 8x_3 + 10x_4\\4x_1 + x_2 + x_3 + 5x_5\\-20x_1 + 7x_2 + 9x_3 - 7x_4\\-21x_1 + 6x_2 + 8x_3 - 9x_4\end{bmatrix}$$

(b) Yes. In the computation above, the standard matrix A is completely determined by the four vectors and their images under T.

103. The given vector $\mathbf{v} = \begin{bmatrix} 2 \\ -1 \\ 0 \\ 3 \end{bmatrix}$ is in the range of T if and only there is a vector $\mathbf{u}$ such that $T(\mathbf{u}) = \mathbf{v}$. If A is the standard matrix of T, then this condition is equivalent to the system $A\mathbf{x} = \mathbf{v}$ being consistent, where

$$A = \begin{bmatrix} 1 & 1 & 1 & 2 \\ 1 & 2 & -3 & 4 \\ 0 & 1 & 0 & 2 \\ 1 & 5 & -1 & 0 \end{bmatrix}.$$

If we solve this system, we obtain

$$\mathbf{u} = \frac{1}{4} \begin{bmatrix} 5 \\ 2 \\ 3 \\ -1 \end{bmatrix}.$$

So $T(\mathbf{u}) = \mathbf{v}$, and thus $\mathbf{v}$ is in the range of T. Alternatively, we can show that the reduced row echelon form of A is I_4 and conclude from (b) and (e) of Theorem 2.6 that the system $A\mathbf{x} = \mathbf{b}$ is consistent for every $\mathbf{b}$ in $\mathcal{R}^4$.

2.8 COMPOSITION AND INVERTIBILITY OF LINEAR TRANSFORMATIONS

1. $\left\{ \begin{bmatrix} 2 \\ 4 \end{bmatrix}, \begin{bmatrix} 3 \\ 5 \end{bmatrix} \right\}$ **2.** $\left\{ \begin{bmatrix} 0 \\ 1 \end{bmatrix}, \begin{bmatrix} 1 \\ 1 \end{bmatrix} \right\}$

3. By the boxed result on page 180, the range of T equals the span of the columns of the standard matrix of T. So the desired set is

$$\{T(\mathbf{e}_1), T(\mathbf{e}_2)\} = \left\{ \begin{bmatrix} 0 \\ 2 \\ 1 \end{bmatrix}, \begin{bmatrix} 3 \\ -1 \\ 1 \end{bmatrix} \right\}.$$

4. $\left\{ \begin{bmatrix} 1 \\ 2 \end{bmatrix}, \begin{bmatrix} 1 \\ 0 \end{bmatrix} \right\}$ **5.** $\left\{ \begin{bmatrix} 2 \\ 2 \\ 4 \end{bmatrix}, \begin{bmatrix} 1 \\ 2 \\ 1 \end{bmatrix}, \begin{bmatrix} 1 \\ 3 \\ 0 \end{bmatrix} \right\}$

6. $\left\{ \begin{bmatrix} 5 \\ 1 \\ 1 \end{bmatrix}, \begin{bmatrix} -4 \\ -2 \\ 0 \end{bmatrix}, \begin{bmatrix} 1 \\ 0 \\ 1 \end{bmatrix} \right\}$ **7.** $\left\{ \begin{bmatrix} 1 \\ 0 \end{bmatrix} \right\}$

8. $\left\{ \begin{bmatrix} 1 \\ 2 \\ 0 \\ 0 \end{bmatrix}, \begin{bmatrix} -4 \\ -3 \\ 0 \\ 1 \end{bmatrix} \right\}$ **9.** $\left\{ \begin{bmatrix} 1 \\ 0 \\ 0 \end{bmatrix}, \begin{bmatrix} 0 \\ 1 \\ 0 \end{bmatrix} \right\}$

10. $\left\{ \begin{bmatrix} 1 \\ 0 \\ 0 \end{bmatrix}, \begin{bmatrix} 0 \\ 1 \\ 0 \end{bmatrix}, \begin{bmatrix} 0 \\ 0 \\ 1 \end{bmatrix} \right\}$ **11.** $\left\{ \begin{bmatrix} 0 \\ 0 \end{bmatrix} \right\}$

12. $\left\{ \begin{bmatrix} 1 \\ 0 \\ 0 \end{bmatrix}, \begin{bmatrix} 0 \\ 1 \\ 0 \end{bmatrix}, \begin{bmatrix} 0 \\ 0 \\ 1 \end{bmatrix} \right\}$

13. The null space of T is the solution set of $A\mathbf{x} = \mathbf{0}$, where

$$A = \begin{bmatrix} 0 & 1 \\ 1 & 1 \end{bmatrix}$$

is the standard matrix of T. Thus the general solution of $A\mathbf{x} = \mathbf{0}$ is

$$x_1 = 0$$
$$x_2 = 0.$$

So a generating set is $\{\mathbf{0}\}$. By Theorem 2.11, T is one-to-one.

14. $\left\{ \begin{bmatrix} 0 \\ 0 \end{bmatrix} \right\}$, T is one-to-one.

15. $\left\{ \begin{bmatrix} 0 \\ -1 \\ 1 \end{bmatrix} \right\}$, T is not one-to-one.

16. $\left\{ \begin{bmatrix} 0 \\ 0 \end{bmatrix} \right\}$, T is one-to-one.

17. The null space of T is the solution set of $A\mathbf{x} = \mathbf{0}$, where

$$A = \begin{bmatrix} 1 & 2 & 1 \\ 1 & 3 & 2 \\ 2 & 5 & 3 \end{bmatrix}$$

is the standard matrix of T. The general solution of $A\mathbf{x} = \mathbf{0}$ is

$$\begin{aligned} x_1 &= x_3 \\ x_2 &= -x_3 \\ x_3 &\text{ free,} \end{aligned}$$

or

$$\begin{bmatrix} x_1 \\ x_2 \\ x_3 \end{bmatrix} = \begin{bmatrix} x_3 \\ -x_3 \\ x_3 \end{bmatrix} = x_3 \begin{bmatrix} 1 \\ -1 \\ 1 \end{bmatrix}.$$

So a generating set is

$$\left\{ \begin{bmatrix} 1 \\ -1 \\ 1 \end{bmatrix} \right\}.$$

By Theorem 2.11, we have that T is not one-to-one.

18. $\left\{ \begin{bmatrix} 0 \\ 0 \\ 0 \end{bmatrix} \right\}$, T is one-to-one.

19. $\{\mathbf{0}\}$, T is one-to-one.

20. $\left\{ \begin{bmatrix} 1 \\ 0 \\ 0 \end{bmatrix}, \begin{bmatrix} 0 \\ 1 \\ 0 \end{bmatrix}, \begin{bmatrix} 0 \\ 0 \\ 1 \end{bmatrix} \right\}$, T is not one-to-one.

21. $\{\mathbf{e}_2\}$, T is not one-to-one.

22. $\left\{\begin{bmatrix} 0 \\ 1 \\ 0 \end{bmatrix}, \begin{bmatrix} -2 \\ 0 \\ 1 \end{bmatrix}\right\}$, T is not one-to-one.

23. $\left\{\begin{bmatrix} 1 \\ -3 \\ 1 \\ 0 \end{bmatrix}, \begin{bmatrix} 3 \\ -5 \\ 0 \\ 1 \end{bmatrix}\right\}$, T is not one-to-one.

24. $\begin{bmatrix} 0 & 1 \\ 1 & 1 \end{bmatrix}$, T is one-to-one.

25. The standard matrix of T is
$$[T(\mathbf{e}_1)\ T(\mathbf{e}_2)] = \begin{bmatrix} 2 & 3 \\ 4 & 5 \end{bmatrix}.$$
The reduced row echelon form of this matrix is
$$\begin{bmatrix} 1 & 0 \\ 0 & 1 \end{bmatrix},$$
which has rank 2. So by Theorem 2.11, T is one-to-one.

26. $\begin{bmatrix} 1 & 1 & 1 \\ 2 & 0 & 0 \end{bmatrix}$, T is not one-to-one.

27. $\begin{bmatrix} 0 & 3 \\ 2 & -1 \\ 1 & 1 \end{bmatrix}$, T is one-to-one.

28. $\begin{bmatrix} 1 & 0 & -2 \\ -3 & 4 & 0 \\ 0 & 0 & 0 \end{bmatrix}$, T is not one-to-one.

29. $\begin{bmatrix} 1 & -1 & 0 \\ 0 & 1 & -1 \\ 1 & 0 & -1 \end{bmatrix}$, T is not one-to-one.

30. $\begin{bmatrix} 1 & -1 & 1 & 1 \\ -2 & 1 & -1 & -1 \\ 2 & 3 & -6 & 5 \\ -1 & 2 & -1 & -5 \end{bmatrix}$, T is one-to-one.

31. $\begin{bmatrix} 1 & 2 & 2 & 1 & 8 \\ 1 & 2 & 1 & 0 & 6 \\ 1 & 1 & 1 & 2 & 5 \\ 3 & 2 & 0 & 5 & 8 \end{bmatrix}$, T is not one-to-one.

32. $\begin{bmatrix} 0 & 1 \\ 1 & 1 \end{bmatrix}$, and T is onto.

33. The standard matrix of T is $A = \begin{bmatrix} 2 & 3 \\ 4 & 5 \end{bmatrix}$. Because rank $A = 2$, we see that T is onto by Theorem 2.10.

34. $\begin{bmatrix} 1 & 1 & 1 \\ 2 & 0 & 0 \end{bmatrix}$, T is onto.

35. The standard matrix is $\begin{bmatrix} 0 & 3 \\ 2 & -1 \\ 1 & 1 \end{bmatrix}$, T is not onto.

36. $\begin{bmatrix} 2 & -5 & 4 \end{bmatrix}$, T is onto.

37. $\begin{bmatrix} 0 & 1 & -2 \\ 1 & 0 & -1 \\ -1 & 2 & -3 \end{bmatrix}$, T is not onto.

38. $\begin{bmatrix} 1 & -1 & 2 & 0 \\ -2 & 1 & -7 & 0 \\ 1 & -1 & 2 & 0 \\ -1 & 2 & 1 & 0 \end{bmatrix}$, T is not onto.

39. $\begin{bmatrix} 1 & -2 & 2 & -1 \\ -1 & 1 & 3 & 2 \\ 1 & -1 & -6 & -1 \\ 1 & -2 & 5 & -5 \end{bmatrix}$, T is onto.

40. $\begin{bmatrix} 1 & 2 & 2 & 1 \\ 1 & 2 & 1 & 0 \\ 1 & 1 & 1 & 2 \\ 3 & 2 & 0 & 5 \end{bmatrix}$, T is not onto.

41. True

42. False, the span of the columns must equal the codomain for the transformation to be onto.

43. False, $A = \begin{bmatrix} 1 & 0 \\ 0 & 1 \\ 0 & 0 \end{bmatrix}$ has linearly independent columns, but the vector $\begin{bmatrix} 0 \\ 0 \\ 1 \end{bmatrix}$ is not in the range of T_A.

44. True   45. True   46. True

47. False, T_A must be onto.

48. True

49. False, the range must equal its codomain.

50. True

51. False, the function must be linear.

52. False, the rank must equal m.

53. True

54. False, the function must be linear.

55. False, the rank must equal n.

56. True   57. True

58. False, the standard matrix of TU is AB.

59. True   60. True

61. (a) The null space is $\{\mathbf{0}\}$. (The only vector that is rotated to $\mathbf{0}$ is the zero vector.)
    (b) Yes, by Theorem 2.11.

(c) $\mathcal{R}^2$. For any vector $\mathbf{v}$ in $\mathcal{R}^2$, let $\mathbf{u}$ be the vector formed by rotating $\mathbf{v}$ clockwise by $90°$. Clearly $T(\mathbf{u}) = \mathbf{v}$. So every vector in $\mathcal{R}^2$ is in the range of T.

(d) Yes, by (c).

62. (a) The null space is $\{\mathbf{0}\}$. The only vector that is reflected to $\mathbf{0}$ is the zero vector.
(b) Yes, by Theorem 2.11.
(c) $\mathcal{R}^2$. For any vector $\mathbf{v}$ in $\mathcal{R}^2$, let $\mathbf{u}$ be the reflection of $\mathbf{v}$. Clearly $T(\mathbf{u}) = \mathbf{v}$. So every vector in $\mathcal{R}^2$ is in the range of T.
(d) Yes, by (c).

63. (a) Span $\{\mathbf{e}_1\}$. The only vectors that are projected to $\mathbf{0}$ are the multiples of $\mathbf{e}_1$.
(b) No, because $T(\mathbf{e}_1) = \mathbf{0}$.
(c) Span $\{\mathbf{e}_2\}$. Clearly every vector is projected onto the y-axis and hence is a multiple of $\mathbf{e}_2$.
(d) No, from (c), it follows that $\mathbf{e}_1$ is not in the range of T.

64. (a) Span $\{\mathbf{e}_1, \mathbf{e}_2\}$ (b) no
(c) Span $\{\mathbf{e}_3\}$ (d) no

65. (a) Span $\{\mathbf{e}_3\}$ (b) no
(c) Span $\{\mathbf{e}_1, \mathbf{e}_2\}$ (d) no

66. (a) $\{\mathbf{0}\}$ (b) yes (c) $\mathcal{R}^3$ (d) yes

67. (a) T is one-to-one. The columns of the standard matrix of T are $T(\mathbf{e}_1)$ and $T(\mathbf{e}_2)$, which are linearly independent because neither is a multiple of the other. So by Theorem 2.11, T is one-to-one.
(b) T is onto. The rank of the standard matrix of T is 2. So, by Theorem 2.10, T is onto.

68. (a) No, the columns of the standard matrix of T are linearly dependent.
(b) No, the rank of the standard matrix of T is 1.

69. The domain and codomain are both $\mathcal{R}^2$. Also
$$(UT)\left(\begin{bmatrix}x_1\\x_2\end{bmatrix}\right) = U\left(T\left(\begin{bmatrix}x_1\\x_2\end{bmatrix}\right)\right)$$
$$= U\left(\begin{bmatrix}x_1 + x_2\\x_1 - 3x_2\\4x_1\end{bmatrix}\right)$$
$$= \begin{bmatrix}(x_1+x_2)-(x_1-3x_2)+4(4x_1)\\(x_1+x_2)+3(x_1-3x_2)\end{bmatrix}$$
$$= \begin{bmatrix}16x_1 + 4x_2\\4x_1 - 8x_2\end{bmatrix}.$$

70. $\begin{bmatrix}16 & 4\\4 & -8\end{bmatrix}$

71. $A = \begin{bmatrix}1 & 1\\1 & -3\\4 & 0\end{bmatrix}$ and $B = \begin{bmatrix}1 & -1 & 4\\1 & 3 & 0\end{bmatrix}$

72. $\begin{bmatrix}16 & 4\\4 & -8\end{bmatrix}$

73. The domain and codomain are $\mathcal{R}^3$. The rule is
$$TU\left(\begin{bmatrix}x_1\\x_2\\x_3\end{bmatrix}\right) = \begin{bmatrix}2x_1 + 2x_2 + 4x_3\\-2x_1 - 10x_2 + 4x_3\\4x_1 - 4x_2 + 16x_3\end{bmatrix}$$

74. $\begin{bmatrix}2 & 2 & 4\\-2 & -10 & 4\\4 & -4 & 16\end{bmatrix}$

75. $\begin{bmatrix}2 & 2 & 4\\-2 & -10 & 4\\4 & -4 & 16\end{bmatrix}$

76. The domain and the codomain are both $\mathcal{R}^2$.
$$UT\left(\begin{bmatrix}x_1\\x_2\end{bmatrix}\right) = \begin{bmatrix}-x_1 + 5x_2\\15x_1 - 5x_2\end{bmatrix}$$

77. $\begin{bmatrix}-1 & 5\\15 & -5\end{bmatrix}$

78. $A = \begin{bmatrix}1 & 2\\3 & -1\end{bmatrix}$, $B = \begin{bmatrix}2 & -1\\0 & 5\end{bmatrix}$

79. $\begin{bmatrix}-1 & 5\\15 & -5\end{bmatrix}$

80. The domain and the codomain are both $\mathcal{R}^2$.
$$TU\left(\begin{bmatrix}x_1\\x_2\end{bmatrix}\right) = \begin{bmatrix}2x_1 + 9x_2\\6x_1 - 8x_2\end{bmatrix}$$

81. $\begin{bmatrix}2 & 9\\6 & -8\end{bmatrix}$ 82. $AB = \begin{bmatrix}2 & 9\\6 & -8\end{bmatrix}$

83. $T^{-1}\left(\begin{bmatrix}x_1\\x_2\end{bmatrix}\right) = \begin{bmatrix}\frac{1}{3}x_1 + \frac{1}{3}x_2\\-\frac{1}{3}x_1 + \frac{2}{3}x_2\end{bmatrix}$

84. $T^{-1}\left(\begin{bmatrix}x_1\\x_2\end{bmatrix}\right) = \begin{bmatrix}-.2x_1 + .6x_2\\.4x_1 - .2x_2\end{bmatrix}$

85. Let A be the standard matrix of T. By Theorem 2.12, $T^{-1} = T_{A^{-1}}$. Now
$$A = \begin{bmatrix}-1 & 1 & 3\\2 & 0 & -1\\-1 & 2 & 5\end{bmatrix} \text{ and }$$
$$A^{-1} = \begin{bmatrix}2 & 1 & -1\\-9 & -2 & 5\\4 & 1 & -2\end{bmatrix}.$$
So
$$T^{-1}\left(\begin{bmatrix}x_1\\x_2\\x_3\end{bmatrix}\right) = \begin{bmatrix}2x_1 + x_2 - x_3\\-9x_1 - 2x_2 + 5x_3\\4x_1 + x_2 - 2x_3\end{bmatrix}.$$

60 Chapter 2 Matrices and Linear Transformations

86. $T^{-1}\left(\begin{bmatrix} x_1 \\ x_2 \\ x_3 \end{bmatrix}\right) = \begin{bmatrix} -2x_1 + 2x_2 + x_2 \\ -3x_1 + 2x_2 + 2x_3 \\ -2x_1 + x_2 + x_3 \end{bmatrix}$

87. $T^{-1}\left(\begin{bmatrix} x_1 \\ x_2 \\ x_3 \end{bmatrix}\right) = \begin{bmatrix} x_1 - 2x_2 + x_3 \\ -x_1 + x_2 - x_3 \\ 2x_1 - 7x_2 + 3x_3 \end{bmatrix}$

88. $T^{-1}\left(\begin{bmatrix} x_1 \\ x_2 \\ x_3 \end{bmatrix}\right) = \begin{bmatrix} -2x_1 - x_2 + x_3 \\ 5x_1 + 3x_2 - x_3 \\ 4x_1 + 2x_2 - x_2 \end{bmatrix}$

89. $T^{-1}\left(\begin{bmatrix} x_1 \\ x_2 \\ x_3 \\ x_4 \end{bmatrix}\right) = \frac{1}{2}\begin{bmatrix} x_1 - 3x_2 - 6x_3 + 3x_4 \\ 3x_1 - 2x_2 - 3x_3 + 3x_4 \\ -3x_1 + 3x_2 + 4x_3 - 3x_4 \\ -3x_1 + 6x_2 + 9x_3 - 5x_4 \end{bmatrix}$

90. $T^{-1}\left(\begin{bmatrix} x_1 \\ x_2 \\ x_3 \\ x_4 \end{bmatrix}\right) = \frac{1}{2}\begin{bmatrix} 7x_1 - 4x_2 + 4x_3 - 2x_4 \\ -9x_1 + 8x_2 - 9x_3 \\ -14x_1 + 10x_2 - 11x_3 + 2x_4 \\ 11x_1 - 7x_2 + 7x_3 - 3x_4 \end{bmatrix}$

91. Yes. Let $f: \mathcal{R}^n \to \mathcal{R}^m$ and $g: \mathcal{R}^m \to \mathcal{R}^p$ be one-to-one functions. To show $gf: \mathcal{R}^n \to \mathcal{R}^p$ is one-to-one, suppose $(gf)(\mathbf{u}) = (gf)(\mathbf{v})$. Then $g(f(\mathbf{u})) = g(f(\mathbf{v}))$. Because g is one-to-one, we have $f(\mathbf{u}) = f(\mathbf{v})$, and since f is also one-to-one, we have $\mathbf{u} = \mathbf{v}$.

92. Yes. Let $f: \mathcal{R}^n \to \mathcal{R}^m$ and $g: \mathcal{R}^m \to \mathcal{R}^p$ both be onto. To show that $gf: \mathcal{R}^n \to \mathcal{R}^p$ is onto, let $\mathbf{w}$ be in $\mathcal{R}^p$. Because g is onto, there exists $\mathbf{v}$ in $\mathcal{R}^m$ such that $g(\mathbf{v}) = \mathbf{w}$. Because f is onto, there exists $\mathbf{u}$ in $\mathcal{R}^n$ such that $f(\mathbf{u}) = \mathbf{v}$. So

$$(gf)(\mathbf{u}) = g(f(\mathbf{u})) = g(\mathbf{v}) = \mathbf{w}.$$

So every vector in $\mathcal{R}^p$ is an image.

93. If T is the reflection about the x-axis, then

$$T\left(\begin{bmatrix} x_1 \\ x_2 \end{bmatrix}\right) = \begin{bmatrix} x_1 \\ -x_2 \end{bmatrix}.$$

So

$$(TT)\left(\begin{bmatrix} x_1 \\ x_2 \end{bmatrix}\right) = T\left(T\left(\begin{bmatrix} x_1 \\ x_2 \end{bmatrix}\right)\right)$$

$$= T\left(\begin{bmatrix} x_1 \\ -x_2 \end{bmatrix}\right)$$

$$= T\left(\begin{bmatrix} x_1 \\ -(-x_2) \end{bmatrix}\right)$$

$$= \begin{bmatrix} x_1 \\ x_2 \end{bmatrix} = I\left(\begin{bmatrix} x_1 \\ x_2 \end{bmatrix}\right).$$

So $TT = I$.

94. Let T be the reflection about the y-axis, and U be the rotation by $180°$. Then

$$T\left(\begin{bmatrix} x_1 \\ x_2 \end{bmatrix}\right) = \begin{bmatrix} -x_1 \\ x_2 \end{bmatrix} \text{ and } U\left(\begin{bmatrix} x_1 \\ x_2 \end{bmatrix}\right) = \begin{bmatrix} -x_1 \\ -x_2 \end{bmatrix}.$$

So

$$(UT)\left(\begin{bmatrix} x_1 \\ x_2 \end{bmatrix}\right) = \begin{bmatrix} x_1 \\ -x_2 \end{bmatrix},$$

which is the rule for the reflection about the x-axis.

95. Let T be the projection on the x-axis, and let U be the reflection about the y-axis. Then

$$T\left(\begin{bmatrix} x_1 \\ x_2 \end{bmatrix}\right) = \begin{bmatrix} x_1 \\ 0 \end{bmatrix} \text{ and } U\left(\begin{bmatrix} x_1 \\ x_2 \end{bmatrix}\right) = \begin{bmatrix} -x_1 \\ x_2 \end{bmatrix}.$$

So

$$(UT)\left(\begin{bmatrix} x_1 \\ x_2 \end{bmatrix}\right) = \begin{bmatrix} -x_1 \\ 0 \end{bmatrix} = (TU)\left(\begin{bmatrix} x_1 \\ x_2 \end{bmatrix}\right).$$

96. Let T and U be shear transformations. Then

$$T\left(\begin{bmatrix} x_1 \\ x_2 \end{bmatrix}\right) = \begin{bmatrix} x_1 + kx_2 \\ x_2 \end{bmatrix}$$

for some k and

$$T\left(\begin{bmatrix} x_1 \\ x_2 \end{bmatrix}\right) = \begin{bmatrix} x_1 + cx_2 \\ x_2 \end{bmatrix}$$

for some c. So

$$UT\left(\begin{bmatrix} x_1 \\ x_2 \end{bmatrix}\right) = \begin{bmatrix} x_1 + (k+c)x_2 \\ x_2 \end{bmatrix},$$

which is a shear transformation.

97. (a) Suppose

$$a_1 T(\mathbf{v}_1) + a_2 T(\mathbf{v}_2) + \cdots + a_k T(\mathbf{v}_k) = \mathbf{0}$$

for some scalars $a_1, a_2, \ldots, a_k$. Because T is linear, we have

$$T(a_1 \mathbf{v}_1 + a_2 \mathbf{v}_2 + \cdots + a_k \mathbf{v}_k) = \mathbf{0}.$$

So $a_1 \mathbf{v}_1 + a_2 \mathbf{v}_2 + \cdots + a_k \mathbf{v}_k$ is in the null space of T. Therefore, by Theorem 2.11, $a_1 \mathbf{v}_1 + a_2 \mathbf{v}_2 + \cdots + a_k \mathbf{v}_k = \mathbf{0}$. Because $\{\mathbf{v}_1, \mathbf{v}_2, \ldots, \mathbf{v}_k\}$ is linearly independent, we have $a_1 = a_2 = \cdots = a_k = 0$. Hence $\{T(\mathbf{v}_1), T(\mathbf{v}_2), \ldots, T(\mathbf{v}_k)\}$ is linearly independent.

(b) Let T be the projection on the x-axis. Now $\{\mathbf{e}_2\}$ is linearly independent, but $\{T(\mathbf{e}_1)\} = \{\mathbf{0}\}$ is not.

98. Let A, B, and C be matrices such that $(AB)C$ and $A(BC)$ are defined. We will show that $(AB)C = A(BC)$. Recall that function composition is associative. (See Appendix B.) Using Theorem 2.12, we have

$$T_{(AB)C} = T_{AB}T_C = (T_AT_B)T_C$$
$$= T_A(T_BT_C) = T_A(T_{BC})$$
$$= T_{A(BC)}.$$

Therefore the uniqueness statement in Theorem 2.9 implies that $(AB)C = A(BC)$.

99. (a) $A = \begin{bmatrix} 1 & 3 & -2 & 1 \\ 3 & 0 & 4 & 1 \\ 2 & -1 & 0 & 2 \\ 0 & 0 & 1 & 1 \end{bmatrix}$ and

$B = \begin{bmatrix} 0 & 1 & 0 & -3 \\ 2 & 0 & 1 & -1 \\ 1 & -2 & 0 & 4 \\ 0 & 5 & 1 & 0 \end{bmatrix}.$

(b) $AB = \begin{bmatrix} 4 & 10 & 4 & -14 \\ 4 & 0 & 1 & 7 \\ -2 & 12 & 1 & -5 \\ 1 & 3 & 1 & 4 \end{bmatrix}$

(c) $TU\left(\begin{bmatrix} x_1 \\ x_2 \\ x_3 \\ x_4 \end{bmatrix}\right) =$

$\begin{bmatrix} 4x_1 + 10x_2 + 4x_3 - 14x_4 \\ 4x_1 \quad\quad\quad + x_3 + 7x_4 \\ -2x_1 + 12x_2 + x_3 - 5x_4 \\ x_1 + 3x_2 + x_3 + 4x_4 \end{bmatrix}$

100. (a) $\begin{bmatrix} 2 & 4 & 1 & 6 \\ 3 & 7 & -1 & 11 \\ 1 & 2 & 0 & 2 \\ 2 & 5 & -1 & 8 \end{bmatrix}$

(b) $A^{-1} = \begin{bmatrix} -2 & 10 & -1 & -12 \\ 1 & -6 & 2 & 7 \\ 1 & -2 & 0 & 2 \\ 0 & 1 & -1 & -1 \end{bmatrix}$

(c) $T^{-1}\left(\begin{bmatrix} x_1 \\ x_2 \\ x_3 \\ x_4 \end{bmatrix}\right) =$

$\begin{bmatrix} -2x_1 + 10x_2 - x_3 - 12x_4 \\ x_1 - 6x_2 + 2x_3 + 7x_4 \\ x_1 - 2x_2 \quad\quad + 2x_4 \\ x_2 - x_3 - x_4 \end{bmatrix}$

CHAPTER 2 REVIEW EXERCISES

1. True
2. False, consider $\begin{bmatrix} 1 & 2 \\ 2 & 4 \end{bmatrix}$.
3. False, the product of a 2×2 and 3×3 matrix is not defined.
4. True
5. False, see page 122.
6. False, consider $I_2 + I_2 = 2I_2 \neq O$.
7. True   8. True   9. True
10. False, $O\mathbf{x} = \mathbf{0}$ is consistent for $\mathbf{b} = \mathbf{0}$, but O is not invertible.
11. True
12. False, the null space is contained in the domain.
13. True
14. False, let T be the projection on the x-axis. Then $\{\mathbf{e}_2\}$ is a linearly independent set but $\{T(\mathbf{e}_2)\} = \{\mathbf{0}\}$ is a linearly dependent set.
15. True   16. True
17. False, the transformation

$$T_A\left(\begin{bmatrix} x_1 \\ x_2 \end{bmatrix}\right) = \begin{bmatrix} x_1 \\ x_2 \\ 0 \end{bmatrix}$$

is one-to-one, but not onto.
18. True
19. False, the null space consists exactly of the zero vector.
20. False, the columns of its standard matrix form a generating set for its codomain.
21. True
22. (a) Misuse. A function, not a matrix, has a range.
    (b) Misuse. Only a linear transformation has a standard matrix.
    (c) Not a misuse.
    (d) Misuse. Only a linear transformation has a null space. A system has solutions.
    (e) Misuse. A function, not a matrix, can be one-to-one.
23. (a) BA is defined if and only if $q = m$.
    (b) $p \times n$
24. $\begin{bmatrix} 10 \\ 8 \end{bmatrix}$ 25. $\begin{bmatrix} 64 & -4 \\ 32 & -2 \end{bmatrix}$
26. incompatible dimensions   27. $\begin{bmatrix} 2 \\ 29 \\ 4 \end{bmatrix}$

62 Chapter 2 Matrices and Linear Transformations

28. $[-4 \quad -9]$ **29.** incompatible dimensions

30. $\begin{bmatrix} 20 & 30 \\ -2 & -3 \end{bmatrix}$ **31.** $\dfrac{1}{6}\begin{bmatrix} 5 & 10 \\ 2 & 4 \end{bmatrix}$

32. A is not invertible.   **33.** $\begin{bmatrix} 30 \\ 42 \end{bmatrix}$

34. $\begin{bmatrix} 128 & 192 \\ 256 & 384 \end{bmatrix}$ **35.** incompatible dimensions

36. $\left[\begin{array}{cc|cc} 1 & -1 & 6 & 2 \\ 2 & 1 & 4 & 8 \\ \hline 0 & 0 & 4 & 2 \\ 0 & 0 & -2 & 6 \end{array}\right]$

37. We have
$$[I_2 \mid -I_2]\begin{bmatrix} 1 \\ 3 \\ -7 \\ -4 \end{bmatrix} = I_2\begin{bmatrix} 1 \\ 3 \end{bmatrix} + (-I_2)\begin{bmatrix} -7 \\ -4 \end{bmatrix}$$
$$= \begin{bmatrix} 1 \\ 3 \end{bmatrix} + \begin{bmatrix} 7 \\ 4 \end{bmatrix} = \begin{bmatrix} 8 \\ 7 \end{bmatrix}.$$

38. $\begin{bmatrix} -11 & 2 & 2 \\ -4 & 0 & 1 \\ 6 & -1 & -1 \end{bmatrix}$

39. $\dfrac{1}{50}\begin{bmatrix} 22 & 14 & -2 \\ -42 & -2 & 11 \\ -5 & -10 & 5 \end{bmatrix}$

40. By item 1 of the boxed result on page 150, the first row of AB is the product of the first row of A and B, which is zero.

41. By the definition of matrix multiplication and Theorem 1.3(f), the first column of AB is $A\mathbf{b}_1 = A\mathbf{0} = \mathbf{0}$.

42. $A = I_2$ and $B = 2I_2$

43. The inverse of the coefficient matrix is
$$\begin{bmatrix} 1 & -1 \\ -1 & 2 \end{bmatrix}.$$
So the solution is
$$\begin{bmatrix} 1 & -1 \\ -1 & 2 \end{bmatrix}\begin{bmatrix} 3 \\ 5 \end{bmatrix} = \begin{bmatrix} -2 \\ 7 \end{bmatrix}.$$

44. The inverse of the coefficient matrix is
$$\dfrac{1}{8}\begin{bmatrix} 13 & -3 & -1 \\ -7 & 1 & 3 \\ 2 & 2 & -2 \end{bmatrix}.$$
Therefore
$$\begin{bmatrix} x_1 \\ x_2 \\ x_3 \end{bmatrix} = \dfrac{1}{8}\begin{bmatrix} 13 & -3 & -1 \\ -7 & 1 & 3 \\ 2 & 2 & -2 \end{bmatrix}\begin{bmatrix} 3 \\ -1 \\ 2 \end{bmatrix}$$
$$= \begin{bmatrix} 5 \\ -2 \\ 0 \end{bmatrix}.$$

45. Because
$$\mathbf{r}_2 = 2\mathbf{r}_1 \quad \text{and} \quad \mathbf{r}_5 = -2\mathbf{r}_1 + 3\mathbf{r}_3 + \mathbf{r}_4,$$
by the column correspondence property we have
$$\mathbf{a}_2 = 2\mathbf{a}_1 = 2\begin{bmatrix} 3 \\ 5 \\ 2 \end{bmatrix} = \begin{bmatrix} 6 \\ 10 \\ 4 \end{bmatrix}$$
and
$$\mathbf{a}_5 = -2\mathbf{a}_1 + 3\mathbf{a}_3 + \mathbf{a}_4$$
$$= -2\begin{bmatrix} 3 \\ 5 \\ 2 \end{bmatrix} + 3\begin{bmatrix} 2 \\ 0 \\ -1 \end{bmatrix} + \begin{bmatrix} 2 \\ -1 \\ 3 \end{bmatrix} = \begin{bmatrix} 2 \\ -11 \\ -4 \end{bmatrix}.$$
So
$$A = \begin{bmatrix} 3 & 6 & 2 & 2 & 2 \\ 5 & 10 & 0 & -1 & -11 \\ 2 & 4 & -1 & 3 & -4 \end{bmatrix}.$$

46. Since $T_A \colon \mathcal{R}^3 \to \mathcal{R}^2$, the codomain is $\mathcal{R}^2$. Because rank $A = 2$, the table on page 188 tells us that T_A is onto, and so its range is $\mathcal{R}^2$.

47. Since $T_B \colon \mathcal{R}^2 \to \mathcal{R}^3$, the codomain is $\mathcal{R}^3$. The range equals
$$\mathrm{Span}\left\{\begin{bmatrix} 4 \\ 1 \\ 0 \end{bmatrix}, \begin{bmatrix} 2 \\ -3 \\ 1 \end{bmatrix}\right\}.$$

48. $T_A\left(\begin{bmatrix} 2 \\ 0 \\ 3 \end{bmatrix}\right) = \begin{bmatrix} 2 & -1 & 3 \\ 4 & 0 & -2 \end{bmatrix}\begin{bmatrix} 2 \\ 0 \\ 3 \end{bmatrix} = \begin{bmatrix} 13 \\ 2 \end{bmatrix}$

49. $T_B\left(\begin{bmatrix} 4 \\ 2 \end{bmatrix}\right) = \begin{bmatrix} 4 & 2 \\ 1 & -3 \\ 0 & 1 \end{bmatrix}\begin{bmatrix} 4 \\ 2 \end{bmatrix} = \begin{bmatrix} 20 \\ -2 \\ 2 \end{bmatrix}$

50. The standard matrix equals
$$[T(\mathbf{e}_1) \ T(\mathbf{e}_2)] = \begin{bmatrix} 3 & -1 \\ 4 & 0 \end{bmatrix}.$$

51. $\begin{bmatrix} 2 & 0 & -1 \\ 4 & 0 & 0 \end{bmatrix}$ **52.** $\begin{bmatrix} 6 & 0 \\ 0 & 6 \end{bmatrix}$

53. The standard matrix of T is
$$A = [T(\mathbf{e}_1) \ T(\mathbf{e}_2)].$$
Now
$$T(\mathbf{e}_1) = 2\mathbf{e}_1 + U(\mathbf{e}_1)$$

$$= 2\begin{bmatrix}1\\0\end{bmatrix} + U\left(\begin{bmatrix}1\\0\end{bmatrix}\right)$$

$$= \begin{bmatrix}2\\0\end{bmatrix} + \begin{bmatrix}2\\3\end{bmatrix} = \begin{bmatrix}4\\3\end{bmatrix}.$$

Also
$$T(\mathbf{e}_2) = 2\mathbf{e}_2 + U(\mathbf{e}_2)$$
$$= 2\begin{bmatrix}0\\1\end{bmatrix} + U\left(\begin{bmatrix}0\\1\end{bmatrix}\right)$$
$$= \begin{bmatrix}0\\2\end{bmatrix} + \begin{bmatrix}1\\0\end{bmatrix} = \begin{bmatrix}1\\2\end{bmatrix}.$$

So
$$A = \begin{bmatrix}4 & 1\\3 & 2\end{bmatrix}.$$

54. not linear because $T(\mathbf{0}) = \begin{bmatrix}1\\0\end{bmatrix} \neq \mathbf{0}$

55. linear

56. not linear because
$$T(\mathbf{e}_1 + \mathbf{e}_2) = T\left(\begin{bmatrix}1\\1\end{bmatrix}\right) = \begin{bmatrix}1\\1\end{bmatrix},$$
but
$$T(\mathbf{e}_1) + T(\mathbf{e}_2) = T\left(\begin{bmatrix}1\\0\end{bmatrix}\right) + T\left(\begin{bmatrix}0\\1\end{bmatrix}\right)$$
$$= \begin{bmatrix}0\\1\end{bmatrix} + \begin{bmatrix}0\\0\end{bmatrix} = \begin{bmatrix}0\\1\end{bmatrix}$$

57. linear   58. $\left\{\begin{bmatrix}1\\0\\2\end{bmatrix}, \begin{bmatrix}1\\0\\-1\end{bmatrix}\right\}$

59. $\left\{\begin{bmatrix}1\\0\end{bmatrix}, \begin{bmatrix}2\\1\end{bmatrix}, \begin{bmatrix}0\\-1\end{bmatrix}\right\}$

60. $\left\{\begin{bmatrix}0\\0\end{bmatrix}\right\}$, T is one-to-one.

61. The null space is the solution set of $A\mathbf{x} = \mathbf{0}$, where A is the standard matrix of T. The general solution is
$$\begin{aligned}x_1 &= -2x_3\\x_2 &= x_3\\x_3 &\text{ free.}\end{aligned}$$
Thus
$$\begin{bmatrix}x_1\\x_2\\x_3\end{bmatrix} = \begin{bmatrix}-2x_3\\x_3\\x_3\end{bmatrix} = x_3\begin{bmatrix}-2\\1\\1\end{bmatrix}.$$
So the generating set is $\left\{\begin{bmatrix}-2\\1\\1\end{bmatrix}\right\}$. By Theorem 2.11, T is not one-to-one.

62. $\begin{bmatrix}1 & 2 & 0\\0 & 1 & -1\end{bmatrix}$ The columns are not linearly independent; so T is not one-to-one by Theorem 2.11.

63. $\begin{bmatrix}1 & 1\\0 & 0\\2 & -1\end{bmatrix}$ The columns are linearly independent; so T is one-to-one by Theorem 2.11.

64. $\begin{bmatrix}2 & 0 & 1\\1 & 1 & -1\end{bmatrix}$ The rank is 2; so T is onto by Theorem 2.10.

65. $\begin{bmatrix}3 & -1\\0 & 1\\1 & 1\end{bmatrix}$ The rank is 2; so T is not onto by Theorem 2.10.

66. The domain and the codomain are both $\mathcal{R}^3$.
$$UT\left(\begin{bmatrix}x_1\\x_2\\x_3\end{bmatrix}\right) = \begin{bmatrix}5x_1 - x_2 + 4x_3\\x_1 + x_2 - x_3\\3x_1 + x_2\end{bmatrix}$$

67. $\begin{bmatrix}5 & -1 & 4\\1 & 1 & -1\\3 & 1 & 0\end{bmatrix}$

68. $A = \begin{bmatrix}2 & 0 & 1\\1 & 1 & -1\end{bmatrix}$ and $B = \begin{bmatrix}3 & -1\\0 & 1\\1 & 1\end{bmatrix}$

69. $\begin{bmatrix}5 & -1 & 4\\1 & 1 & -1\\3 & 1 & 0\end{bmatrix}$

70. The domain and the codomain are both $\mathcal{R}^2$.
$$TU\left(\begin{bmatrix}x_1\\x_2\end{bmatrix}\right) = \begin{bmatrix}7x_1 - x_2\\2x_1 - x_2\end{bmatrix}$$

71. $\begin{bmatrix}7 & -1\\2 & -1\end{bmatrix}$ 72. $AB = \begin{bmatrix}7 & -1\\2 & -1\end{bmatrix}$

73. $T^{-1}\left(\begin{bmatrix}x_1\\x_2\end{bmatrix}\right) = \frac{1}{5}\begin{bmatrix}3x_1 - 2x_2\\x_1 + x_2\end{bmatrix}$

74. $T^{-1}\left(\begin{bmatrix}x_1\\x_2\\x_3\end{bmatrix}\right) = \frac{1}{8}\begin{bmatrix}13x_1 - 3x_2 - x_3\\-7x_1 + x_2 + 3x_3\\2x_1 + 2x_2 - 2x_3\end{bmatrix}$

CHAPTER 2 MATLAB EXERCISES

1. (a) $AD = \begin{bmatrix}4 & 10 & 9\\1 & 2 & 9\\5 & 8 & 15\\5 & 8 & -8\\-4 & -8 & 1\end{bmatrix}$

64 Chapter 2 Matrices and Linear Transformations

(b) $DB = \begin{bmatrix} 6 & -2 & 5 & 11 & 9 \\ -3 & -1 & 10 & 7 & -3 \\ -3 & 1 & 2 & -1 & -3 \\ 2 & -2 & 7 & 9 & 3 \\ 0 & -1 & 10 & 10 & 2 \end{bmatrix}$

(c), (d) $(AB^T)C = A(B^TC)$
$= \begin{bmatrix} 38 & -22 & 14 & 38 & 57 \\ 10 & -4 & 4 & 10 & 11 \\ -12 & -9 & -11 & -12 & 12 \\ 9 & -5 & 4 & 9 & 14 \\ 28 & 10 & 20 & 28 & -9 \end{bmatrix}$

(e) $D(B - 2C)$
$= \begin{bmatrix} -2 & 10 & 5 & 3 & -17 \\ -31 & -7 & -8 & -21 & -1 \\ -11 & -5 & -4 & -9 & 7 \\ -14 & 2 & -1 & -7 & -11 \\ -26 & -1 & -4 & -16 & -6 \end{bmatrix}$

(f) $\begin{bmatrix} 11 \\ 8 \\ 20 \\ -3 \\ -9 \end{bmatrix}$

(g), (h) $C(A\mathbf{v}) = (CA)\mathbf{v} = \begin{bmatrix} 1 \\ -18 \\ 81 \end{bmatrix}$

(i) $A^3 = \begin{bmatrix} 23 & 14 & 9 & -7 & 46 \\ 2 & 11 & 6 & -2 & 10 \\ 21 & 26 & -8 & -17 & 11 \\ -6 & 18 & 53 & 24 & -36 \\ -33 & -6 & 35 & 25 & -12 \end{bmatrix}$

2. (a) The entries of the following matrices are rounded to four places after the decimal.

$A^{10} =$
$\begin{bmatrix} .2056 & .2837 & .2240 & .1380 & .0589 & 0 \\ .1375 & .2056 & .1749 & .1101 & .0471 & 0 \\ .1414 & .1767 & .1584 & .1083 & .0475 & 0 \\ .1266 & .1616 & .1149 & .0793 & .0356 & 0 \\ .0356 & .0543 & .0420 & .0208 & .0081 & 0 \\ .0027 & .0051 & .0051 & .0036 & .0016 & 0 \end{bmatrix}$

$A^{100} =$
$\begin{bmatrix} .0045 & .0062 & .0051 & .0033 & .0014 & 0 \\ .0033 & .0045 & .0037 & .0024 & .0010 & 0 \\ .0031 & .0043 & .0035 & .0023 & .0010 & 0 \\ .0026 & .0036 & .0029 & .0019 & .0008 & 0 \\ .0008 & .0011 & .0009 & .0006 & .0003 & 0 \\ .0001 & .0001 & .0001 & .0001 & .0000 & 0 \end{bmatrix}$

$A^{500} =$
$c\begin{bmatrix} .2126 & .2912 & .2393 & .1539 & .0665 & 0 \\ .1552 & .2126 & .1747 & .1124 & .0486 & 0 \\ .1457 & .1996 & .1640 & .1055 & .0456 & 0 \\ .1216 & .1665 & .1369 & .0880 & .0381 & 0 \\ .0381 & .0521 & .0428 & .0275 & .0119 & 0 \\ .0040 & .0054 & .0045 & .0029 & .0012 & 0 \end{bmatrix}$,

where $c = \dfrac{1}{10^9}$.

The colony will disappear.

(b) (i) Given a population distribution $\mathbf{x}_{n-1}$ of females in year $n-1$, the population distribution of females in year n without immigration is given by $A\mathbf{x}_{n-1}$. Since this is augmented by $\mathbf{b}$, the final population distribution in year n is given by $\mathbf{x}_n = A\mathbf{x}_{n-1} + \mathbf{b}$.

(ii) The entries of the vectors that follow are rounded to four places after the decimal.

$\mathbf{x}_1 = \begin{bmatrix} 4.7900 \\ 3.2700 \\ 4.0800 \\ 3.4400 \\ 0.7200 \\ 0.1800 \end{bmatrix}, \quad \mathbf{x}_2 = \begin{bmatrix} 5.3010 \\ 4.4530 \\ 5.0430 \\ 3.2640 \\ 1.0320 \\ 0.0720 \end{bmatrix},$

$\mathbf{x}_3 = \begin{bmatrix} 6.1254 \\ 4.8107 \\ 6.1077 \\ 4.0344 \\ 0.9792 \\ 0.1032 \end{bmatrix}, \quad \mathbf{x}_4 = \begin{bmatrix} 7.2115 \\ 5.3878 \\ 6.4296 \\ 4.8862 \\ 1.2103 \\ 0.0979 \end{bmatrix},$

and

$\mathbf{x}_5 = \begin{bmatrix} 8.1259 \\ 6.1480 \\ 6.9490 \\ 5.1437 \\ 1.4658 \\ 0.1210 \end{bmatrix}$

(iii) Observe that
$$\mathbf{x}_1 = A\mathbf{x}_0 + \mathbf{b},$$
$$\mathbf{x}_2 = A\mathbf{x}_1 + \mathbf{b} = A(A\mathbf{x}_0 + \mathbf{b}) + \mathbf{b}$$
$$= A^2\mathbf{x}_0 + (A + I_6)\mathbf{b},$$
$$\mathbf{x}_3 = A\mathbf{x}_2 + \mathbf{b}$$
$$= A((A^2\mathbf{x}_0 + (A + I_6)\mathbf{b}) + \mathbf{b}$$
$$= A^3\mathbf{x}_0 + (A^2 + A + I_6)\mathbf{b}.$$

Continuing this process, we obtain, for any positive integer n,

$$\mathbf{x}_n = A^n \mathbf{x}_0 + (A^{n-1} + \cdots + A + I_6)\mathbf{b}.$$

(For those familiar with mathematical induction, this proof can be given using induction.)
Let $S = A^{n-1} + \cdots + A + I_6$. Then

$$(A - I_6)S = AS - S = (A^n - I_6),$$

and hence

$$S = (A - I_6)^{-1}(A^n - I_6).$$

We conclude that

$$\mathbf{x}_n = A^n \mathbf{x}_0 + (A - I_6)^{-1}(A^n - I_6)\mathbf{b}.$$

(iv) Assuming that $\mathbf{x}_{n+1} = \mathbf{x}_n$, we have $\mathbf{x}_n = A\mathbf{x}_n + \mathbf{b}$, and hence

$$(I_6 - A)\mathbf{x}_n = \mathbf{x}_n - A\mathbf{x}_n = \mathbf{b}.$$

So $\mathbf{x}_n = (I_6 - A)^{-1}\mathbf{b}$. Using the given $\mathbf{b}$, we obtain (with entries rounded to four places after the decimal)

$$\mathbf{x}_n = \begin{bmatrix} 28.1412 \\ 20.7988 \\ 20.8189 \\ 16.6551 \\ 4.9965 \\ 0.4997 \end{bmatrix}.$$

3. The eight airports divide up into two subsets: $\{1, 2, 6, 8\}$ and $\{3, 4, 5, 7\}$.

4. (a) $R = \begin{bmatrix} 1 & 2 & 0 & -1 & 0 & 0 & 0 \\ 0 & 0 & 1 & 1 & 0 & 0 & 1 \\ 0 & 0 & 0 & 0 & 1 & 0 & -1 \\ 0 & 0 & 0 & 0 & 0 & 1 & -1 \\ 0 & 0 & 0 & 0 & 0 & 0 & 0 \end{bmatrix}$

(b) $S = \begin{bmatrix} -2 & 1 & 0 \\ 1 & 0 & 0 \\ 0 & -1 & -1 \\ 0 & 1 & 0 \\ 0 & 0 & 1 \\ 0 & 0 & 1 \\ 0 & 0 & 1 \end{bmatrix}$

(c) As in the discussion following Theorem 2.3 in Section 2.3, $A\mathbf{x} = \mathbf{0}$ and $R\mathbf{x} = \mathbf{0}$ have the same solutions. Furthermore, we may infer from the description of null$(A, \,'r')$ in Table D.2 of Appendix D that the columns of S are the vectors in the vector form of the general solution of $R\mathbf{x} = \mathbf{0}$ and their coefficients are the free variables.

For the matrix R in (a), the free variables (x_2, x_4, and x_7) and the basic variables (x_1, x_3, x_5, and x_6) are obtained by solving the homogeneous system of linear equations with coefficient matrix R. This results in the general solution $\mathbf{x} = x_2\mathbf{s}_1 + x_4\mathbf{s}_2 + x_7\mathbf{s}_3$, where $\mathbf{s}_i$ is the ith column of the matrix S in (b).

In general, consider any $m \times n$ matrix A with reduced row echelon form R, and let S be the matrix whose columns form a basis for Null A obtained by applying the MATLAB function null$(A, \,'r')$. Then S is an $n \times (n - r)$ matrix, where r is the rank of A, and so $n - r$ is equal to the number of free variables in the vector solution of $R\mathbf{x} = \mathbf{0}$. Each column of S corresponds to a free variable. Consider a free variable x_j, and let $\mathbf{v}$ be the column of S that corresponds to it. We can obtain $\mathbf{v}$ from R as follows.

(i) The jth entry of $\mathbf{v}$ is 1.
(ii) For each i, if the ith row of R is nonzero and has its leading entry in column k, then the kth entry of $\mathbf{v}$ is $-r_{ij}$.
(iii) All other entries of $\mathbf{v}$ equal 0.

For example, consider the matrix R in (a). We apply the rules above to compute the third column of S in (b). This column corresponds to the free variable x_7. By (i), we have $s_{73} = 1$. Applying (ii), we obtain $s_{13} = 0$, $s_{33} = -1$, $s_{53} = 1$, and $s_{63} = 1$. The other entries of this column equal 0.

5. (a) First, observe that if C is an $m \times p$ matrix in reduced row echelon form, then for any $q \leq p$, the $n \times q$ matrix consisting of the first q columns of C is also in reduced row echelon form.

Given an $m \times n$ matrix A, let $B = [A \; I_m]$, the $m \times (m + n)$ whose first n columns are the columns of A, and whose last m columns are those of I_m. By Theorem 2.3, there is an invertible $m \times m$ matrix P such that PB is in reduced row echelon form. Furthermore, $PB = P[A \; I_m] = [PA \; P]$. Thus PA is in reduced echelon form, and the final m columns of PB are the columns of P.

(b) $P = \begin{bmatrix} 0.0 & -0.8 & -2.2 & -1.8 & 1.0 \\ 0.0 & -0.8 & -1.2 & -1.8 & 1.0 \\ 0.0 & 0.4 & 1.6 & 2.4 & -1.0 \\ 0.0 & 1.0 & 2.0 & 2.0 & -1.0 \\ 1.0 & 0.0 & -1.0 & -1.0 & 0.0 \end{bmatrix}$

66 Chapter 2 Matrices and Linear Transformations

6. (a) $M = \begin{bmatrix} 1 & 2 & 1 & 2 \\ 2 & 0 & -1 & 3 \\ -1 & 1 & 0 & 0 \\ 2 & 1 & 1 & 2 \\ 4 & 4 & 1 & 6 \end{bmatrix}$

(b) $S = \begin{bmatrix} 1 & 0 & 1 & 0 & 1 & 0 \\ 0 & 1 & 1 & 0 & 0 & 0 \\ 0 & 0 & 0 & 1 & 2 & 0 \\ 0 & 0 & 0 & 0 & 0 & 1 \end{bmatrix}$

(c) The sizes of M and S are 5×4 and 4×6, respectively, and hence the number of columns of M is equal to the number of rows of S. Hence the product MS is defined and has size 5×6.

(e) Let A be an $m \times n$ matrix of rank r, and let R be the reduced row echelon form of A. Observe that M is an $m \times r$ matrix because A has r pivot columns, and S is an $r \times n$ matrix because the first r rows of R are nonzero and the final $m-r$ rows of R are zero. Thus the product MS is defined and is an $m \times n$ matrix.

Consider any j, $1 \le j \le n$. Suppose that $\mathbf{r}_j$ is a linear combination of the first k pivot columns, which are the first k standard vectors of $\mathcal{R}^m$,

$$\mathbf{r}_j = c_1 \mathbf{e}_1 + c_2 \mathbf{e}_2 + \cdots + c_k \mathbf{e}_k.$$

By the column correspondence property, the jth column of A is a linear combination of the first k pivot columns of A, with the same corresponding coefficients. That is,

$$\mathbf{a}_j = c_1 \mathbf{m}_1 + c_2 \mathbf{m}_2 + \cdots + c_k \mathbf{m}_k.$$

Observe also that the first k entries of $\mathbf{r}_j$, and hence of $\mathbf{s}_j$, are the coefficients $c_1, c_2, \ldots, c_k$, respectively, and that the rmaining entries equal 0. It follows that

$$M\mathbf{s}_j = c_1 \mathbf{m}_1 + c_2 \mathbf{m}_2 + \cdots + c_k \mathbf{m}_k = \mathbf{a}_j$$

Since $M\mathbf{s}_j$ is the jth column of MS, it follows that $MS = A$.

7. $A^{-1}B = \begin{bmatrix} 6 & -4 & 3 & 19 & 5 & -2 & -5 \\ -1 & 2 & -4 & -1 & 4 & -3 & -2 \\ -2 & 0 & 2 & 6 & -1 & 6 & 3 \\ 0 & 1 & -3 & -8 & 2 & -3 & 1 \\ -1 & 0 & 2 & -6 & -5 & 2 & 2 \end{bmatrix}$

8. (a) We have

$$L = \frac{1}{3}\begin{bmatrix} 3 & 0 & 0 & 0 \\ 6 & 3 & 0 & 0 \\ 3 & 1 & 3 & 0 \\ 6 & 1 & 3 & 3 \end{bmatrix} \text{ and }$$

$$U = \frac{1}{3}\begin{bmatrix} 3 & -3 & 6 & 0 & -6 & 12 \\ 0 & 9 & -9 & -6 & 15 & -15 \\ 0 & 0 & -6 & 11 & -8 & -1 \\ 0 & 0 & 0 & -6 & 21 & -9 \end{bmatrix}.$$

(b) $\mathbf{x} = \begin{bmatrix} \frac{17}{4} \\ -\frac{5}{4} \\ -\frac{1}{4} \\ \frac{3}{2} \\ 0 \\ 0 \end{bmatrix} + x_5 \begin{bmatrix} -\frac{29}{12} \\ \frac{23}{4} \\ \frac{61}{12} \\ \frac{7}{2} \\ 1 \\ 0 \end{bmatrix} + x_6 \begin{bmatrix} -\frac{5}{12} \\ -\frac{9}{4} \\ -\frac{35}{12} \\ -\frac{3}{2} \\ 0 \\ 1 \end{bmatrix}$

9. (a) $A = \begin{bmatrix} 1 & 2 & 0 & 1 & -3 & -2 \\ 0 & 1 & 0 & -1 & 0 & 0 \\ 1 & 0 & 1 & 0 & 0 & 3 \\ 2 & 4 & 0 & 3 & -6 & -4 \\ 3 & 2 & 2 & 1 & -2 & -4 \\ 4 & 4 & 2 & 2 & -5 & 3 \end{bmatrix}$

(b) $A^{-1} =$

$\frac{1}{9}\begin{bmatrix} 10 & -18 & -54 & -27 & 1 & 26 \\ -18 & 9 & 0 & 9 & 0 & 0 \\ -7 & 18 & 63 & 27 & 2 & -29 \\ -18 & 0 & 0 & 9 & 0 & 0 \\ -17 & 0 & -18 & 0 & 1 & 8 \\ -1 & 0 & 0 & 0 & -1 & 1 \end{bmatrix}$

(c) $T^{-1}\left(\begin{bmatrix} x_1 \\ x_2 \\ x_3 \\ x_4 \\ x_5 \\ x_6 \end{bmatrix}\right) = A^{-1}\begin{bmatrix} x_1 \\ x_2 \\ x_3 \\ x_4 \\ x_5 \\ x_6 \end{bmatrix} =$

$\begin{bmatrix} \frac{10}{9}x_1 - 2x_2 - 6x_3 - 3x_4 + \frac{1}{9}x_5 + \frac{26}{9}x_6 \\ -2x_1 + x_2 + x_4 \\ -\frac{7}{9}x_1 + 2x_2 + 7x_3 + 3x_4 + \frac{2}{9}x_5 - \frac{29}{9}x_6 \\ -2x_1 + x_4 \\ -\frac{17}{9}x_1 - 2x_3 + \frac{1}{9}x_5 + \frac{8}{9}x_6 \\ -\frac{1}{9}x_1 - \frac{1}{9}x_5 + \frac{1}{9}x_6 \end{bmatrix}$

10. (a) $B = \begin{bmatrix} 1 & 0 & 2 & 0 & 0 & 1 \\ 2 & -1 & 0 & 1 & 0 & 0 \\ 0 & 3 & 0 & 0 & -1 & 0 \\ 2 & 1 & -1 & 0 & 0 & 1 \end{bmatrix}$

(b) The standard matrix of UT is

$$BA = \begin{bmatrix} 7 & 6 & 4 & 3 & -8 & 7 \\ 4 & 7 & 0 & 6 & -12 & -8 \\ -3 & 1 & -2 & -4 & 2 & 4 \\ 5 & 9 & 1 & 3 & -11 & -4 \end{bmatrix}.$$

(c) $UT\left(\begin{bmatrix} x_1 \\ x_2 \\ x_3 \\ x_4 \\ x_5 \\ x_6 \end{bmatrix}\right) =$

$\begin{bmatrix} 7x_1 + 6x_2 + 4x_3 + 3x_4 - 8x_5 + 7x_6 \\ 4x_1 + 7x_2 + 6x_4 - 12x_5 - 8x_6 \\ -3x_1 + x_2 - 2x_3 - 4x_4 + 2x_5 + 4x_6 \\ 5x_1 + 9x_2 + x_3 + 3x_4 - 11x_5 - 4x_6 \end{bmatrix}$

(d) The standard matrix of UT^{-1} is
$BA^{-1} =$
$\begin{bmatrix} -\frac{5}{9} & 2 & 8 & 3 & \frac{4}{9} & -\frac{31}{9} \\ \frac{20}{9} & -5 & -12 & -6 & \frac{2}{9} & \frac{52}{9} \\ -\frac{37}{9} & 3 & 2 & 3 & -\frac{1}{9} & -\frac{8}{9} \\ \frac{8}{9} & -5 & -19 & -8 & -\frac{1}{9} & \frac{82}{9} \end{bmatrix},$

and hence
$UT^{-1}\left(\begin{bmatrix} x_1 \\ x_2 \\ x_3 \\ x_4 \\ x_5 \\ x_6 \end{bmatrix}\right) =$

$\begin{bmatrix} -\frac{5}{9}x_1 + 2x_2 + 8x_3 + 3x_4 + \frac{4}{9}x_5 - \frac{31}{9}x_6 \\ \frac{20}{9}x_1 - 5x_2 - 12x_3 - 6x_4 + \frac{2}{9}x_5 + \frac{52}{9}x_6 \\ -\frac{37}{9}x_1 + 3x_2 + 2x_3 + 3x_4 - \frac{1}{9}x_5 - \frac{8}{9}x_6 \\ \frac{8}{9}x_1 - 5x_2 - 19x_3 - 8x_4 - \frac{1}{9}x_5 + \frac{82}{9}x_6 \end{bmatrix}.$

Chapter 3

Determinants

3.1 COFACTOR EXPANSION

1. $\det \begin{bmatrix} 6 & 2 \\ -3 & -1 \end{bmatrix} = 6(-1) - 2(-3) = 0$

2. $\det \begin{bmatrix} 4 & 5 \\ 3 & -7 \end{bmatrix} = 4(-7) - 5(3) = -43$

3. $\det \begin{bmatrix} -2 & 9 \\ 1 & 8 \end{bmatrix} = (-2)(8) - 9(1) = -25$

4. 42 **5.** 0 **6.** 3 **7.** 2

8. 2 **9.** 16 **10.** −86

11. The $(3,1)$-cofactor of A is
$$(-1)^{3+1} \det \begin{bmatrix} -2 & 4 \\ 6 & 3 \end{bmatrix} = 1[(-2)(3) - 4(6)]$$
$$= 1(-30) = -30.$$

12. 52

13. We have
$$2\det \begin{bmatrix} 4 & -2 \\ 0 & 1 \end{bmatrix} - (-1)\det \begin{bmatrix} 1 & -2 \\ -1 & 1 \end{bmatrix}$$
$$+ 3\det \begin{bmatrix} 1 & 4 \\ -1 & 0 \end{bmatrix}$$
$$= 2(4) + 1(-1) + 3(4) = 19.$$

14. We have
$$(-2)\det \begin{bmatrix} -2 & 2 \\ 1 & -1 \end{bmatrix} + (-1)\det \begin{bmatrix} 1 & 2 \\ 0 & -1 \end{bmatrix}$$
$$- 3\det \begin{bmatrix} 1 & -2 \\ 0 & 1 \end{bmatrix}$$
$$= -2(0) - 1(-1) - 3(1)$$
$$= -2.$$

15. We have
$$0\det \begin{bmatrix} -2 & 2 \\ -1 & 3 \end{bmatrix} - 1\det \begin{bmatrix} 1 & 2 \\ 2 & 3 \end{bmatrix}$$
$$+ (-1)\det \begin{bmatrix} 1 & -2 \\ 2 & -1 \end{bmatrix}$$
$$= 0(-4) - 1(-1) - 1(3) = -2.$$

16. We have
$$(-1)\det \begin{bmatrix} -1 & 3 \\ 4 & -2 \end{bmatrix} - 0\det \begin{bmatrix} 2 & 3 \\ 1 & -2 \end{bmatrix}$$
$$+ 1\det \begin{bmatrix} 2 & -1 \\ 1 & 4 \end{bmatrix}$$
$$= (-1)(-10) - 0(-7) + 1(9)$$
$$= 19.$$

17. We have
$$(-5)\det \begin{bmatrix} 4 & -3 \\ 0 & -1 \end{bmatrix} + 0\det \begin{bmatrix} 1 & -3 \\ 2 & -1 \end{bmatrix}$$
$$- 0\det \begin{bmatrix} 1 & 4 \\ 2 & 0 \end{bmatrix}$$
$$= (-5)(-4) + 0(5) - 0(-8) = 20$$

18. We have
$$4\det \begin{bmatrix} 3 & -2 \\ 0 & 5 \end{bmatrix} - 1\det \begin{bmatrix} 0 & -2 \\ 2 & 5 \end{bmatrix}$$
$$+ 0\det \begin{bmatrix} 0 & 3 \\ 2 & 0 \end{bmatrix}$$
$$= 4(15) - 1(4) + 0(-6) = 56$$

19. We have
$$(-0)\det \begin{bmatrix} 2 & 1 & -1 \\ -3 & 2 & -1 \\ 3 & 0 & -2 \end{bmatrix} + (-1)\det \begin{bmatrix} 1 & 1 & -1 \\ 4 & 2 & -1 \\ 0 & 0 & -2 \end{bmatrix}$$
$$- 0\det \begin{bmatrix} 1 & 2 & -1 \\ 4 & -3 & -1 \\ 0 & 3 & -2 \end{bmatrix} + 1\det \begin{bmatrix} 1 & 2 & 1 \\ 4 & -3 & 2 \\ 0 & 3 & 0 \end{bmatrix}$$
$$= (-0)(-11) + (-1)(4) - 0(13) + 1(6) = 2.$$

20. We have
$$(-0)\det \begin{bmatrix} -1 & 0 & 1 \\ 3 & 1 & 4 \\ -2 & 2 & 3 \end{bmatrix} + 1\det \begin{bmatrix} 0 & 0 & 1 \\ -2 & 1 & 4 \\ 1 & 2 & 3 \end{bmatrix}$$
$$- 0\det \begin{bmatrix} 0 & -1 & 1 \\ -2 & 3 & 4 \\ 1 & -2 & 3 \end{bmatrix} + (-2)\det \begin{bmatrix} 0 & -1 & 0 \\ -2 & 3 & 1 \\ 1 & -2 & 2 \end{bmatrix}$$
$$= (-0)(13) + 1(-5) - 0(-9) + (-2)(-5) = 5$$

21. 60 **22.** −48 **23.** 180 **24.** 42
25. −147 **26.** 18 **27.** −24 **28.** 78

29. The area of the parallelogram determined by **u** and **v** is

$$|\det [\mathbf{u}\ \mathbf{v}]| = \left|\det \begin{bmatrix} 3 & -2 \\ 5 & 7 \end{bmatrix}\right| = |3(7) - (-2)(5)|$$
$$= |21 + 10| = 31.$$

30. The area of the parallelogram determined by **u** and **v** is

$$|\det [\mathbf{u}\ \mathbf{v}]| = \left|\det \begin{bmatrix} -3 & 8 \\ 6 & -5 \end{bmatrix}\right|$$
$$= |-3(-5) - 8(6)| = |15 - 48| = 33.$$

31. The area of the parallelogram determined by **u** and **v** is

$$|\det [\mathbf{u}\ \mathbf{v}]| = \left|\det \begin{bmatrix} 6 & 3 \\ 4 & 2 \end{bmatrix}\right|$$
$$= |6(2) - 3(4)| = |0| = 0.$$

32. 13 **33.** 22 **34.** 16 **35.** 22 **36.** 16

37. We have

$$\det \begin{bmatrix} 3 & 6 \\ c & 4 \end{bmatrix} = 3(4) - 6c = 12 - 6c.$$

The given matrix is not invertible when its determinant equals 0, that is, when $c = 2$.

38. −8 **39.** −9 **40.** $-\frac{2}{5}$ **41.** ±4
42. no c **43.** no c **44.** ±6

45. False, the determinant of a matrix is a *scalar*.

46. False, $\det \begin{bmatrix} a & b \\ c & d \end{bmatrix} = ad - bc$.

47. False, if the determinant of a 2 × 2 matrix is *nonzero*, then the matrix is invertible.

48. False, if a 2 × 2 matrix is invertible, then its determinant is *nonzero*.

49. True

50. False, the (i,j)-cofactor of A equals $(-1)^{i+j}$ times the determinant of the $(n-1) \times (n-1)$ matrix obtained by deleting row i and column j from A.

51. True **52.** True

53. False, cofactor expansion is very inefficient. (See pages 204–205.)

54. True

55. False, consider $\begin{bmatrix} 1 & 2 \\ 2 & 1 \end{bmatrix}$.

56. True

57. False, see Example 1 on page 154.

58. False, a matrix in which all the entries to the left and below the diagonal entries equal zero is called an *upper* triangular matrix.

59. True **60.** True

61. False, the determinant of an upper triangular or a lower triangular square matrix equals the *product* of its diagonal entries.

62. True

63. False, the area of the parallelogram determined by **u** and **v** is $|\det[\mathbf{u}\ \mathbf{v}]|$.

64. False, $|\det[T(\mathbf{u})\ T(\mathbf{v})]| = |\det A| \cdot |\det[\mathbf{u}\ \mathbf{v}]|$, where A is the standard matrix of T.

65. We have

$$\det A_\theta = \det \begin{bmatrix} \cos \theta & -\sin \theta \\ \sin \theta & \cos \theta \end{bmatrix}$$
$$= \cos^2 \theta - (-\sin^2 \theta)$$
$$= \cos^2 \theta + \sin^2 \theta = 1.$$

66. If $a_{ij} = 0$ or $a_{ji} = 1$ for every i and j, then each term in

$$\det A = a_{11}a_{22} - a_{12}a_{21}$$

is 0 or 1. Hence $\det A$ equals 0, 1, or −1.

67. The determinant of the given matrix is 2.

68. $\det \begin{bmatrix} a & b \\ a & b \end{bmatrix} = ab - ba = 0$

69. Let $A = \begin{bmatrix} a & b \\ c & d \end{bmatrix}$. Then

$$\det A^T = \det \begin{bmatrix} a & c \\ b & d \end{bmatrix} = ad - cb$$
$$= \det \begin{bmatrix} a & b \\ c & d \end{bmatrix} = \det A.$$

70. $\det kA = k^2(\det A)$

71. We have

$$\det AB = \det \left(\begin{bmatrix} a_{11} & a_{12} \\ a_{21} & a_{22} \end{bmatrix} \begin{bmatrix} b_{11} & b_{12} \\ b_{21} & b_{22} \end{bmatrix} \right)$$
$$= \det \begin{bmatrix} a_{11}b_{11} + a_{12}b_{21} & a_{11}b_{12} + a_{12}b_{22} \\ a_{21}b_{11} + a_{22}b_{21} & a_{21}b_{12} + a_{22}b_{22} \end{bmatrix}$$
$$= (a_{11}b_{11} + a_{12}b_{21})(a_{21}b_{12} + a_{22}b_{22})$$
$$\quad - (a_{11}b_{12} + a_{12}b_{22})(a_{21}b_{11} + a_{22}b_{21})$$

$$= a_{12}b_{21}a_{21}b_{12} + a_{11}b_{11}a_{22}b_{22}$$
$$- a_{12}b_{22}a_{21}b_{11} - a_{11}b_{12}a_{22}b_{21}$$
$$= (a_{11}a_{22} - a_{12}a_{21})(b_{11}b_{22} - b_{12}b_{21})$$
$$= (\det A)(\det B).$$

72. Evaluating the determinant by a cofactor expansion along the zero row, we see that its value is 0.

73. We have
$$\det EA = \det \begin{bmatrix} a & b \\ kc & kd \end{bmatrix}$$
$$= k(ad - bc)$$
$$= (\det E)(\det A).$$

74. We have
$$\det EA = \det \begin{bmatrix} c & d \\ a & b \end{bmatrix}$$
$$= cb - da$$
$$= -1(ad - bc)$$
$$= (\det E)(\det A).$$

75. We have
$$\det EA = \det \begin{bmatrix} a & b \\ ka+c & kb+d \end{bmatrix}$$
$$= ad - bc$$
$$= 1(ad - bc)$$
$$= (\det E)(\det A).$$

76. We have
$$\det EA = \det \begin{bmatrix} a+kc & b+kd \\ c & d \end{bmatrix}$$
$$= 1(ad - bc) = (\det E)(\det A).$$

77. We have
$$\det \begin{bmatrix} a & b \\ c+kp & d+kq \end{bmatrix}$$
$$= ad + akq - bc - bkp$$
$$= (ad - bc) + k(aq - bp)$$
$$= \det \begin{bmatrix} a & b \\ c & d \end{bmatrix} + k \det \begin{bmatrix} a & b \\ p & q \end{bmatrix}.$$

78. The determinant of a square matrix with integer entries is an integer.

79. $\frac{1}{2}|\det[\mathbf{u}\ \mathbf{v}]|$ 80. $(-1)^{mn}$ 81. (c) no

82. (c) yes   83. (c) yes

84. (a) $\det A = -40$, $\det A^{-1} = -0.25$
    (c) $(\det A)(\det A^{-1}) = 1$

3.2 PROPERTIES OF DETERMINANTS

1. We have
$$-0 \det \begin{bmatrix} -1 & 4 \\ 2 & -2 \end{bmatrix} + 0 \det \begin{bmatrix} 1 & -1 \\ 2 & -2 \end{bmatrix}$$
$$- 3 \det \begin{bmatrix} 1 & -1 \\ -1 & 4 \end{bmatrix}$$
$$= 0(-6) + 0(0) - 3(3) = -9.$$

2. We have
$$1 \det \begin{bmatrix} -1 & 3 \\ 1 & -1 \end{bmatrix} - 2 \det \begin{bmatrix} -2 & 2 \\ 1 & -1 \end{bmatrix}$$
$$+ 0 \det \begin{bmatrix} -2 & 2 \\ -1 & 3 \end{bmatrix}$$
$$= 1(-2) - 2(0) + 0(-4) = -2.$$

3. We have
$$1 \det \begin{bmatrix} 1 & -2 \\ -1 & 1 \end{bmatrix} + 4 \det \begin{bmatrix} 2 & 3 \\ -1 & 1 \end{bmatrix}$$
$$- 0 \det \begin{bmatrix} 2 & 3 \\ 1 & -2 \end{bmatrix}$$
$$= 1(-1) + 4(5) - 0(-7) = 19.$$

4. We have
$$1 \det \begin{bmatrix} 5 & -9 \\ 3 & -1 \end{bmatrix} - (-2) \det \begin{bmatrix} -1 & 2 \\ 3 & -1 \end{bmatrix}$$
$$+ 2 \det \begin{bmatrix} -1 & 2 \\ 5 & -9 \end{bmatrix}$$
$$= 1(22) + 2(-5) + 2(-1) = 10.$$

5. We have
$$2 \det \begin{bmatrix} 2 & 2 \\ 3 & 1 \end{bmatrix} - 3 \det \begin{bmatrix} 1 & 3 \\ 3 & 1 \end{bmatrix}$$
$$+ 1 \det \begin{bmatrix} 1 & 3 \\ 2 & 2 \end{bmatrix}$$
$$= 2(-4) - 3(-8) + 1(-4) = 12.$$

6. We have
$$1 \det \begin{bmatrix} 2 & 3 \\ 1 & 1 \end{bmatrix} - 2 \det \begin{bmatrix} 3 & 2 \\ 1 & 1 \end{bmatrix}$$
$$+ 3 \det \begin{bmatrix} 3 & 2 \\ 2 & 3 \end{bmatrix}$$
$$= 1(-1) - 2(1) + 3(5) = 12.$$

3.2 Properties of Determinants

7. We have
$$0\det\begin{bmatrix}1 & 2\\-1 & 1\end{bmatrix} - 1\det\begin{bmatrix}2 & 0\\-1 & 1\end{bmatrix}$$
$$+ 0\det\begin{bmatrix}2 & 0\\1 & 2\end{bmatrix}$$
$$= 0 - 1(2) + 0 = -2.$$

8. We have
$$0\det\begin{bmatrix}1 & 1\\0 & -1\end{bmatrix} - 2\det\begin{bmatrix}0 & 2\\0 & -1\end{bmatrix}$$
$$+ 1\det\begin{bmatrix}0 & 2\\1 & 1\end{bmatrix}$$
$$= 0 - 2(0) + 1(-2) = -2.$$

9. We have
$$-2\det\begin{bmatrix}1 & -1\\-2 & 1\end{bmatrix} + 0\det\begin{bmatrix}3 & 1\\-2 & 1\end{bmatrix}$$
$$+ 1\det\begin{bmatrix}3 & 1\\1 & -1\end{bmatrix}$$
$$= -2(-1) + 0 + 1(-4) = -2.$$

10. We have
$$1\det\begin{bmatrix}1 & 0\\-2 & -1\end{bmatrix} + 1\det\begin{bmatrix}3 & 2\\-2 & -1\end{bmatrix}$$
$$+ 1\det\begin{bmatrix}3 & 2\\1 & 0\end{bmatrix}$$
$$= 1(-1) + 1(1) + 1(-2) = -2.$$

11.
$$\det\begin{bmatrix}0 & 0 & 5\\0 & 3 & 7\\4 & -1 & -2\end{bmatrix} = -\det\begin{bmatrix}4 & -1 & 2\\0 & 3 & 7\\0 & 0 & 5\end{bmatrix}$$
$$= -(4)(3)(5) = -60$$

12. -72 13. -15 14. -30

15. We have
$$\det\begin{bmatrix}3 & -2 & 1\\0 & 0 & 5\\-9 & 4 & 2\end{bmatrix} = -\det\begin{bmatrix}3 & -2 & 1\\-9 & 4 & 2\\0 & 0 & 5\end{bmatrix}$$
$$= -\det\begin{bmatrix}3 & -2 & 1\\0 & -2 & 5\\0 & 0 & 5\end{bmatrix}$$
$$= -(3)(-2)(5) = 30.$$

16. 66 17. -20 18. 10

19. We have
$$\det\begin{bmatrix}1 & 2 & 1\\1 & 1 & 2\\3 & 4 & 8\end{bmatrix} = \det\begin{bmatrix}1 & 2 & 1\\0 & -1 & 1\\0 & -2 & 5\end{bmatrix}$$
$$= \det\begin{bmatrix}1 & 2 & 1\\0 & -1 & 1\\0 & 0 & 3\end{bmatrix}$$
$$= 1(-1)(3) = -3.$$

20. -40 21. 18 22. -36

23. We have
$$\det\begin{bmatrix}0 & 4 & -1 & 1\\-3 & 1 & 1 & 2\\1 & 0 & -2 & 3\\2 & 3 & 0 & 1\end{bmatrix} = -\det\begin{bmatrix}1 & 0 & -2 & 3\\-3 & 1 & 1 & 2\\0 & 4 & -1 & 1\\2 & 3 & 0 & 1\end{bmatrix}$$
$$= -\det\begin{bmatrix}1 & 0 & -2 & 3\\0 & 1 & -5 & 11\\0 & 4 & -1 & 1\\0 & 3 & 4 & -5\end{bmatrix}$$
$$= -\det\begin{bmatrix}1 & 0 & -2 & 3\\0 & 1 & -5 & 11\\0 & 0 & 19 & -43\\0 & 0 & 19 & -38\end{bmatrix}$$
$$= -\det\begin{bmatrix}1 & 0 & -2 & 3\\0 & 1 & -5 & 11\\0 & 0 & 19 & -43\\0 & 0 & 0 & 5\end{bmatrix}$$
$$= -1(1)(19)(5) = -95.$$

24. -56 25. -8 26. 15

27. We have
$$\det\begin{bmatrix}c & 6\\2 & c+4\end{bmatrix} = c(c+4) - 12$$
$$= c^2 + 4c - 12 = (c+6)(c-2).$$

A matrix is not invertible if its determinant equals 0; so this matrix is not invertible if $c = -6$ or $c = 2$.

28. $-4, 2$ 29. 5 30. -12

31. We have
$$\det\begin{bmatrix}1 & -1 & 2\\-1 & 0 & 4\\2 & 1 & c\end{bmatrix} = \det\begin{bmatrix}1 & -1 & 2\\0 & -1 & 6\\0 & 3 & c-4\end{bmatrix}$$
$$= \det\begin{bmatrix}1 & -1 & 2\\0 & -1 & 6\\0 & 0 & c+14\end{bmatrix}$$
$$= 1(-1)(c+14) = -(c+14).$$

So the matrix is not invertible if $c = -14$.

32. -4 **33.** -5 and 3 **34.** $-2, 5$

35. -1 **36.** every c **37.** $c = \frac{1}{2}$

38. $c = 0$

39. False, $\det \begin{bmatrix} 1 & 2 \\ 3 & 4 \end{bmatrix} \neq 1 \cdot 4$.

40. True

41. False, multiplying a row of a square matrix by a scalar c changes its determinant by a factor of c.

42. True

43. False, consider $A = [\mathbf{e}_1 \ \mathbf{0}]$ and $B = [\mathbf{0} \ \mathbf{e}_2]$.

44. True

45. False, $\det A \neq 0$ if A is an invertible matrix.

46. False, for any square matrix A, $\det A^T = \det A$.

47. True

48. False, the determinant of $2I_2$ is 4, but its reduced row echelon form is I_2.

49. True   **50.** True

51. False, $\det cA = c^n \det A$ if A is an $n \times n$ matrix.

52. False, Cramer's rule can be used to solve only systems with an invertible coefficient matrix.

53. True   **54.** True   **55.** True

56. False, $\det(-A) = -\det A$ if A is a 5×5 matrix.

57. True

58. False, if an $n \times n$ matrix A is transformed into an upper triangular matrix U using only row interchanges and row addition operations, then $\det A = (-1)^r u_{11} u_{22} \cdots u_{nn}$, where r is the number of row interchanges performed.

59. We have
$$x_1 = \frac{\det \begin{bmatrix} 6 & 2 \\ -3 & 4 \end{bmatrix}}{\det \begin{bmatrix} 1 & 2 \\ 3 & 4 \end{bmatrix}} = \frac{6(4) - 2(-3)}{1(4) - 2(3)} = -15$$

and

$$x_2 = \frac{\det \begin{bmatrix} 1 & 6 \\ 3 & -3 \end{bmatrix}}{\det \begin{bmatrix} 1 & 2 \\ 3 & 4 \end{bmatrix}} = \frac{1(-3) - 6(3)}{-2} = 10.5.$$

60. $x_1 = -10, x_2 = 9$ **61.** $x_1 = 11, x_2 = -6$

62. $x_1 = -16, x_2 = 21$

63. We have

$$x_1 = \frac{\det \begin{bmatrix} 6 & 0 & -2 \\ -5 & 1 & 3 \\ 4 & 2 & 1 \end{bmatrix}}{\det \begin{bmatrix} 1 & 0 & -2 \\ -1 & 1 & 3 \\ 0 & 2 & 1 \end{bmatrix}} = \frac{\det \begin{bmatrix} 6 & 0 & -2 \\ -5 & 1 & 3 \\ 14 & 0 & -5 \end{bmatrix}}{\det \begin{bmatrix} 1 & 0 & -2 \\ 0 & 1 & 1 \\ 0 & 2 & 1 \end{bmatrix}}$$

$$= \frac{6(-5) - (-2)(14)}{1(1) - 1(2)} = 2,$$

$$x_2 = \frac{\det \begin{bmatrix} 1 & 6 & -2 \\ -1 & -5 & 3 \\ 0 & 4 & 1 \end{bmatrix}}{\det \begin{bmatrix} 1 & 0 & -2 \\ -1 & 1 & 3 \\ 0 & 2 & 1 \end{bmatrix}} = \frac{\det \begin{bmatrix} 1 & 6 & -2 \\ 0 & 1 & 1 \\ 0 & 4 & 1 \end{bmatrix}}{-1}$$

$$= \frac{1(1) - 1(4)}{-1} = 3,$$

and

$$x_3 = \frac{\det \begin{bmatrix} 1 & 0 & 6 \\ -1 & 1 & -5 \\ 0 & 2 & 4 \end{bmatrix}}{\det \begin{bmatrix} 1 & 0 & -2 \\ -1 & 1 & 3 \\ 0 & 2 & 1 \end{bmatrix}} = \frac{\det \begin{bmatrix} 1 & 0 & 6 \\ 0 & 1 & 1 \\ 0 & 2 & 4 \end{bmatrix}}{-1}$$

$$= \frac{1(4) - 1(2)}{-1} = -2.$$

64. $x_1 = -2, x_2 = -3, x_3 = 1$

65. $x_1 = -0.4, x_2 = 1.8, x_3 = -2.4$

66. $x_1 = 0.6, x_2 = -0.4, x_3 = 0.4$

67. Take $A = I_2$ and $k = 2$. Then
$$\det kA = \det \begin{bmatrix} 2 & 0 \\ 0 & 2 \end{bmatrix} = 2 \cdot 2 = 4,$$
but $k \cdot \det A = 2 \cdot 1 = 2$.

68. By repeated use of Theorem 3.3(b), we see that $\det kA = k^n (\det A)$.

69. Let A be an invertible $n \times n$ matrix. Then $1 = \det I_n = \det AA^{-1} = (\det A)(\det A^{-1})$. So $\det A \neq 0$, and hence $\det A^{-1} = \frac{1}{\det A}$.

70. We have $\det(-A) = -\det A$ if and only if A is an $n \times n$ matrix for some odd positive integer n. To evaluate $\det(-A)$, remove a factor of -1 from each row. Then by Theorem 3.3(b), we have $\det(-A) = (-1)^n \det A = -\det A$ if and only if n is odd.

71. We have
$$\det(B^{-1}AB) = (\det B^{-1})(\det A)(\det B)$$
$$= \left(\frac{1}{\det B}\right)(\det A)(\det B)$$
$$= \det A.$$

72. If $A^k = O$, then $\det A^k = 0$. But $\det A^k = (\det A)^k$ by Theorem 3.4(b). So $(\det A)^k = 0$, and therefore $\det A = 0$.

73. If $QQ^T = I_n$, then by Theorem 3.4(c) we have
$$1 = \det I_n = \det QQ^T$$
$$= (\det Q)(\det Q^T) = (\det Q)^2.$$

Hence $\det Q = \pm 1$.

74. Let A be a skew-symmetric $n \times n$ matrix, where n is odd. Then, by Exercise 70,
$$\det A = \det A^T = \det(-A) = -\det A.$$

So $\det A = 0$, and hence A is not invertible.

For even n, a skew-symmetric $n \times n$ matrix may be invertible or not invertible. Consider, for example,
$$\begin{bmatrix} 0 & 1 \\ -1 & 0 \end{bmatrix} \quad \text{and} \quad \begin{bmatrix} 0 & 0 \\ 0 & 0 \end{bmatrix}.$$

75. We have
$$\det \begin{bmatrix} 1 & a & a^2 \\ 1 & b & b^2 \\ 1 & c & c^2 \end{bmatrix} = \det \begin{bmatrix} 1 & a & a^2 \\ 0 & b-a & b^2-a^2 \\ 0 & c-a & c^2-a^2 \end{bmatrix}$$
$$= \det \begin{bmatrix} b-a & b^2-a^2 \\ c-a & c^2-a^2 \end{bmatrix}$$
$$= (b-a)(c-a) \det \begin{bmatrix} 1 & b+a \\ 1 & c+a \end{bmatrix}$$
$$= (b-a)(c-a)[(c+a)-(b+a)]$$
$$= (b-a)(c-a)(c-b).$$

76. We have
$$\det \begin{bmatrix} 1 & x_1 & y_1 \\ 1 & x_2 & y_2 \\ 1 & x & y \end{bmatrix} = \det \begin{bmatrix} 1 & x_1 & y_1 \\ 0 & x_2-x_1 & y_2-y_1 \\ 0 & x-x_1 & y-y_1 \end{bmatrix}$$
$$= \det \begin{bmatrix} x_2-x_1 & y_2-y_1 \\ x-x_1 & y-y_1 \end{bmatrix}$$
$$= (x_2-x_1)(y-y_1) - (x-x_1)(y_2-y_1).$$

Hence the given equation reduces to
$$(x_2-x_1)(y-y_1) - (x-x_1)(y_2-y_1) = 0,$$

that is,
$$y - y_1 = \frac{y_2-y_1}{x_2-x_1}(x-x_1) \quad \text{if } x_1 \neq x_2,$$

or to $x = x_1$ if $x_1 = x_2$.

77. By the Corollary to Theorem 2.5, $\mathcal{B}$ is linearly independent if and only if B is invertible. So $\mathcal{B}$ is linearly independent if and only if $\det B \neq 0$.

78. Interchange rows 1 and n of B, then rows 2 and $n-1$ of B, etc. It follows from Theorem 3.3(a) that $\det B = -\det A$ if $n = 2, 3, 6, 7, 10, 11, \ldots$ and $\det B = \det A$ otherwise.

79. Suppose that B is a matrix obtained by multiplying each entry of row r of A by a scalar k. Let A_{rj} and B_{rj} be the $(n-1) \times (n-1)$ matrices obtained by deleting row r and column j from A and B, respectively. Then $A_{rj} = B_{rj}$ for $j = 1, 2, \ldots, n$. Since $b_{rj} = ka_{rj}$ for $j = 1, 2, \ldots, n$, the conclusion follows from the cofactor expansions of A and B along row r.

80. If E is obtained by performing an interchange operation on I_n, the result follows from Theorem 3.3(a). Let E be the elementary matrix obtained by multiplying row r of I_n by k. Then EA is obtained by multiplying row r of A by k, and so Theorem 3.3(b) implies that
$$\det EA = k(\det A) = (\det E)(\det A).$$

A similar proof establishes the result if E is obtained by adding k times row r of I_n to row s.

81. Suppose that E is an elementary matrix obtained by interchanging two rows of I_n or by multiplying a row of I_n by some scalar. Then $E^T = E$, and so $\det E^T = \det E$. If E is an elementary matrix obtained by adding a multiple of some row to another, then E is an upper triangular or lower triangular $n \times n$ matrix with every diagonal entry equal to 1. Thus $\det E = 1$. But E^T has the same form, and so $\det E^T = 1$ also.

82. (a) Let B denote the matrix whose (i,j)-entry is b_{ij}, the cofactor of the (j,i)-entry of A. By performing a cofactor expansion along column k of B, we see that $\det B = b_{kj}$.

(b) Let B_k be the $n \times n$ matrix obtained by replacing column k of A by $\mathbf{e}_j$. Applying Cramer's rule to $A\mathbf{x} = \mathbf{e}_j$, we see that
$$x_k = \frac{\det B_k}{\det A} = \frac{b_{kj}}{\det A}.$$

74 Chapter 3 Determinants

So

$$(\det A)\mathbf{e}_j = (\det A)A \begin{bmatrix} x_1 \\ x_2 \\ \vdots \\ x_n \end{bmatrix} = A \begin{bmatrix} b_{1j} \\ b_{2j} \\ \vdots \\ b_{nj} \end{bmatrix}.$$

(c) From (b), we see that

$$AB = A[\mathbf{b}_1 \ \mathbf{b}_2 \ \cdots \ \mathbf{b}_n]$$
$$= [(\det A)\mathbf{e}_1 \ (\det A)\mathbf{e}_2 \ \cdots \ (\det A)\mathbf{e}_n]$$
$$= (\det A)[\mathbf{e}_1 \ \mathbf{e}_2 \ \cdots \ \mathbf{e}_n]$$
$$= (\det A)I_n.$$

(d) If $\det A \neq 0$, it follows from (c) that $A\left(\frac{1}{\det A}B\right) = I_n$. Hence

$$A^{-1} = \frac{1}{\det A}B.$$

83. (a) We have

$$A \xrightarrow{r_1 \leftrightarrow r_2} \begin{bmatrix} 2.4 & 3.0 & -6 & 9 \\ 0.0 & -3.0 & -2 & -5 \\ -4.8 & 6.3 & 4 & -2 \\ 9.6 & 1.5 & 5 & 9 \end{bmatrix}$$

$$\xrightarrow[-4r_1 + r_4 \to r_4]{2r_1 + r_3 \to r_3} \begin{bmatrix} 2.4 & 3.0 & -6 & 9 \\ 0.0 & -3.0 & -2 & -5 \\ 0.0 & 12.3 & -8 & 16 \\ 0.0 & -10.5 & 29 & -27 \end{bmatrix}$$

$$\xrightarrow[-3.5r_2 + r_4 \to r_4]{4.1r_2 + r_3 \to r_3} \begin{bmatrix} 2.4 & 3 & -6.0 & 9.0 \\ 0.0 & -3 & -2.0 & -5.0 \\ 0.0 & 0 & -16.2 & -4.5 \\ 0.0 & 0 & 36.0 & -9.5 \end{bmatrix}$$

$$\xrightarrow{-\frac{20}{9}r_3 + r_4 \to r_4} \begin{bmatrix} 2.4 & 3 & -6.0 & 9.0 \\ 0.0 & -3 & -2.0 & -5.0 \\ 0.0 & 0 & -16.2 & -4.5 \\ 0.0 & 0 & 0.0 & -19.5 \end{bmatrix}.$$

(b) Because there is one interchange operation used in (a),

$$\det A = -(2.4)(-3)(-16.2)(-19.5)$$
$$= 2274.48.$$

84. (a) $x_1 = 119$, $x_2 = -262$, $x_3 = 61$, $x_4 = -164$
(b) 5

85. $\begin{bmatrix} 13 & -8 & -3 & 6 \\ -28 & 20 & 8 & -12 \\ 7 & -4 & -1 & 2 \\ -18 & 12 & 6 & -8 \end{bmatrix}$

CHAPTER 3 REVIEW EXERCISES

1. False, $\det \begin{bmatrix} a & b \\ c & d \end{bmatrix} = ad - bc$.

2. False, for $n \geq 2$, the (i,j)-cofactor of an $n \times n$ matrix A equals $(-1)^{i+j}$ times the determinant of the $(n-1) \times (n-1)$ matrix obtained by deleting row i and column j from A.

3. True

4. False, consider $A = [\mathbf{e}_1 \ \mathbf{0}]$ and $B = [\mathbf{0} \ \mathbf{e}_2]$.

5. True

6. False, if B is obtained by interchanging two rows of an $n \times n$ matrix A, then $\det B = -\det A$.

7. False, an $n \times n$ matrix is invertible if and only if its determinant is *nonzero*.

8. True

9. False, for any invertible matrix A,

$$\det A^{-1} = \frac{1}{\det A}.$$

10. False, for any $n \times n$ matrix A and scalar c, $\det cA = c^n(\det A)$.

11. False, the determinant of an upper triangular or a lower triangular square matrix equals the *product* of its diagonal entries.

12. 1 **13.** 5 **14.** −3

15. The $(3,1)$-cofactor of the matrix is

$$(-1)^{3+1}\det\begin{bmatrix} -1 & 2 \\ 2 & -1 \end{bmatrix} = 1[(-1)(-1) - 2(2)]$$
$$= -3.$$

16. $1(7) - (-1)(-1) + 2(-5)$

17. $2(-3) - 1(1) + 3(1)$

18. $-(-1)(-1) + 2(-1) - 1(1)$

19. $1(7) - (-1)(-5) + 2(-3)$

20. −8 **21.** 0 **22.** 0 **23.** 3

24. (a) We have

$$\begin{bmatrix} 0 & 3 & -6 & 1 \\ -2 & -2 & 2 & 6 \\ 1 & 1 & -1 & -1 \\ 2 & -1 & 2 & -2 \end{bmatrix} \xrightarrow{r_1 \leftrightarrow r_3}$$

$$\begin{bmatrix} 1 & 1 & -1 & -1 \\ -2 & -2 & 2 & 6 \\ 0 & 3 & -6 & 1 \\ 2 & -1 & 2 & -2 \end{bmatrix} \xrightarrow[-2r_1 + r_4 \to r_4]{2r_1 + r_2 \to r_2}$$

$$\begin{bmatrix} 1 & 1 & -1 & -1 \\ 0 & 0 & 0 & 4 \\ 0 & 3 & -6 & 1 \\ 0 & -3 & 4 & 0 \end{bmatrix} \xrightarrow{r_2 \leftrightarrow r_4}$$

$$\begin{bmatrix} 1 & 1 & -1 & -1 \\ 0 & -3 & 4 & 0 \\ 0 & 3 & -6 & 1 \\ 0 & 0 & 0 & 4 \end{bmatrix} \xrightarrow{r_2 + r_3 \to r_3}$$

$$\begin{bmatrix} 1 & 1 & -1 & -1 \\ 0 & -3 & 4 & 0 \\ 0 & 0 & -2 & 1 \\ 0 & 0 & 0 & 4 \end{bmatrix}.$$

(b) $(-1)^2(1)(-3)(-2)(4) = 24$

25. -3 and 4 **26.** -2

27. We have

$$\det \begin{bmatrix} c+4 & -1 & c+5 \\ -3 & 3 & -4 \\ c+6 & -3 & c+7 \end{bmatrix}$$

$$= \det \begin{bmatrix} c+4 & -1 & c+5 \\ 3c+9 & 0 & 3c+11 \\ -2c-6 & 0 & -2c-8 \end{bmatrix}$$

$$= (-1)(-1)^{1+2} \det \begin{bmatrix} 3c+9 & 3c+11 \\ -2c-6 & -2c-8 \end{bmatrix}$$

$$= (c+3)[3(-2c-8) - (3c+11)(-2)]$$

$$= -2(c+3).$$

Because a matrix is not invertble if and only if its determinant equals 0, this matrix is not invertible if and only if $c = -3$.

28. $-1, 2$ **29.** 25

30. The volume is

$$\left| \det \begin{bmatrix} 1 & -1 & 3 \\ 0 & 2 & 1 \\ 2 & 1 & -1 \end{bmatrix} \right| = |-17| = 17.$$

31. We have

$$x_1 = \frac{\det \begin{bmatrix} 5 & 1 \\ -6 & 3 \end{bmatrix}}{\det \begin{bmatrix} 2 & 1 \\ -4 & 3 \end{bmatrix}} = \frac{5(3) - 1(-6)}{2(3) - 1(-4)} = \frac{21}{10}$$

and

$$x_2 = \frac{\det \begin{bmatrix} 2 & 5 \\ -4 & -6 \end{bmatrix}}{\det \begin{bmatrix} 2 & 1 \\ -4 & 3 \end{bmatrix}} = \frac{2(-6) - 5(-4)}{10} = \frac{8}{10}.$$

32. $x_1 = -2.2$, $x_2 = -0.4$, $x_3 = 4.4$

33. 5 **34.** 0.2

35. If $\det A = 5$, then

$$\det 2A = 2^3 (\det A) = 8(5) = 40$$

because we can remove a factor of 2 from each row of $2A$ and apply Theorem 3.3(b).

36. 125 **37.** 5 **38.** -10

39. We have

$$\det \begin{bmatrix} a_{11} + 5a_{31} & a_{12} + 5a_{32} & a_{13} + 5a_{33} \\ 4a_{21} & 4a_{22} & 4a_{23} \\ a_{31} - 2a_{21} & a_{32} - 2a_{22} & a_{33} - 2a_{23} \end{bmatrix}$$

$$= 4 \cdot \det \begin{bmatrix} a_{11} + 5a_{31} & a_{12} + 5a_{32} & a_{13} + 5a_{33} \\ a_{21} & a_{22} & a_{23} \\ a_{31} - 2a_{21} & a_{32} - 2a_{22} & a_{33} - 2a_{22} \end{bmatrix}$$

$$= 4 \cdot \det \begin{bmatrix} a_{11} + 5a_{31} & a_{12} + 5a_{32} & a_{13} + 5a_{33} \\ a_{21} & a_{22} & a_{23} \\ a_{31} & a_{32} & a_{33} \end{bmatrix}$$

$$= 4 \cdot \det \begin{bmatrix} a_{11} & a_{12} & a_{13} \\ a_{21} & a_{22} & a_{23} \\ a_{31} & a_{32} & a_{33} \end{bmatrix}$$

$$= 4(\det A) = 4(5) = 20.$$

40. -5

41. If $B^2 = B$, then

$$\det B = \det B^2 = \det BB = (\det B)(\det B)$$
$$= (\det B)^2.$$

Hence $\det B$ is a solution of the equation $x = x^2$, and so $\det B = 0$ or $\det B = 1$.

42. We have

$$\det A = \det(PDP^{-1})$$
$$= (\det P)(\det D) \left(\frac{1}{\det P} \right)$$
$$= \det D.$$

Since D is a diagonal matrix, $\det D$ is the product of its diagonal entries.

43. We have

$$\det \begin{bmatrix} 1 & x & y \\ 1 & x_1 & y_1 \\ 0 & 1 & m \end{bmatrix} = \det \begin{bmatrix} 1 & x & y \\ 0 & x_1 - x & y_1 - y \\ 0 & 1 & m \end{bmatrix}$$

$$= m(x_1 - x) - (y_1 - y).$$

So the given equation can be written

$$y_1 - y = m(x_1 - x) \quad \text{or} \quad y - y_1 = m(x - x_1).$$

CHAPTER 3 MATLAB EXERCISES

1. Matrix A can be transformed into an upper triangular matrix U using only row addition operations. The diagonal entries of U (rounded to 4 places after the decimal point) are -0.8000, -30.4375, 1.7865, -0.3488, -1.0967, and 0.3749. Thus the determinant of A equals $(-1)^0$ times the product of these numbers, which is 6.2400.

2. This sequence of elementary row operations transforms A into an upper triangular matrix: $r_1 \leftrightarrow r_2$, $-2r_1 + r_3 \to r_3$, $-2r_1 + r_4 \to r_4$, $2r_1 + r_5 \to r_5$, $-r_1 + r_6 \to r_6$, $-r_2 + r_4 \to r_4$, $4r_2 + r_5 \to r_5$, $-2r_2 + r_6 \to r_6$, $r_3 \leftrightarrow r_6$, $17r_3 + r_5 \to r_5$, $r_4 \leftrightarrow r_6$, $r_5 \leftrightarrow r_6$, and $-\frac{152}{9}r_5 + r_6 \to r_6$.
(Other sequences are possible.) This matrix is

$$\begin{bmatrix} 1 & 1 & 2 & -2 & 1 & 2 \\ 0 & 1 & 2 & -2 & 3 & 1 \\ 0 & 0 & -1 & 1 & -10 & 2 \\ 0 & 0 & 0 & -1 & 4 & -1 \\ 0 & 0 & 0 & 0 & -9 & 0 \\ 0 & 0 & 0 & 0 & 0 & 46 \end{bmatrix}.$$

Thus
$$\det A = (-1)^4 (1)(1)(-1)(-1)(-9)(46) = -414.$$

3. (a) $\det \begin{bmatrix} \mathbf{v} \\ A \end{bmatrix} = 2$ and $\det \begin{bmatrix} \mathbf{w} \\ A \end{bmatrix} = -10$.

   (b) $\det \begin{bmatrix} \mathbf{v} + \mathbf{w} \\ A \end{bmatrix} = \det \begin{bmatrix} \mathbf{v} \\ A \end{bmatrix} + \det \begin{bmatrix} \mathbf{w} \\ A \end{bmatrix} = -8$.

   (c) $\det \begin{bmatrix} 3\mathbf{v} - 2\mathbf{w} \\ A \end{bmatrix} = 3 \det \begin{bmatrix} \mathbf{v} \\ A \end{bmatrix} - 2 \det \begin{bmatrix} \mathbf{w} \\ A \end{bmatrix} = 26$.

   (d) Any such function is a linear transformation.

   (e) Define $T \colon \mathcal{R}^n \to R$ by
   $$T(\mathbf{x}) = \det \begin{bmatrix} \mathbf{x} \\ C \end{bmatrix},$$
   where C is an $(n-1) \times n$ matrix. Let $\mathbf{u} = \begin{bmatrix} u_1 & u_2 & \ldots & u_n \end{bmatrix}$ and $\mathbf{v} = \begin{bmatrix} v_1 & v_2 & \ldots & v_n \end{bmatrix}$ be $1 \times n$ row vectors. Then $T(\mathbf{u} + \mathbf{v}) = \det A$, where
   $$A = \begin{bmatrix} \mathbf{u} + \mathbf{v} \\ C \end{bmatrix}.$$
   Using the cofactor expansion of A along the first row, we see that
   $$\det A = (u_1 + v_1)c_1 - \cdots + (-1)^{n+1}(u_n + v_n)c_n,$$
   where c_i denotes the determinant of the $(n-1) \times (n-1)$ matrix obtained by deleting column i from C. Similarly, we see that
   $$T(\mathbf{u}) = u_1 c_1 - \cdots + (-1)^{n+1} u_n c_n,$$
   and
   $$T(\mathbf{v}) = v_1 c_1 - \cdots + (-1)^{n+1} v_n c_n.$$
   Hence $T(\mathbf{u} + \mathbf{v}) = T(\mathbf{u}) + T(\mathbf{v})$, and so T preserves vector addition. Moreover, $T(c\mathbf{u}) = cT(\mathbf{u})$ by Theorem 3.3(b), and so T preserves scalar multiplication. Thus T is a linear transformation.

   (f) Any such function is a linear transformation. The proof follows from Theorem 3.3(a) and the result in (e) above.

   (g) Any such function is a linear transformation. The proof follows from Theorem 3.4(c) and the result in (f) above.

Chapter 4

Subspaces and Their Properties

4.1 SUBSPACES

1. $\left\{\begin{bmatrix} 0 \\ 1 \end{bmatrix}\right\}$ 2. $\left\{\begin{bmatrix} 2 \\ -3 \end{bmatrix}\right\}$

3. $\left\{\begin{bmatrix} 4 \\ -1 \end{bmatrix}\right\}$ 4. $\left\{\begin{bmatrix} 0 \\ 1 \\ -3 \end{bmatrix}, \begin{bmatrix} 4 \\ 1 \\ 1 \end{bmatrix}\right\}$

5. Since
$$\begin{bmatrix} -s+t \\ 2s-t \\ s+3t \end{bmatrix} = s\begin{bmatrix} -1 \\ 2 \\ 1 \end{bmatrix} + t\begin{bmatrix} 1 \\ -1 \\ 3 \end{bmatrix},$$
a generating set for the subspace is
$$\left\{\begin{bmatrix} -1 \\ 2 \\ 1 \end{bmatrix}, \begin{bmatrix} 1 \\ -1 \\ 3 \end{bmatrix}\right\}.$$

6. $\left\{\begin{bmatrix} -1 \\ 0 \\ 0 \\ 1 \end{bmatrix}, \begin{bmatrix} 3 \\ 0 \\ 1 \\ 0 \end{bmatrix}, \begin{bmatrix} 0 \\ 0 \\ -1 \\ -2 \end{bmatrix}\right\}$

7. $\left\{\begin{bmatrix} -1 \\ 0 \\ 0 \\ 3 \end{bmatrix}, \begin{bmatrix} 1 \\ 4 \\ 0 \\ 0 \end{bmatrix}, \begin{bmatrix} 0 \\ -3 \\ 0 \\ -1 \end{bmatrix}\right\}$

8. $\left\{\begin{bmatrix} 1 \\ 2 \\ -1 \\ -2 \end{bmatrix}, \begin{bmatrix} -1 \\ 0 \\ 3 \\ 1 \end{bmatrix}, \begin{bmatrix} 3 \\ -1 \\ 2 \\ 1 \end{bmatrix}\right\}$

9. Since
$$\begin{bmatrix} 2s-5t \\ 3r+s-2t \\ r-4s+3t \\ -r+2s \end{bmatrix} = r\begin{bmatrix} 0 \\ 3 \\ 1 \\ -1 \end{bmatrix} + s\begin{bmatrix} 2 \\ 1 \\ -4 \\ 2 \end{bmatrix} + t\begin{bmatrix} -5 \\ -2 \\ 3 \\ 0 \end{bmatrix},$$

a generating set for the subspace is
$$\left\{\begin{bmatrix} 0 \\ 3 \\ 1 \\ -1 \end{bmatrix}, \begin{bmatrix} 2 \\ 1 \\ -4 \\ 2 \end{bmatrix}, \begin{bmatrix} -5 \\ -2 \\ 3 \\ 0 \end{bmatrix}\right\}.$$

10. $\left\{\begin{bmatrix} -1 \\ 1 \\ 0 \\ 1 \end{bmatrix}, \begin{bmatrix} 0 \\ -1 \\ 0 \\ 0 \end{bmatrix}, \begin{bmatrix} 4 \\ 2 \\ 3 \\ -1 \end{bmatrix}\right\}$

11. For the given vector $\mathbf{v}$, we have $A\mathbf{v} = \mathbf{0}$. Hence $\mathbf{v}$ is in Null A.

12. For the given vector $\mathbf{v}$, we have $A\mathbf{v} \ne \mathbf{0}$. Hence $\mathbf{v}$ is not in Null A.

13. For the given vector $\mathbf{v}$, we have $A\mathbf{v} \ne \mathbf{0}$. Hence $\mathbf{v}$ is not in Null A.

14. yes 15. yes 16. no 17. yes
18. no 19. no 20. yes

21. Vector $\mathbf{u} = \begin{bmatrix} 1 \\ -4 \\ 2 \end{bmatrix}$ belongs to Col A if and only if $A\mathbf{x} = \mathbf{u}$ has a solution. Since $\begin{bmatrix} -7 \\ -4 \\ 0 \\ 0 \end{bmatrix}$ is a solution of this system, $\mathbf{u}$ is in Col A.

22. no 23. yes 24. no
25. yes 26. no

27. $\left\{\begin{bmatrix} 7 \\ 5 \\ 1 \end{bmatrix}\right\}$ 28. $\left\{\begin{bmatrix} -2 \\ 1 \\ 1 \end{bmatrix}\right\}$

29. $\left\{\begin{bmatrix} 2 \\ -1 \\ 1 \\ 0 \end{bmatrix}, \begin{bmatrix} -1 \\ -3 \\ 0 \\ 1 \end{bmatrix}\right\}$ 30. $\left\{\begin{bmatrix} 2 \\ -3 \\ 1 \end{bmatrix}\right\}$

78 Chapter 4 Subspaces and Their Properties

31. $\left\{ \begin{bmatrix} -5 \\ 3 \\ 1 \\ 0 \end{bmatrix}, \begin{bmatrix} 3 \\ -4 \\ 0 \\ 1 \end{bmatrix} \right\}$

32. The reduced row echelon form of the given matrix is
$$R = \begin{bmatrix} 1 & 0 & 1 & 0 & -1 \\ 0 & 1 & -1 & 0 & 0 \\ 0 & 0 & 0 & 1 & 1 \end{bmatrix}.$$

Since the vector form of the general solution of $R\mathbf{x} = \mathbf{0}$ is
$$\begin{bmatrix} x_1 \\ x_2 \\ x_3 \\ x_4 \\ x_5 \end{bmatrix} = x_3 \begin{bmatrix} -1 \\ 1 \\ 1 \\ 0 \\ 0 \end{bmatrix} + x_5 \begin{bmatrix} 1 \\ 0 \\ 0 \\ -1 \\ 1 \end{bmatrix},$$

a generating set for Null A = Null R is
$$\left\{ \begin{bmatrix} -1 \\ 1 \\ 1 \\ 0 \\ 0 \end{bmatrix}, \begin{bmatrix} 1 \\ 0 \\ 0 \\ -1 \\ 1 \end{bmatrix} \right\}.$$

33. $\left\{ \begin{bmatrix} 3 \\ 1 \\ 0 \\ 0 \\ 0 \\ 0 \end{bmatrix}, \begin{bmatrix} -1 \\ 0 \\ -2 \\ 1 \\ 0 \\ 0 \end{bmatrix}, \begin{bmatrix} -2 \\ 0 \\ -3 \\ 0 \\ -2 \\ 1 \end{bmatrix} \right\}$

34. $\left\{ \begin{bmatrix} 3 \\ -2 \\ 0 \\ 1 \\ 0 \\ 0 \end{bmatrix}, \begin{bmatrix} -1 \\ 1 \\ 0 \\ 0 \\ 1 \\ 0 \end{bmatrix}, \begin{bmatrix} -3 \\ 2 \\ 1 \\ 0 \\ 0 \\ 1 \end{bmatrix} \right\}$

35. The standard matrix of T is
$$A = \begin{bmatrix} 1 & 2 & -1 \end{bmatrix}.$$

The range of T equals the column space of A; so $\{1, 2, -1\}$ is a generating set for the range of T.

Note that A is in reduced row echelon form. The general solution of $A\mathbf{x} = \mathbf{0}$ is
$$\begin{aligned} x_1 &= -2x_2 + x_3 \\ x_2 &\quad \text{free} \\ x_3 &\quad \text{free,} \end{aligned}$$

and its vector form is
$$\begin{bmatrix} x_1 \\ x_2 \\ x_3 \end{bmatrix} = x_2 \begin{bmatrix} -2 \\ 1 \\ 0 \end{bmatrix} + x_3 \begin{bmatrix} 1 \\ 0 \\ 1 \end{bmatrix}.$$

Hence
$$\left\{ \begin{bmatrix} -2 \\ 1 \\ 0 \end{bmatrix}, \begin{bmatrix} 1 \\ 0 \\ 1 \end{bmatrix} \right\}$$

is a generating set for the null spaces of both A and T.

36. $\left\{ \begin{bmatrix} 1 \\ 2 \end{bmatrix}, \begin{bmatrix} 2 \\ 4 \end{bmatrix} \right\}$, $\left\{ \begin{bmatrix} -2 \\ 1 \end{bmatrix} \right\}$

37. $\left\{ \begin{bmatrix} 1 \\ 1 \\ 1 \\ 0 \end{bmatrix}, \begin{bmatrix} 1 \\ -1 \\ 0 \\ 1 \end{bmatrix} \right\}$, $\left\{ \begin{bmatrix} 0 \\ 0 \end{bmatrix} \right\}$

38. $\left\{ \begin{bmatrix} 1 \\ -2 \end{bmatrix}, \begin{bmatrix} -2 \\ 4 \end{bmatrix}, \begin{bmatrix} 3 \\ -6 \end{bmatrix} \right\}$, $\left\{ \begin{bmatrix} 2 \\ 1 \\ 0 \end{bmatrix}, \begin{bmatrix} -3 \\ 0 \\ 1 \end{bmatrix} \right\}$

39. $\left\{ \begin{bmatrix} 1 \\ 0 \\ 2 \end{bmatrix}, \begin{bmatrix} 1 \\ 0 \\ 0 \end{bmatrix}, \begin{bmatrix} -1 \\ 0 \\ -1 \end{bmatrix} \right\}$, $\left\{ \begin{bmatrix} 1 \\ 1 \\ 2 \end{bmatrix} \right\}$

40. $\left\{ \begin{bmatrix} 1 \\ 0 \\ 1 \\ 1 \end{bmatrix}, \begin{bmatrix} 1 \\ 1 \\ 0 \\ 2 \end{bmatrix}, \begin{bmatrix} 0 \\ 1 \\ -1 \\ 1 \end{bmatrix} \right\}$, $\left\{ \begin{bmatrix} 1 \\ -1 \\ 1 \end{bmatrix} \right\}$

41. $\left\{ \begin{bmatrix} 1 \\ -1 \\ 2 \\ 0 \end{bmatrix}, \begin{bmatrix} -1 \\ 2 \\ -1 \\ 2 \end{bmatrix}, \begin{bmatrix} -5 \\ 7 \\ -8 \\ 4 \end{bmatrix} \right\}$, $\left\{ \begin{bmatrix} 3 \\ -2 \\ 1 \end{bmatrix} \right\}$

42. $\left\{ \begin{bmatrix} 1 \\ -1 \\ 1 \\ 1 \end{bmatrix}, \begin{bmatrix} -1 \\ 2 \\ 0 \\ 1 \end{bmatrix}, \begin{bmatrix} -3 \\ 4 \\ -2 \\ -1 \end{bmatrix}, \begin{bmatrix} -2 \\ 5 \\ 1 \\ 4 \end{bmatrix} \right\}$, $\left\{ \begin{bmatrix} 2 \\ -1 \\ 1 \\ 0 \end{bmatrix}, \begin{bmatrix} -1 \\ -3 \\ 0 \\ 1 \end{bmatrix} \right\}$

43. True 44. True
45. False, $\{\mathbf{0}\}$ is called the *zero subspace*.
46. True 47. True
48. False, the column space of an $m \times n$ matrix is contained in $\mathcal{R}^m$.
49. False, the row space of an $m \times n$ matrix is contained in $\mathcal{R}^n$.
50. False, the *column space* of an $m \times n$ matrix equals $\{A\mathbf{v} \colon \mathbf{v} \text{ is in } \mathcal{R}^n\}$.
51. True 52. True 53. True 54. True
55. False, the range of a linear transformation equals the *column space* of its standard matrix.
56. True 57. True 58. True 59. True

60. True **61.** True **62.** True

63. The columns of a matrix form a generating set for the column space of the matrix. For the matrix in Exercise 27, we have

$$-7\begin{bmatrix}-1\\1\end{bmatrix} - 5\begin{bmatrix}1\\-2\end{bmatrix} = \begin{bmatrix}2\\3\end{bmatrix}.$$

Thus, by Theorem 1.7,

$$\operatorname{Col} A = \operatorname{Span}\left\{\begin{bmatrix}-1\\1\end{bmatrix}, \begin{bmatrix}1\\-2\end{bmatrix}, \begin{bmatrix}2\\3\end{bmatrix}\right\}$$

$$= \operatorname{Span}\left\{\begin{bmatrix}-1\\1\end{bmatrix}, \begin{bmatrix}1\\-2\end{bmatrix}\right\}.$$

So $\left\{\begin{bmatrix}-1\\1\end{bmatrix}, \begin{bmatrix}1\\-2\end{bmatrix}\right\}$ is a generating set for Col A containing exactly 2 vectors.

64. As in Exercise 63, $\left\{\begin{bmatrix}1\\0\\1\end{bmatrix}, \begin{bmatrix}2\\-1\\0\end{bmatrix}\right\}$ is a generating set for Col A.

65. From the matrix R in the solution to Exercise 32, we see that the pivot columns of the given matrix A are columns 1, 2, and 4. Choosing each of these columns and exactly one of the other columns gives a generating set for the column space of A that contains exactly four vectors. (See Theorems 2.4(b) and 1.7.) One such generating set is the set containing the first four columns of A.

66. $\left\{\begin{bmatrix}1\\2\\-1\end{bmatrix}, \begin{bmatrix}-3\\-6\\3\end{bmatrix}, \begin{bmatrix}0\\-1\\2\end{bmatrix}, \begin{bmatrix}-2\\2\\-1\end{bmatrix}\right\}$

67. $\left\{\begin{bmatrix}1\\-2\end{bmatrix}, \begin{bmatrix}-3\\4\end{bmatrix}\right\}$

68. $\left\{\begin{bmatrix}-1\\5\\4\end{bmatrix}, \begin{bmatrix}6\\-3\\-2\end{bmatrix}, \begin{bmatrix}-7\\8\\3\end{bmatrix}\right\}$

69. $\left\{\begin{bmatrix}-2\\4\\5\\-1\end{bmatrix}, \begin{bmatrix}-1\\1\\2\\0\end{bmatrix}, \begin{bmatrix}3\\-4\\-5\\1\end{bmatrix}\right\}$

70. $\left\{\begin{bmatrix}1\\1\\0\\1\end{bmatrix}, \begin{bmatrix}0\\-1\\1\\1\end{bmatrix}\right\}$

71. $\mathcal{R}^n$, the zero subspace of $\mathcal{R}^m$, the zero subspace of $\mathcal{R}^n$

72. Yes, $A\mathbf{x} = \mathbf{0}$ is equivalent to $R\mathbf{x} = \mathbf{0}$.

73. Consider $A = \begin{bmatrix}1 & 1\\2 & 2\end{bmatrix}$. The reduced row echelon form of A is $R = \begin{bmatrix}1 & 1\\0 & 0\end{bmatrix}$. Clearly $\begin{bmatrix}1\\2\end{bmatrix}$ belongs to Col A but not to Col R; so Col $A \neq$ Col R.

74. See Theorem 4.8 in Section 4.3.

75. $\begin{bmatrix}1 & -1\\-1 & 1\end{bmatrix}$ **76.** $\begin{bmatrix}1 & -1\\1 & -1\end{bmatrix}$

77. Let V and W be subspaces of $\mathcal{R}^n$. Since $\mathbf{0}$ is contained in both V and W, $\mathbf{0}$ is contained in $V \cap W$. Let $\mathbf{v}$ and $\mathbf{w}$ be contained in $V \cap W$. Then $\mathbf{v}$ and $\mathbf{w}$ are contained in both V and W, and so $\mathbf{v} + \mathbf{w}$ is contained in both V and W. Thus $\mathbf{v} + \mathbf{w}$ is contained in $V \cap W$. Finally, for any scalar c, $c\mathbf{v}$ is contained in both V and W; so $c\mathbf{v}$ is in $V \cap W$. It follows that $V \cap W$ is a subspace of $\mathcal{R}^n$.

78. (a) Because

$$V = \operatorname{Span}\{\mathbf{e}_2\} \text{ and } W = \operatorname{Span}\{\mathbf{e}_1\},$$

the result follows from Theorem 4.1.

(b) Note that $\mathbf{e}_2$ is in V and $\mathbf{e}_1$ is in W, but their sum is not in $V \cup W$.

79. Suppose that $\mathcal{S}$ has the property that $\mathbf{u}+c\mathbf{v}$ is in $\mathcal{S}$ for all vectors $\mathbf{u}$ and $\mathbf{v}$ in $\mathcal{S}$ and all scalars c. If $\mathbf{w}$ is in $\mathcal{S}$, then so is $\mathbf{w}+(-1)\mathbf{w} = \mathbf{0}$. Taking $c = 1$ shows that $\mathcal{S}$ is closed under vector addition, and taking $\mathbf{u} = \mathbf{0}$ shows that $\mathcal{S}$ is closed under scalar multiplication. So $\mathcal{S}$ is a subspace of $\mathcal{R}^n$. The proof of the converse is straightforward.

80. Since V is closed under vector addition and scalar multiplication, V must contain every linear combination of its vectors. So if $\mathbf{u}_1, \mathbf{u}_2, \ldots, \mathbf{u}_k$ are in V, every vector in the span of $\{\mathbf{u}_1, \mathbf{u}_2, \ldots, \mathbf{u}_k\}$ is in V.

81. The vectors $\begin{bmatrix}1\\0\end{bmatrix}$ and $\begin{bmatrix}0\\1\end{bmatrix}$ are in the set, but $\begin{bmatrix}1\\0\end{bmatrix} + \begin{bmatrix}0\\1\end{bmatrix}$ is not.

82. The vector $\begin{bmatrix}0\\2\end{bmatrix}$ is in the set, but $3\begin{bmatrix}0\\2\end{bmatrix}$ is not.

83. The zero vector is not in the set.

84. The vector $\begin{bmatrix}1\\0\end{bmatrix}$ is in the set, but $3\begin{bmatrix}1\\0\end{bmatrix}$ is not.

85. The zero vector is not in the set.

86. The vector $\begin{bmatrix}3\\2\\1\end{bmatrix}$ is in the set, but $(-1)\begin{bmatrix}3\\2\\1\end{bmatrix}$ is not.

80 Chapter 4 Subspaces and Their Properties

87. The vector $\begin{bmatrix} 6 \\ 2 \\ 3 \end{bmatrix}$ is in the set, but $(-1)\begin{bmatrix} 6 \\ 2 \\ 3 \end{bmatrix}$ is not.

88. The vectors $\begin{bmatrix} 1 \\ 0 \\ 0 \end{bmatrix}$ and $\begin{bmatrix} 0 \\ 1 \\ 0 \end{bmatrix}$ are in the set, but their sum is not.

89. Let V denote the given set, and let $\mathbf{u}$ and $\mathbf{v}$ be in V. Then $u_1 - 3u_2 = 0$ and $v_1 - 3v_2 = 0$. Consider
$$\mathbf{u} + \mathbf{v} = \begin{bmatrix} u_1 + v_1 \\ u_2 + v_2 \end{bmatrix}.$$
Since
$$(u_1 + v_1) - 3(u_2 + v_2) = (u_1 - 3u_2) + (v_1 - 3v_2)$$
$$= 0 + 0 = 0,$$
$\mathbf{u} + \mathbf{v}$ is in V. Likewise $c\mathbf{u}$ is in V because
$$cu_1 - 3(cu_2) = c(u_1 - 3u_2) = c(0) = 0.$$
Since $\mathbf{0}$ clearly belongs to V, V is a subspace of $\mathcal{R}^2$.

90. Imitate the arguments in Exercise 89 and Example 3.

91. Imitate the arguments in Exercise 89 and Example 3.

92. Imitate the arguments in Exercise 89 and Example 3.

93. Imitate the arguments in Exercise 89 and Example 3.

94. Imitate the arguments in Exercise 89 and Example 3.

95. Let V denote the null space of T. Since $T(\mathbf{0}) = \mathbf{0}$ by Theorem 2.8(a), $\mathbf{0}$ is in V. If $\mathbf{u}$ and $\mathbf{v}$ are in V, then $T(\mathbf{u}) = T(\mathbf{v}) = \mathbf{0}$. Hence
$$T(\mathbf{u} + \mathbf{v}) = T(\mathbf{u}) + T(\mathbf{v}) = \mathbf{0} + \mathbf{0} = \mathbf{0}.$$
So $\mathbf{u} + \mathbf{v}$ is in V. Finally, for any scalar c and any vector $\mathbf{u}$ in V, we have
$$T(c\mathbf{u}) = cT(\mathbf{u}) = c(\mathbf{0}) = \mathbf{0}.$$
So $c\mathbf{u}$ is in V. Thus V is a subspace of $\mathcal{R}^n$.

96. Let W denote the range of T. Since $T(\mathbf{0}) = \mathbf{0}$ by Theorem 2.8(a), $\mathbf{0}$ is in W. Let $\mathbf{w}_1$ and $\mathbf{w}_2$ belong to W. Then there exist $\mathbf{v}_1$ and $\mathbf{v}_2$ in $\mathcal{R}^n$ such that $T(\mathbf{v}_1) = \mathbf{w}_1$ and $T(\mathbf{v}_2) = \mathbf{w}_2$. Now $\mathbf{v}_1 + \mathbf{v}_2$ is in $\mathcal{R}^n$ and
$$T(\mathbf{v}_1 + \mathbf{v}_2) = T(\mathbf{v}_1) + T(\mathbf{v}_2) = \mathbf{w}_1 + \mathbf{w}_2.$$
So $\mathbf{w}_1 + \mathbf{w}_2$ is in W. Finally, for any scalar c, $T(c\mathbf{v}_1) = cT(\mathbf{v}_1) = c\mathbf{w}_1$. Hence $c\mathbf{w}_1$ is in W. Therefore W is a subspace of $\mathcal{R}^m$.

97. By changing $\mathcal{R}^n$ to V, we can use the same argument as in Exercise 96.

98. Let $V = \{\mathbf{u} : T(\mathbf{u}) \text{ is in } W\}$. Since $T(\mathbf{0}) = \mathbf{0}$ is in W, we see that $\mathbf{0}$ is in V. Let $\mathbf{u}_1$ and $\mathbf{u}_2$ be in V. Then $T(\mathbf{u}_1)$ and $T(\mathbf{u}_2)$ are in W, and hence $T(\mathbf{u}_1) + T(\mathbf{u}_2)$ is in W. Since $T(\mathbf{u}_1 + \mathbf{u}_2) = T(\mathbf{u}_1) + T(\mathbf{u}_2)$, it follows that $\mathbf{u}_1 + \mathbf{u}_2$ is in V. Finally, for any scalar c, $cT(\mathbf{u}_1)$ is in W. So $c\mathbf{u}_1$ is in V because $T(c\mathbf{u}_1) = cT(\mathbf{u}_1)$. Thus V is a subspace of $\mathcal{R}^n$.

99. Because $A\mathbf{0} = \mathbf{0} = B\mathbf{0}$, the zero vector is in V. Assume that $\mathbf{u}$ and $\mathbf{v}$ are in V. Then $A\mathbf{u} = B\mathbf{u}$ and $A\mathbf{v} = B\mathbf{v}$. Hence
$$A(\mathbf{u} + \mathbf{v}) = A\mathbf{u} + A\mathbf{v} = B\mathbf{u} + B\mathbf{v} = B(\mathbf{u} + \mathbf{v}).$$
Thus $\mathbf{u} + \mathbf{v}$ is in V, and so V is closed under vector addition. Also, for any scalar c,
$$A(c\mathbf{u}) = cA\mathbf{u} = cB\mathbf{u} = B(c\mathbf{u}).$$
Hence $c\mathbf{u}$ is in V, and V is closed under scalar multiplication. Since V is a subset of $\mathcal{R}^n$ that contains $\mathbf{0}$ and is closed under both vector addition and scalar multiplication, V is a subspace of $\mathcal{R}^n$.

100. Because $\mathbf{0}$ is in both V and W, $\mathbf{0} + \mathbf{0} = \mathbf{0}$ is in S. Let $\mathbf{s}_1$ and $\mathbf{s}_2$ be in S. Then $\mathbf{s}_1 = \mathbf{v}_1 + \mathbf{w}_1$ and $\mathbf{s}_2 = \mathbf{v}_2 + \mathbf{w}_2$ for some $\mathbf{v}_1$ and $\mathbf{v}_2$ in V and some $\mathbf{w}_1$ and $\mathbf{w}_2$ in W. Hence
$$\mathbf{s}_1 + \mathbf{s}_2 = (\mathbf{v}_1 + \mathbf{w}_1) + (\mathbf{v}_2 + \mathbf{w}_2)$$
$$= (\mathbf{v}_1 + \mathbf{v}_2) + (\mathbf{w}_1 + \mathbf{w}_2),$$
and $\mathbf{v}_1 + \mathbf{v}_2$ is in V and $\mathbf{w}_1 + \mathbf{w}_2$ is in W because V and W are closed under vector addition. Thus $\mathbf{s}_1 + \mathbf{s}_2$ is in S. Also, for any scalar c,
$$c\mathbf{s}_1 = c(\mathbf{v}_1 + \mathbf{w}_1) = c\mathbf{v}_1 + c\mathbf{w}_1,$$
and $c\mathbf{v}_1$ is in V and $c\mathbf{w}_1$ is in W because V and W are closed under scalar multiplication. Thus $c\mathbf{s}_1$ is in S. Therefore S is a subspace of $\mathcal{R}^n$.

101. (a) The system $A\mathbf{x} = \mathbf{u}$ is consistent since the reduced row echelon form of $[A \ \mathbf{u}]$ contains no row whose only nonzero entry lies in the last column. Hence $\mathbf{u}$ belongs to Col A.

(b) On the other hand, $A\mathbf{x} = \mathbf{v}$ is not consistent, and so $\mathbf{v}$ does not belong to Col A.

102. (a) no, because $A\mathbf{u} \neq \mathbf{0}$
(b) yes, because $A\mathbf{v} = \mathbf{0}$

103. (a) yes, because $A^T\mathbf{u} = \mathbf{0}$
(b) no, because $A^T\mathbf{v} \neq \mathbf{0}$

4.2 BASIS AND DIMENSION

1. (a) $\left\{\begin{bmatrix}1\\-1\end{bmatrix}\right\}$ (b) $\left\{\begin{bmatrix}3\\1\\0\\0\end{bmatrix}, \begin{bmatrix}-4\\0\\1\\0\end{bmatrix}, \begin{bmatrix}2\\0\\0\\1\end{bmatrix}\right\}$

2. (a) $\left\{\begin{bmatrix}1\\2\end{bmatrix}, \begin{bmatrix}0\\-1\end{bmatrix}\right\}$ (b) $\left\{\begin{bmatrix}2\\1\\1\\0\end{bmatrix}, \begin{bmatrix}-1\\2\\0\\1\end{bmatrix}\right\}$

3. (a) $\left\{\begin{bmatrix}1\\-1\\-1\end{bmatrix}, \begin{bmatrix}2\\-1\\0\end{bmatrix}\right\}$ (b) $\left\{\begin{bmatrix}2\\-3\\1\end{bmatrix}\right\}$

4. (a) $\left\{\begin{bmatrix}1\\-1\\2\end{bmatrix}\right\}$ (b) $\left\{\begin{bmatrix}-3\\1\\0\end{bmatrix}, \begin{bmatrix}2\\0\\1\end{bmatrix}\right\}$

5. (a) $\left\{\begin{bmatrix}1\\-1\\2\end{bmatrix}, \begin{bmatrix}0\\1\\3\end{bmatrix}\right\}$ (b) $\left\{\begin{bmatrix}2\\1\\0\\0\end{bmatrix}, \begin{bmatrix}-2\\0\\1\\1\end{bmatrix}\right\}$

6. (a) $\left\{\begin{bmatrix}1\\-1\\2\end{bmatrix}, \begin{bmatrix}1\\-2\\3\end{bmatrix}, \begin{bmatrix}-1\\1\\1\end{bmatrix}\right\}$ (b) $\left\{\begin{bmatrix}-2\\1\\-3\\1\end{bmatrix}\right\}$

7. The reduced row echelon form of the given matrix A is
$$R = \begin{bmatrix}1 & 0 & 0 & 4\\0 & 1 & 0 & 4\\0 & 0 & 1 & 1\\0 & 0 & 0 & 0\end{bmatrix}.$$

(a) Hence the first three columns of the given matrix are its pivot columns, and so
$$\left\{\begin{bmatrix}-1\\2\\1\\0\end{bmatrix}, \begin{bmatrix}1\\0\\-1\\1\end{bmatrix}, \begin{bmatrix}2\\-5\\-1\\-2\end{bmatrix}\right\}$$
is a basis for the column space of A.

(b) The null space of A is the solution space of $R\mathbf{x} = \mathbf{0}$. Since the vector form of the general solution of $R\mathbf{x} = \mathbf{0}$ is
$$\begin{bmatrix}x_1\\x_2\\x_3\\x_4\end{bmatrix} = x_4 \begin{bmatrix}-4\\-4\\-1\\1\end{bmatrix},$$
the set
$$\left\{\begin{bmatrix}-4\\-4\\-1\\1\end{bmatrix}\right\}$$
is a basis for the null space of A.

8. (a) $\left\{\begin{bmatrix}1\\3\\0\\2\end{bmatrix}, \begin{bmatrix}2\\5\\3\\1\end{bmatrix}\right\}$ (b) $\left\{\begin{bmatrix}1\\1\\0\\0\end{bmatrix}, \begin{bmatrix}-3\\0\\1\\1\end{bmatrix}\right\}$

9. (a) $\left\{\begin{bmatrix}1\\2\\1\end{bmatrix}, \begin{bmatrix}2\\3\\2\end{bmatrix}, \begin{bmatrix}1\\3\\4\end{bmatrix}\right\}$
(b) The null space of T is $\{\mathbf{0}\}$.

10. (a) $\left\{\begin{bmatrix}1\\1\\0\end{bmatrix}, \begin{bmatrix}2\\1\\1\end{bmatrix}\right\}$ (b) $\left\{\begin{bmatrix}-1\\1\\1\end{bmatrix}\right\}$

11. The standard matrix of T is
$$A = \begin{bmatrix}1 & -2 & 1 & 1\\2 & -5 & 1 & 3\\1 & -3 & 0 & 2\end{bmatrix}.$$

(a) The range of T equals the column space of A; so we proceed as in Exercise 7. The reduced row echelon form of A is
$$R = \begin{bmatrix}1 & 0 & 3 & -1\\0 & 1 & 1 & -1\\0 & 0 & 0 & 0\end{bmatrix}.$$

Hence the set of pivot columns of A,
$$\left\{\begin{bmatrix}1\\2\\1\end{bmatrix}, \begin{bmatrix}-2\\-5\\-3\end{bmatrix}\right\},$$
is a basis for the range of T.

(b) Since the null space of T is the same as the null space of A, we must determine the vector form of the general solution of $A\mathbf{x} = \mathbf{0}$. This representation is:
$$\begin{bmatrix}x_1\\x_2\\x_3\\x_4\end{bmatrix} = x_3\begin{bmatrix}-3\\-1\\1\\0\end{bmatrix} + x_4\begin{bmatrix}1\\1\\0\\1\end{bmatrix}.$$

Hence
$$\left\{\begin{bmatrix}-3\\-1\\1\\0\end{bmatrix}, \begin{bmatrix}1\\1\\0\\1\end{bmatrix}\right\}$$
is a basis for the null space of T.

12. (a) $\left\{\begin{bmatrix}1\\1\\-1\end{bmatrix}, \begin{bmatrix}2\\3\\1\end{bmatrix}, \begin{bmatrix}1\\2\\0\end{bmatrix}\right\}$ (b) $\left\{\begin{bmatrix}0\\1\\0\\0\end{bmatrix}\right\}$

13. (a) $\left\{\begin{bmatrix}1\\2\\0\\3\end{bmatrix}, \begin{bmatrix}1\\1\\0\\1\end{bmatrix}\right\}$ (b) $\left\{\begin{bmatrix}1\\-3\\1\\0\end{bmatrix}, \begin{bmatrix}-1\\2\\0\\1\end{bmatrix}\right\}$

82 Chapter 4 Subspaces and Their Properties

14. (a) $\left\{ \begin{bmatrix} -2 \\ 0 \\ 1 \\ 2 \end{bmatrix}, \begin{bmatrix} -1 \\ 0 \\ 2 \\ 3 \end{bmatrix} \right\}$ (b) $\left\{ \begin{bmatrix} 1 \\ -2 \\ 1 \\ 0 \end{bmatrix}, \begin{bmatrix} 2 \\ -3 \\ 0 \\ 1 \end{bmatrix} \right\}$

15. (a) $\left\{ \begin{bmatrix} 1 \\ 3 \\ 7 \end{bmatrix}, \begin{bmatrix} 2 \\ 1 \\ 4 \end{bmatrix} \right\}$

    (b) $\left\{ \begin{bmatrix} 1 \\ -2 \\ 1 \\ 0 \\ 0 \end{bmatrix}, \begin{bmatrix} 0 \\ 0 \\ 0 \\ 1 \\ 0 \end{bmatrix}, \begin{bmatrix} 2 \\ -3 \\ 0 \\ 0 \\ 1 \end{bmatrix} \right\}$

16. (a) $\left\{ \begin{bmatrix} -1 \\ 1 \\ 3 \\ 1 \end{bmatrix}, \begin{bmatrix} 1 \\ 1 \\ 1 \\ 2 \end{bmatrix} \right\}$

    (b) $\left\{ \begin{bmatrix} 1 \\ -3 \\ 1 \\ 0 \\ 0 \end{bmatrix}, \begin{bmatrix} 1 \\ -5 \\ 0 \\ 1 \\ 0 \end{bmatrix}, \begin{bmatrix} 3 \\ -6 \\ 0 \\ 0 \\ 1 \end{bmatrix} \right\}$

17. Since
$$\begin{bmatrix} s \\ -2s \end{bmatrix} = s \begin{bmatrix} 1 \\ -2 \end{bmatrix}$$
and $\left\{ \begin{bmatrix} 1 \\ -2 \end{bmatrix} \right\}$ is linearly independent, this set is a basis for the given subspace.

18. $\left\{ \begin{bmatrix} 2 \\ -1 \\ 1 \end{bmatrix}, \begin{bmatrix} 0 \\ 4 \\ -3 \end{bmatrix} \right\}$ 19. $\left\{ \begin{bmatrix} 5 \\ 2 \\ 0 \\ 0 \end{bmatrix}, \begin{bmatrix} -3 \\ 0 \\ 0 \\ -4 \end{bmatrix} \right\}$

20. $\left\{ \begin{bmatrix} 5 \\ 2 \\ 0 \\ 3 \end{bmatrix}, \begin{bmatrix} -3 \\ 6 \\ 4 \\ -1 \end{bmatrix}, \begin{bmatrix} 0 \\ 0 \\ -7 \\ 9 \end{bmatrix} \right\}$

21. $\left\{ \begin{bmatrix} 3 \\ 1 \\ 0 \end{bmatrix}, \begin{bmatrix} -5 \\ 0 \\ 1 \end{bmatrix} \right\}$ 22. $\left\{ \begin{bmatrix} 10 \\ 8 \\ 1 \end{bmatrix} \right\}$

23. $\left\{ \begin{bmatrix} 2 \\ 1 \\ 0 \\ 0 \end{bmatrix}, \begin{bmatrix} -3 \\ 0 \\ 1 \\ 0 \end{bmatrix}, \begin{bmatrix} 4 \\ 0 \\ 0 \\ 1 \end{bmatrix} \right\}$

24. $\left\{ \begin{bmatrix} -11 \\ -9 \\ 1 \\ 0 \end{bmatrix}, \begin{bmatrix} -4 \\ -3 \\ 0 \\ 1 \end{bmatrix} \right\}$

25. Let
$$A = \begin{bmatrix} 1 & 2 & 1 \\ 2 & 1 & -4 \\ 1 & 3 & 3 \end{bmatrix}.$$

Then the given subspace is Col A, and so a basis for the given subspace can be obtained by choosing the pivot columns of A. This basis is

$$\left\{ \begin{bmatrix} 1 \\ 2 \\ 1 \end{bmatrix}, \begin{bmatrix} 2 \\ 1 \\ 3 \end{bmatrix} \right\}.$$

26. $\left\{ \begin{bmatrix} 1 \\ 1 \\ -1 \end{bmatrix}, \begin{bmatrix} 1 \\ 2 \\ 0 \end{bmatrix} \right\}$

27. $\left\{ \begin{bmatrix} 1 \\ -1 \\ 3 \end{bmatrix}, \begin{bmatrix} 0 \\ -1 \\ 1 \end{bmatrix}, \begin{bmatrix} 1 \\ -2 \\ 0 \end{bmatrix} \right\}$

28. $\left\{ \begin{bmatrix} 2 \\ 3 \\ -5 \end{bmatrix}, \begin{bmatrix} 8 \\ -12 \\ 20 \end{bmatrix}, \begin{bmatrix} 1 \\ 0 \\ -2 \end{bmatrix} \right\}$

29. $\left\{ \begin{bmatrix} 1 \\ 0 \\ -1 \\ 2 \end{bmatrix}, \begin{bmatrix} 1 \\ 1 \\ -2 \\ 1 \end{bmatrix}, \begin{bmatrix} 0 \\ 1 \\ -1 \\ 2 \end{bmatrix} \right\}$

30. $\left\{ \begin{bmatrix} 0 \\ 2 \\ 3 \\ 1 \end{bmatrix}, \begin{bmatrix} 1 \\ 1 \\ 1 \\ 3 \end{bmatrix}, \begin{bmatrix} 1 \\ 0 \\ 1 \\ -1 \end{bmatrix} \right\}$

31. $\left\{ \begin{bmatrix} -2 \\ 4 \\ 5 \\ -1 \end{bmatrix}, \begin{bmatrix} 3 \\ -4 \\ -5 \\ 1 \end{bmatrix}, \begin{bmatrix} 1 \\ 5 \\ 4 \\ -2 \end{bmatrix} \right\}$

32. $\left\{ \begin{bmatrix} 1 \\ 3 \\ 3 \\ 1 \end{bmatrix}, \begin{bmatrix} 1 \\ -1 \\ -1 \\ 1 \end{bmatrix} \right\}$

33. False, every nonzero subspace of $\mathcal{R}^n$ has infinitely many bases.

34. True

35. False, a basis for a subspace is a generating set that is as *small* as possible.

36. True 37. True 38. True 39. True

40. False, the *pivot columns* of any matrix form a basis for its column space.

41. False, if $A = \begin{bmatrix} 1 & 2 \\ 1 & 2 \end{bmatrix}$, then the pivot columns of the reduced row echelon form of A do not form a basis for Col A.

42. True

43. False, every generating set for V contains *at least* k vectors.

44. True   45. True   46. True
47. True   48. True   49. True
50. False, neither standard vector is in the subspace $\left\{ \begin{bmatrix} u_1 \\ u_2 \end{bmatrix} \in \mathcal{R}^2 : u_1 + u_2 = 0 \right\}$.
51. True   52. True
53. A generating set for $\mathcal{R}^4$ must contain at least 4 vectors.
54. No linearly independent subset of $\mathcal{R}^3$ can contain more than 3 vectors.
55. A basis for $\mathcal{R}^3$ must contain exactly 3 vectors.
56. A generating set for $\mathcal{R}^3$ must contain at least 3 vectors.
57. No linearly independent subset of $\mathcal{R}^2$ can contain more than 2 vectors.
58. A basis for $\mathcal{R}^2$ must contain exactly 2 vectors.
59. The dimension of the subspace V in Exercise 21 is 2. Since the set given in Exercise 59 is a linearly independent subset of V containing 2 vectors, it is a basis for V by Theorem 4.7.
60. The dimension of the subspace V in Exercise 24 is 2. Since the set given in Exercise 60 is a linearly independent subset of V containing 2 vectors, it is a basis for V by Theorem 4.7.
61. The dimension of the subspace V in Exercise 29 is 3. Since the set given in Exercise 61 is a linearly independent subset of V containing 3 vectors, it is a basis for V.
62. The dimension of the subspace V in Exercise 30 is 3. Since the set given in Exercise 62 is a linearly independent subset of V containing 3 vectors, it is a basis for V.
63. The dimension of the null space of the matrix in Exercise 5 is 2. Since the set given in Exercise 63 is a linearly independent subset of this null space containing 2 vectors, it is a basis for the null space.
64. The dimension of the null space of the matrix in Exercise 8 is 2. Since the set given in Exercise 64 is a linearly independent subset of this null space containing 2 vectors, it is a basis for the null space.
65. Let S denote the set of vectors given in Exercise 65 and A denote the matrix given in Exercise 7. For each $\mathbf{v}$ in S, the matrix equation $A\mathbf{x} = \mathbf{v}$ is consistent; so S is contained in the subspace V. Moreover, S is linearly independent. Because the rank of A equals 3, there are 3 pivot columns of A, and so, by Theorem 2.4, the dimension of V is 3. Hence S is a basis for V by Theorem 4.7.
66. The argument is similar to that of the preceding exercise except that the rank of the matrix (and hence the dimension of the subspace) is 2.
67. If $\mathbf{v} \neq \mathbf{0}$, then $\{\mathbf{v}\}$ is a basis for Span $\{\mathbf{v}\}$. Hence the dimension of Span $\{\mathbf{v}\}$ is 1.
68. The given subspace has $\{\mathbf{e}_2, \mathbf{e}_3, \ldots, \mathbf{e}_n\}$ as a basis; so its dimension is $n - 1$.
69. The given subspace has $\{\mathbf{e}_3, \mathbf{e}_4, \ldots, \mathbf{e}_n\}$ as a basis; so its dimension is $n - 2$.
70. The set
$$\left\{ \begin{bmatrix} -1 \\ 1 \\ 0 \\ 0 \\ \vdots \\ 0 \\ 0 \end{bmatrix}, \begin{bmatrix} -1 \\ 0 \\ 1 \\ 0 \\ \vdots \\ 0 \\ 0 \end{bmatrix}, \ldots, \begin{bmatrix} -1 \\ 0 \\ 0 \\ 0 \\ \vdots \\ 0 \\ 1 \end{bmatrix} \right\}$$
is a basis for the given subspace. Thus its dimension is $n - 1$.
71. By Exercise 89 in Section 1.7, $\mathcal{B}$ is a linearly independent subset of V. Since it contains exactly k vectors, it is a basis for V.
72. It follows from Exercise 98 in Section 1.7 that $\mathcal{B}$ is a linearly independent subset of V. Since it contains exactly k vectors, it is a basis for V.
73. Clearly $\mathbf{v}$ belongs to $V = \operatorname{Span} \mathcal{A}$. Thus $\mathcal{B} = \{\mathbf{v}, \mathbf{u}_2, \mathbf{u}_3, \ldots, \mathbf{u}_k\}$ is a subset of V, because $\mathbf{u}_2, \mathbf{u}_3, \ldots, \mathbf{u}_k$ belong to $\mathcal{A}$, which is a subset of V.

We claim that $\mathcal{B}$ is linearly independent. Suppose that $c_1, c_2, \ldots, c_k$ are scalars such that
$$c_1 \mathbf{v} + c_2 \mathbf{u}_2 + \cdots + c_k \mathbf{u}_k = \mathbf{0}.$$
Then
$$c_1(\mathbf{u}_1 + \mathbf{u}_2 + \cdots + \mathbf{u}_k) + c_2 \mathbf{u}_2 + \cdots + c_k \mathbf{u}_k = \mathbf{0}$$
$$c_1 \mathbf{u}_1 + (c_1 + c_2)\mathbf{u}_2 + \cdots + (c_1 + c_k)\mathbf{u}_k = \mathbf{0}.$$
Since $\mathcal{A}$ is linearly independent, it follows that $c_1 = 0, c_1 + c_2 = 0, \cdots, c_1 + c_k = 0$. Hence $c_1 = c_2 = \cdots = c_k = 0$, proving that $\mathcal{B}$ is linearly independent.

Note that the vectors $\mathbf{u}_2, \mathbf{u}_3, \ldots, \mathbf{u}_k$ are distinct because $\mathcal{A}$ contains k vectors. Suppose that $\mathbf{v} = \mathbf{u}_j$ for some $j = 2, 3, \ldots, k$. Then
$$\mathbf{u}_1 + \mathbf{u}_2 + \cdots + \mathbf{u}_k = \mathbf{u}_j.$$

Subtracting $\mathbf{u}_j$ from both sides of this equation gives a linear combination of the vectors in $\mathcal{A}$ equal to $\mathbf{0}$. Since not all of the coefficients in this linear combination are 0, we have contradicted that $\mathcal{A}$ is linearly independent. Thus the vectors $\mathbf{v}, \mathbf{u}_2, \ldots, \mathbf{u}_k$ are distinct, and so $\mathcal{B}$ contains k vectors. It follows from Theorem 4.7 that $\mathcal{B}$ is a basis for V.

74. Set $\mathcal{B}$ is a linearly independent subset of V containing exactly k elements.

75. (a) Let $\mathbf{w}$ be in the range of T. Then $T(\mathbf{v}) = \mathbf{w}$ for some $\mathbf{v}$ in $\mathcal{R}^n$, and $\mathbf{v}$ can be written as $\mathbf{v} = c_1\mathbf{u}_1 + c_2\mathbf{u}_2 + \cdots + c_n\mathbf{u}_n$ for some scalars $c_1, c_2, \ldots, c_n$. Thus

$$\mathbf{w} = T(\mathbf{v})$$
$$= c_1 T(\mathbf{u}_1) + c_2 T(\mathbf{u}_2) + \cdots + c_n T(\mathbf{u}_n),$$

proving that $\mathcal{S}$ is a generating set for the range of T.

(b) If T is the zero transformation, then $\mathcal{S} = \{\mathbf{0}\}$ is not a basis for the range of T (because it is not linearly independent).

76. (a) The set $\{T(\mathbf{u}_1), T(\mathbf{u}_2), \ldots, T(\mathbf{u}_k)\}$ is a generating set for W by Exercise 75(a), and it is a linearly independent subset of W by Exercise 97 in Section 1.7.

(b) Both V and W have bases containing exactly k vectors, so $\dim V = \dim W = k$.

77. (a) Because V and W are subspaces of $\mathcal{R}^n$, $\mathbf{0}$ is in both V and W. Suppose that $\mathbf{u}$ is in both V and W. Then $\mathbf{u} = \mathbf{v}_1 + \mathbf{w}_1$, where $\mathbf{v}_1 = \mathbf{u}$ and $\mathbf{w}_1 = \mathbf{0}$, and also $\mathbf{u} = \mathbf{v}_2 + \mathbf{w}_2$, where $\mathbf{v}_2 = \mathbf{0}$ and $\mathbf{w}_2 = \mathbf{u}$. It follows from the uniqueness of the representation of $\mathbf{u}$ in the form $\mathbf{v} + \mathbf{w}$ for some $\mathbf{v}$ in V and some $\mathbf{w}$ in W that $\mathbf{v}_1 = \mathbf{v}_2 = \mathbf{0}$ and $\mathbf{w}_2 = \mathbf{w}_1 = \mathbf{0}$. Hence $\mathbf{u} = \mathbf{0} + \mathbf{0} = \mathbf{0}$, and so $\mathbf{0}$ is the only vector in both V and W.

(b) Let $\mathcal{B}_1 = \{\mathbf{v}_1, \mathbf{v}_2, \ldots, \mathbf{v}_k\}$ be a basis for V and $\mathcal{B}_2 = \{\mathbf{w}_1, \mathbf{w}_2, \ldots, \mathbf{w}_m\}$ be a basis for W. Thus $\dim V = k$ and $\dim W = m$. Let $\mathcal{B} = \{\mathbf{v}_1, \mathbf{v}_2, \ldots, \mathbf{v}_k, \mathbf{w}_1, \mathbf{w}_2, \ldots, \mathbf{w}_m\}$. We will show that $\mathcal{B}$ is a basis for $\mathcal{R}^n$, so that, by Theorem 4.5, the number of vectors in $\mathcal{B}$ must be n, that is,

$$\dim V + \dim W = k + m = n.$$

First we show that $\mathcal{B}$ is linearly independent. Let $a_1, a_2, \ldots, a_k, b_1, b_2, \ldots, b_m$ be scalars such that

$$a_1\mathbf{v}_1 + a_2\mathbf{v}_2 + \cdots + a_k\mathbf{v}_k$$
$$+ b_1\mathbf{w}_1 + b_2\mathbf{w}_2 + \cdots + b_m\mathbf{w}_m = \mathbf{0}.$$

Define $\mathbf{v} = a_1\mathbf{v}_1 + a_2\mathbf{v}_2 + \cdots + a_k\mathbf{v}_k$ and $\mathbf{w} = b_1\mathbf{w}_1 + b_2\mathbf{w}_2 + \cdots + b_m\mathbf{w}_m$. Then $\mathbf{v} = -\mathbf{w}$. Because $\mathbf{0}$ is the only vector in both V and W, we have $\mathbf{v} = \mathbf{0}$ and $\mathbf{w} = \mathbf{0}$. But if $\mathbf{v} = a_1\mathbf{v}_1 + a_2\mathbf{v}_2 + \cdots + a_k\mathbf{v}_k = \mathbf{0}$, then $a_1 = a_2 = \ldots = a_k = 0$ because $\mathcal{B}_1$ is linearly independent. Similarly, $b_1 = b_2 = \ldots = b_m = 0$. Thus $\mathcal{B}$ is linearly independent.

Next, we show that $\mathcal{B}$ is a generating set for $\mathcal{R}^n$. For any vector $\mathbf{u}$ in $\mathcal{R}^n$, there exist $\mathbf{v}$ in V and $\mathbf{w}$ in W with $\mathbf{u} = \mathbf{v} + \mathbf{w}$. Because $\mathcal{B}_1$ is a basis for V, there exist scalars $a_1, a_2, \ldots, a_k$ such that $\mathbf{v} = a_1\mathbf{v}_1 + a_2\mathbf{v}_2 + \cdots + a_k\mathbf{v}_k$. Similarly, there exist scalars $b_1, b_2, \ldots, b_m$ such that $\mathbf{w} = b_1\mathbf{w}_1 + b_2\mathbf{w}_2 + \cdots + b_m\mathbf{w}_m$. Hence $\mathbf{u} = \mathbf{v} + \mathbf{w}$ is a linear combination of the vectors in $\mathcal{B}$, and so $\mathcal{B}$ is a generating set for $\mathcal{R}^n$. Because $\mathcal{B}$ is a linearly independent generating set for $\mathcal{R}^n$, $\mathcal{B}$ is a basis for $\mathcal{R}^n$, completing the proof.

78. The columns of the given matrix form a generating set for V, because $\mathcal{S}$ is a generating set for V and $\mathcal{L}$ is a subset of V. So the pivot columns of this matrix must be a basis for V. Necessarily this basis includes $\mathcal{L}$ because $\mathcal{L}$ is linearly independent.

79. $\left\{ \begin{bmatrix} 2 \\ 3 \\ 0 \end{bmatrix}, \begin{bmatrix} 1 \\ 0 \\ 0 \end{bmatrix}, \begin{bmatrix} 0 \\ 0 \\ 1 \end{bmatrix} \right\}$

80. $\left\{ \begin{bmatrix} -1 \\ -1 \\ 6 \\ -7 \end{bmatrix}, \begin{bmatrix} 5 \\ -9 \\ -2 \\ -1 \end{bmatrix}, \begin{bmatrix} 1 \\ -2 \\ 0 \\ 1 \end{bmatrix} \right\}$

81. Let
$$A = \begin{bmatrix} 1 & -1 & 2 & 1 \\ 2 & -2 & 4 & 2 \\ -3 & 3 & -6 & -3 \end{bmatrix}.$$

Since the reduced row echelon form of A is

$$\begin{bmatrix} 1 & -1 & 2 & 1 \\ 0 & 0 & 0 & 0 \\ 0 & 0 & 0 & 0 \end{bmatrix},$$

the vector form of the general solution of $A\mathbf{x} = \mathbf{0}$ is

$$\begin{bmatrix} x_1 \\ x_2 \\ x_3 \\ x_4 \end{bmatrix} = x_2 \begin{bmatrix} 1 \\ 1 \\ 0 \\ 0 \end{bmatrix} + x_3 \begin{bmatrix} -2 \\ 0 \\ 1 \\ 0 \end{bmatrix} + x_4 \begin{bmatrix} -1 \\ 0 \\ 0 \\ 1 \end{bmatrix}.$$

Hence
$$\left\{ \begin{bmatrix} 1 \\ 1 \\ 0 \\ 0 \end{bmatrix}, \begin{bmatrix} -2 \\ 0 \\ 1 \\ 0 \end{bmatrix}, \begin{bmatrix} -1 \\ 0 \\ 0 \\ 1 \end{bmatrix} \right\}$$
is a basis for Null A.

Since the reduced row echelon form of
$$\begin{bmatrix} 0 & 1 & -2 & -1 \\ 2 & 1 & 0 & 0 \\ 1 & 0 & 1 & 0 \\ 0 & 0 & 0 & 1 \end{bmatrix} \text{ is } \begin{bmatrix} 1 & 0 & 1 & 0 \\ 0 & 1 & -2 & 0 \\ 0 & 0 & 0 & 1 \\ 0 & 0 & 0 & 0 \end{bmatrix},$$
it follows from Exercise 78 that
$$\left\{ \begin{bmatrix} 0 \\ 2 \\ 1 \\ 0 \end{bmatrix}, \begin{bmatrix} 1 \\ 1 \\ 0 \\ 0 \end{bmatrix}, \begin{bmatrix} -1 \\ 0 \\ 0 \\ 1 \end{bmatrix} \right\}$$
is a basis for Null A that contains $\mathcal{L}$.

82. $\left\{ \begin{bmatrix} 0 \\ 0 \\ 1 \\ 0 \end{bmatrix}, \begin{bmatrix} 1 \\ -1 \\ -3 \\ 2 \end{bmatrix}, \begin{bmatrix} 1 \\ 2 \\ -1 \\ 1 \end{bmatrix} \right\}$

83. (a) The reduced row echelon form of the matrix whose columns are the vectors in $\mathcal{S}$ is I_3; so $\mathcal{S}$ is linearly independent.

    (b) Simplification of the left side of the given equation gives
    $$\begin{bmatrix} v_1 \\ v_2 \\ -v_1 + v_2 \end{bmatrix} = \begin{bmatrix} v_1 \\ v_2 \\ v_3 \end{bmatrix},$$
    because $v_1 - v_2 + v_3 = 0$.

    (c) Because $\mathcal{S}$ is not a subset of V, it is not a basis for V.

84. (a) The reduced row echelon form of the matrix whose columns are the vectors in $\mathcal{S}$ is I_3; so $\mathcal{S}$ is linearly independent.

    (b) Simplification of the left side of the given equation gives
    $$\begin{bmatrix} v_1 \\ v_2 \\ v_3 \\ 3v_1 - v_3 \end{bmatrix} = \begin{bmatrix} v_1 \\ v_2 \\ v_3 \\ v_4 \end{bmatrix},$$
    because $3v_1 - v_3 = 0 = v_4$.

    (c) Because $\mathcal{S}$ is not a subset of V, it is not a basis for V.

85. The reduced row echelon form of A is
$$\begin{bmatrix} 1 & 0 & -1.2 & 0 & 1.4 \\ 0 & 1 & 2.3 & 0 & -2.9 \\ 0 & 0 & 0.0 & 1 & 0.7 \end{bmatrix}.$$

    (a) As in Exercise 7,
    $$\left\{ \begin{bmatrix} 0.1 \\ 0.7 \\ -0.5 \end{bmatrix}, \begin{bmatrix} 0.2 \\ 0.9 \\ 0.5 \end{bmatrix}, \begin{bmatrix} 0.5 \\ -0.5 \\ -0.5 \end{bmatrix} \right\}$$
    is a basis for the column space of A.

    (b) The vector form of the general solution of $A\mathbf{x} = \mathbf{0}$ is
    $$\begin{bmatrix} x_1 \\ x_2 \\ x_3 \\ x_4 \\ x_5 \end{bmatrix} = x_3 \begin{bmatrix} 1.2 \\ -2.3 \\ 1.0 \\ 0.0 \\ 0.0 \end{bmatrix} + x_5 \begin{bmatrix} -1.4 \\ 2.9 \\ 0.0 \\ -0.7 \\ 1.0 \end{bmatrix}.$$

    Hence
    $$\left\{ \begin{bmatrix} 1.2 \\ -2.3 \\ 1.0 \\ 0.0 \\ 0.0 \end{bmatrix}, \begin{bmatrix} -1.4 \\ 2.9 \\ 0.0 \\ -0.7 \\ 1.0 \end{bmatrix} \right\}$$
    is a basis for Null A.

86. By Exercise 85(b), the dimension of Null A is 2. Since the given set is a linearly independent subset of Null A containing exactly two vectors, it is a basis for Null A.

87. By Exercise 85(a), the dimension of Col A is 3. Since the given set is a linearly independent subset of Col A containing exactly three vectors, it is a basis for Col A.

88. (a) We have rank $A = 3$, dim (Col A) = 3, and dim (Row A) = 3.

    (b) All three numbers are equal.

4.3 THE DIMENSION OF SUBSPACES ASSOCIATED WITH A MATRIX

1. (a) The dimension of Col A equals the rank of A, which is 2.

   (b) The dimension of Null A equals the nullity of A, which is $4 - 2 = 2$.

   (c) The dimension of Row A equals the rank of A, which is 2.

   (d) The dimension of Null A^T equals the nullity of A^T. Since A^T is a 4×3 matrix, the nullity of A^T equals
   $$3 - \text{rank } A^T = 3 - \text{rank } A = 3 - 2 = 1.$$

86 Chapter 4 Subspaces and Their Properties

2. (a) 3 (b) 1 (c) 3 (d) 0
3. (a) 3 (b) 2 (c) 3 (d) 0
4. (a) 3 (b) 2 (c) 3 (d) 1
5. (a) 1 (b) 3 (c) 1 (d) 0
6. (a) 2 (b) 1 (c) 2 (d) 1
7. (a) 2 (b) 1 (c) 2 (d) 0
8. (a) 1 (b) 2 (c) 1 (d) 1
9. The reduced row echelon form of A is
$$\begin{bmatrix} 1 & 0 & 6 & 0 \\ 0 & 1 & -4 & 1 \\ 0 & 0 & 0 & 0 \end{bmatrix}.$$
Hence rank $A = 2$. As in Exercise 1, the answers are:
(a) 2 (b) 2 (c) 2 (d) 1.
10. (a) 2 (b) 3 (c) 2 (d) 1
11. (a) 2 (b) 1 (c) 2 (d) 2
12. (a) 3 (b) 0 (c) 3 (d) 1
13. A basis for this subspace is
$$\left\{ \begin{bmatrix} -2 \\ 1 \end{bmatrix} \right\}.$$
Thus the dimension of the subspace is 1.
14. A basis for this subspace is
$$\left\{ \begin{bmatrix} 1 \\ 0 \\ 2 \end{bmatrix} \right\}.$$
Thus the dimension of the subspace is 1.
15. The vectors in the given subspace have the form
$$\begin{bmatrix} -3s + 4t \\ s - 2t \\ 2s \end{bmatrix} = s \begin{bmatrix} -3 \\ 1 \\ 2 \end{bmatrix} + t \begin{bmatrix} 4 \\ -2 \\ 0 \end{bmatrix}.$$
So a generating set for the given subspace is
$$\mathcal{B} = \left\{ \begin{bmatrix} -3 \\ 1 \\ 2 \end{bmatrix}, \begin{bmatrix} 4 \\ -2 \\ 0 \end{bmatrix} \right\}.$$
Since neither vector in $\mathcal{B}$ is a multiple of the other, $\mathcal{B}$ is also linearly independent. Therefore $\mathcal{B}$ is a basis for the given subspace. It follows that $\dim V = 2$.
16. 2
17. By Theorem 4.8, $\left\{ \begin{bmatrix} 1 \\ 0 \\ 3 \end{bmatrix}, \begin{bmatrix} 0 \\ 1 \\ 2 \end{bmatrix} \right\}$ is a basis for Row A.

18. $\left\{ \begin{bmatrix} 1 \\ -1 \\ 0 \\ -2 \end{bmatrix}, \begin{bmatrix} 0 \\ 0 \\ 1 \\ 3 \end{bmatrix} \right\}$ 19. $\left\{ \begin{bmatrix} 1 \\ 0 \\ 0 \\ 1 \end{bmatrix}, \begin{bmatrix} 0 \\ 1 \\ 1 \\ -1 \end{bmatrix} \right\}$

20. $\left\{ \begin{bmatrix} 1 \\ -2 \\ 0 \\ 2 \\ 3 \end{bmatrix}, \begin{bmatrix} 0 \\ 0 \\ 1 \\ -3 \\ -5 \end{bmatrix} \right\}$

21. $\left\{ \begin{bmatrix} 1 \\ 0 \\ 0 \\ -3 \\ 1 \\ 3 \end{bmatrix}, \begin{bmatrix} 0 \\ 1 \\ 0 \\ 2 \\ -1 \\ -2 \end{bmatrix}, \begin{bmatrix} 0 \\ 0 \\ 1 \\ 0 \\ 0 \\ -1 \end{bmatrix} \right\}$

22. $\left\{ \begin{bmatrix} 1 \\ 0 \\ 0 \\ -3 \\ 1 \\ 3 \end{bmatrix}, \begin{bmatrix} 0 \\ 1 \\ 0 \\ 2 \\ -1 \\ -2 \end{bmatrix}, \begin{bmatrix} 0 \\ 0 \\ 1 \\ 0 \\ 0 \\ -1 \end{bmatrix} \right\}$

23. $\left\{ \begin{bmatrix} 1 \\ 0 \\ 0 \\ 1 \\ 0 \end{bmatrix}, \begin{bmatrix} 0 \\ 1 \\ 0 \\ -1 \\ 0 \end{bmatrix}, \begin{bmatrix} 0 \\ 0 \\ 1 \\ 0 \\ 0 \end{bmatrix}, \begin{bmatrix} 0 \\ 0 \\ 0 \\ 0 \\ 1 \end{bmatrix} \right\}$

24. $\left\{ \begin{bmatrix} 1 \\ 0 \\ 0 \\ -3 \\ 1 \\ 3 \end{bmatrix}, \begin{bmatrix} 0 \\ 1 \\ 0 \\ 2 \\ -1 \\ -2 \end{bmatrix}, \begin{bmatrix} 0 \\ 0 \\ 1 \\ 0 \\ 0 \\ -1 \end{bmatrix} \right\}$

25. $\left\{ \begin{bmatrix} 1 \\ -1 \\ 1 \end{bmatrix}, \begin{bmatrix} 0 \\ 1 \\ 2 \end{bmatrix} \right\}$ 26. $\left\{ \begin{bmatrix} 1 \\ -1 \\ 0 \\ -2 \end{bmatrix}, \begin{bmatrix} 1 \\ -1 \\ 2 \\ 4 \end{bmatrix} \right\}$

27. $\left\{ \begin{bmatrix} -1 \\ 1 \\ 1 \\ -2 \end{bmatrix}, \begin{bmatrix} 2 \\ -1 \\ -1 \\ 3 \end{bmatrix} \right\}$ 28. $\left\{ \begin{bmatrix} 1 \\ -2 \\ 1 \\ -1 \\ -2 \end{bmatrix}, \begin{bmatrix} 2 \\ -4 \\ 1 \\ 1 \\ 1 \end{bmatrix} \right\}$

29. $\left\{ \begin{bmatrix} 1 \\ 0 \\ -1 \\ -3 \\ 1 \\ 4 \end{bmatrix}, \begin{bmatrix} 2 \\ -1 \\ -1 \\ -8 \\ 3 \\ 9 \end{bmatrix}, \begin{bmatrix} 0 \\ 1 \\ 1 \\ 2 \\ -1 \\ -3 \end{bmatrix} \right\}$

4.3 The Dimension of Subspaces Associated with a Matrix

30. $\left\{ \begin{bmatrix} -1 \\ 1 \\ 1 \\ 5 \\ -2 \\ -6 \end{bmatrix}, \begin{bmatrix} 2 \\ -1 \\ -1 \\ -8 \\ 3 \\ 9 \end{bmatrix}, \begin{bmatrix} 0 \\ 1 \\ -1 \\ 2 \\ -1 \\ -1 \end{bmatrix} \right\}$

31. $\left\{ \begin{bmatrix} 1 \\ 0 \\ -1 \\ 1 \\ 3 \end{bmatrix}, \begin{bmatrix} 2 \\ -1 \\ -1 \\ 3 \\ -8 \end{bmatrix}, \begin{bmatrix} 0 \\ 1 \\ -1 \\ -1 \\ 2 \end{bmatrix}, \begin{bmatrix} -1 \\ 1 \\ 1 \\ -2 \\ 5 \end{bmatrix} \right\}$

32. $\left\{ \begin{bmatrix} 1 \\ 0 \\ -1 \\ -3 \\ 1 \\ 4 \end{bmatrix}, \begin{bmatrix} 2 \\ -1 \\ -1 \\ -8 \\ 3 \\ 9 \end{bmatrix}, \begin{bmatrix} 0 \\ 1 \\ 1 \\ 2 \\ -1 \\ -3 \end{bmatrix} \right\}$

33. The standard matrix of T is $A = \begin{bmatrix} 1 & 2 \\ 2 & 1 \end{bmatrix}$, and its reduced row echelon form is I_2.

    (a) Since the range of T equals the column space of A, the dimension of the range of T equals the rank of A, which is 2. Thus, by Theorem 2.10, T is onto.

    (b) The null space of T equals the null space of A. Hence the dimension of the null space of T equals the nullity of A, which is 0. Thus T is one-to-one by Theorem 2.11.

34. (a) 1   (b) 1   neither one-to-one nor onto
35. (a) 1   (b) 2   neither one-to-one nor onto
36. (a) 2   (b) 1   onto, not one-to-one
37. The standard matrix of T is
    $$\begin{bmatrix} 1 & 0 \\ 2 & 1 \\ 0 & -1 \end{bmatrix},$$
    and its reduced row echelon form is
    $$\begin{bmatrix} 1 & 0 \\ 0 & 1 \\ 0 & 0 \end{bmatrix}.$$

    (a) As in Exercise 33, the dimension of the range of T is 2. Since the codomain of T is $\mathcal{R}^3$, T is not onto.

    (b) As in Exercise 33, the dimension of the null space of T is 0. Hence T is one-to-one.

38. (a) 2   (b) 0   one-to-one, not onto
39. (a) 2   (b) 1   onto, not one-to-one
40. (a) 3   (b) 0   both one-to-one and onto

41. False, the dimensions of the subspaces $V = \text{Span } \{\mathbf{e}_1\}$ and $W = \text{Span } \{\mathbf{e}_2\}$ of $\mathcal{R}^2$ are both 1, but $V \neq W$.
42. True   43. True
44. False, the dimension of the null space of a matrix equals the *nullity* of the matrix.
45. False, the dimension of the column space of a matrix equals the *rank* of the matrix.
46. True   47. True
48. False, consider $A = \begin{bmatrix} 1 & 2 \\ 1 & 2 \end{bmatrix}$ and the reduced row echelon form of A, which is $\begin{bmatrix} 1 & 2 \\ 0 & 0 \end{bmatrix}$.
49. True
50. False, the nonzero rows of *the reduced row echelon form* of a matrix form a basis for its row space.
51. False, consider $A = \begin{bmatrix} 1 & 0 & 0 \\ 0 & 0 & 0 \\ 0 & 1 & 0 \end{bmatrix}$.
52. False, consider $A = \begin{bmatrix} 1 & 0 & 0 \\ 0 & 1 & 0 \end{bmatrix}$.
53. True   54. True
55. False, consider any nonsquare matrix.
56. False, the dimension of the null space of any $m \times n$ matrix A plus the dimension of its column space equals n.
57. True   58. True   59. True   60. True
61. We have
    $$\begin{bmatrix} 2s_1 - t_1 \\ s_1 + 3t_1 \end{bmatrix} = \begin{bmatrix} 1 \\ 0 \end{bmatrix} \quad \text{and} \quad \begin{bmatrix} 2s_2 - t_2 \\ s_2 + 3t_2 \end{bmatrix} = \begin{bmatrix} 0 \\ 1 \end{bmatrix}$$
    for $s_1 = \frac{3}{7}$, $t_1 = -\frac{1}{7}$, $s_2 = \frac{1}{7}$, and $t_2 = \frac{2}{7}$. Hence $\mathcal{B}$ is contained in V, which has dimension 2. Since $\mathcal{B}$ is linearly independent, it is a basis for V by Theorem 4.7.

62. Imitate the argument in Exercise 61.

63. Note that $V = \text{Span} \left\{ \begin{bmatrix} 0 \\ 1 \\ -3 \end{bmatrix}, \begin{bmatrix} 4 \\ 1 \\ 1 \end{bmatrix} \right\}$ is a 2-dimensional subspace of $\mathcal{R}^3$. Since
    $$\frac{1}{4}\begin{bmatrix} 0 \\ 1 \\ -3 \end{bmatrix} + \frac{3}{4}\begin{bmatrix} 4 \\ 1 \\ 1 \end{bmatrix} = \begin{bmatrix} 3 \\ 1 \\ 0 \end{bmatrix}$$
    and
    $$\frac{1}{2}\begin{bmatrix} 0 \\ 1 \\ -3 \end{bmatrix} + \frac{1}{2}\begin{bmatrix} 4 \\ 1 \\ 1 \end{bmatrix} = \begin{bmatrix} 2 \\ 1 \\ -1 \end{bmatrix},$$

$\mathcal{B}$ is contained in V. Moreover, $\mathcal{B}$ is linearly independent because neither vector in $\mathcal{B}$ is a multiple of the other. Thus $\mathcal{B}$ is a basis for V by Theorem 4.7.

64. Imitate the argument in Exercise 63.
65. Imitate the argument in Exercise 63.
66. Imitate the argument in Exercise 63.
67. Imitate the argument in Exercise 63.
68. Imitate the argument in Exercise 63.
69. (a) Refer to the solution to Exercise 9. By Theorem 4.8,
$$\left\{ \begin{bmatrix} 1 \\ 0 \\ 6 \\ 0 \end{bmatrix}, \begin{bmatrix} 0 \\ 1 \\ -4 \\ 1 \end{bmatrix} \right\}$$
is a basis for Row A. Also, the vector form of the general solution of $A\mathbf{x} = \mathbf{0}$ is
$$\begin{bmatrix} x_1 \\ x_2 \\ x_3 \\ x_4 \end{bmatrix} = x_3 \begin{bmatrix} -6 \\ 4 \\ 1 \\ 0 \end{bmatrix} + x_4 \begin{bmatrix} 0 \\ -1 \\ 0 \\ 1 \end{bmatrix}.$$
Thus
$$\left\{ \begin{bmatrix} -6 \\ 4 \\ 1 \\ 0 \end{bmatrix}, \begin{bmatrix} 0 \\ -1 \\ 0 \\ 1 \end{bmatrix} \right\}$$
is a basis for Null A.

(b) The set
$$\left\{ \begin{bmatrix} 1 \\ 0 \\ 6 \\ 0 \end{bmatrix}, \begin{bmatrix} 0 \\ 1 \\ -4 \\ 1 \end{bmatrix}, \begin{bmatrix} -6 \\ 4 \\ 1 \\ 0 \end{bmatrix}, \begin{bmatrix} 0 \\ -1 \\ 0 \\ 1 \end{bmatrix} \right\}$$
is a linearly independent subset of $\mathcal{R}^4$. Since it contains 4 vectors, it is a basis for $\mathcal{R}^4$ by Theorem 4.7.

70. (a) $\left\{ \begin{bmatrix} 1 \\ 0 \\ 3 \end{bmatrix}, \begin{bmatrix} 0 \\ 1 \\ -2 \end{bmatrix} \right\}, \left\{ \begin{bmatrix} -3 \\ 2 \\ 1 \end{bmatrix} \right\}$

(b) The union of the two bases in (a) is a linearly independent subset of $\mathcal{R}^3$ containing three vectors.

71. Let A be the standard matrix of T. The null space of T equals Null A, whose dimension is the nullity of A. The range of T equals Col A, whose dimension is the rank of A. Hence the sum of the dimensions of the null space and range of T is nullity A + rank $A = n$.

72. No, their dimensions cannot be equal and have a sum of 3.
73. Let $\mathbf{v}$ be in the column space of AB. Then $\mathbf{v} = (AB)\mathbf{u}$ for some $\mathbf{u}$ in $\mathcal{R}^p$. Consider $\mathbf{w} = B\mathbf{u}$. Since $A\mathbf{w} = A(B\mathbf{u}) = (AB)\mathbf{u} = \mathbf{v}$, $\mathbf{v}$ is in the column space of A.
74. Let $\mathbf{u}$ be in the null space of B. Then
$$(AB)\mathbf{u} = A(B\mathbf{u}) = A\mathbf{0} = \mathbf{0}.$$
So $\mathbf{u}$ is in the null space of AB.
75. Since the dimension of the column space of a matrix equals its rank, it follows from Exercise 73 and Theorem 4.9 that rank $AB \leq$ rank A.
76. Note that both A and AB are $m \times n$ matrices. Since rank $AB \leq$ rank A by Exercise 75, it follows that
$$n - \text{rank } A \leq n - \text{rank } AB.$$
So nullity $A \leq$ nullity AB.
77. By Exercise 75, rank $AB =$ rank $(AB)^T =$ rank $B^T A^T \leq$ rank $B^T =$ rank B.
78. By Exercise 77,
$$\text{nullity } B = p - \text{rank } B$$
$$\leq p - \text{rank } AB = \text{nullity } AB.$$
79. Let $V =$ Span $\{\mathbf{e}_1, \mathbf{e}_2\}$ and $W =$ Span $\{\mathbf{e}_4, \mathbf{e}_5\}$.
80. Let $\mathcal{A} = \{\mathbf{v}_1, \mathbf{v}_2, \mathbf{v}_3\}$ be a basis for V and $\mathcal{B} = \{\mathbf{w}_1, \mathbf{w}_2, \mathbf{w}_3\}$ be a basis for W. If $\mathcal{A}$ and $\mathcal{B}$ contain a common vector, that vector is in both V and W. Suppose that $\mathcal{A}$ and $\mathcal{B}$ do not contain a common vector. Then the six vectors $\mathbf{v}_1, \mathbf{v}_2, \mathbf{v}_3, \mathbf{w}_1, \mathbf{w}_2, \mathbf{w}_3$ from $\mathcal{R}^5$ must be linearly dependent. Thus
$$a_1\mathbf{v}_1 + a_2\mathbf{v}_2 + a_3\mathbf{v}_3 + b_1\mathbf{w}_1 + b_2\mathbf{w}_2 + b_3\mathbf{w}_3 = \mathbf{0}$$
for some scalars, not all zero. Then
$$a_1\mathbf{v}_1 + a_2\mathbf{v}_2 + a_3\mathbf{v}_3 = -b_1\mathbf{w}_1 - b_2\mathbf{w}_2 - b_3\mathbf{w}_3$$
is a nonzero vector in both V and W.
81. (a) Let $\mathbf{v}$ and $\mathbf{w}$ be in $\mathcal{R}^k$. Then
$$T(\mathbf{v} + \mathbf{w})$$
$$= (v_1 + w_1)\mathbf{u}_1 + \cdots + (v_k + w_k)\mathbf{u}_k$$
$$= (v_1\mathbf{u}_1 + \cdots + v_k\mathbf{u}_k)$$
$$\quad + (w_1\mathbf{u}_1 + \cdots + w_k\mathbf{u}_k)$$
$$= T(\mathbf{v}) + T(\mathbf{w}),$$
and, for any scalar c,
$$T(c\mathbf{v}) = (cv_1)\mathbf{u}_1 + (cv_2)\mathbf{u}_2 + \cdots + (cv_k)\mathbf{u}_k$$
$$= c(v_1\mathbf{u}_1 + v_2\mathbf{u}_2 + \cdots + v_k\mathbf{u}_k)$$
$$= cT(\mathbf{v}).$$

(b) Since $\{\mathbf{u}_1, \mathbf{u}_2, \ldots, \mathbf{u}_k\}$ is linearly independent,
$$x_1\mathbf{u}_1 + x_2\mathbf{u}_2 + \cdots + x_k\mathbf{u}_k = \mathbf{0}$$
implies $x_1 = x_2 = \cdots = x_k = 0$. Thus $T(\mathbf{x}) = \mathbf{0}$ implies $\mathbf{x} = \mathbf{0}$. Therefore T is one-to-one by Theorem 2.11.

(c) If $\mathbf{v}$ is in V, then
$$\mathbf{v} = a_1\mathbf{u}_1 + a_2\mathbf{u}_2 + \cdots + a_k\mathbf{u}_k$$
for some scalars $a_1, a_2, \ldots, a_k$. For
$$\mathbf{a} = \begin{bmatrix} a_1 \\ a_2 \\ \vdots \\ a_k \end{bmatrix},$$
we have $T(\mathbf{a}) = \mathbf{v}$. Thus every vector in V is an image.

82. (a) Let $\mathbf{w}_1$ and $\mathbf{w}_2$ be in W. For $i = 1, 2$, $\mathbf{w}_i = \mathbf{v}_i + c_i\mathbf{u}$ for some $\mathbf{v}_i$ in V and some scalar c_i. So
$$\mathbf{w}_1 + \mathbf{w}_2 = (\mathbf{v}_1 + \mathbf{v}_2) + (c_1 + c_2)\mathbf{u}$$
is in W because $\mathbf{v}_1 + \mathbf{v}_2$ is in V and $c_1 + c_2$ is a scalar. Also, for any scalar a,
$$a\mathbf{w}_1 = a\mathbf{v}_1 + (ac_1)\mathbf{u}$$
is in W because $a\mathbf{v}_1$ is in V and ac_1 is a scalar. Finally, $\mathbf{0} = \mathbf{0} + 0\mathbf{u}$ is in W because $\mathbf{0}$ is in V. Hence W is a subspace of $\mathcal{R}^n$.

(b) If $\{\mathbf{v}_1, \mathbf{v}_2, \ldots, \mathbf{v}_k\}$ is a basis for V, then $\mathcal{B} = \{\mathbf{v}_1, \mathbf{v}_2, \ldots, \mathbf{v}_k, \mathbf{u}\}$ is a generating set for W. In addition, $\mathcal{B}$ is linearly independent by property 3 of linearly dependent and independent sets in Section 1.7. Therefore $\mathcal{B}$ is a basis for W, and so
$$\dim W = k + 1 = \dim V + 1.$$

83. (a) We have $\mathbf{u}^T\mathbf{u} = u_1^2 + u_2^2 + \cdots + u_n^2$ for any $\mathbf{u}$ in $\mathcal{R}^n$. Hence $\mathbf{u}^T\mathbf{u} = 0$ if and only if $\mathbf{u} = \mathbf{0}$.

(b) Let $\mathbf{u}$ be in Row A = Col A^T and $\mathbf{v}$ be in Null A. Then there exists $\mathbf{w}$ such that $A^T\mathbf{w} = \mathbf{u}$. Hence
$$\mathbf{u}^T\mathbf{v} = (A^T\mathbf{w})^T\mathbf{v} = \mathbf{w}^T(A\mathbf{v}) = \mathbf{w}^T\mathbf{0} = 0.$$

(c) By applying (b) with $\mathbf{v} = \mathbf{u}$, we obtain $\mathbf{u}^T\mathbf{u} = 0$. Hence $\mathbf{u} = \mathbf{0}$ by (a).

84. Let $\mathcal{B}_1$ be a basis for Row A and $\mathcal{B}_2$ be a basis for Null A. The number of vectors in $\mathcal{B}_1$ is rank A, and the number of vectors in $\mathcal{B}_2$ is $n - $ rank A. Thus $\mathcal{B}_1 \cup \mathcal{B}_2$ contains n vectors by Exercise 83(c). An argument similar to that in Exercise 80 shows that $\mathcal{B}_1 \cup \mathcal{B}_2$ is linearly independent. Hence $\mathcal{B}_1 \cup \mathcal{B}_2$ is a basis for $\mathcal{R}^n$ by Theorem 4.7.

85. (a) Let B be a 4×4 matrix such that $AB = O$. Then
$$O = AB = A[\mathbf{b}_1 \; \mathbf{b}_2 \; \mathbf{b}_3 \; \mathbf{b}_4]$$
$$= [A\mathbf{b}_1 \; A\mathbf{b}_2 \; A\mathbf{b}_3 \; A\mathbf{b}_4].$$

Hence each column of B is a solution of $A\mathbf{x} = \mathbf{0}$. Because the reduced row echelon form of A is
$$\begin{bmatrix} 1 & 0 & -1 & -2 \\ 0 & 1 & 1 & -1 \\ 0 & 0 & 0 & 0 \\ 0 & 0 & 0 & 0 \end{bmatrix},$$
the vector form of the general solution of $A\mathbf{x} = \mathbf{0}$ is
$$\begin{bmatrix} x_1 \\ x_2 \\ x_3 \\ x_4 \end{bmatrix} = x_3 \begin{bmatrix} 1 \\ -1 \\ 1 \\ 0 \end{bmatrix} + x_4 \begin{bmatrix} 2 \\ 1 \\ 0 \\ 1 \end{bmatrix}.$$

Therefore
$$B = \begin{bmatrix} 1 & 2 & 0 & 0 \\ -1 & 1 & 0 & 0 \\ 1 & 0 & 0 & 0 \\ 0 & 1 & 0 & 0 \end{bmatrix}$$
is a 4×4 matrix with rank 2 such that $AB = O$.

(b) If C is a 4×4 matrix with $AC = O$, then the preceding argument shows that each column of C is a vector in Null A, a 2-dimensional subspace. Hence C can have at most two linearly independent columns, so that rank $C \leq 2$.

86. The reduced row echelon form of the matrix whose columns are the vectors in $\mathcal{B}$ is $[\mathbf{e}_1 \; \mathbf{e}_2 \; \mathbf{e}_3]$.

87. (a) No, the first vector in $\mathcal{A}_1$ is not in W.

(b) Yes, $\mathcal{A}_2$ is a linearly independent subset of W that contains $\dim W = 3$ vectors.

(c) $[\mathbf{e}_1 \; \mathbf{e}_2 \; \mathbf{e}_3], [\mathbf{e}_1 \; \mathbf{e}_2 \; \mathbf{e}_3], [\mathbf{e}_1 \; \mathbf{e}_2 \; \mathbf{e}_3],$
$$\begin{bmatrix} 1 & 0 & 0 & -.4 & -.2 \\ 0 & 1 & 0 & .8 & .4 \\ 0 & 0 & 1 & -.2 & -.6 \end{bmatrix},$$

$$\begin{bmatrix} 1 & 0 & 0 & -.4 & -.2 \\ 0 & 1 & 0 & .8 & .4 \\ 0 & 0 & 1 & -.2 & -.5 \end{bmatrix},$$

$$\begin{bmatrix} 1 & 0 & 0 & -.4 & -.2 \\ 0 & 1 & 0 & .8 & .4 \\ 0 & 0 & 1 & -.2 & -.6 \end{bmatrix}$$

88. We claim that $\mathcal{A}$ is a basis for W if and only if the reduced row echelon forms of A^T and B^T are equal.

    Suppose that the reduced row echelon forms of A^T and B^T are equal. If $\dim W = k$, then B is an $n \times k$ matrix. Since $A^T = B^T$, it follows that A must also be an $n \times k$ matrix. There is an invertible matrix P such that $B^T = PA^T$ (Exercise 89 in Section 2.3). Thus $B = (B^T)^T = (PA^T)^T = (A^T)^T P^T = AP^T$, and so $\operatorname{Col} B = \operatorname{Col}(AP^T)$, which is contained in $\operatorname{Col} A$. It follows that W is a k-dimensional subspace of $\operatorname{Col} A$. Hence $W = \operatorname{Col} A$ by Theorem 4.9, and so $\mathcal{A}$ is a basis for W by Theorem 4.7.

    Conversely, suppose that $\mathcal{A}$ is a basis for W and $\mathbf{x}$ is in $\mathcal{R}^n$. Then $\mathbf{x}^T \mathbf{w} = 0$ for every $\mathbf{w}$ in W if and only if $\mathbf{x}^T \mathbf{a}_j = 0$ for every column $\mathbf{a}_j$ of A, which is equivalent to the condition that $A^T \mathbf{x} = \mathbf{0}$. Since the same is true for $\mathcal{B}$, we have $A^T \mathbf{x} = \mathbf{0}$ if and only if $B^T \mathbf{x} = \mathbf{0}$. Thus the reduced row echelon forms of A^T and B^T are equal by Exercise 92 in Section 2.3.

4.4 COORDINATE SYSTEMS

1. $4 \begin{bmatrix} 1 \\ -1 \end{bmatrix} + 3 \begin{bmatrix} -1 \\ 2 \end{bmatrix} = \begin{bmatrix} 1 \\ 2 \end{bmatrix}$

2. $\begin{bmatrix} -5 \\ 7 \end{bmatrix}$ 3. $\begin{bmatrix} -5 \\ 11 \end{bmatrix}$ 4. $\begin{bmatrix} -5 \\ 9 \end{bmatrix}$

5. $\mathbf{v} = 2 \begin{bmatrix} 1 \\ -1 \end{bmatrix} + 5 \begin{bmatrix} -1 \\ 2 \end{bmatrix} = \begin{bmatrix} -3 \\ 8 \end{bmatrix}$ 6. $\begin{bmatrix} 3 \\ -1 \end{bmatrix}$

7. $\begin{bmatrix} 4 \\ 5 \\ 4 \end{bmatrix}$ 8. $\begin{bmatrix} -5 \\ -1 \\ 0 \end{bmatrix}$

9. $\mathbf{v} = (-1)\begin{bmatrix} 0 \\ 1 \\ 1 \end{bmatrix} + 5\begin{bmatrix} -1 \\ 0 \\ 1 \end{bmatrix} + (-2)\begin{bmatrix} 1 \\ 1 \\ 1 \end{bmatrix} = \begin{bmatrix} -7 \\ -3 \\ 2 \end{bmatrix}$

10. $\begin{bmatrix} 6 \\ 5 \\ 1 \end{bmatrix}$

11. (a) $\mathcal{B}$ is a linearly independent subset of $\mathcal{R}^2$ containing 2 vectors.

    (b) $\begin{bmatrix} 5 \\ -3 \end{bmatrix}$

12. (a) $\mathcal{B}$ is a linearly independent subset of $\mathcal{R}^2$ containing 2 vectors.

    (b) $\begin{bmatrix} -2 \\ 4 \end{bmatrix}$

13. (a) Let B be the matrix whose columns are the vectors in $\mathcal{B}$. Since the reduced row echelon form of B is I_3, $\mathcal{B}$ is linearly independent. So $\mathcal{B}$ is a linearly independent set of 3 vectors from $\mathcal{R}^3$, and hence $\mathcal{B}$ is a basis for $\mathcal{R}^3$ by Theorem 4.7.

    (b) The components of $[\mathbf{v}]_\mathcal{B}$ are the coefficients that express $\mathbf{v}$ as a linear combination of the vectors in $\mathcal{B}$. Thus
    $$[\mathbf{v}]_\mathcal{B} = \begin{bmatrix} 3 \\ 0 \\ -1 \end{bmatrix}.$$

14. (a) $\mathcal{B}$ is a linearly independent subset of $\mathcal{R}^3$ containing 3 vectors.

    (b) $\begin{bmatrix} 0 \\ 1 \\ -4 \end{bmatrix}$

15. $\begin{bmatrix} -5 \\ -1 \end{bmatrix}$ 16. $\begin{bmatrix} 0 \\ 1 \end{bmatrix}$

17. $[\mathbf{v}]_\mathcal{B} = \begin{bmatrix} 1 & -1 \\ -1 & 2 \end{bmatrix}^{-1} \begin{bmatrix} 5 \\ -3 \end{bmatrix} = \begin{bmatrix} 7 \\ 2 \end{bmatrix}$ 18. $\begin{bmatrix} 8 \\ 5 \end{bmatrix}$

19. $\begin{bmatrix} 0 \\ -1 \\ 3 \end{bmatrix}$ 20. $\begin{bmatrix} 11 \\ -3 \\ -5 \end{bmatrix}$

21. $[\mathbf{v}]_\mathcal{B} = \begin{bmatrix} 0 & -1 & 1 \\ 1 & 0 & 1 \\ 1 & 1 & 1 \end{bmatrix}^{-1} \begin{bmatrix} 1 \\ -3 \\ -2 \end{bmatrix} = \begin{bmatrix} -5 \\ 1 \\ 2 \end{bmatrix}$

22. $\begin{bmatrix} 9 \\ -3 \\ -4 \end{bmatrix}$

23. The unique representation of $\mathbf{u}$ as a linear combination of $\mathcal{B} = \{\mathbf{b}_1, \mathbf{b}_2\}$ is $\mathbf{u} = c_1 \mathbf{b}_1 + c_2 \mathbf{b}_2$ if and only if $\begin{bmatrix} c_1 \\ c_2 \end{bmatrix}$ is the coordinate vector of $\mathbf{u}$ relative to $\mathcal{B}$. Computing $[\mathbf{u}]_\mathcal{B} = [\mathbf{b}_1 \ \mathbf{b}_2]^{-1} \mathbf{u}$, we find that $\mathbf{u} = (a + 2b)\mathbf{b}_1 + (a + 3b)\mathbf{b}_2$.

24. $(a + b)\mathbf{b}_1 + (a + 2b)\mathbf{b}_2 = \mathbf{u}$

25. $(-5a - 3b)\mathbf{b}_1 + (-3a - 2b)\mathbf{b}_2 = \mathbf{u}$

26. $(a - 2b)\mathbf{b}_1 + (-a + 3b)\mathbf{b}_2 = \mathbf{u}$

27. $(-4a - 3b + 2c)\mathbf{b}_1 + (-2a - b + c)\mathbf{b}_2 + (3a + 2b - c)\mathbf{b}_3 = \mathbf{u}$

28. $(-2a - b + c)\mathbf{b}_1 + (3a + b - c)\mathbf{b}_2 + (-2a + c)\mathbf{b}_3 = \mathbf{u}$

29. $(-a - b + 2c)\mathbf{b}_1 + b\mathbf{b}_2 + (-a - b + c)\mathbf{b}_3 = \mathbf{u}$
30. $(-a - 4b + 2c)\mathbf{b}_1 + (-2b + c)\mathbf{b}_2 + (a + 3b - c)\mathbf{b}_3 = \mathbf{u}$
31. False, every vector in V can be *uniquely* represented as a linear combination of the vectors in $\mathcal{S}$ if and only if $\mathcal{S}$ is a basis for V.
32. True   33. True   34. True   35. True
36. True   37. True   38. True   39. True
40. True   41. True   42. True   43. True
44. False, $\begin{bmatrix} x' \\ y' \end{bmatrix} = A_\theta^T \begin{bmatrix} x \\ y \end{bmatrix}$.
45. True   46. True   47. True
48. False, the graph of such an equation is a *hyperbola*.
49. True   50. True
51. (a) Since the reduced row echelon form of $\begin{bmatrix} 1 & 2 \\ 2 & 3 \end{bmatrix}$ is $\begin{bmatrix} 1 & 0 \\ 0 & 1 \end{bmatrix}$, $\mathcal{B}$ is a linearly independent subset of $\mathcal{R}^2$ containing 2 vectors. Hence $\mathcal{B}$ is a basis for $\mathcal{R}^2$ by Theorem 4.7.

    (b) Let $B = [\mathbf{b}_1 \ \mathbf{b}_2]$. Then
    $$[\mathbf{e}_1]_\mathcal{B} = B^{-1}\mathbf{e}_1 = \begin{bmatrix} -3 \\ 2 \end{bmatrix}$$
    and
    $$[\mathbf{e}_2]_\mathcal{B} = B^{-1}\mathbf{e}_2 = \begin{bmatrix} 2 \\ -1 \end{bmatrix}.$$
    Hence
    $$A = \begin{bmatrix} -3 & 2 \\ 2 & -1 \end{bmatrix}.$$

    (c) From (b), we see that
    $$A = [B^{-1}\mathbf{e}_1 \ B^{-1}\mathbf{e}_2]$$
    $$= B^{-1}[\mathbf{e}_1 \ \mathbf{e}_2] = B^{-1}I_2 = B^{-1};$$
    so the matrices A and B are inverses of each other.

52. (a) $\mathcal{B}$ is a linearly independent subset of $\mathcal{R}^2$ containing 2 vectors.

    (b) $\begin{bmatrix} -2 & 1 \\ -3 & 1 \end{bmatrix}$ (c) $A = B^{-1}$

53. (a) $\mathcal{B}$ is a linearly independent subset of $\mathcal{R}^3$ containing 3 vectors.

    (b) $\begin{bmatrix} 1 & 0 & 1 \\ 1 & 1 & 3 \\ 0 & -1 & -1 \end{bmatrix}$ (c) $A = B^{-1}$

54. (a) $\mathcal{B}$ is a linearly independent subset of $\mathcal{R}^3$ containing 3 vectors.

    (b) $\begin{bmatrix} -3 & -1 & 2 \\ -4 & -1 & 2 \\ 3 & 1 & -1 \end{bmatrix}$ (c) $A = B^{-1}$

55. Let $\mathbf{v} = \begin{bmatrix} x \\ y \end{bmatrix}$ and $[\mathbf{v}]_\mathcal{B} = \begin{bmatrix} x' \\ y' \end{bmatrix}$, where $\mathcal{B}$ is the basis obtained by rotating the vectors in the standard basis by $30°$. As in the example on pages 269–270, we have
    $$\begin{bmatrix} x' \\ y' \end{bmatrix} = [\mathbf{v}]_\mathcal{B} = (A_{30°})^{-1}\mathbf{v}$$
    $$= A_{30°}^T \mathbf{v} = \begin{bmatrix} \frac{\sqrt{3}}{2} & \frac{1}{2} \\ -\frac{1}{2} & \frac{\sqrt{3}}{2} \end{bmatrix} \begin{bmatrix} x \\ y \end{bmatrix}.$$
    Hence
    $$x' = \tfrac{\sqrt{3}}{2}x + \tfrac{1}{2}y$$
    $$y' = -\tfrac{1}{2}x + \tfrac{\sqrt{3}}{2}y.$$

56. $x' = \tfrac{1}{2}x + \tfrac{\sqrt{3}}{2}y$
    $y' = -\tfrac{\sqrt{3}}{2}x + \tfrac{1}{2}y$

57. $x' = -\tfrac{\sqrt{2}}{2}x + \tfrac{\sqrt{2}}{2}y$
    $y' = -\tfrac{\sqrt{2}}{2}x - \tfrac{\sqrt{2}}{2}y$

58. $x' = \tfrac{\sqrt{3}}{2}x - \tfrac{1}{2}y$
    $y' = \tfrac{1}{2}x + \tfrac{\sqrt{3}}{2}y$

59. $x' = -5x - 3y$
    $y' = -2x - y$

60. $x' = 3x - 2y$
    $y' = -4x + 3y$

61. $x' = -x - y$
    $y' = -2x - y$

62. $x' = -3x - 2y$
    $y' = 5x + 3y$

63. Let
    $$B = \begin{bmatrix} 1 & 1 & 0 \\ 0 & 1 & -2 \\ 1 & 0 & 1 \end{bmatrix}.$$
    Then, as in Exercise 55,
    $$\begin{bmatrix} x' \\ y' \\ z' \end{bmatrix} = [\mathbf{v}]_\mathcal{B} = B^{-1}\mathbf{v} = \begin{bmatrix} -1 & 1 & 2 \\ 2 & -1 & -2 \\ 1 & -1 & -1 \end{bmatrix} \begin{bmatrix} x \\ y \\ z \end{bmatrix}.$$
    Hence
    $$x' = -x + y + 2z$$
    $$y' = 2x - y - 2z$$
    $$z' = x - y - z.$$

64. $x' = -2x + 3y + 2z$
    $y' = x - 2y - z$
    $z' = x - y$

65. $x' = x - y + z$
    $y' = -3x + 4y - 2z$
    $z' = x - 2y + z$

66. $x' = -x + z$
    $y' = 2x + y - 3z$
    $z' = -2x - y + 4z$

67. Let $\mathbf{v} = \begin{bmatrix} x \\ y \end{bmatrix}$ and $[\mathbf{v}]_\mathcal{B} = \begin{bmatrix} x' \\ y' \end{bmatrix}$, where $\mathcal{B}$ is the basis obtained by rotating the vectors in the standard basis by $60°$. As in Example 4,
$$\begin{bmatrix} x \\ y \end{bmatrix} = \mathbf{v} = A_{60°}[\mathbf{v}]_\mathcal{B} = \begin{bmatrix} \frac{1}{2} & -\frac{\sqrt{3}}{2} \\ \frac{\sqrt{3}}{2} & \frac{1}{2} \end{bmatrix} \begin{bmatrix} x' \\ y' \end{bmatrix}.$$

Hence
$$x = \tfrac{1}{2}x' - \tfrac{\sqrt{3}}{2}y'$$
$$y = \tfrac{\sqrt{3}}{2}x' + \tfrac{1}{2}y'.$$

68. $\begin{aligned} x &= \tfrac{\sqrt{2}}{2}x' - \tfrac{\sqrt{2}}{2}y' \\ y &= \tfrac{\sqrt{2}}{2}x' + \tfrac{\sqrt{2}}{2}y' \end{aligned}$ **69.** $\begin{aligned} x &= -\tfrac{\sqrt{2}}{2}x' - \tfrac{\sqrt{2}}{2}y' \\ y &= \tfrac{\sqrt{2}}{2}x' - \tfrac{\sqrt{2}}{2}y' \end{aligned}$

70. $\begin{aligned} x &= \tfrac{\sqrt{3}}{2}x' + \tfrac{1}{2}y' \\ y &= -\tfrac{1}{2}x' + \tfrac{\sqrt{3}}{2}y' \end{aligned}$ **71.** $\begin{aligned} x &= x' + 3y' \\ y &= 2x' + 4y' \end{aligned}$

72. $\begin{aligned} x &= 2x' + y' \\ y &= -x' + 3y' \end{aligned}$ **73.** $\begin{aligned} x &= -x' + 3y' \\ y &= 3x' + 5y' \end{aligned}$

74. $\begin{aligned} x &= 3x' + 2y' \\ y &= 2x' + 4y' \end{aligned}$

75. Let
$$\begin{bmatrix} 1 & -1 & 0 \\ 3 & 1 & -1 \\ 0 & 1 & 1 \end{bmatrix}.$$
As in Exercise 67, we have
$$\begin{bmatrix} x \\ y \\ z \end{bmatrix} = \mathbf{v} = B[\mathbf{v}]_\mathcal{B} = \begin{bmatrix} 1 & -1 & 0 \\ 3 & 1 & -1 \\ 0 & 1 & 1 \end{bmatrix} \begin{bmatrix} x' \\ y' \\ z' \end{bmatrix}.$$

Thus
$$\begin{aligned} x &= x' - y' \\ y &= 3x' + y' - z' \\ z &= y' + z'. \end{aligned}$$

76. $\begin{aligned} x &= 2x' + z' \\ y &= -x' - y' - z' \\ z &= x' + y' + 2z' \end{aligned}$

77. $\begin{aligned} x &= x' - y' - z' \\ y &= -x' + 3y' + z' \\ z &= x' + 2y' + z' \end{aligned}$

78. $\begin{aligned} x &= -x' - y' + z' \\ y &= x' + 2y' - z' \\ z' &= x' + 2y' + z' \end{aligned}$

79. As in Exercise 55, we have
$$\begin{bmatrix} x' \\ y' \end{bmatrix} = A_{60°}^T \begin{bmatrix} x \\ y \end{bmatrix} = \begin{bmatrix} \frac{1}{2} & \frac{\sqrt{3}}{2} \\ -\frac{\sqrt{3}}{2} & \frac{1}{2} \end{bmatrix} \begin{bmatrix} x \\ y \end{bmatrix}.$$

Thus
$$x' = \tfrac{1}{2}x + \tfrac{\sqrt{3}}{2}y$$
$$y' = -\tfrac{\sqrt{3}}{2}x + \tfrac{1}{2}y.$$

Rewrite the given equation in the form
$$25(x')^2 + 16(y')^2 = 400.$$

Then substitute the expressions for x' and y' into this equation to obtain
$$\frac{73}{4}x^2 + \frac{9\sqrt{3}}{2}xy + \frac{91}{4}y^2 = 400,$$
that is,
$$73x^2 + 18\sqrt{3}xy + 91y^2 = 1600.$$

80. $5x^2 + 26xy + 5y^2 = 72$

81. $8x^2 - 34xy + 8y^2 = 225$

82. $9x^2 + 2\sqrt{3}xy + 11y^2 = 48$

83. $-23x^2 - 26\sqrt{3}xy + 3y^2 = 144$

84. $79x^2 - 42\sqrt{3}xy + 37y^2 = 400$

85. $-11x^2 + 50\sqrt{3}xy + 39y^2 = 576$

86. $31x^2 + 10\sqrt{3}xy + 21y^2 = 144$

87. As in Exercise 67, we have
$$\begin{bmatrix} x \\ y \end{bmatrix} = A_{45°}\begin{bmatrix} x' \\ y' \end{bmatrix} = \begin{bmatrix} \frac{\sqrt{2}}{2} & -\frac{\sqrt{2}}{2} \\ \frac{\sqrt{2}}{2} & \frac{\sqrt{2}}{2} \end{bmatrix} \begin{bmatrix} x' \\ y' \end{bmatrix}.$$

Thus
$$x = \tfrac{\sqrt{2}}{2}x' - \tfrac{\sqrt{2}}{2}y'$$
$$y = \tfrac{\sqrt{2}}{2}x' + \tfrac{\sqrt{2}}{2}y'.$$

Substituting these expressions for x and y into the given equation produces
$$4(x')^2 - 10(y')^2 = 20,$$
that is,
$$2(x')^2 - 5(y')^2 = 10.$$

88. $3(x')^2 + 7(y')^2 = 21$ **89.** $4(x')^2 + 3(y')^2 = 12$

90. $2(x')^2 - (y')^2 = 6$ **91.** $4(x')^2 - 3(y')^2 = 60$

92. $8(x')^2 + 9(y')^2 = 180$ **93.** $5(x')^2 + 2(y')^2 = 10$

94. $5(x')^2 + 3(y')^2 = 24$

95. By definition of $[\mathbf{v}]_\mathcal{A}$, we have
$$\begin{aligned} \mathbf{v} &= a_1\mathbf{u}_1 + a_2\mathbf{u}_2 + \cdots + a_n\mathbf{u}_n \\ &= \frac{a_1}{c_1}(c_1\mathbf{u}_1) + \frac{a_2}{c_2}(c_2\mathbf{u}_2) + \cdots + \frac{a_n}{c_n}(c_n\mathbf{u}_n). \end{aligned}$$

Hence
$$[\mathbf{v}]_\mathcal{B} = \begin{bmatrix} \frac{a_1}{c_1} \\ \frac{a_2}{c_2} \\ \vdots \\ \frac{a_n}{c_n} \end{bmatrix}.$$

96. $\begin{bmatrix} a_1 - a_2 - \cdots - a_n \\ a_2 \\ a_3 \\ \vdots \\ a_n \end{bmatrix}$ 97. $\begin{bmatrix} a_1 \\ a_2 - a_1 \\ \vdots \\ a_n - a_1 \end{bmatrix}$

98. $\begin{bmatrix} a_1 \\ a_2 - a_1 \\ a_3 - a_2 \\ \vdots \\ a_n - a_{n-1} \end{bmatrix}$

99. No, consider $\mathcal{A} = \{\mathbf{e}_1, \mathbf{e}_2\}$ and $\mathcal{B} = \{\mathbf{e}_1, \mathbf{e}_1 + \mathbf{e}_2\}$, which are both bases for $\mathcal{R}^2$. For $\mathbf{v} = \mathbf{e}_1$, we have $[\mathbf{v}]_\mathcal{A} = [\mathbf{v}]_\mathcal{B} = \mathbf{e}_1$, but $\mathcal{A} \neq \mathcal{B}$.

100. Let $\mathcal{A} = \{\mathbf{u}_1, \mathbf{u}_2, \ldots, \mathbf{u}_n\}$. Then $[\mathbf{u}_i]_\mathcal{A} = \mathbf{e}_i$ for each i. So if $[\mathbf{v}]_\mathcal{A} = [\mathbf{v}]_\mathcal{B}$ for each $\mathbf{v}$ in $\mathcal{R}^n$, we must have $[\mathbf{u}_i]_\mathcal{B} = \mathbf{e}_i$ for each i, so that $\mathcal{B} = \mathcal{A}$.

101. Let $\mathcal{S} = \{\mathbf{u}_1, \mathbf{u}_2, \ldots, \mathbf{u}_k\}$ be linearly dependent. There exist scalars $c_1, c_2, \ldots, c_k$, not all zero, such that $c_1\mathbf{u}_1 + c_2\mathbf{u}_2 + \cdots + c_k\mathbf{u}_k = \mathbf{0}$. If $\mathbf{v}$ is in the span of $\mathcal{S}$ and $\mathbf{v} = a_1\mathbf{u}_1 + a_2\mathbf{u}_2 + \cdots + a_k\mathbf{u}_k$, then we also have
$$\mathbf{v} = (a_1 + c_1)\mathbf{u}_1 + (a_2 + c_2)\mathbf{u}_2 + \cdots + (a_k + c_k)\mathbf{u}_k.$$

102. Let A and B be the matrices whose columns are the vectors in $\mathcal{A}$ and $\mathcal{B}$, respectively. Then
$$[\mathbf{v}]_\mathcal{A} = A^{-1}\mathbf{v} = A^{-1}(B[\mathbf{v}]_\mathcal{B}) = A^{-1}B[\mathbf{v}]_\mathcal{B}.$$

103. (a) Let B be the matrix whose columns are the vectors in $\mathcal{B}$. Then T is the matrix transformation induced by B^{-1} (Theorem 4.11), and so T is linear by Theorem 2.6.

 (b) Since the standard matrix of T is the invertible matrix B^{-1}, T is one-to-one and onto by Theorems 2.11 and 2.10.

104. B^{-1}, where B is the matrix whose columns are the vectors in $\mathcal{B}$

105. If there are unique scalars such that
$$a_1\mathbf{u}_1 + a_2\mathbf{u}_2 + \cdots + a_k\mathbf{u}_k = \mathbf{0}$$
$$= 0\mathbf{u}_1 + 0\mathbf{u}_2 + \cdots + 0\mathbf{u}_k,$$

then $a_1 = a_2 = \cdots = a_k = 0$. Hence $\mathcal{B}$ is linearly independent. Moreover, every vector in V is in the span of $\mathcal{B}$; so $\mathcal{B}$ is a basis for V.

106. (a) The reduced row echelon form of the matrix whose columns are the vectors in $\mathcal{S}$ is I_3.

 (b) The third component of the vector on the left side of the given equation is $2v_1 - v_2 = v_3$.

 (c) No, $\mathcal{S}$ is not a subset of V.

107. Suppose that $\mathcal{A} = \{\mathbf{u}_1, \mathbf{u}_2, \ldots, \mathbf{u}_k\}$ is a linearly independent subset of $\mathcal{R}^n$, and let $c_1, c_2, \ldots, c_k$ be scalars such that
$$c_1[\mathbf{u}_1]_\mathcal{B} + c_2[\mathbf{u}_2]_\mathcal{B} + \cdots + c_k[\mathbf{u}_k]_\mathcal{B} = \mathbf{0}.$$

Define $T: \mathcal{R}^n \to \mathcal{R}^n$ by $T(\mathbf{v}) = [\mathbf{v}]_\mathcal{B}$ for all $\mathbf{v}$ in $\mathcal{R}^n$. Then T is a linear transformation by Exercise 103(a), and so
$$c_1 T(\mathbf{u}_1) + c_2 T(\mathbf{u}_2) + \cdots + c_k T(\mathbf{u}_k) = \mathbf{0}$$
$$T(c_1\mathbf{u}_1 + c_2\mathbf{u}_2 + \cdots + c_k\mathbf{u}_k) = \mathbf{0}.$$

Thus $c_1\mathbf{u}_1 + c_2\mathbf{u}_2 + \cdots + c_k\mathbf{u}_k$ is in the null space of T. Since T is one-to-one by Exercise 103(b), it follows that
$$c_1\mathbf{u}_1 + c_2\mathbf{u}_2 + \cdots + c_k\mathbf{u}_k = \mathbf{0}.$$

Because $\{\mathbf{u}_1, \mathbf{u}_2, \ldots, \mathbf{u}_k\}$ is linearly independent, it follows that $c_1 = c_2 = \cdots = c_k = 0$. Therefore $\{[\mathbf{u}_1]_\mathcal{B}, [\mathbf{u}_2]_\mathcal{B}, \ldots, [\mathbf{u}_k]_\mathcal{B}\}$ is linearly independent. The proof of the converse is similar.

108. The argument is similar to that in Exercise 107.

109. (a) The matrix whose columns are the vectors in $\mathcal{B}$ has I_5 as its reduced row echelon form. Hence $\mathcal{B}$ is a linearly independent subset of $\mathcal{R}^5$ containing five vectors.

 (b) $\begin{bmatrix} 29 \\ 44 \\ -52 \\ 33 \\ 39 \end{bmatrix}$

110. $\mathbf{u} = \begin{bmatrix} 2 \\ 0 \\ 0 \\ -1 \\ 1 \end{bmatrix}$

111. Let B be the matrix whose columns are the vectors in $\mathcal{B}$. Since $[\mathbf{v}]_\mathcal{B} = B^{-1}\mathbf{v}$, we must find a nonzero vector $\mathbf{v}$ in $\mathcal{R}^5$ such that
$$B^{-1}\mathbf{v} = .5\mathbf{v}$$

94 Chapter 4 Subspaces and Their Properties

$$B^{-1}\mathbf{v} - .5\mathbf{v} = \mathbf{0}$$
$$(B^{-1} - .5I_5)\mathbf{v} = \mathbf{0}.$$

The reduced row echelon form of $B^{-1} - .5I_5$ is

$$\begin{bmatrix} 1 & 0 & 0 & 0 & 0 \\ 0 & 1 & 0 & 0 & -2 \\ 0 & 0 & 1 & 0 & 2 \\ 0 & 0 & 0 & 1 & -2 \\ 0 & 0 & 0 & 0 & 0 \end{bmatrix}.$$

Thus the vector form of the general solution of $(B^{-1} - .5I_5)\mathbf{x} = \mathbf{0}$ is

$$\begin{bmatrix} x_1 \\ x_2 \\ x_3 \\ x_4 \\ x_5 \end{bmatrix} = x_5 \begin{bmatrix} 0 \\ 2 \\ -2 \\ 2 \\ 1 \end{bmatrix}.$$

So by taking

$$\mathbf{v} = \begin{bmatrix} 0 \\ 2 \\ -2 \\ 2 \\ 1 \end{bmatrix},$$

we have $[\mathbf{v}]_\mathcal{B} = .5\mathbf{v}$.

112. (a) $\mathbf{v} = 5\mathbf{u}_1 - 3\mathbf{u}_2 - \mathbf{u}_3$

(b) $[\mathbf{v}]_\mathcal{B} = 5[\mathbf{u}_1]_\mathcal{B} - 3[\mathbf{u}_2]_\mathcal{B} - [\mathbf{u}_3]_\mathcal{B}$

(c) Let $\mathcal{B} = \{\mathbf{u}_1, \mathbf{u}_2, \ldots, \mathbf{u}_n\}$ be a basis for $\mathcal{R}^n$, and let $\mathbf{v}$ be in $\mathcal{R}^n$. The conjecture is: There exist scalars $c_1, c_2, \ldots, c_n$ such that

$$\mathbf{v} = c_1\mathbf{u}_1 + c_2\mathbf{u}_2 + \cdots + c_n\mathbf{u}_n$$

if and only if

$$[\mathbf{v}]_\mathcal{B} = c_1[\mathbf{u}_1]_\mathcal{B} + c_2[\mathbf{u}_2]_\mathcal{B} + \cdots + c_n[\mathbf{u}_n]_\mathcal{B}.$$

4.5 MATRIX REPRESENTATIONS OF LINEAR OPERATORS

1. The standard matrix of T is $A = \begin{bmatrix} 2 & 1 \\ 1 & -1 \end{bmatrix}$. If B is the matrix whose columns are the vectors in $\mathcal{B}$, then, by Theorem 4.12, we have

$$[T]_\mathcal{B} = B^{-1}AB = \begin{bmatrix} 1 & 1 \\ 3 & 0 \end{bmatrix}.$$

2. As in Exercise 1, $[T]_\mathcal{B} = \begin{bmatrix} 4 & 1 \\ -9 & -2 \end{bmatrix}$.

3. As in Exercise 1, $[T]_\mathcal{B} = \begin{bmatrix} 1 & 2 \\ 1 & 1 \end{bmatrix}$.

4. $\begin{bmatrix} 15 & 29 \\ -7 & -14 \end{bmatrix}$ **5.** $\begin{bmatrix} 10 & 19 & 16 \\ -5 & -8 & -8 \\ 2 & 2 & 3 \end{bmatrix}$

6. $\begin{bmatrix} 2 & 2 & 1 \\ 1 & -3 & -3 \\ 0 & 3 & 3 \end{bmatrix}$ **7.** $\begin{bmatrix} 0 & -19 & 28 \\ 3 & 34 & -47 \\ 3 & 23 & -31 \end{bmatrix}$

8. $\begin{bmatrix} 29 & 17 & 70 \\ 39 & 21 & 92 \\ -20 & -11 & -47 \end{bmatrix}$

9. $\begin{bmatrix} -10 & -12 & -9 & 1 \\ 20 & 26 & 20 & -7 \\ -10 & -15 & -12 & 7 \\ 7 & 7 & 5 & 1 \end{bmatrix}$

10. $\begin{bmatrix} -104.5 & -191 & -140.5 & -537 \\ 44.5 & 81 & 59.5 & 230 \\ -33.5 & -59 & -40.5 & -168 \\ 13.5 & 24 & 16.5 & 67 \end{bmatrix}$

11. Let A be the standard matrix of T and B be the matrix whose columns are the vectors in $\mathcal{B}$. Then, by Theorem 4.12, we have

$$A = B[T]_\mathcal{B} B^{-1} = \begin{bmatrix} 10 & -19 \\ 3 & -4 \end{bmatrix}.$$

12. $\begin{bmatrix} -4 & -2 \\ 9 & 5 \end{bmatrix}$ **13.** $\begin{bmatrix} 45 & 25 \\ -79 & -44 \end{bmatrix}$

14. $\begin{bmatrix} 9 & -4 \\ 8 & -2 \end{bmatrix}$ **15.** $\begin{bmatrix} 2 & 5 & 10 \\ -6 & 1 & -7 \\ 2 & -2 & 0 \end{bmatrix}$

16. $\begin{bmatrix} -5 & -3 & 3 \\ 4 & 1 & -1 \\ -5 & -4 & 5 \end{bmatrix}$ **17.** $\begin{bmatrix} -1 & -1 & 0 \\ 1 & 3 & -1 \\ -1 & 0 & 1 \end{bmatrix}$

18. $\begin{bmatrix} 4 & 9 & -3 \\ -3 & -4 & 1 \\ -3 & 2 & -2 \end{bmatrix}$

19. False, a linear operator on $\mathcal{R}^n$ is a linear transformation whose domain and codomain both equal $\mathcal{R}^n$.

20. True **21.** True

22. False, the matrix representation of T with respect to $\mathcal{B}$ is

$$[[T(\mathbf{b}_1)]_\mathcal{B} \ [T(\mathbf{b}_2)]_\mathcal{B} \ \cdots \ [T(\mathbf{b}_n)]_\mathcal{B}].$$

23. True

24. False, $[T]_\mathcal{B} = B^{-1}AB$.

25. False, $[T]_\mathcal{B} = B^{-1}AB$.

26. True

27. False, $T(\mathbf{v}) = \mathbf{v}$ for every vector $\mathbf{v}$ on L.
28. False, $T(\mathbf{v}) = \mathbf{v}$ for every vector $\mathbf{v}$ on L.
29. False, there exists a basis $\mathcal{B}$ for $\mathcal{R}^n$ such that $[T]_\mathcal{B} = \begin{bmatrix} 1 & 0 \\ 0 & -1 \end{bmatrix}$.
30. True
31. False, $\mathcal{B}$ consists of one vector on L and one vector perpendicular to L.
32. False, an $n \times n$ matrix A is said to be similar to an $n \times n$ matrix B if $B = P^{-1}AP$ for some invertible matrix P.
33. True   34. True   35. True   36. True
37. False, $[T]_\mathcal{B}[\mathbf{v}]_\mathcal{B} = [T(\mathbf{v})]_\mathcal{B}$.
38. True
39. $\begin{bmatrix} 1 & -3 \\ 4 & 0 \end{bmatrix}$ 40. $\begin{bmatrix} 2 & -1 \\ -5 & 3 \end{bmatrix}$ 41. $\begin{bmatrix} 3 & 2 \\ -5 & 4 \end{bmatrix}$
42. $\begin{bmatrix} 1 & 0 & 5 \\ -2 & 6 & 2 \\ 3 & -1 & -4 \end{bmatrix}$
43. Since $T(\mathbf{b}_1) = 0\mathbf{b}_1 - 5\mathbf{b}_2 + 4\mathbf{b}_3$, we have

$$[T(\mathbf{b}_1)]_\mathcal{B} = \begin{bmatrix} 0 \\ -5 \\ 4 \end{bmatrix}.$$

Likewise

$$[T(\mathbf{b}_2)]_\mathcal{B} = \begin{bmatrix} 2 \\ 0 \\ -7 \end{bmatrix} \quad \text{and} \quad [T(\mathbf{b}_2)]_\mathcal{B} = \begin{bmatrix} 3 \\ 0 \\ 1 \end{bmatrix}.$$

Hence

$$[T]_\mathcal{B} = \begin{bmatrix} 0 & 2 & 3 \\ -5 & 0 & 0 \\ 4 & -7 & 1 \end{bmatrix}.$$

44. $\begin{bmatrix} 2 & -1 & 0 \\ 5 & 3 & 1 \\ 0 & 0 & -2 \end{bmatrix}$ 45. $\begin{bmatrix} 1 & 0 & -3 & 0 \\ -1 & 2 & 0 & 4 \\ 1 & 0 & 5 & -1 \\ -1 & -1 & 0 & 3 \end{bmatrix}$

46. $\begin{bmatrix} 0 & 1 & 2 & 0 \\ -1 & 0 & 0 & -1 \\ 0 & -2 & 0 & 2 \\ 1 & 0 & -3 & 1 \end{bmatrix}$

47. (a) $\begin{bmatrix} 0 & 3 \\ 1 & 0 \end{bmatrix}$ (b) $\begin{bmatrix} -1 & 2 \\ 1 & 1 \end{bmatrix}$

    (c) $T\left(\begin{bmatrix} x_1 \\ x_2 \end{bmatrix}\right) = \begin{bmatrix} -x_1 + 2x_2 \\ x_1 + x_2 \end{bmatrix}$

48. (a) $\begin{bmatrix} 1 & 2 \\ -2 & -1 \end{bmatrix}$ (b) $\frac{1}{3}\begin{bmatrix} 3 & -2 \\ 18 & -3 \end{bmatrix}$

    (c) $T\left(\begin{bmatrix} x_1 \\ x_2 \end{bmatrix}\right) = \frac{1}{3}\begin{bmatrix} 3x_1 - 2x_2 \\ 18x_1 - 3x_2 \end{bmatrix}$

49. (a) $\begin{bmatrix} 3 & 2 \\ -1 & 0 \end{bmatrix}$ (b) $\begin{bmatrix} -8 & -6 \\ 15 & 11 \end{bmatrix}$

    (c) $T\left(\begin{bmatrix} x_1 \\ x_2 \end{bmatrix}\right) = \begin{bmatrix} -8x_1 - 6x_2 \\ 15x_1 + 11x_2 \end{bmatrix}$

50. (a) $\begin{bmatrix} -1 & 3 \\ 4 & -2 \end{bmatrix}$ (b) $\begin{bmatrix} 7 & -2 \\ 30 & -10 \end{bmatrix}$

    (c) $T\left(\begin{bmatrix} x_1 \\ x_2 \end{bmatrix}\right) = \begin{bmatrix} 7x_1 - 2x_2 \\ 30x_1 - 10x_2 \end{bmatrix}$

51. Let

$$\mathbf{b}_1 = \begin{bmatrix} 1 \\ 0 \\ 1 \end{bmatrix}, \quad \mathbf{b}_2 = \begin{bmatrix} 0 \\ 1 \\ 0 \end{bmatrix}, \quad \text{and} \quad \mathbf{b}_3 = \begin{bmatrix} 1 \\ 1 \\ 0 \end{bmatrix}.$$

(a) Since $T(\mathbf{b}_1) = 0\mathbf{b}_1 - \mathbf{b}_2 + 0\mathbf{b}_3$, we have

$$[T(\mathbf{b}_1)]_\mathcal{B} = \begin{bmatrix} 0 \\ -1 \\ 0 \end{bmatrix}.$$

Likewise

$$[T(\mathbf{b}_2)]_\mathcal{B} = \begin{bmatrix} 0 \\ 0 \\ 2 \end{bmatrix} \quad \text{and} \quad [T(\mathbf{b}_3)]_\mathcal{B} = \begin{bmatrix} 1 \\ 2 \\ 0 \end{bmatrix}.$$

Hence

$$[T]_\mathcal{B} = \begin{bmatrix} 0 & 0 & 1 \\ -1 & 0 & 2 \\ 0 & 2 & 0 \end{bmatrix}.$$

(b) The standard matrix A of T is given by

$$A = B[T]_\mathcal{B} B^{-1} = \begin{bmatrix} -1 & 2 & 1 \\ 0 & 2 & -1 \\ 1 & 0 & -1 \end{bmatrix},$$

where $B = [\mathbf{b}_1 \ \mathbf{b}_2 \ \mathbf{b}_3]$.

(c) For any vector $\mathbf{x}$ in $\mathcal{R}^3$, we have

$$T(\mathbf{x}) = A\mathbf{x}$$
$$= \begin{bmatrix} -1 & 2 & 1 \\ 0 & 2 & -1 \\ 1 & 0 & -1 \end{bmatrix} \begin{bmatrix} x_1 \\ x_2 \\ x_3 \end{bmatrix}$$
$$= \begin{bmatrix} -x_1 + 2x_2 + x_3 \\ 2x_2 - x_3 \\ x_1 - x_3 \end{bmatrix}.$$

52. (a) $\begin{bmatrix} 0 & 4 & -1 \\ 1 & 0 & 3 \\ 2 & -1 & 2 \end{bmatrix}$ (b) $\frac{1}{3}\begin{bmatrix} -11 & 13 & -4 \\ 7 & 7 & -1 \\ 62 & -31 & 10 \end{bmatrix}$

    (c) $T\left(\begin{bmatrix} x_1 \\ x_2 \\ x_3 \end{bmatrix}\right) = \frac{1}{3}\begin{bmatrix} -11x_1 + 13x_2 - 4x_3 \\ 7x_1 + 7x_2 - x_3 \\ 62x_1 - 31x_2 + 10x_3 \end{bmatrix}$

53. (a) $\begin{bmatrix} 0 & -1 & 2 \\ 3 & 0 & 5 \\ -2 & 4 & 0 \end{bmatrix}$ (b) $\begin{bmatrix} 2 & -7 & -1 \\ -8 & -8 & 11 \\ -4 & -9 & 6 \end{bmatrix}$

(c) $T\left(\begin{bmatrix} x_1 \\ x_2 \\ x_3 \end{bmatrix}\right) = \begin{bmatrix} 2x_1 - 7x_2 - x_3 \\ -8x_1 - 8x_2 + 11x_3 \\ -4x_1 - 9x_2 + 6x_3 \end{bmatrix}$

54. (a) $\begin{bmatrix} 3 & -1 & 5 \\ -2 & 3 & -2 \\ 1 & 0 & -1 \end{bmatrix}$ (b) $\begin{bmatrix} 17 & 6 & -7 \\ -19 & -7 & 9 \\ 15 & 6 & -5 \end{bmatrix}$

(c) $T\left(\begin{bmatrix} x_1 \\ x_2 \\ x_3 \end{bmatrix}\right) = \begin{bmatrix} 17x_1 + 6x_2 - 7x_3 \\ -19x_1 - 7x_2 + 9x_3 \\ 15x_1 + 6x_2 - 5x_3 \end{bmatrix}$

55. We can calculate the image of $3\mathbf{b}_1 - 2\mathbf{b}_2$ directly using the linearity properties of T. Alternatively, we can use the comment before Example 1: $[T(\mathbf{v})]_\mathcal{B} = [T]_\mathcal{B}[\mathbf{v}]_\mathcal{B}$ for every vector $\mathbf{v}$ in $\mathcal{R}^2$. From Exercise 39, we have

$$[T]_\mathcal{B} = \begin{bmatrix} 1 & -3 \\ 4 & 0 \end{bmatrix}.$$

Therefore

$$[T(3\mathbf{b}_1 - 2\mathbf{b}_2)]_\mathcal{B} = [T]_\mathcal{B}[3\mathbf{b}_1 - 2\mathbf{b}_2]_\mathcal{B}$$
$$= [T]_\mathcal{B}\begin{bmatrix} 3 \\ -2 \end{bmatrix} = \begin{bmatrix} 9 \\ 12 \end{bmatrix}.$$

So $T(3\mathbf{b}_1 - 2\mathbf{b}_2) = 9\mathbf{b}_1 + 12\mathbf{b}_2$.

56. $-6\mathbf{b}_1 + 17\mathbf{b}_2$ **57.** $-3\mathbf{b}_1 - 17\mathbf{b}_2$

58. $-10\mathbf{b}_1 + 2\mathbf{b}_2 + 7\mathbf{b}_3$ **59.** $-2\mathbf{b}_1 - 10\mathbf{b}_2 + 15\mathbf{b}_3$

60. $-\mathbf{b}_1 + 12\mathbf{b}_2 + 4\mathbf{b}_3$ **61.** $8\mathbf{b}_1 + 5\mathbf{b}_2 - 16\mathbf{b}_3 - \mathbf{b}_4$

62. $-2\mathbf{b}_1 - 3\mathbf{b}_2 + 4\mathbf{b}_3 + 6\mathbf{b}_4$

63. For any $\mathbf{v}$ in $\mathcal{R}^n$, we have $I(\mathbf{v}) = \mathbf{v}$. Hence if $\mathcal{B} = \{\mathbf{b}_1, \mathbf{b}_2, \ldots, \mathbf{b}_n\}$, then

$$[I]_\mathcal{B} = [[I(\mathbf{b}_1)]_\mathcal{B} \ [I(\mathbf{b}_2)]_\mathcal{B} \ \cdots \ [I(\mathbf{b}_n)]_\mathcal{B}]$$
$$= [[\mathbf{b}_1]_\mathcal{B} \ [\mathbf{b}_2]_\mathcal{B} \ \cdots \ [\mathbf{b}_n]_\mathcal{B}]$$
$$= [\mathbf{e}_1 \ \mathbf{e}_2 \ \cdots \ \mathbf{e}_n]$$
$$= I_n.$$

64. the $n \times n$ zero matrix

65. $T\left(\begin{bmatrix} x_1 \\ x_2 \end{bmatrix}\right) = \begin{bmatrix} .8x_1 + .6x_2 \\ .6x_1 - .8x_2 \end{bmatrix}$

66. $T\left(\begin{bmatrix} x_1 \\ x_2 \end{bmatrix}\right) = \begin{bmatrix} -.6x_1 + .8x_2 \\ .8x_1 + .6x_2 \end{bmatrix}$

67. Take

$$\mathbf{b}_1 = \begin{bmatrix} 1 \\ -2 \end{bmatrix} \quad \text{and} \quad \mathbf{b}_2 = \begin{bmatrix} 2 \\ 1 \end{bmatrix}.$$

Then $\mathbf{b}_1$ lies on the line with equation $y = -2x$, and $\mathbf{b}_2$ is perpendicular to this line. Hence $T(\mathbf{b}_1) = \mathbf{b}_1$ and $T(\mathbf{b}_2) = -\mathbf{b}_2$, so that for $\mathcal{B} = \{\mathbf{b}_1, \mathbf{b}_2\}$, we have

$$[T]_\mathcal{B} = \begin{bmatrix} 1 & 0 \\ 0 & -1 \end{bmatrix}.$$

Thus the standard matrix of T is

$$B[T]_\mathcal{B} B^{-1} = \begin{bmatrix} -.6 & -.8 \\ -.8 & .6 \end{bmatrix},$$

where $B = [\mathbf{b}_1 \ \mathbf{b}_2]$. Therefore

$$T\left(\begin{bmatrix} x_1 \\ x_2 \end{bmatrix}\right) = \begin{bmatrix} -.6x_1 - .8x_2 \\ -.8x_1 + .6x_2 \end{bmatrix}.$$

68. $T\left(\begin{bmatrix} x_1 \\ x_2 \end{bmatrix}\right) = \frac{1}{m^2+1}\begin{bmatrix} (1-m^2)x_1 + 2mx_2 \\ 2mx_1 + (m^2-1)x_2 \end{bmatrix}$

69. $U\left(\begin{bmatrix} x_1 \\ x_2 \end{bmatrix}\right) = \begin{bmatrix} .5x_1 + .5x_2 \\ .5x_1 + .5x_2 \end{bmatrix}$

70. $U\left(\begin{bmatrix} x_1 \\ x_2 \end{bmatrix}\right) = \begin{bmatrix} .8x_1 - .4x_2 \\ -.4x_1 + .2x_2 \end{bmatrix}$

71. Take

$$\mathbf{b}_1 = \begin{bmatrix} 1 \\ -3 \end{bmatrix} \quad \text{and} \quad \mathbf{b}_2 = \begin{bmatrix} 3 \\ 1 \end{bmatrix}.$$

Then $\mathbf{b}_1$ lies on the line with equation $y = -3x$, and $\mathbf{b}_2$ is perpendicular to this line. Hence $U(\mathbf{b}_1) = \mathbf{b}_1$ and $U(\mathbf{b}_2) = \mathbf{0}$, so that for the basis $\mathcal{B} = \{\mathbf{b}_1, \mathbf{b}_2\}$, we have

$$[U]_\mathcal{B} = \begin{bmatrix} 1 & 0 \\ 0 & 0 \end{bmatrix}.$$

It follows that the standard matrix of U is

$$B[U]_\mathcal{B} B^{-1} = \begin{bmatrix} .1 & -.3 \\ -.3 & .9 \end{bmatrix},$$

where $B = [\mathbf{b}_1 \ \mathbf{b}_2]$. Therefore

$$U\left(\begin{bmatrix} x_1 \\ x_2 \end{bmatrix}\right) = \begin{bmatrix} .1x_1 - .3x_2 \\ -.3x_1 + .9x_2 \end{bmatrix}.$$

72. $U\left(\begin{bmatrix} x_1 \\ x_2 \end{bmatrix}\right) = \frac{1}{m^2+1}\begin{bmatrix} x_1 + mx_2 \\ mx_1 + m^2x_2 \end{bmatrix}$

73. (a) $T_W\left(\begin{bmatrix} -2 \\ 1 \\ 0 \end{bmatrix}\right) = \begin{bmatrix} -2 \\ 1 \\ 0 \end{bmatrix},$

$T_W\left(\begin{bmatrix} 3 \\ 0 \\ 1 \end{bmatrix}\right) = \begin{bmatrix} 3 \\ 0 \\ 1 \end{bmatrix},$

and

$T_W\left(\begin{bmatrix} 1 \\ 2 \\ -3 \end{bmatrix}\right) = -\begin{bmatrix} 1 \\ 2 \\ -3 \end{bmatrix}$

(b) $\mathcal{B}$ is a linearly independent set of 3 vectors from $\mathcal{R}^3$.

(c) $\begin{bmatrix} 1 & 0 & 0 \\ 0 & 1 & 0 \\ 0 & 0 & -1 \end{bmatrix}$ (d) $\dfrac{1}{7}\begin{bmatrix} 6 & -2 & 3 \\ -2 & 3 & 6 \\ 3 & 6 & -2 \end{bmatrix}$

(e) $T_W\left(\begin{bmatrix} x_1 \\ x_2 \\ x_3 \end{bmatrix}\right) = \dfrac{1}{7}\begin{bmatrix} 6x_1 - 2x_2 + 3x_3 \\ -2x_1 + 3x_2 + 6x_3 \\ 3x_1 + 6x_2 - 2x_3 \end{bmatrix}$

74. $T_W\left(\begin{bmatrix} x_1 \\ x_2 \\ x_3 \end{bmatrix}\right) = \dfrac{1}{3}\begin{bmatrix} -x_1 + 2x_2 - 2x_3 \\ 2x_1 + 2x_2 + x_3 \\ -2x_1 + x_2 + 2x_3 \end{bmatrix}$

75. $T_W\left(\begin{bmatrix} x_1 \\ x_2 \\ x_3 \end{bmatrix}\right) = \dfrac{1}{13}\begin{bmatrix} 12x_1 + 4x_2 - 3x_3 \\ 4x_1 - 3x_2 + 12x_3 \\ -3x_1 + 12x_2 + 4x_3 \end{bmatrix}$

76. $T_W\left(\begin{bmatrix} x_1 \\ x_2 \\ x_3 \end{bmatrix}\right) = \dfrac{1}{15}\begin{bmatrix} 14x_1 - 2x_2 + 5x_3 \\ -2x_1 + 11x_2 + 10x_3 \\ 5x_1 + 10x_2 - 10x_3 \end{bmatrix}$

77. $T_W\left(\begin{bmatrix} x_1 \\ x_2 \\ x_3 \end{bmatrix}\right) = \dfrac{1}{41}\begin{bmatrix} 39x_1 - 12x_2 + 4x_3 \\ -12x_1 - 31x_2 + 24x_3 \\ 4x_1 + 24x_2 + 33x_3 \end{bmatrix}$

78. $T_W\left(\begin{bmatrix} x_1 \\ x_2 \\ x_3 \end{bmatrix}\right) = \dfrac{1}{35}\begin{bmatrix} 33x_1 + 6x_2 - 10x_3 \\ 6x_1 + 17x_2 + 30x_3 \\ -10x_1 + 30x_2 - 15x_3 \end{bmatrix}$

79. $T_W\left(\begin{bmatrix} x_1 \\ x_2 \\ x_3 \end{bmatrix}\right) = \dfrac{1}{21}\begin{bmatrix} 19x_1 + 4x_2 + 8x_3 \\ 4x_1 + 13x_2 - 16x_3 \\ 8x_1 - 16x_2 - 11x_3 \end{bmatrix}$

80. $T_W\left(\begin{bmatrix} x_1 \\ x_2 \\ x_3 \end{bmatrix}\right) = \dfrac{1}{75}\begin{bmatrix} 73x_1 - 10x_2 - 14x_3 \\ -10x_1 + 25x_2 - 70x_3 \\ -14x_1 - 70x_2 - 23x_3 \end{bmatrix}$

81. Let
$$\mathbf{b}_1 = \begin{bmatrix} -2 \\ 1 \\ 0 \end{bmatrix},\ \mathbf{b}_2 = \begin{bmatrix} 3 \\ 0 \\ 1 \end{bmatrix},\ \text{and } \mathbf{b}_3 = \begin{bmatrix} 1 \\ 2 \\ -3 \end{bmatrix}.$$

(a) Since $\mathbf{b}_1$ and $\mathbf{b}_2$ lie in W, we have $U_W(\mathbf{b}_1) = \mathbf{b}_1$ and $U_W(\mathbf{b}_2) = \mathbf{b}_2$. Moreover, since $\mathbf{b}_3$ is perpendicular to W, $U_W(\mathbf{b}_3) = \mathbf{0}$.

(b) $[U_W]_\mathcal{B} =$
$[[U_W(\mathbf{b}_1)]_\mathcal{B}\ [U_W(\mathbf{b}_2)]_\mathcal{B}\ [U_W(\mathbf{b}_3)]_\mathcal{B}] =$
$[\mathbf{e}_1\ \mathbf{e}_2\ \mathbf{0}] = \begin{bmatrix} 1 & 0 & 0 \\ 0 & 1 & 0 \\ 0 & 0 & 0 \end{bmatrix}$.

(c) Let $B = [\mathbf{b}_1\ \mathbf{b}_2\ \mathbf{b}_3]$. By Theorem 4.12, the standard matrix of U_W is
$$B[U_W]_\mathcal{B} B^{-1} = \dfrac{1}{14}\begin{bmatrix} 13 & -2 & 3 \\ -2 & 10 & 6 \\ 3 & 6 & 5 \end{bmatrix}.$$

(d) Using the preceding standard matrix of U_W, we have
$$U_W\left(\begin{bmatrix} x_1 \\ x_2 \\ x_3 \end{bmatrix}\right) = \dfrac{1}{14}\begin{bmatrix} 13x_1 - 2x_2 + 3x_3 \\ -2x_1 + 10x_2 + 6x_3 \\ 3x_1 + 6x_2 + 5x_3 \end{bmatrix}.$$

82. $U_W\left(\begin{bmatrix} x_1 \\ x_2 \\ x_3 \end{bmatrix}\right) = \dfrac{1}{6}\begin{bmatrix} 5x_1 - x_2 + 2x_3 \\ -x_1 + 5x_2 + 2x_3 \\ 2x_1 + 2x_2 + 2x_3 \end{bmatrix}$

83. $U_W\left(\begin{bmatrix} x_1 \\ x_2 \\ x_3 \end{bmatrix}\right) = \dfrac{1}{30}\begin{bmatrix} 29x_1 + 2x_2 - 5x_3 \\ 2x_1 + 26x_2 + 10x_3 \\ -5x_1 + 10x_2 + 5x_3 \end{bmatrix}$

84. $U_W\left(\begin{bmatrix} x_1 \\ x_2 \\ x_3 \end{bmatrix}\right) = \dfrac{1}{26}\begin{bmatrix} 25x_1 - 4x_2 + 3x_3 \\ -4x_1 + 10x_2 + 12x_3 \\ 3x_1 + 12x_2 + 17x_3 \end{bmatrix}$

85. $U_W\left(\begin{bmatrix} x_1 \\ x_2 \\ x_3 \end{bmatrix}\right) = \dfrac{1}{35}\begin{bmatrix} 34x_1 + 3x_2 + 5x_3 \\ 3x_1 + 26x_2 - 15x_3 \\ 5x_1 - 15x_2 + 10x_3 \end{bmatrix}$

86. $U_W\left(\begin{bmatrix} x_1 \\ x_2 \\ x_3 \end{bmatrix}\right) = \dfrac{1}{41}\begin{bmatrix} 40x_1 - 6x_2 - 2x_3 \\ -6x_1 + 5x_2 - 12x_3 \\ -2x_1 - 12x_2 + 37x_3 \end{bmatrix}$

87. $U_W\left(\begin{bmatrix} x_1 \\ x_2 \\ x_3 \end{bmatrix}\right) = \dfrac{1}{75}\begin{bmatrix} 74x_1 + 5x_2 - 7x_3 \\ 5x_1 + 50x_2 + 35x_3 \\ -7x_1 + 35x_2 + 26x_3 \end{bmatrix}$

88. $U_W\left(\begin{bmatrix} x_1 \\ x_2 \\ x_3 \end{bmatrix}\right) = \dfrac{1}{21}\begin{bmatrix} 20x_1 - 2x_2 + 4x_3 \\ -2x_1 + 17x_2 + 8x_3 \\ 4x_1 + 8x_2 + 5x_3 \end{bmatrix}$

89. Let A be the standard matrix of T. By Theorem 2.13, T is invertible if and only if A is invertible. So if B is the matrix whose columns are the vectors in $\mathcal{B}$, then T is invertible if and only if $B^{-1}AB = [T]_\mathcal{B}$ is invertible.

90. Let B be the matrix whose columns are the vectors in $\mathcal{B}$, and let A and C be the standard matrices of T and U, respectively. Then by Theorems 4.12 and 2.12,
$$[UT]_\mathcal{B} = B^{-1}(CA)B$$
$$= (B^{-1}CB)(B^{-1}AB) = [U]_\mathcal{B}[T]_\mathcal{B}.$$

91. Let B be the matrix whose columns are the vectors in $\mathcal{B}$, and let A be the standard matrix of T. Then, by Theorem 4.12 and Exercises 68 and 70 in Section 2.4,
$$\operatorname{rank}[T]_\mathcal{B} = \operatorname{rank}(B^{-1}AB) = \operatorname{rank}A.$$

Since the range of T equals $\operatorname{Col}A$, the rank of $[T]_\mathcal{B}$ equals the dimension of $\operatorname{Col}A$, which is the dimension of the range of T.

92. Let B be the matrix whose columns are the vectors in $\mathcal{B}$, and let A be the standard matrix of T. The null space of T equals Null A, and so the dimension of the null space of T equals $n - \operatorname{rank} A = n - \operatorname{rank} [T]_\mathcal{B}$ by the displayed line in the solution of Exercise 91.

93. Let B be the matrix whose columns are the vectors in $\mathcal{B}$, and let A and C be the standard matrices of T and U, respectively. By Exercise 86 of Section 2.7, we have
$$[T+U]_\mathcal{B} = B^{-1}(A+C)B$$
$$= B^{-1}AB + B^{-1}CB = [T]_\mathcal{B} + [U]_\mathcal{B}.$$

94. Let B be the matrix whose columns are the vectors in $\mathcal{B}$, and let A be the standard matrix of T. By Exercise 85 in Section 2.7, for any scalar c we have
$$[cT]_\mathcal{B} = B^{-1}(cA)B = c(B^{-1}AB) = c[T]_\mathcal{B}.$$

95. Let A and B be the matrices whose columns are the vectors in $\mathcal{A}$ and $\mathcal{B}$, respectively, and let C be the standard matrix of T. Then
$$[T]_\mathcal{A} = A^{-1}CA \quad \text{and} \quad [T]_\mathcal{B} = B^{-1}CB.$$
So
$$[T]_\mathcal{A} = A^{-1}(B[T]_\mathcal{B}B^{-1})A$$
$$= (B^{-1}A)^{-1}[T]_\mathcal{B}(B^{-1}A),$$
and thus $[T]_\mathcal{A}$ and $[T]_\mathcal{B}$ are similar.

96. If $B = P^{-1}AP$, take $\mathcal{A} = \{\mathbf{e}_1, \mathbf{e}_2, \ldots, \mathbf{e}_n\}$ and $\mathcal{B} = \{\mathbf{p}_1, \mathbf{p}_2, \ldots, \mathbf{p}_n\}$.

97. Let T be a reflection of $\mathcal{R}^2$ about a line, and let A be the standard matrix of T. As shown in Section 4.5, there is a basis $\mathcal{B}$ for $\mathcal{R}^2$ such that
$$[T]_\mathcal{B} = \begin{bmatrix} 1 & 0 \\ 0 & -1 \end{bmatrix}.$$
Let B be the matrix whose columns are the vectors in $\mathcal{B}$. By Theorem 4.12, we have $A = B[T]_\mathcal{B}B^{-1}$, and so
$$\det A = (\det B)\left(\det \begin{bmatrix} 1 & 0 \\ 0 & -1 \end{bmatrix}\right)\left(\frac{1}{\det B}\right)$$
$$= -1.$$

98. (a) Let B be the matrix whose columns are the vectors in $\mathcal{B}$, and let $A = CB^{-1}$. The matrix transformation T induced by A satisfies
$$[T(\mathbf{b}_1) \quad T(\mathbf{b}_2) \quad \cdots \quad T(\mathbf{b}_n)]$$

$$= [A\mathbf{b}_1 \quad A\mathbf{b}_2 \quad \cdots \quad A\mathbf{b}_n]$$
$$= A[\mathbf{b}_1 \quad \mathbf{b}_2 \quad \cdots \quad \mathbf{b}_n]$$
$$= AB$$
$$= (CB^{-1})B$$
$$= C$$
$$= [\mathbf{c}_1 \quad \mathbf{c}_2 \quad \cdots \quad \mathbf{c}_n].$$
Thus $T(\mathbf{b}_j) = \mathbf{c}_j$ for $j = 1, 2, \ldots, n$.

(b) To prove uniqueness, assume that U is a linear operator on $\mathcal{R}^n$ such that $U(\mathbf{b}_j) = \mathbf{c}_j$ for $j = 1, 2, \ldots, n$. For every $\mathbf{v}$ in $\mathcal{R}^n$, there are unique scalars $a_1, a_2, \ldots, a_n$ such that $\mathbf{v} = a_1\mathbf{b}_1 + a_2\mathbf{b}_2 \cdots + a_n\mathbf{b}_n$. So
$$U(\mathbf{v}) = U(a_1\mathbf{b}_1 + \cdots + a_n\mathbf{b}_n)$$
$$= a_1U(\mathbf{b}_1) + \cdots + a_nU(\mathbf{b}_n)$$
$$= a_1T(\mathbf{b}_1) + \cdots + a_nT(\mathbf{b}_n)$$
$$= T(a_1\mathbf{b}_1 + \cdots + a_n\mathbf{b}_n)$$
$$= T(\mathbf{v})$$
for every $\mathbf{v}$ in $\mathcal{R}^n$. Thus $U = T$.

(c) The desired generalization is as follows. Let $\mathcal{B} = \{\mathbf{b}_1, \mathbf{b}_2, \ldots, \mathbf{b}_n\}$ be a basis for $\mathcal{R}^n$ and $\mathbf{c}_1, \mathbf{c}_2, \ldots, \mathbf{c}_n$ be (not necessarily distinct) vectors in $\mathcal{R}^m$. A unique linear transformation $T\colon \mathcal{R}^n \to \mathcal{R}^m$ exists such that $T(\mathbf{b}_j) = \mathbf{c}_j$ for each j. Let B and C be the $n \times n$ and $m \times n$ matrices whose jth columns are $\mathbf{b}_j$ and $\mathbf{c}_j$, respectively. If A is the standard matrix of T, then $A = CB^{-1}$. The proof of this generalization is similar to the proofs in (a) and (b).

99. Consider column j of $[T]_\mathcal{B}$. Let
$$[T(\mathbf{b}_j)]_\mathcal{B} = \begin{bmatrix} c_1 \\ c_2 \\ \vdots \\ c_n \end{bmatrix}.$$
Because the jth column of $[T]_\mathcal{B}$ is $[T(\mathbf{b}_j)]_\mathcal{B}$, if $[T]_\mathcal{B}$ is an upper triangular matrix, then $c_i = 0$ for $i > j$. Thus we have
$$T(\mathbf{b}_j) = c_1\mathbf{b}_1 + c_2\mathbf{b}_2 + \cdots + c_n\mathbf{b}_n$$
$$= c_1\mathbf{b}_1 + c_2\mathbf{b}_2 + \cdots + c_j\mathbf{b}_j,$$
which is a linear combination of $\mathbf{b}_1, \mathbf{b}_2, \ldots, \mathbf{b}_j$. Conversely, if $T(\mathbf{b}_j)$ is a linear combination of $\mathbf{b}_1, \mathbf{b}_2, \ldots, \mathbf{b}_j$ for each j, then the (i, j)-entry of the matrix $[T]_\mathcal{B}$ equals 0 for $i > j$. Hence $[T]_\mathcal{B}$ is an upper triangular matrix.

100. (a) Let A be the standard matrix of T and B be the matrix whose columns are the vectors in $\mathcal{B}$. Then $[T]_\mathcal{B} = B^{-1}AB$ by Theorem 4.12. Because $B[\mathbf{v}]_\mathcal{B} = \mathbf{v}$, $B^{-1}\mathbf{v} = [\mathbf{v}]_\mathcal{B}$, and $T(\mathbf{v}) = A\mathbf{v}$ for all $\mathbf{v}$ in $\mathcal{R}^n$, we have

$$[T(\mathbf{v})]_\mathcal{B} = B^{-1}(T(\mathbf{v}))$$
$$= B^{-1}(A\mathbf{v}) = (B^{-1}A)\mathbf{v}$$
$$= (B^{-1}A)(B[\mathbf{v}]_\mathcal{B})$$
$$= (B^{-1}AB)[\mathbf{v}]_\mathcal{B}$$
$$= [T]_\mathcal{B}[\mathbf{v}]_\mathcal{B}.$$

(b) In order to show uniqueness, suppose that C is another $n \times n$ matrix such that $[T(\mathbf{v})]_\mathcal{B} = C[\mathbf{v}]_\mathcal{B}$ for all $\mathbf{v}$ in $\mathcal{R}^n$. Then $C[\mathbf{v}]_\mathcal{B} = [T]_\mathcal{B}[\mathbf{v}]_\mathcal{B}$ for all $\mathbf{v}$ in $\mathcal{R}^n$. Let $\mathcal{B} = \{\mathbf{b}_1, \mathbf{b}_2, \ldots, \mathbf{b}_n\}$. Since $[\mathbf{b}_j]_\mathcal{B} = \mathbf{e}_j$ for each j, we have

$$C\mathbf{e}_j = C[\mathbf{b}_j]_\mathcal{B} = [T]_\mathcal{B}[\mathbf{b}_j]_\mathcal{B} = [T]_\mathcal{B}\mathbf{e}_j$$

for every j. But $C\mathbf{e}_j$ is the jth column of C and $[T]_\mathcal{B}\mathbf{e}_j$ is the jth column of $[T]_\mathcal{B}$, and so the corresponding columns of C and $[T]_\mathcal{B}$ are equal. Thus $C = [T]_\mathcal{B}$.

101. (a) Since $\mathcal{B}$ is a linearly independent set of three vectors from $\mathcal{R}^3$, it is a basis for $\mathcal{R}^3$ by Theorem 4.7. Likewise $\mathcal{C}$ is a basis for $\mathcal{R}^2$.

(b) Column j of $[T]_\mathcal{B}^\mathcal{C}$ is the $\mathcal{C}$-coordinate vector of $T(\mathbf{b}_j)$. Let C denote the matrix whose columns are the vectors in $\mathcal{C}$. Now

$$T\left(\begin{bmatrix}1\\1\\1\end{bmatrix}\right) = \begin{bmatrix}2\\2\end{bmatrix}, \quad T\left(\begin{bmatrix}1\\-1\\1\end{bmatrix}\right) = \begin{bmatrix}-2\\4\end{bmatrix},$$

and

$$T\left(\begin{bmatrix}1\\1\\-1\end{bmatrix}\right) = \begin{bmatrix}4\\-2\end{bmatrix}.$$

So the respective $\mathcal{C}$-coordinate vectors are

$$C^{-1}\begin{bmatrix}2\\2\end{bmatrix} = \begin{bmatrix}-2\\2\end{bmatrix}, \quad C^{-1}\begin{bmatrix}-2\\4\end{bmatrix} = \begin{bmatrix}14\\-8\end{bmatrix},$$

and

$$C^{-1}\begin{bmatrix}4\\-2\end{bmatrix} = \begin{bmatrix}-16\\10\end{bmatrix}.$$

Thus $[T]_\mathcal{B}^\mathcal{C} = \begin{bmatrix}-2 & 14 & -16\\ 2 & -8 & 10\end{bmatrix}$.

102. (a) Let A denote the standard matrix of T. Column j of $[T]_\mathcal{B}^\mathcal{C}$ is the $\mathcal{C}$-coordinate vector of $T(\mathbf{b}_j)$, which is $C^{-1}T(\mathbf{b}_j) = C^{-1}A\mathbf{b}_j$. Thus

$$[T]_\mathcal{B}^\mathcal{C} = [C^{-1}A\mathbf{b}_1 \; C^{-1}A\mathbf{b}_2 \; \cdots \; C^{-1}A\mathbf{b}_n]$$
$$= C^{-1}A[\mathbf{b}_1 \; \mathbf{b}_2 \; \cdots \; \mathbf{b}_n] = C^{-1}AB.$$

(b) Let $U: \mathcal{R}^n \to \mathcal{R}^m$ be a linear transformation with standard matrix P.

(i) By Exercise 86 in Section 2.7, the standard matrix of $T + U$ is $A + P$. Thus, by (a), we have

$$[T + U]_\mathcal{B}^\mathcal{C} = C^{-1}(A + P)B$$
$$= C^{-1}AB + C^{-1}PB$$
$$= [T]_\mathcal{B}^\mathcal{C} + [U]_\mathcal{B}^\mathcal{C}.$$

(ii) By Exercise 85 in Section 2.7, the standard matrix of sT is sA. Thus, by (a), we have

$$[sT]_\mathcal{B}^\mathcal{C} = C^{-1}(sA)B$$
$$= s(C^{-1}AB) = s[T]_\mathcal{B}^\mathcal{C}.$$

(iii) For any vector $\mathbf{v}$ in $\mathcal{R}^n$, we have

$$[T(\mathbf{v})]_\mathcal{C} = C^{-1}T(\mathbf{v}) = C^{-1}A\mathbf{v}$$
$$= (C^{-1}AB)B^{-1}\mathbf{v} = [T]_\mathcal{B}^\mathcal{C}[\mathbf{v}]_\mathcal{B}.$$

(c) Let D be the matrix whose columns are the vectors in $\mathcal{D}$, and let P be the standard matrix of U. By Theorem 2.12, the standard matrix of the composition UT is PA. Thus, by (a), we have

$$[U]_\mathcal{C}^\mathcal{D}[T]_\mathcal{B}^\mathcal{C} = (D^{-1}PC)(C^{-1}AB)$$
$$= D^{-1}(PA)B = [UT]_\mathcal{B}^\mathcal{D}.$$

103. (a) Let $B = [\mathbf{b}_1 \; \mathbf{b}_2 \; \mathbf{b}_3 \; \mathbf{b}_4]$. The standard matrices of T and U are

$$A = \begin{bmatrix} 1 & -2 & 0 & 0 \\ 0 & 0 & 1 & 0 \\ -1 & 0 & 3 & 0 \\ 0 & 2 & 0 & -1 \end{bmatrix}$$

and

$$C = \begin{bmatrix} 0 & 1 & -1 & 2 \\ -2 & 0 & 0 & 3 \\ 0 & 2 & -1 & 0 \\ 3 & 0 & 0 & 1 \end{bmatrix},$$

respectively. Hence, by Theorem 2.12, the standard matrix of UT is

$$CA = \begin{bmatrix} 1 & 4 & -2 & -2 \\ -2 & 10 & 0 & -3 \\ 1 & 0 & -1 & 0 \\ 3 & -4 & 0 & -1 \end{bmatrix}.$$

Thus, by Theorem 4.12, we have

$$[T]_\mathcal{B} = B^{-1}AB$$
$$= \begin{bmatrix} 11 & 5 & 13 & 1 \\ -2 & 0 & -5 & -3 \\ -8 & -3 & -9 & 0 \\ 6 & 1 & 8 & 1 \end{bmatrix},$$

$$[U]_\mathcal{B} = B^{-1}CB$$
$$= \begin{bmatrix} -5 & 10 & -38 & -31 \\ 2 & -3 & 9 & 6 \\ 6 & -10 & 27 & 17 \\ -4 & 7 & -25 & -19 \end{bmatrix},$$

and

$$[UT]_\mathcal{B} = B^{-1}(CA)B$$
$$= \begin{bmatrix} 43 & 58 & -21 & -66 \\ -8 & -11 & 8 & 17 \\ -28 & -34 & 21 & 53 \\ 28 & 36 & -14 & -44 \end{bmatrix}.$$

(b) From (a), we see that

$$[T]_\mathcal{B}[U]_\mathcal{B} = (B^{-1}AB)(B^{-1}CB)$$
$$= B^{-1}AI_4CB$$
$$= B^{-1}ACB$$
$$= [UT]_\mathcal{B}.$$

104. See the solution to Exercise 90.

105. (a) $\begin{bmatrix} 0 & 0 & 0 & 1 \\ 1 & 0 & 0 & 0 \\ 0 & 1 & 0 & 0 \\ 0 & 0 & 1 & 0 \end{bmatrix}$,

$$T\left(\begin{bmatrix} x_1 \\ x_2 \\ x_3 \\ x_4 \end{bmatrix}\right) = \begin{bmatrix} 8x_1 - 4x_2 + 3x_3 + x_4 \\ -11x_1 + 7x_2 - 4x_3 - 2x_4 \\ -35x_1 + 20x_2 - 13x_3 - 5x_4 \\ -9x_1 + 4x_2 - 3x_3 - 2x_4 \end{bmatrix}$$

106. (a) $T^{-1}\left(\begin{bmatrix} x_1 \\ x_2 \\ x_3 \\ x_4 \end{bmatrix}\right)$

$$= \begin{bmatrix} 6x_1 - x_2 - 4x_3 - 2x_4 \\ 19x_1 - 3x_2 - 13x_3 - 4x_4 \\ 17x_1 - 3x_2 - 12x_3 - 4x_4 \\ -4x_1 + x_2 + 3x_3 + x_4 \end{bmatrix}$$

(b) $[T]_\mathcal{B} = \begin{bmatrix} 83 & 25 & -114 & -256 \\ -20 & -7 & 29 & 63 \\ -55 & -18 & 78 & 172 \\ 52 & 16 & -72 & -161 \end{bmatrix}$

and

$$[T^{-1}]_\mathcal{B} = \begin{bmatrix} 8 & -6 & 1 & -14 \\ -13 & -12 & 15 & 32 \\ -15 & -7 & 13 & 35 \\ 8 & 0 & -4 & -17 \end{bmatrix}$$

(c) $[T^{-1}]_\mathcal{B} = [T]_\mathcal{B}^{-1}$

107. We claim that $[T^{-1}]_\mathcal{B} = [T]_\mathcal{B}^{-1}$. Let A be the standard matrix of T and B be the matrix whose columns are the vectors in $\mathcal{B}$. By Theorem 2.13, A is invertible because T is invertible. Moreover, the standard matrix of T^{-1} is A^{-1}. Thus

$$[T^{-1}]_\mathcal{B} = B^{-1}A^{-1}B = (B^{-1}AB)^{-1} = [T]_\mathcal{B}^{-1}.$$

CHAPTER 4 REVIEW EXERCISES

1. True 2. True
3. False, the null space of an $m \times n$ matrix is contained in $\mathcal{R}^n$.
4. False, the column space of an $m \times n$ matrix is contained in $\mathcal{R}^m$.
5. False, the row space of an $m \times n$ matrix is contained in $\mathcal{R}^n$.
6. True 7. True
8. False, the range of every linear transformation equals the *column space* of its standard matrix.
9. False, a nonzero subspace of $\mathcal{R}^n$ has infinitely many bases.
10. False, every basis for a particular subspace contains the same number of vectors.
11. True 12. True 13. True
14. True 15. True 16. True
17. False, the dimension of the null space of a matrix equals the *nullity* of the matrix.
18. True
19. False, the dimension of the row space of a matrix equals the *rank* of the matrix.
20. False, consider $\begin{bmatrix} 1 & 2 \\ 1 & 2 \end{bmatrix}$.
21. True 22. True
23. False, $[T]_\mathcal{B} = B^{-1}AB$.
24. True 25. True

26. (a) Misuse, the term "basis" applies only to a subspace.

(b) Misuse, the term "rank" applies to a matrix.

(c) Misuse, the term "dimension" applies only to a subspace.

(d) Correct use

(e) Misuse, the term "dimension" applies only to a subspace.

(f) Misuse, the term "column space" applies only to a matrix.

(g) Misuse, the term "dimension" applies only to a subspace.

(h) Misuse, the term "coordinate vector" applies only to a vector.

27. (a) There are at most k vectors in a linearly independent subset of V.

(b) No conclusions can be drawn about the values of k and m in this case.

(c) There are at least k vectors in a generating set for V.

28. (a) 2 (b) 3 (c) 2 (d) 5 (e) 3

29. No, $\begin{bmatrix} -1 \\ 0 \\ 1 \\ 0 \end{bmatrix}$ is in the set, but $(-1)\begin{bmatrix} -1 \\ 0 \\ 1 \\ 0 \end{bmatrix}$ is not.

30. Yes, the given set is the null space of
$$\begin{bmatrix} 1 & 0 & 0 & 0 \\ 0 & 1 & -5 & 0 \\ 0 & 0 & 0 & 1 \end{bmatrix}.$$

31. (a) $\left\{ \begin{bmatrix} -3 \\ 2 \\ 1 \end{bmatrix} \right\}$ (b) $\left\{ \begin{bmatrix} 1 \\ -1 \\ 2 \\ 1 \end{bmatrix}, \begin{bmatrix} 2 \\ -1 \\ 1 \\ 4 \end{bmatrix} \right\}$

(c) $\left\{ \begin{bmatrix} 1 \\ 0 \\ 3 \end{bmatrix}, \begin{bmatrix} 0 \\ 1 \\ -2 \end{bmatrix} \right\}$

32. (a) $\left\{ \begin{bmatrix} 1 \\ 1 \\ 0 \\ 0 \\ 0 \end{bmatrix}, \begin{bmatrix} -2 \\ 0 \\ -3 \\ 1 \\ 2 \end{bmatrix} \right\}$

(b) $\left\{ \begin{bmatrix} -1 \\ 2 \\ 1 \\ 1 \end{bmatrix}, \begin{bmatrix} 2 \\ -1 \\ 1 \\ 4 \end{bmatrix}, \begin{bmatrix} 2 \\ -3 \\ 1 \\ 8 \end{bmatrix} \right\}$

(c) $\left\{ \begin{bmatrix} 1 \\ -1 \\ 0 \\ 0 \\ 1 \end{bmatrix}, \begin{bmatrix} 0 \\ 0 \\ 2 \\ 0 \\ 3 \end{bmatrix}, \begin{bmatrix} 0 \\ 0 \\ 0 \\ 2 \\ -1 \end{bmatrix} \right\}$

33. The standard matrix of T is
$$A = \begin{bmatrix} 0 & 1 & -2 \\ -1 & 3 & 1 \\ 1 & -4 & 1 \\ 2 & -1 & 3 \end{bmatrix},$$

and its reduced row echelon form is $[\mathbf{e}_1 \ \mathbf{e}_2 \ \mathbf{e}_3]$.

(a) The set
$$\left\{ \begin{bmatrix} 0 \\ -1 \\ 1 \\ 2 \end{bmatrix}, \begin{bmatrix} 1 \\ 3 \\ -4 \\ -1 \end{bmatrix}, \begin{bmatrix} -2 \\ 1 \\ 1 \\ 3 \end{bmatrix} \right\}$$

of pivot columns of A is a basis for the range of T.

(b) The only solution of $A\mathbf{x} = \mathbf{0}$ is $\mathbf{x} = \mathbf{0}$; so the null space of T is the zero subspace.

34. (a) $\left\{ \begin{bmatrix} 1 \\ -2 \end{bmatrix}, \begin{bmatrix} -2 \\ 3 \end{bmatrix} \right\}$ (b) $\left\{ \begin{bmatrix} -3 \\ -1 \\ 1 \\ 0 \end{bmatrix}, \begin{bmatrix} -5 \\ -4 \\ 0 \\ 1 \end{bmatrix} \right\}$

35. From Exercise 34, we see that the null space of T is 2-dimensional, and the given set is a linearly independent subset of the null space that contains 2 vectors.

36. From Exercise 32, we see that the column space of the matrix is 3-dimensional, and the given set is a linearly independent subset of the column space that contains 3 vectors.

37. Let B be the matrix whose columns are the vectors in $\mathcal{B}$.

(a) Since the reduced row echelon form of B is I_3, $\mathcal{B}$ is a linearly independent subset of $\mathcal{R}^3$. Since $\mathcal{B}$ contains exactly 3 vectors, $\mathcal{B}$ is a basis for $\mathcal{R}^3$ by Theorem 4.7.

(b) We have
$$\mathbf{v} = 4 \begin{bmatrix} 0 \\ -1 \\ 1 \end{bmatrix} - 3 \begin{bmatrix} 1 \\ 0 \\ -1 \end{bmatrix} - 2 \begin{bmatrix} -1 \\ -1 \\ 1 \end{bmatrix} = \begin{bmatrix} -1 \\ -2 \\ 5 \end{bmatrix}.$$

(c) By Theorem 4.11, we have
$$[\mathbf{w}]_{\mathcal{B}} = B^{-1}\mathbf{w} = \begin{bmatrix} 1 \\ -8 \\ -6 \end{bmatrix}.$$

38. (a) $\begin{bmatrix} 0 & 4 & 5 \\ -2 & 0 & -4 \\ 1 & -3 & 2 \end{bmatrix}$ (b) $\frac{19}{30}\mathbf{b}_1 + \frac{5}{6}\mathbf{b}_2 - \frac{1}{15}\mathbf{b}_3$

39. (a) $\begin{bmatrix} -17 & 1 \\ -10 & 1 \end{bmatrix}$ (b) $\begin{bmatrix} -7 & -5 \\ -14 & -9 \end{bmatrix}$

(c) $T\left(\begin{bmatrix} x_1 \\ x_2 \end{bmatrix}\right) = \begin{bmatrix} -7x_1 - 5x_2 \\ -14x_1 - 9x_2 \end{bmatrix}$

40. $\begin{bmatrix} 9 & 26 \\ -3 & -9 \end{bmatrix}$

41. Let B be the matrix whose columns are the vectors in $\mathcal{B}$. By Theorem 4.12, the standard matrix of T is

$$B[T]_\mathcal{B} B^{-1} = \begin{bmatrix} 1 & 6 & -5 \\ -4 & 4 & 5 \\ -1 & 3 & 1 \end{bmatrix}.$$

Hence

$$T\left(\begin{bmatrix} x_1 \\ x_2 \\ x_3 \end{bmatrix}\right) = \begin{bmatrix} x_1 + 6x_2 - 5x_3 \\ -4x_1 + 4x_2 + 5x_3 \\ -x_1 + 3x_2 + x_3 \end{bmatrix}.$$

42. Proceeding as in Example 3 in Section 4.5, we obtain $T\left(\begin{bmatrix} x_1 \\ x_2 \\ x_3 \end{bmatrix}\right) = \begin{bmatrix} x_1 + x_3 \\ -4x_1 + 2x_2 + 5x_3 \\ -2x_1 + x_2 \end{bmatrix}$.

43. Let $\mathbf{v} = \begin{bmatrix} x \\ y \end{bmatrix}$ and $[\mathbf{v}]_\mathcal{B} = \begin{bmatrix} x' \\ y' \end{bmatrix}$, where $\mathcal{B}$ is the basis obtained by rotating the vectors in the standard basis by 120°. As in Section 4.4, we have

$$\begin{bmatrix} x' \\ y' \end{bmatrix} = [\mathbf{v}]_\mathcal{B} = (A_{120°})^{-1}\mathbf{v}$$

$$= (A_{120°})^T \mathbf{v} = \begin{bmatrix} -\frac{1}{2} & \frac{\sqrt{3}}{2} \\ -\frac{\sqrt{3}}{2} & -\frac{1}{2} \end{bmatrix} \begin{bmatrix} x \\ y \end{bmatrix}.$$

Hence

$$x' = -\tfrac{1}{2}x + \tfrac{\sqrt{3}}{2}y$$
$$y' = -\tfrac{\sqrt{3}}{2}x - \tfrac{1}{2}y.$$

Rewrite the given equation in the form

$$9(x')^2 + 4(y')^2 = 36,$$

and substitute the expressions above for x' and y'. The resulting equation is

$$\frac{21}{4}x^2 - \frac{5\sqrt{3}}{2}xy + \frac{31}{4}y^2 = 36,$$

that is,

$$21x^2 - 10\sqrt{3}xy + 31y^2 = 144.$$

44. $xy = 3$

45. Let

$$\mathbf{v} = \begin{bmatrix} x \\ y \end{bmatrix} \text{ and } [\mathbf{v}]_\mathcal{B} = \begin{bmatrix} x' \\ y' \end{bmatrix},$$

where $\mathcal{B}$ is the basis obtained by rotating $\mathbf{e}_1$ and $\mathbf{e}_2$ through 315°. Then

$$\begin{bmatrix} x \\ y \end{bmatrix} = \mathbf{v} = A_{315°}[\mathbf{v}]_\mathcal{B} = \begin{bmatrix} \frac{\sqrt{2}}{2} & \frac{\sqrt{2}}{2} \\ -\frac{\sqrt{2}}{2} & \frac{\sqrt{2}}{2} \end{bmatrix} \begin{bmatrix} x' \\ y' \end{bmatrix}.$$

Hence

$$x = \tfrac{\sqrt{2}}{2}x' + \tfrac{\sqrt{2}}{2}y'$$
$$y = -\tfrac{\sqrt{2}}{2}x' + \tfrac{\sqrt{2}}{2}y'.$$

Substituting these expressions for x and y transforms $29x^2 - 42xy + 29y^2 = 200$ into the form

$$50(x')^2 + 8(y')^2 = 200,$$

that is,

$$\frac{(x')^2}{4} + \frac{(y')^2}{25} = 1.$$

46. $\dfrac{(y')^2}{16} - \dfrac{(x')^2}{9} = 1$

47. $T\left(\begin{bmatrix} x_1 \\ x_2 \end{bmatrix}\right) = \dfrac{1}{13}\begin{bmatrix} -5x_1 - 12x_2 \\ -12x_1 + 5x_2 \end{bmatrix}$

48. $U\left(\begin{bmatrix} x_1 \\ x_2 \end{bmatrix}\right) = \dfrac{1}{13}\begin{bmatrix} 4x_1 - 6x_2 \\ -6x_1 + 9x_2 \end{bmatrix}$

49. Let $B = \begin{bmatrix} \mathbf{v}_1 & \mathbf{v}_2 & \ldots & \mathbf{v}_n \end{bmatrix}$. Because the columns of B form a basis for $\mathcal{R}^n$, they are linearly independent. Hence B is invertible by Theorem 2.6. Moreover, by the definition of matrix multiplication, $A\mathbf{v}_1, A\mathbf{v}_2, \ldots, A\mathbf{v}_n$ are the columns of AB. Since both A and B are invertible, it follows from Theorem 2.2(b) that AB is invertible. Therefore the columns of AB are linearly independent and a generating set for $\mathcal{R}^n$ by (d) and (g) of Theorem 2.6. It follows that $\{A\mathbf{v}_1, A\mathbf{v}_2, \ldots, A\mathbf{v}_n\}$ is a basis for $\mathcal{R}^n$.

50. If V is contained in W, then $V \cup W = W$ is a subspace of $\mathcal{R}^n$; and if W is contained in V, then $V \cup W = V$ is a subspace of $\mathcal{R}^n$. If neither V is contained in W nor W is contained in V, then there exist vectors $\mathbf{v}$ in V and $\mathbf{w}$ in W such that $\mathbf{v}$ is not in W and $\mathbf{w}$ is not in V. Then $\mathbf{v} + \mathbf{w}$ is not in V (lest $\mathbf{w} = (\mathbf{v} + \mathbf{w}) - \mathbf{v}$ be in V) and, similarly, $\mathbf{v} + \mathbf{w}$ is not in W. So $\mathbf{v} + \mathbf{w}$ is not in $V \cup W$.

51. Let A be the standard matrix of T and B be the matrix whose columns are the vectors in $\mathcal{B}$. By

Theorem 2.13, A is invertible because T is invertible. Moreover, the standard matrix of T^{-1} is A^{-1}. Thus

$$[T^{-1}]_\mathcal{B} = B^{-1}A^{-1}B = (B^{-1}AB)^{-1} = [T]_\mathcal{B}^{-1}.$$

52. Let V and W be subspaces of $\mathcal{R}^n$. Then $\mathbf{0}+\mathbf{0} = \mathbf{0}$ is in $V+W$. Let $\mathbf{u}_1$ and $\mathbf{u}_2$ be in $V+W$. For $i=1,2$, there exist $\mathbf{v}_i$ in V and $\mathbf{w}_i$ in W such that $\mathbf{u}_i = \mathbf{v}_i + \mathbf{w}_i$. So

$$\mathbf{u}_1 + \mathbf{u}_2 = (\mathbf{v}_1 + \mathbf{w}_1) + (\mathbf{v}_2 + \mathbf{w}_2)$$
$$= (\mathbf{v}_1 + \mathbf{v}_2) + (\mathbf{w}_1 + \mathbf{w}_2)$$

is in $V+W$ because $\mathbf{v}_1+\mathbf{v}_2$ is in V and $\mathbf{w}_1+\mathbf{w}_2$ is in W. Also, for any scalar c,

$$c\mathbf{u}_1 = c(\mathbf{v}_1 + \mathbf{w}_1) = c\mathbf{v}_1 + c\mathbf{w}_1$$

is in $V+W$ because $c\mathbf{v}_1$ is in V and $c\mathbf{w}_1$ is in W. Hence $V+W$ is a subspace of $\mathcal{R}^n$.

53. A vector in V is a solution of the system of linear equations
$$\begin{array}{rcl} v_1 + v_2 & = & 0 \\ 2v_1 \quad - v_3 & = & 0. \end{array}$$
The vector form of the general solution of this system is
$$\begin{bmatrix} v_1 \\ v_2 \\ v_3 \end{bmatrix} = v_3 \begin{bmatrix} 1 \\ -1 \\ 2 \end{bmatrix}.$$

Moreover, a vector in W is a solution of
$$\begin{array}{rcl} w_1 \quad - 2w_3 & = & 0 \\ w_2 + w_3 & = & 0. \end{array}$$

The vector form of the general solution of this system is
$$\begin{bmatrix} w_1 \\ w_2 \\ w_3 \end{bmatrix} = w_3 \begin{bmatrix} 2 \\ -1 \\ 1 \end{bmatrix}.$$

It follows that a vector in $V+W$ has the form

$$r \begin{bmatrix} 1 \\ -1 \\ 2 \end{bmatrix} + s \begin{bmatrix} 2 \\ -1 \\ 1 \end{bmatrix}$$

for some scalars r and s. So

$$\mathcal{B} = \left\{ \begin{bmatrix} 1 \\ -1 \\ 2 \end{bmatrix}, \begin{bmatrix} 2 \\ -1 \\ 1 \end{bmatrix} \right\}$$

is a generating set for $V+W$. Moreover, $\mathcal{B}$ is linearly independent since neither vector in $\mathcal{B}$ is a multiple of the other. So $\mathcal{B}$ is a basis for $V+W$.

54. Let the subsets be $\mathcal{S}_1 = \{\mathbf{v}_1, \mathbf{v}_2, \ldots, \mathbf{v}_k\}$ and $\mathcal{S}_2 = \{\mathbf{w}_1, \mathbf{w}_2, \ldots, \mathbf{w}_m\}$. Note that each vector $\mathbf{v}_i$ in $\mathcal{S}_1$ is in $V+W$ because it can be written as $\mathbf{v}_i + \mathbf{0}$, and similarly each vector $\mathbf{w}_j$ in $\mathcal{S}_2$ is in $V+W$. If $\mathbf{u}$ is in $V+W$, then $\mathbf{u} = \mathbf{v}+\mathbf{w}$ for some $\mathbf{v}$ in V and $\mathbf{w}$ in W. So there exist scalars $a_1, a_2, \ldots, a_k$ and $b_1, b_2, \ldots, b_m$ such that

$$\mathbf{v} = a_1\mathbf{v}_1 + a_2\mathbf{v}_2 + \cdots + a_k\mathbf{v}_k$$

and

$$\mathbf{w} = b_1\mathbf{w}_1 + b_2\mathbf{w}_2 + \cdots + b_m\mathbf{w}_m.$$

Hence

$$\mathbf{u} = \mathbf{v} + \mathbf{w}$$
$$= a_1\mathbf{v}_1 + a_2\mathbf{v}_2 + \cdots + a_k\mathbf{v}_k$$
$$\quad + b_1\mathbf{w}_1 + b_2\mathbf{w}_2 + \cdots + b_m\mathbf{w}_m.$$

Since $\mathcal{S}$ is contained in $V+W$, it is a generating set for $V+W$.

CHAPTER 4 MATLAB EXERCISES

1. Recall that a vector $\mathbf{v}$ is in the column space of A if and only if the equation $A\mathbf{x} = \mathbf{v}$ is consistent.
   (a) yes   (b) no   (c) no   (d) yes

2. Recall that a vector $\mathbf{u}$ is in the null space of A if and only if $A\mathbf{u} = \mathbf{0}$.
   (a) yes   (b) yes   (c) no   (d) yes

3. (a) Choose the set of pivot columns of A:
$$\left\{ \begin{bmatrix} 1.2 \\ -1.1 \\ 2.3 \\ -1.2 \\ 1.1 \\ 0.1 \end{bmatrix}, \begin{bmatrix} 2.3 \\ 3.2 \\ 1.1 \\ 1.4 \\ -4.1 \\ -2.1 \end{bmatrix}, \begin{bmatrix} 1.2 \\ -3.1 \\ 2.1 \\ -1.4 \\ 5.1 \\ 1.2 \end{bmatrix} \right\}.$$

(b) $\left\{ \begin{bmatrix} 1.2 \\ -1.1 \\ 2.3 \\ -1.2 \\ 1.1 \\ 0.1 \end{bmatrix}, \begin{bmatrix} 2.3 \\ 3.2 \\ 1.1 \\ 1.4 \\ -4.1 \\ -2.1 \end{bmatrix}, \begin{bmatrix} 1.2 \\ -3.1 \\ 2.1 \\ -1.4 \\ 5.1 \\ 1.2 \end{bmatrix}, \mathbf{e}_1, \mathbf{e}_2, \mathbf{e}_3 \right\}$

(c) $\left\{ \begin{bmatrix} -1 \\ -1 \\ -1 \\ 1 \\ 0 \end{bmatrix}, \begin{bmatrix} 0 \\ 2 \\ 1 \\ 0 \\ 1 \end{bmatrix} \right\}$

(d) $\left\{ \begin{bmatrix} 1.2 \\ 2.3 \\ 1.2 \\ 4.7 \\ -5.8 \end{bmatrix}, \begin{bmatrix} -1.1 \\ 3.2 \\ -3.1 \\ -1.0 \\ -3.3 \end{bmatrix}, \begin{bmatrix} 2.3 \\ 1.1 \\ 2.1 \\ 5.5 \\ -4.3 \end{bmatrix} \right\}$

104 Chapter 4 Subspaces and Their Properties

4. (a) $\left\{ \begin{bmatrix} 1.3 \\ 2.2 \\ -1.2 \\ 4.0 \\ 1.7 \\ -3.1 \end{bmatrix}, \begin{bmatrix} 2.1 \\ -1.4 \\ 1.3 \\ 2.7 \\ 4.1 \\ 1.0 \end{bmatrix}, \begin{bmatrix} 2.9 \\ -3.0 \\ 3.8 \\ 1.4 \\ 6.5 \\ 5.1 \end{bmatrix} \right\}$

(b) $\left\{ \begin{bmatrix} 1.3 \\ 2.2 \\ -1.2 \\ 4.0 \\ 1.7 \\ -3.1 \end{bmatrix}, \begin{bmatrix} 2.1 \\ -1.4 \\ 1.3 \\ 2.7 \\ 4.1 \\ 1.0 \end{bmatrix}, \begin{bmatrix} 2.9 \\ -3.0 \\ 3.8 \\ 1.4 \\ 6.5 \\ 5.1 \end{bmatrix}, \mathbf{e}_1, \mathbf{e}_3, \mathbf{e}_4 \right\}$

(c) $\left\{ \begin{bmatrix} -2 \\ 1 \\ 1 \\ 0 \end{bmatrix} \right\}$ (d) $\left\{ \begin{bmatrix} 1 \\ 0 \\ 2 \\ 0 \end{bmatrix}, \begin{bmatrix} 0 \\ 1 \\ -1 \\ 0 \end{bmatrix}, \begin{bmatrix} 0 \\ 0 \\ 0 \\ 1 \end{bmatrix} \right\}$

5. Let $\mathbf{b}_i$ ($1 \leq i \leq 6$) denote the ith vector in $\mathcal{B}$.

   (a) $\mathcal{B}$ is a linearly independent set of 6 vectors from $\mathcal{R}^6$.

   (b) For any vector $\mathbf{v}$, the coefficients that express $\mathbf{v}$ as a linear combination of the vectors in $\mathcal{B}$ are the components of $[\mathbf{v}]_\mathcal{B}$. By computing $[\mathbf{v}]_\mathcal{B}$, we obtain the following linear combinations.

   (i) $2\mathbf{b}_1 - \mathbf{b}_2 - 3\mathbf{b}_3 + 2\mathbf{b}_5 - \mathbf{b}_6$
   (ii) $\mathbf{b}_1 - \mathbf{b}_2 + \mathbf{b}_3 + 2\mathbf{b}_4 - 3\mathbf{b}_5 + \mathbf{b}_6$
   (iii) $-3\mathbf{b}_2 + \mathbf{b}_3 + 2\mathbf{b}_4 - 4\mathbf{b}_5$

   (c) (i) $\begin{bmatrix} 2 \\ -1 \\ -3 \\ 0 \\ 2 \\ -1 \end{bmatrix}$ (ii) $\begin{bmatrix} 1 \\ -1 \\ 1 \\ 2 \\ -3 \\ 1 \end{bmatrix}$ (iii) $\begin{bmatrix} 0 \\ -3 \\ 1 \\ 2 \\ -4 \\ 0 \end{bmatrix}$

6. Proceeding as in Example 3 in Section 4.5, we obtain

$$\begin{bmatrix} -47.6 & 0.6 & 3.4 & -44.6 & 23.5 \\ -30.9 & 1.4 & 2.1 & -28.9 & 12.5 \\ 22.2 & -0.2 & -1.8 & 21.2 & -10.5 \\ 0.7 & -1.2 & 1.7 & -0.3 & 0.0 \\ -38.5 & -1.0 & 4.5 & -38.5 & 21.5 \end{bmatrix}.$$

7. Proceeding as in Example 3 in Section 4.5, we obtain

$$\begin{bmatrix} -1 & 2 & 1 & 0 & -1 & -2 \\ -8 & -4 & -9 & 3 & 4 & -10 \\ 6 & 1 & 6 & 0 & -1 & 7 \\ 1 & 1 & -1 & -1 & -2 & 2 \end{bmatrix}.$$

8. First suppose that $\mathbf{v}$ is in W. Then for some scalars $c_1, c_2, \ldots, c_k$, we can write

$$\mathbf{v} = c_1\mathbf{b}_1 + c_2\mathbf{b}_2 + \cdots + c_k\mathbf{b}_k.$$

Hence
$$\begin{aligned} A\mathbf{v} &= T(\mathbf{v}) \\ &= c_1T(\mathbf{b}_1) + c_2T(\mathbf{b}_2) + \cdots + c_kT(\mathbf{b}_k) \\ &= c_1\mathbf{0} + c_2\mathbf{0} + \cdots + c_k\mathbf{0} \\ &= \mathbf{0}. \end{aligned}$$

Therefore $\mathbf{v}$ is in Null A.

Conversely, suppose that $\mathbf{v}$ is in Null A. Let $c_1, c_2, \ldots, c_n$ be the unique scalars such that

$$\mathbf{v} = c_1\mathbf{b}_1 + c_2\mathbf{b}_2 + \cdots + c_k\mathbf{b}_k + \cdots + c_n\mathbf{b}_n.$$

Then
$$\begin{aligned} \mathbf{0} = A\mathbf{v} &= T(\mathbf{v}) \\ &= c_1T(\mathbf{b}_1) + \cdots + c_kT(\mathbf{b}_k) + \cdots + c_nT(\mathbf{b}_n) \\ &= c_{k+1}\mathbf{b}_{k+1} + \cdots + c_n\mathbf{b}_n. \end{aligned}$$

But $\{\mathbf{b}_{k+1}, \ldots, \mathbf{b}_n\}$ is linearly independent, and hence $c_{k+1} = \cdots = c_n = 0$. Thus we have

$$\mathbf{v} = c_1\mathbf{b}_1 + c_2\mathbf{b}_2 + \cdots + c_k\mathbf{b}_k,$$

so it follows that $\mathbf{v}$ is in W.

Since each of W and Null A is a subset of the other, $W = $ Null A.

9. Extend the given basis for W by including $\mathbf{e}_1$ and $\mathbf{e}_3$, and then use the method described in the preceding exercise to obtain

$$A = \begin{bmatrix} 1.00 & 0.00 & 0.00 & 0.75 & -0.50 \\ 0.00 & 0.00 & 0.00 & 0.00 & 0.00 \\ 0.00 & 0.00 & 1.00 & -0.75 & 0.50 \\ 0.00 & 0.00 & 0.00 & 0.00 & 0.00 \\ 0.00 & 0.00 & 0.00 & 0.00 & 0.00 \end{bmatrix}.$$

10. (a) For any vector $\mathbf{v}$ in $\mathcal{R}^n$, we may use block multiplication to compute

$$C\mathbf{v} = \begin{bmatrix} A \\ B \end{bmatrix} \mathbf{v} = \begin{bmatrix} A\mathbf{v} \\ B\mathbf{v} \end{bmatrix}.$$

Thus
$$C\mathbf{v} = O = \begin{bmatrix} \mathbf{0} \\ \mathbf{0} \end{bmatrix}$$

if and only if $A\mathbf{v} = \mathbf{0}$ and $B\mathbf{v} = \mathbf{0}$. This occurs if and only if $\mathbf{v}$ is in both V and W, that is, if and only if $\mathbf{v}$ is in $V \cap W$. Hence

$$\text{Null } C = V \cap W.$$

(b) Let
$$\mathbf{v}_1 = \begin{bmatrix} 1 \\ 2 \\ 1 \\ -1 \end{bmatrix}, \mathbf{v}_2 = \begin{bmatrix} 2 \\ 1 \\ 0 \\ 1 \end{bmatrix}, \text{ and } \mathbf{v}_3 = \begin{bmatrix} 1 \\ 3 \\ 1 \\ 0 \end{bmatrix}$$

and
$$\mathbf{w}_1 = \begin{bmatrix} 1 \\ -1 \\ 1 \\ 1 \end{bmatrix}, \ \mathbf{w}_2 = \begin{bmatrix} 0 \\ 1 \\ 1 \\ 1 \end{bmatrix}, \text{ and } \mathbf{w}_3 = \begin{bmatrix} 1 \\ 0 \\ 1 \\ 2 \end{bmatrix}.$$

Let
$$A = \begin{bmatrix} 0 & 0 & 0 & 0 \\ 0 & 1 & -3 & -1 \\ 0 & 0 & 0 & 0 \\ 0 & 0 & 0 & 0 \end{bmatrix}$$
and
$$B = \begin{bmatrix} 1 & \frac{1}{2} & 0 & -\frac{1}{2} \\ 0 & 0 & 0 & 0 \\ 0 & 0 & 0 & 0 \\ 0 & 0 & 0 & 0 \end{bmatrix}.$$

Then $\{\mathbf{v}_1, \mathbf{v}_2, \mathbf{v}_3, \mathbf{e}_2\}$ and $\{\mathbf{w}_1, \mathbf{w}_2, \mathbf{w}_3, \mathbf{e}_1\}$ are bases for $\mathcal{R}^4$. Proceeding as in Exercise 8, we see that $V = \text{Null } A$ and $W = \text{Null } B$. Thus $V \cap W = \text{Null} \begin{bmatrix} A \\ B \end{bmatrix}$ by (a). The zero rows in A and B can be ignored, however, and so
$$V \cap W = \text{Null} \begin{bmatrix} 0 & 1 & -3 & -1 \\ 1 & \frac{1}{2} & 0 & -\frac{1}{2} \end{bmatrix}.$$

Therefore
$$\left\{ \begin{bmatrix} -3 \\ 6 \\ 2 \\ 0 \end{bmatrix}, \begin{bmatrix} 0 \\ 1 \\ 0 \\ 1 \end{bmatrix} \right\}$$

is a basis for $V \cap W$.

Chapter 5

Eigenvalues, Eigenvectors, and Diagonalization

5.1 EIGENVALUES AND EIGENVECTORS

1. 6   2. −2   3. 3   4. 3
5. −2   6. −2   7. −3   8. −5
9. The eigenvalue is −4 because

$$\begin{bmatrix} 2 & -6 & 6 \\ 1 & 9 & -6 \\ -2 & 16 & -13 \end{bmatrix} \begin{bmatrix} -1 \\ 1 \\ 2 \end{bmatrix} = \begin{bmatrix} 4 \\ -4 \\ -8 \end{bmatrix}$$

$$= (-4) \begin{bmatrix} -1 \\ 1 \\ 2 \end{bmatrix}.$$

10. −1   11. 2   12. −4   13. $\left\{ \begin{bmatrix} -1 \\ 1 \end{bmatrix} \right\}$

14. $\left\{ \begin{bmatrix} 2 \\ 1 \end{bmatrix} \right\}$ 15. $\left\{ \begin{bmatrix} -3 \\ 1 \end{bmatrix} \right\}$ 16. $\left\{ \begin{bmatrix} 1 \\ 2 \end{bmatrix} \right\}$

17. Let A denote the given matrix. The reduced row echelon form of $A - 3I_3$ is

$$\begin{bmatrix} 1 & 1 & 0 \\ 0 & 0 & 1 \\ 0 & 0 & 0 \end{bmatrix},$$

and so

$$\left\{ \begin{bmatrix} -1 \\ 1 \\ 0 \end{bmatrix} \right\}$$

is a basis for the eigenspace of A corresponding to eigenvalue 3.

18. $\left\{ \begin{bmatrix} 1 \\ 1 \\ 1 \end{bmatrix} \right\}$ 19. $\left\{ \begin{bmatrix} -2 \\ -1 \\ 1 \end{bmatrix} \right\}$ 20. $\left\{ \begin{bmatrix} -2 \\ 2 \\ 3 \end{bmatrix} \right\}$

21. Let A denote the given matrix. The reduced row echelon form of $A - (-1)I_3 = A + I_3$ is

$$\begin{bmatrix} 1 & \frac{1}{3} & -\frac{2}{3} \\ 0 & 0 & 0 \\ 0 & 0 & 0 \end{bmatrix},$$

and so

$$\left\{ \begin{bmatrix} -1 \\ 3 \\ 0 \end{bmatrix}, \begin{bmatrix} 2 \\ 0 \\ 3 \end{bmatrix} \right\}$$

is a basis for the eigenspace corresponding to eigenvalue −1.

22. $\left\{ \begin{bmatrix} -1 \\ 1 \\ 0 \end{bmatrix}, \begin{bmatrix} -2 \\ 0 \\ 1 \end{bmatrix} \right\}$ 23. $\left\{ \begin{bmatrix} 1 \\ 1 \\ 0 \end{bmatrix}, \begin{bmatrix} 1 \\ 0 \\ 1 \end{bmatrix} \right\}$

24. $\left\{ \begin{bmatrix} -1 \\ 1 \\ 0 \end{bmatrix}, \begin{bmatrix} -3 \\ 0 \\ 1 \end{bmatrix} \right\}$

25. The eigenvalue is 6 because

$$T\left(\begin{bmatrix} -2 \\ 3 \end{bmatrix} \right) = \begin{bmatrix} -12 \\ 18 \end{bmatrix} = 6 \begin{bmatrix} -2 \\ 3 \end{bmatrix}.$$

26. 4   27. 4   28. 5   29. −3
30. −2   31. 5   32. −4

33. $\left\{ \begin{bmatrix} 2 \\ 3 \end{bmatrix} \right\}$ 34. $\left\{ \begin{bmatrix} -2 \\ 3 \end{bmatrix} \right\}$

35. $\left\{ \begin{bmatrix} -2 \\ 3 \end{bmatrix} \right\}$ 36. $\left\{ \begin{bmatrix} 1 \\ 2 \end{bmatrix} \right\}$

37. The standard matrix of T is

$$A = \begin{bmatrix} 1 & -1 & -3 \\ -3 & -1 & -9 \\ 1 & 1 & 5 \end{bmatrix},$$

and the reduced row echelon form of $A - 2I_3$ is
$$\begin{bmatrix} 1 & 1 & 3 \\ 0 & 0 & 0 \\ 0 & 0 & 0 \end{bmatrix}.$$
Hence
$$\left\{ \begin{bmatrix} -1 \\ 1 \\ 0 \end{bmatrix}, \begin{bmatrix} -3 \\ 0 \\ 1 \end{bmatrix} \right\}$$
is a basis for the eigenspace corresponding to eigenvalue 2.

38. $\left\{ \begin{bmatrix} 1 \\ 1 \\ 1 \end{bmatrix} \right\}$ 39. $\left\{ \begin{bmatrix} 1 \\ -2 \\ 2 \end{bmatrix} \right\}$

40. $\left\{ \begin{bmatrix} -1 \\ 2 \\ 0 \end{bmatrix}, \begin{bmatrix} 1 \\ 0 \\ 1 \end{bmatrix} \right\}$

41. False, if $A\mathbf{v} = \lambda\mathbf{v}$ for some *nonzero* vector $\mathbf{v}$, then λ is an eigenvalue of A.

42. False, if $A\mathbf{v} = \lambda\mathbf{v}$ for some *nonzero* vector $\mathbf{v}$, then $\mathbf{v}$ is an eigenvector of A.

43. True   44. True   45. True

46. False, the eigenspace of A corresponding to eigenvalue λ is the null space of $A - \lambda I_n$.

47. True   48. True

49. False, the linear operator on $\mathcal{R}^2$ that rotates a vector by 90° has no real eigenvalues. (See pages 298–299.)

50. True

51. False, the exception is the zero vector.

52. True   53. True   54. True

55. False, the exception is $c = 0$.

56. True

57. False, if $A = B = I_n$, then $\lambda = 1$ is an eigenvalue of I_n, but not of $A + B = 2I_n$.

58. True

59. False, if $A = B = 2I_n$, then $\lambda = 2$ is an eigenvalue of $2I_n$, but not of $AB = 4I_n$.

60. True

61. The only eigenvalue is 1; its eigenspace is $\mathcal{R}^n$.

62. The only eigenvalue is 0, and the eigenspace corresponding to 0 is $\mathcal{R}^n$.

63. Let λ be the eigenvalue of A corresponding to $\mathbf{v}$. Then $\mathbf{v} \neq \mathbf{0}$, and
$$A(c\mathbf{v}) = c(A\mathbf{v}) = c(\lambda\mathbf{v}) = \lambda(c\mathbf{v}).$$

64. Let $\mathbf{v}$ be an eigenvector of A such that $A\mathbf{v} = \lambda\mathbf{v}$ and $A\mathbf{v} = \mu\mathbf{v}$. Then $\lambda\mathbf{v} = \mu\mathbf{v}$; so $(\lambda - \mu)\mathbf{v} = \mathbf{0}$. Since $\mathbf{v} \neq \mathbf{0}$, we must have $\lambda = \mu$.

65. Null A

66. If 0 is an eigenvalue, then $A\mathbf{v} = \mathbf{0}$ for some nonzero $\mathbf{v}$. Hence A is not invertible by Theorem 2.6. If 0 is not an eigenvalue, then $A\mathbf{x} = \mathbf{0}$ has no solution but $\mathbf{x} = \mathbf{0}$. Thus A is invertible by Theorem 2.6.

67. Let λ be an eigenvalue of an invertible matrix A. By Exercise 66, $\lambda \neq 0$. Then $A\mathbf{v} = \lambda\mathbf{v}$ for some $\mathbf{v} \neq \mathbf{0}$. Thus
$$\mathbf{v} = A^{-1}(\lambda\mathbf{v}) = \lambda(A^{-1}\mathbf{v})$$
and hence $\frac{1}{\lambda}\mathbf{v} = A^{-1}\mathbf{v}$. So $\frac{1}{\lambda}$ is an eigenvalue of A^{-1}.

68. $\begin{bmatrix} 1 \\ 1 \\ \vdots \\ 1 \end{bmatrix}$

69. If $\mathbf{v}$ is an eigenvector of A with λ as the corresponding eigenvalue, then $A\mathbf{v} = \lambda\mathbf{v}$. So
$$A^2\mathbf{v} = A(A\mathbf{v}) = A(\lambda\mathbf{v}) = \lambda(A\mathbf{v})$$
$$= \lambda(\lambda\mathbf{v}) = \lambda^2\mathbf{v}.$$
Hence λ^2 is an eigenvalue of A^2.

70. By a repetition of the argument in the solution of Exercise 69, we see that if λ is an eigenvalue of A, then λ^k is an eigenvalue of A^k for every positive integer k.

71. Either $\mathbf{v} = \mathbf{0}$ or $\mathbf{v}$ is an eigenvector of A.

72. Suppose λ is an eigenvalue of a nilpotent matrix A. Then $A\mathbf{v} = \lambda\mathbf{v}$ for some $\mathbf{v} \neq \mathbf{0}$. Multiplying both sides by A^{k-1} and using the result of Exercise 46, we obtain $\mathbf{0} = O\mathbf{v} = A^k\mathbf{v} = \lambda^k\mathbf{v}$. Since $\mathbf{v} \neq \mathbf{0}$, we must have $\lambda = 0$.

73. Suppose that $c_1\mathbf{v}_1 + c_2\mathbf{v}_2 = \mathbf{0}$ for some scalars c_1 and c_2. Then
$$\mathbf{0} = T(\mathbf{0}) = T(c_1\mathbf{v}_1 + c_2\mathbf{v}_2) = c_1\lambda_1\mathbf{v}_1 + c_2\lambda_2\mathbf{v}_2$$
$$= \lambda_1(-c_2\mathbf{v}_2) + c_2\lambda_2\mathbf{v}_2 = (\lambda_2 - \lambda_1)c_2\mathbf{v}_2.$$

Since $\lambda_1 \neq \lambda_2$ and $\mathbf{v}_2 \neq \mathbf{0}$, we have $c_2 = 0$. Thus we also have $c_1 = 0$, and so $\{\mathbf{v}_1, \mathbf{v}_2\}$ is linearly independent.

74. Applying the hint, we have
$$c_1\lambda_1\mathbf{v}_1 + c_2\lambda_2\mathbf{v}_2 + c_3\lambda_1\mathbf{v}_3 =$$

$$c_1\lambda_3\mathbf{v}_1 + c_2\lambda_3\mathbf{v}_2 + c_3\lambda_3\mathbf{v}_3,$$

and hence

$$c_1(\lambda_1 - \lambda_3)\mathbf{v}_1 + c_2(\lambda_2 - \lambda_3)\mathbf{v}_2 = \mathbf{0}.$$

Apply Exercise 73 and the fact that the eigenvalues are distinct to obtain that $c_1 = c_2 = 0$. It now follows that $c_3 = 0$, and hence the set is linearly independent.

75. Let $\mathcal{B} = \{\mathbf{v}_1, \mathbf{v}_2\}$ be a basis for the two-dimensional eigenspace of T, and let λ be the eigenvalue corresponding to $\mathbf{v}_1$ and $\mathbf{v}_2$. Then

$$[T]_\mathcal{B} = \begin{bmatrix} \lambda & 0 \\ 0 & \lambda \end{bmatrix} = \lambda I_2.$$

If $B = [\mathbf{v}_1 \ \mathbf{v}_2]$, then the standard matrix of T is

$$B[T]_\mathcal{B}B^{-1} = \lambda I_2$$

by Theorem 4.12. Hence $T = \lambda I$.

76. $-2.7, 2.3, -1.1, -1.1$

77. The eigenvalues of A are $-2.7, 2.3$, and -1.1 (with multiplicity 2), but the eigenvalues of $3A$ are $-8.1, 6.9$, and -3.3 (with multiplicity 2).

78. yes, $-8.1, 6.9, -3.3, -3.3$

79. (a) A scalar λ is an eigenvalue of B if and only if $c\lambda$ is an eigenvalue of cB.
    (b) A vector $\mathbf{v}$ is an eigenvector of B if and only if $\mathbf{v}$ is an eigenvector of cB.
    (c) Suppose that $\mathbf{v}$ is an eigenvector of B with corresponding eigenvalue λ. Then

    $$(cB)\mathbf{v} = c(B\mathbf{v}) = c(\lambda\mathbf{v}) = (c\lambda)\mathbf{v},$$

    and thus $\mathbf{v}$ is an eigenvector of cB with corresponding eigenvalue λ.
    Conversely, suppose that $\mathbf{v}$ is an eigenvector of cB with corresponding eigenvalue $c\lambda$. Then $(cB)\mathbf{v} = (c\lambda)\mathbf{v}$, and therefore $c(B\mathbf{v}) = c(\lambda\mathbf{v})$. Since $c \neq 0$, we can divide both sides of the last equation by c to obtain $B\mathbf{v} = \lambda\mathbf{v}$. So $\mathbf{v}$ is an eigenvector of B with corresponding eigenvalue λ.

80. no

81. Yes, the eigenvalues of A^T are the same as those of A. Eigenvectors of A^T are found by solving $(A^T - \lambda I_4)\mathbf{x} = \mathbf{0}$ for each eigenvalue λ. Four eigenvectors of A^T are

$$\begin{bmatrix} -1 \\ 1 \\ -2 \\ 1 \end{bmatrix}, \begin{bmatrix} 2 \\ 0 \\ 3 \\ 3 \end{bmatrix}, \begin{bmatrix} 1 \\ -1 \\ 2 \\ 0 \end{bmatrix}, \text{ and } \begin{bmatrix} 0 \\ -1 \\ 0 \\ 1 \end{bmatrix}.$$

82. The eigenvalues of an $n \times n$ matrix B are the same as those of B^T, but an eigenvector of B need not be an eigenvector of B^T.

5.2 THE CHARACTERISTIC POLYNOMIAL

1. The eigenvalues of the matrix are the roots of its characteristic polynomial, which are 5 and 6. The vector form of the general solution of $(A - 5I_2)\mathbf{x} = \mathbf{0}$ is

$$\begin{bmatrix} x_1 \\ x_2 \end{bmatrix} = x_2 \begin{bmatrix} -1.5 \\ 1 \end{bmatrix}.$$

So

$$\left\{ \begin{bmatrix} -1.5 \\ 1 \end{bmatrix} \right\} \quad \text{or} \quad \left\{ \begin{bmatrix} -3 \\ 2 \end{bmatrix} \right\}$$

is a basis for the eigenspace of A corresponding to eigenvalue 5. Also, the vector form of the general solution of $(A - 6I_2)\mathbf{x} = \mathbf{0}$ is

$$\begin{bmatrix} x_1 \\ x_2 \end{bmatrix} = x_2 \begin{bmatrix} -1 \\ 1 \end{bmatrix}.$$

Thus

$$\left\{ \begin{bmatrix} -1 \\ 1 \end{bmatrix} \right\}$$

is a basis for the eigenspace of A corresponding to eigenvalue 6.

2. $-5, \left\{ \begin{bmatrix} 1 \\ 2 \end{bmatrix} \right\}, -4, \left\{ \begin{bmatrix} 1 \\ 3 \end{bmatrix} \right\}$

3. $0, \left\{ \begin{bmatrix} 3 \\ 5 \end{bmatrix} \right\}, -1, \left\{ \begin{bmatrix} 2 \\ 3 \end{bmatrix} \right\}$

4. $-2, \left\{ \begin{bmatrix} -1 \\ 1 \end{bmatrix} \right\}, 5, \left\{ \begin{bmatrix} -1 \\ 2 \end{bmatrix} \right\}$

5. $-3, \left\{ \begin{bmatrix} 1 \\ 1 \\ 1 \end{bmatrix} \right\}, 2, \left\{ \begin{bmatrix} 1 \\ 0 \\ 1 \end{bmatrix} \right\}$

6. $-2, \left\{ \begin{bmatrix} -2 \\ -1 \\ 1 \end{bmatrix} \right\}, 4, \left\{ \begin{bmatrix} 0 \\ -1 \\ 1 \end{bmatrix} \right\}$

7. $6, \left\{ \begin{bmatrix} 1 \\ -1 \\ 1 \end{bmatrix} \right\}, -2, \left\{ \begin{bmatrix} 1 \\ 2 \\ 0 \end{bmatrix}, \begin{bmatrix} 1 \\ 0 \\ 2 \end{bmatrix} \right\}$

8. $-4, \left\{ \begin{bmatrix} 1 \\ 0 \\ 1 \end{bmatrix} \right\}, 2, \left\{ \begin{bmatrix} 1 \\ 1 \\ 1 \end{bmatrix} \right\}, 3, \left\{ \begin{bmatrix} 1 \\ 1 \\ 2 \end{bmatrix} \right\}$

9. $-3, \left\{ \begin{bmatrix} -1 \\ 1 \\ 1 \end{bmatrix} \right\}, -2, \left\{ \begin{bmatrix} -1 \\ 1 \\ 0 \end{bmatrix} \right\}, 1, \left\{ \begin{bmatrix} 1 \\ 0 \\ 1 \end{bmatrix} \right\}$

5.2 The Characteristic Polynomial

10. $3, \left\{\begin{bmatrix} 1 \\ -1 \\ 1 \end{bmatrix}\right\}, 1, \left\{\begin{bmatrix} -1 \\ 1 \\ 0 \end{bmatrix}, \begin{bmatrix} -1 \\ 0 \\ 1 \end{bmatrix}\right\}$

11. $3, \left\{\begin{bmatrix} 1 \\ 1 \\ 0 \\ 0 \end{bmatrix}\right\}, 4, \left\{\begin{bmatrix} 0 \\ 1 \\ 0 \\ 1 \end{bmatrix}\right\}, -1, \left\{\begin{bmatrix} 0 \\ 1 \\ 1 \\ 0 \end{bmatrix}, \begin{bmatrix} -1 \\ 1 \\ 0 \\ 1 \end{bmatrix}\right\}$

12. $-5, \left\{\begin{bmatrix} 1 \\ 2 \\ 1 \\ 2 \end{bmatrix}\right\}, -2, \left\{\begin{bmatrix} 0 \\ 2 \\ 1 \\ 1 \end{bmatrix}\right\}, 1, \left\{\begin{bmatrix} 0 \\ 1 \\ 0 \\ 0 \end{bmatrix}, \begin{bmatrix} 1 \\ 1 \\ 0 \\ 1 \end{bmatrix}\right\}$

13. $-4, \left\{\begin{bmatrix} 1 \\ 2 \end{bmatrix}\right\}, 1, \left\{\begin{bmatrix} 1 \\ 1 \end{bmatrix}\right\}$

14. $2, \left\{\begin{bmatrix} -1 \\ 3 \end{bmatrix}\right\}, 4, \left\{\begin{bmatrix} -1 \\ 2 \end{bmatrix}\right\}$

15. $3, \left\{\begin{bmatrix} -2 \\ 3 \end{bmatrix}\right\}, 5, \left\{\begin{bmatrix} -1 \\ 2 \end{bmatrix}\right\}$

16. $-2, \left\{\begin{bmatrix} -2 \\ 5 \end{bmatrix}\right\}, -1, \left\{\begin{bmatrix} -1 \\ 2 \end{bmatrix}\right\}$

17. $-3, \left\{\begin{bmatrix} 1 \\ 0 \\ 1 \end{bmatrix}\right\}, 1, \left\{\begin{bmatrix} 1 \\ 0 \\ 2 \end{bmatrix}\right\}$

18. $1, \left\{\begin{bmatrix} 0 \\ 0 \\ 1 \end{bmatrix}\right\}, -3, \left\{\begin{bmatrix} 1 \\ 0 \\ 1 \end{bmatrix}\right\}, 3, \left\{\begin{bmatrix} -2 \\ 1 \\ 0 \end{bmatrix}\right\}$

19. $-1, \left\{\begin{bmatrix} 0 \\ 0 \\ 1 \end{bmatrix}\right\}, 5, \left\{\begin{bmatrix} 0 \\ -3 \\ 1 \end{bmatrix}\right\}$

20. $-2, \left\{\begin{bmatrix} 0 \\ -2 \\ 1 \end{bmatrix}\right\}, 3, \left\{\begin{bmatrix} 0 \\ 1 \\ 0 \end{bmatrix}\right\}$

21. $-6, \left\{\begin{bmatrix} -1 \\ 1 \\ 1 \end{bmatrix}\right\}, -2, \left\{\begin{bmatrix} 1 \\ 1 \\ 1 \end{bmatrix}\right\}, 4, \left\{\begin{bmatrix} 0 \\ 1 \\ 0 \end{bmatrix}\right\}$

22. $3, \left\{\begin{bmatrix} 1 \\ 1 \\ 0 \end{bmatrix}\right\}, -4, \left\{\begin{bmatrix} 0 \\ 1 \\ 0 \end{bmatrix}, \begin{bmatrix} -1 \\ 0 \\ 1 \end{bmatrix}\right\}$

23. $-1, \left\{\begin{bmatrix} 1 \\ 0 \\ 0 \\ 0 \end{bmatrix}\right\}, 1, \left\{\begin{bmatrix} -1 \\ 1 \\ 0 \\ 0 \end{bmatrix}\right\},$
$-2, \left\{\begin{bmatrix} -1 \\ -2 \\ 3 \\ 0 \end{bmatrix}\right\}, 2, \left\{\begin{bmatrix} 7 \\ -2 \\ -1 \\ 4 \end{bmatrix}\right\}$

24. $-2, \left\{\begin{bmatrix} 0 \\ 1 \\ 0 \\ 0 \end{bmatrix}, \begin{bmatrix} 0 \\ 0 \\ 1 \\ 1 \end{bmatrix}\right\}, 1, \left\{\begin{bmatrix} 0 \\ -1 \\ 1 \\ 0 \end{bmatrix}, \begin{bmatrix} -1 \\ -1 \\ 0 \\ 2 \end{bmatrix}\right\}$

25. The standard matrix of T is

$$\begin{bmatrix} -1 & 6 \\ -8 & 13 \end{bmatrix}.$$

The eigenvalues of T are the roots of its characteristic polynomial, which are 5 and 7. The eigenvectors of T are the same as the eigenvectors of A; so we must find bases for the null spaces of $A - 5I_2$ and $A - 7I_2$. As in Exercise 1, we obtain the following bases for the eigenspaces corresponding to 5 and 7:

$$\left\{\begin{bmatrix} 1 \\ 1 \end{bmatrix}\right\} \quad \text{and} \quad \left\{\begin{bmatrix} 3 \\ 4 \end{bmatrix}\right\}.$$

26. $-5, \left\{\begin{bmatrix} -1 \\ 2 \end{bmatrix}\right\}, -3, \left\{\begin{bmatrix} -1 \\ 1 \end{bmatrix}\right\}$

27. $2, \left\{\begin{bmatrix} -2 \\ 1 \end{bmatrix}\right\}, 6, \left\{\begin{bmatrix} -3 \\ 2 \end{bmatrix}\right\}$

28. $4, \left\{\begin{bmatrix} 2 \\ 5 \end{bmatrix}\right\}, 3, \left\{\begin{bmatrix} 1 \\ 2 \end{bmatrix}\right\}$

29. $-2, \left\{\begin{bmatrix} 1 \\ 1 \\ 0 \end{bmatrix}\right\}, 4, \left\{\begin{bmatrix} 1 \\ 0 \\ 1 \end{bmatrix}\right\}$

30. $-3, \left\{\begin{bmatrix} -1 \\ 1 \\ 0 \end{bmatrix}\right\}, -2, \left\{\begin{bmatrix} -91 \\ 98 \\ 8 \end{bmatrix}\right\}, -9, \left\{\begin{bmatrix} 7 \\ -7 \\ 6 \end{bmatrix}\right\}$

31. $1, \left\{\begin{bmatrix} 1 \\ 2 \\ 3 \end{bmatrix}\right\}, 2, \left\{\begin{bmatrix} -2 \\ 1 \\ 0 \end{bmatrix}, \begin{bmatrix} 2 \\ 0 \\ 1 \end{bmatrix}\right\}$

32. $-1, \left\{\begin{bmatrix} -1 \\ 1 \\ 0 \end{bmatrix}, \begin{bmatrix} 1 \\ 0 \\ 1 \end{bmatrix}\right\}, 3, \left\{\begin{bmatrix} 1 \\ 2 \\ 2 \end{bmatrix}\right\}$

33. $-3, \left\{\begin{bmatrix} 1 \\ 1 \end{bmatrix}\right\}, -2, \left\{\begin{bmatrix} 1 \\ 2 \end{bmatrix}\right\}$

34. $3, \left\{\begin{bmatrix} 1 \\ 3 \end{bmatrix}\right\}, 4, \left\{\begin{bmatrix} 1 \\ 2 \end{bmatrix}\right\}$

35. $5, \left\{\begin{bmatrix} 2 \\ 5 \end{bmatrix}\right\}, 4, \left\{\begin{bmatrix} 1 \\ 2 \end{bmatrix}\right\}$

36. $3, \left\{\begin{bmatrix} -1 \\ 1 \end{bmatrix}\right\}, 7, \left\{\begin{bmatrix} -2 \\ 3 \end{bmatrix}\right\}$

37. The standard matrix of T is

$$\begin{bmatrix} 7 & -10 & 0 \\ 5 & -8 & 0 \\ -1 & 1 & 2 \end{bmatrix},$$

and its characteristic polynomial is

$$-t^3 + t^2 + 8t - 12 = -(t+3)(t-2)^2.$$

So the eigenvalues of T are -3 and 2 (with multiplicity 2). As in Exercise 25, we find the following bases for the eigenspaces:

$$\left\{\begin{bmatrix}1\\1\\0\end{bmatrix}\right\} \quad \text{and} \quad \left\{\begin{bmatrix}0\\0\\1\end{bmatrix}\right\}.$$

38. $4, \left\{\begin{bmatrix}1\\0\\2\end{bmatrix}\right\}, -1, \left\{\begin{bmatrix}-1\\1\\0\end{bmatrix}, \begin{bmatrix}1\\0\\1\end{bmatrix}\right\}$

39. $-3, \left\{\begin{bmatrix}1\\2\\3\end{bmatrix}\right\}, 1, \left\{\begin{bmatrix}0\\1\\0\end{bmatrix}, \begin{bmatrix}0\\0\\1\end{bmatrix}\right\}$

40. $-4, \left\{\begin{bmatrix}1\\0\\1\end{bmatrix}\right\}, 1, \left\{\begin{bmatrix}0\\0\\1\end{bmatrix}\right\}, 2, \left\{\begin{bmatrix}1\\1\\0\end{bmatrix}\right\}$

41. The characteristic polynomial is $t^2 - 3t + 10$, which has no (real) roots.

42. The characteristic polynomial is $t^2 - 2t + 7$, which has no (real) roots.

43. The standard matrix of T is $\begin{bmatrix}1 & 3\\-2 & 5\end{bmatrix}$, which has characteristic polynomial $t^2 - 6t + 11$. This polynomial has no (real) roots.

44. The standard matrix of T is $\begin{bmatrix}2 & -3\\2 & 4\end{bmatrix}$, which has characteristic polynomial $t^2 - 6t + 14$. This polynomial has no (real) roots.

45. Let A denote this matrix. The characteristic polynomial of A is $t^2 - 2t + 5$. So the eigenvalues of A, which are the roots of this polynomial, are $1 - 2i$ and $1 + 2i$. The vector form of the general solution of $(A - (1-2i)I_2)\mathbf{x} = \mathbf{0}$ is

$$\begin{bmatrix}x_1\\x_2\end{bmatrix} = x_2\begin{bmatrix}-\frac{1}{2}\\1\end{bmatrix}$$

So

$$\left\{\begin{bmatrix}-\frac{1}{2}\\1\end{bmatrix}\right\} \quad \text{or} \quad \left\{\begin{bmatrix}1\\-2\end{bmatrix}\right\}$$

is a basis for the eigenspace of A corresponding to the eigenvalue $1 - 2i$.

Likewise, the vector form of the general solution of $(A - (1+2i)I_2)\mathbf{x} = \mathbf{0}$ is

$$\begin{bmatrix}x_1\\x_2\end{bmatrix} = x_2\begin{bmatrix}-\frac{1}{3}\\1\end{bmatrix}$$

So

$$\left\{\begin{bmatrix}-\frac{1}{3}\\1\end{bmatrix}\right\} \quad \text{or} \quad \left\{\begin{bmatrix}-1\\3\end{bmatrix}\right\}$$

is a basis for the eigenspace of A corresponding to the eigenvalue $1 + 2i$.

46. $1 - 3i, \left\{\begin{bmatrix}2i\\-1\end{bmatrix}\right\}, 1 + i, \left\{\begin{bmatrix}2i\\1\end{bmatrix}\right\}$

47. $8 - 12i, \left\{\begin{bmatrix}1+4i\\1\end{bmatrix}\right\}, 8 + 12i, \left\{\begin{bmatrix}1-4i\\1\end{bmatrix}\right\}$

48. $2 - i, \left\{\begin{bmatrix}-i\\1\end{bmatrix}\right\}, 2 + i, \left\{\begin{bmatrix}i\\1\end{bmatrix}\right\}$

49. $2i, \left\{\begin{bmatrix}1\\0\\0\end{bmatrix}\right\}, 4, \left\{\begin{bmatrix}i\\2\\0\end{bmatrix}\right\}, 1, \left\{\begin{bmatrix}2\\1\\i\end{bmatrix}\right\}$

50. $2, \left\{\begin{bmatrix}1\\0\\0\end{bmatrix}\right\}, 0, \left\{\begin{bmatrix}i\\2\\0\end{bmatrix}\right\}, i, \left\{\begin{bmatrix}-4+3i\\5\\5\end{bmatrix}\right\}$

51. $i, \left\{\begin{bmatrix}1\\0\\0\end{bmatrix}\right\}, 1, \left\{\begin{bmatrix}0\\1\\0\end{bmatrix}\right\}, 2, \left\{\begin{bmatrix}1\\1\\2\end{bmatrix}\right\}$

52. $2i, \left\{\begin{bmatrix}1\\0\\0\end{bmatrix}\right\}, 0, \left\{\begin{bmatrix}0\\1\\i\end{bmatrix}\right\}, 1, \left\{\begin{bmatrix}0\\0\\1\end{bmatrix}\right\}$

53. False, consider the matrix A in Example 1 and $B = \begin{bmatrix}-3 & 0\\0 & 5\end{bmatrix}$, which both have $(t+3)(t-5)$ as their characteristic polynomial.

54. True   55. True

56. False, see page 303.

57. False, see page 303.

58. False, consider I_n.

59. False, the rotation matrix $A_{90°}$ has no eigenvectors in $\mathcal{R}^2$.

60. True

61. False, $\begin{bmatrix}0 & -1\\1 & 0\end{bmatrix}$ has a characteristic polynomial of $t^2 + 1$.

62. True

63. False, consider $4I_3$; here 4 is an eigenvalue of multiplicity 3.

64. False, see Example 4.

65. True

66. False, consider the matrix given in Exercise 49.

67. True

68. False, see Example 3 of a matrix with no (real) eigenvalues.

69. True   70. True

71. False, it has the eigenvalue 0.

72. True
73. c is not an eigenvalue of A.
74. $\det A$
75. Let the characteristic polynomial of $-A$ be
$$b_n t^n + b_{n-1} t^{n-1} + \cdots + b_1 t + b_0.$$
By Theorem 3.3(b),
$$\begin{aligned}\det(-A - tI_n) &= (-1)^n \det(A + tI_n) \\ &= (-1)^n \det(A - (-t)I_n).\end{aligned}$$
Thus the characteristic polynomial of $-A$ is
$$(-1)^n \left[(-1)^n a_n(-t)^n + (-1)^{n-1} a_{n-1}(-t)^{n-1} \right.$$
$$\left. + \cdots + a_0 \right]$$
$$= a_n t^n - a_{n-1} t^{n-1} + \cdots + (-1)^{n+1} a_1 t + (-1)^n a_0.$$
Hence, for $0 \le k \le n$, $b_{n-k} = a_{n-k}$ if k is even, and $b_{n-k} = -a_{n-k}$ if k is odd.

76. $(-1)^n$

77. (a) By Theorem 5.1, the eigenvalue 5 must have a multiplicity of 3 or more. In addition, the eigenvalue -9 must have a multiplicity of 1 or more. Since A is a 4×4 matrix, the sum of the multiplicities of its two eigenvalues must be 4. Hence eigenvalue 5 must have multiplicity 3, and eigenvalue -9 must have multiplicity 1. Thus the characteristic polynomial of A must be $(t-5)^3(t+9)$.
   (b) By Theorem 5.1, the eigenvalue -9 must have a multiplicity of 1 or more. As in (a), the sum of the multiplicities of the two eigenvalues of A must be 4. Since eigenvalue 5 must have a multiplicity of at least one, there are three possibilities:
   (i) Eigenvalue 5 has multiplicity 1, and eigenvalue -9 has multiplicity 3, in which case the characteristic polynomial of A is $(t-5)(t+9)^3$.
   (ii) Eigenvalue 5 has multiplicity 2, and eigenvalue -9 has multiplicity 2, in which case the characteristic polynomial of A is $(t-5)^2(t+9)^2$.
   (iii) Eigenvalue 5 has multiplicity 3, and eigenvalue -9 has multiplicity 1, in which case the characteristic polynomial of A is $(t-5)^3(t+9)$.
   (c) If $\dim W_1 = 2$, then eigenvalue 5 must have a multiplicity of 2 or more. This leads to the two cases described in (ii) and (iii) of (b).

78. (a) $-(t-4)(t-6)^3(t-7)$
    (b) $-(t-4)^3(t-6)(t-7)$,
        $-(t-4)^2(t-6)^2(t-7)$, or
        $-(t-4)^2(t-6)(t-7)^2$
    (c) $-(t-4)^2(t-6)^2(t-7)$,
        $-(t-4)(t-6)^3(t-7)$, or
        $-(t-4)(t-6)^2(t-7)^2$
    (d) $-(t-4)(t-6)^2(t-7)^2$

79. Let A be an upper triangular or lower triangular $n \times n$ matrix. Then $A - tI_n$ is a matrix of the same form, and so
$$\det(A - tI_n) = (a_{11} - t)(a_{22} - t)\cdots(a_{nn} - t)$$
by Theorem 3.2. Thus λ is an eigenvalue of multiplicity k if and only if exactly k of the diagonal entries of A equal λ.

80. The characteristic polynomial of A_θ is
$$t^2 - (2\cos\theta)t + 1,$$
which has no (real) roots if $0° < \theta < 180°$.

81. (a) $\left\{ \begin{bmatrix} -1 \\ 1 \end{bmatrix} \right\}, \left\{ \begin{bmatrix} -2 \\ 1 \end{bmatrix} \right\}$
    (b) $\left\{ \begin{bmatrix} -1 \\ 1 \end{bmatrix} \right\}, \left\{ \begin{bmatrix} -2 \\ 1 \end{bmatrix} \right\}$
    (c) $\left\{ \begin{bmatrix} -1 \\ 1 \end{bmatrix} \right\}, \left\{ \begin{bmatrix} -2 \\ 1 \end{bmatrix} \right\}$
    (d) If c is a nonzero scalar, then $\mathbf{v}$ is an eigenvector of B if and only if $\mathbf{v}$ is an eigenvector of cB because
$$(cB)\mathbf{v} = c(B\mathbf{v}) = c(\lambda \mathbf{v}) = (c\lambda)\mathbf{v}.$$
    (e) If c is a nonzero scalar, then λ is an eigenvalue of B if and only if $c\lambda$ is an eigenvalue of cB because
$$(cB)\mathbf{v} = c(B\mathbf{v}) = c(\lambda \mathbf{v}) = (c\lambda)\mathbf{v}.$$

82. (a) $6, \left\{ \begin{bmatrix} -2 \\ 1 \end{bmatrix} \right\}, 7, \left\{ \begin{bmatrix} -1 \\ 1 \end{bmatrix} \right\}$
    (b) $10, \left\{ \begin{bmatrix} -2 \\ 1 \end{bmatrix} \right\}, 11, \left\{ \begin{bmatrix} -1 \\ 1 \end{bmatrix} \right\}$
    (c) $0, \left\{ \begin{bmatrix} -2 \\ 1 \end{bmatrix} \right\}, 1, \left\{ \begin{bmatrix} -1 \\ 1 \end{bmatrix} \right\}$
    (d) For any scalar c, $\mathbf{v}$ is an eigenvector of B if and only if $\mathbf{v}$ is an eigenvector of $B + cI_n$ because
$$(B + cI_n)\mathbf{v} = B\mathbf{v} + cI_n\mathbf{v}$$
$$= \lambda \mathbf{v} + c\mathbf{v} = (\lambda + c)\mathbf{v}.$$

(e) For any scalar c, λ is an eigenvalue of B if and only if $\lambda + c$ is an eigenvalue of $B + cI_n$ because
$$(B + cI_n)\mathbf{v} = B\mathbf{v} + cI_n\mathbf{v}$$
$$= \lambda\mathbf{v} + c\mathbf{v} = (\lambda + c)\mathbf{v}.$$

83. (a) The characteristic polynomial of A^T is
$$t^2 - 13t + 42 = (t-6)(t-7).$$

(b) The characteristic polynomial of B^T equals that of B since
$$\det(B^T - tI_n) = \det(B - tI_n)^T$$
$$= \det(B - tI_n).$$

(c) Because the characteristic polynomials of B and B^T are equal, the eigenvalues of B and B^T are the same.

(d) There is no relationship between the eigenvectors of an arbitrary matrix B and those of B^T.

84. (a) Suppose that $\mathbf{v}$ is in the eigenspace of A corresponding to λ. Then
$$B(P^{-1}\mathbf{v}) = P^{-1}APP^{-1}\mathbf{v} = P^{-1}A\mathbf{v}$$
$$= P^{-1}\lambda\mathbf{v} = \lambda(P^{-1}\mathbf{v}),$$

and hence $P^{-1}\mathbf{v}$ is in the eigenspace of B corresponding to λ.
Conversely, suppose that $P^{-1}\mathbf{v}$ is in the eigenspace of B corresponding to λ. Then
$$A\mathbf{v} = PBP^{-1}\mathbf{v} = P(BP^{-1}\mathbf{v})$$
$$= P\lambda P^{-1}\mathbf{v} = \lambda\mathbf{v},$$

and hence $\mathbf{v}$ is in the eigenspace of A corresponding to λ.

(b) Suppose that $\{\mathbf{v}_1, \mathbf{v}_2, \ldots, \mathbf{v}_k\}$ is a basis for the eigenspace of A corresponding to λ. Since P^{-1} is invertible, $T_{P^{-1}}$, the matrix transformation induced by P^{-1}, is invertible and hence, by Exercise 97 of Section 5.1, takes linearly independent sets to linearly independent sets. It follows from this and (a) that
$$\{P^{-1}\mathbf{v}_1, P^{-1}\mathbf{v}_2, \ldots, P^{-1}\mathbf{v}_k\}$$
is a linearly independent subset of the eigenspace of B corresponding to λ.
Let $\mathbf{w}$ be any vector in the eigenspace of B corresponding to λ. Since $\mathbf{w} = P^{-1}(P\mathbf{w})$, it follows from (a) that $P\mathbf{w}$ is in the eigenspace of A corresponding to λ. So there exist scalars $c_1, c_2, \ldots, c_k$ such that $P\mathbf{w} = c_1\mathbf{v}_1 + c_2\mathbf{v}_2 + \cdots + c_k\mathbf{v}_k$, and hence $\mathbf{w} = c_1 P^{-1}\mathbf{v}_1 + c_2 P^{-1}\mathbf{v}_2 + \cdots + c_k P^{-1}\mathbf{v}_k$. Thus $\{P^{-1}\mathbf{v}_1, P^{-1}\mathbf{v}_2, \ldots, P^{-1}\mathbf{v}_k\}$ is a generating set for the eigenspace of B corresponding to λ, and we conclude that it is a basis for this subspace.

(c) By (b) the eigenspaces of A and B corresponding to λ have bases with equal numbers of vectors, and hence the dimensions of these two subspaces are equal.

85. If
$$A = \begin{bmatrix} a & b \\ b & c \end{bmatrix},$$
then the characteristic polynomial of A is
$$t^2 - (a+c)t + (ac - b^2).$$
Since the discriminant of this quadratic polynomial is
$$(a-c)^2 + 4b^2 \geq 0,$$
A has real eigenvalues.

86. (a) $r = -(a+d)$, $s = ad - bc$

(b) $A^2 + rA + sI_2 = \begin{bmatrix} a^2 + bc & ab + bd \\ ca + dc & cb + d^2 \end{bmatrix}$
$- (a+d)\begin{bmatrix} a & b \\ c & d \end{bmatrix} + (ad - bc)\begin{bmatrix} 1 & 0 \\ 0 & 1 \end{bmatrix}$
$= \begin{bmatrix} 0 & 0 \\ 0 & 0 \end{bmatrix}.$

87. $-t^3 + \dfrac{23}{15}t^2 - \dfrac{127}{720}t + \dfrac{1}{2160}$

88. $t^4 + 20t^3 + 19t^2 + 18t + 17$

89. $\begin{bmatrix} 0 & 0 & 0 & 5 \\ 1 & 0 & 0 & -7 \\ 0 & 1 & 0 & -23 \\ 0 & 0 & 1 & 11 \end{bmatrix}$

90. (b) The characteristic polynomials of B and B^T are equal.

(c) $\det(B^T - tI_n) = \det(B - tI_n)^T$
$= \det(B - tI_n)$

91. (a) The characteristic polynomial of A is
$$t^2 - 2.5t - 1.5 = (t-3)(t+0.5);$$
so the eigenvalues of A are 3 and -0.5. Corresponding eigenvectors are
$$\begin{bmatrix} 1 \\ 1 \end{bmatrix} \quad \text{and} \quad \begin{bmatrix} 1 \\ 2 \end{bmatrix}.$$

(b) Matrix A is invertible because its rank is 2; in fact,
$$A^{-1} = \frac{1}{3}\begin{bmatrix} 8 & -7 \\ 14 & -13 \end{bmatrix}.$$

The characteristic polynomial of A^{-1} is
$$t^2 + \frac{5}{3}t - \frac{2}{3} = \frac{1}{3}(3t-1)(t+2),$$

and so the corresponding eigenvalues of A^{-1} are $\frac{1}{3}$ and -2. Corresponding eigenvectors are
$$\begin{bmatrix} 1 \\ 1 \end{bmatrix} \quad \text{and} \quad \begin{bmatrix} 1 \\ 2 \end{bmatrix}.$$

(c) The eigenvalues of an invertible matrix A are the reciprocals of the eigenvalues of its inverse, and an eigenvector of A corresponding to eigenvalue λ is also an eigenvector of A^{-1} corresponding to the eigenvalue $\frac{1}{\lambda}$.

(d) The eigenvalues of the given matrix are -2, 4, and 5 with corresponding eigenvectors
$$\mathbf{v}_1 = \begin{bmatrix} 0 \\ 1 \\ 1 \end{bmatrix}, \quad \mathbf{v}_2 = \begin{bmatrix} -2 \\ 3 \\ 2 \end{bmatrix},$$
and
$$\mathbf{v}_3 = \begin{bmatrix} -1 \\ 2 \\ 1 \end{bmatrix}.$$

The inverse of the given matrix is
$$\begin{bmatrix} .30 & .10 & -.10 \\ -.85 & .05 & -.55 \\ -.80 & -.10 & -.40 \end{bmatrix}.$$

Its eigenvalues are $-.50$, $.25$, and $.20$ with corresponding eigenvectors $\mathbf{v}_1$, $\mathbf{v}_2$, and $\mathbf{v}_3$.

(e) Let $\mathbf{v}$ be an eigenvector of an invertible matrix A that corresponds to eigenvalue λ. Then
$$A\mathbf{v} = \lambda\mathbf{v}.$$

Since A is invertible and $\mathbf{v} \neq \mathbf{0}$, we have $A\mathbf{v} \neq \mathbf{0}$. Hence $\lambda \neq 0$. So
$$\mathbf{v} = A^{-1}(\lambda\mathbf{v})$$
$$\mathbf{v} = \lambda(A^{-1}\mathbf{v})$$
$$\frac{1}{\lambda}\mathbf{v} = A^{-1}\mathbf{v}.$$

Thus $\mathbf{v}$ is an eigenvector of A^{-1} corresponding to eigenvalue $\frac{1}{\lambda}$.

5.3 DIAGONALIZATION OF MATRICES

1. The eigenvalues of A are 4 and 5. Eigenvectors corresponding to eigenvalue 4 are solutions of $(A - 4I_2)\mathbf{x} = \mathbf{0}$. Since the reduced row echelon form of $A - 4I_2$ is
$$\begin{bmatrix} 1 & 2 \\ 0 & 0 \end{bmatrix},$$
these solutions have the form
$$\begin{bmatrix} x_1 \\ x_2 \end{bmatrix} = x_2 \begin{bmatrix} -2 \\ 1 \end{bmatrix}.$$
Hence
$$\left\{ \begin{bmatrix} -2 \\ 1 \end{bmatrix} \right\}$$
is a basis for the eigenspace of A corresponding to eigenvalue 4. Likewise, the reduced row echelon form of $A - 5I_2$ is
$$\begin{bmatrix} 1 & 3 \\ 0 & 0 \end{bmatrix},$$
and so
$$\left\{ \begin{bmatrix} -3 \\ 1 \end{bmatrix} \right\}$$
is a basis for the eigenspace of A corresponding to eigenvalue 5. Let P be the matrix whose columns are the vectors in the bases for the eigenspaces, and let D be the diagonal matrix whose diagonal entries are the corresponding eigenvalues:
$$P = \begin{bmatrix} -2 & -3 \\ 1 & 1 \end{bmatrix} \quad \text{and} \quad D = \begin{bmatrix} 4 & 0 \\ 0 & 5 \end{bmatrix}.$$
Then $A = PDP^{-1}$.

2. There are no (real) eigenvalues.

3. The eigenspace corresponding to eigenvalue 2 is 1-dimensional.

4. $P = \begin{bmatrix} -5 & -3 \\ 3 & 2 \end{bmatrix}$, $D = \begin{bmatrix} 0 & 0 \\ 0 & -1 \end{bmatrix}$

5. $P = \begin{bmatrix} 0 & -2 & -1 \\ 1 & 3 & 1 \\ 1 & 2 & 1 \end{bmatrix}$, $D = \begin{bmatrix} -5 & 0 & 0 \\ 0 & 2 & 0 \\ 0 & 0 & 3 \end{bmatrix}$

6. $P = \begin{bmatrix} -4 & 1 & -1 \\ -2 & 1 & 0 \\ 1 & 0 & 1 \end{bmatrix}$, $D = \begin{bmatrix} -3 & 0 & 0 \\ 0 & -1 & 0 \\ 0 & 0 & -1 \end{bmatrix}$

7. There is only one real eigenvalue, and its multiplicity is one.

8. $P = \begin{bmatrix} -1 & 0 & 1 \\ 0 & -1 & -1 \\ 1 & 2 & 2 \end{bmatrix}$, $D = \begin{bmatrix} -5 & 0 & 0 \\ 0 & -4 & 0 \\ 0 & 0 & -2 \end{bmatrix}$

9. $P = \begin{bmatrix} -1 & -1 & 1 \\ 4 & 1 & 0 \\ 2 & 0 & 1 \end{bmatrix}, D = \begin{bmatrix} 5 & 0 & 0 \\ 0 & 3 & 0 \\ 0 & 0 & 3 \end{bmatrix}$

10. The eigenvalue 3 has multiplicity 3, but its eigenspace has dimension 1.

11. $P = \begin{bmatrix} 0 & 1 & 0 & 0 \\ 0 & 0 & 1 & 0 \\ 1 & 0 & 0 & 1 \\ 0 & 1 & 1 & 1 \end{bmatrix}, D = \begin{bmatrix} 4 & 0 & 0 & 0 \\ 0 & -1 & 0 & 0 \\ 0 & 0 & -1 & 0 \\ 0 & 0 & 0 & -1 \end{bmatrix}$

12. $P = \begin{bmatrix} 2 & 0 & -1 & 0 \\ 1 & 1 & 0 & 0 \\ -1 & 0 & 1 & 0 \\ 1 & 0 & 0 & 1 \end{bmatrix}, D = \begin{bmatrix} -3 & 0 & 0 & 0 \\ 0 & 2 & 0 & 0 \\ 0 & 0 & 2 & 0 \\ 0 & 0 & 0 & 2 \end{bmatrix}$

13. The characteristic polynomial of A is
$$t^2 - 2t + 1 = (t-1)^2.$$
Since the rank of $A - I_2$ is 1, the eigenspace of A corresponding to eigenvalue 1 is 1-dimensional. Hence A is not diagonalizable because the eigenvalue 1 has multiplicity 2 and its eigenspace is 1-dimensional.

14. $P = \begin{bmatrix} -2 & 1 \\ 1 & 3 \end{bmatrix}, D = \begin{bmatrix} -2 & 0 \\ 0 & 5 \end{bmatrix}$

15. $P = \begin{bmatrix} -2 & -3 \\ 1 & 2 \end{bmatrix}, D = \begin{bmatrix} 3 & 0 \\ 0 & 2 \end{bmatrix}$

16. There are no (real) eigenvalues.

17. Since A is upper triangular, its eigenvalues are its diagonal entries, which are $-1, -3,$ and 2. Bases for the corresponding eigenspaces are
$$\left\{\begin{bmatrix} 1 \\ 0 \\ 0 \end{bmatrix}\right\}, \left\{\begin{bmatrix} -1 \\ 1 \\ 0 \end{bmatrix}\right\}, \text{ and } \left\{\begin{bmatrix} -1 \\ 1 \\ 5 \end{bmatrix}\right\},$$
respectively. Hence we may take
$$P = \begin{bmatrix} 1 & -1 & -1 \\ 0 & 1 & 1 \\ 0 & 0 & 5 \end{bmatrix} \text{ and } D = \begin{bmatrix} -1 & 0 & 0 \\ 0 & -3 & 0 \\ 0 & 0 & 2 \end{bmatrix}.$$

18. The sum of the multiplicities of the eigenvalues of A is not 3.

19. The eigenvalue 0 has multiplicity 2, but its eigenspace is 1-dimensional.

20. The eigenvalue 3 has multiplicity 2, but its eigenspace is 1-dimensional.

21. The characteristic polynomial of A is $t^2 - 4t + 5$, and hence the eigenvalues of A are $2-i$ and $2+i$. Bases for the corresponding eigenspaces are
$$\left\{\begin{bmatrix} -i \\ 1 \end{bmatrix}\right\} \text{ and } \left\{\begin{bmatrix} i \\ 1 \end{bmatrix}\right\},$$
respectively. Hence we may take
$$P = \begin{bmatrix} -i & i \\ 1 & 1 \end{bmatrix} \text{ and } D = \begin{bmatrix} 2-i & 0 \\ 0 & 2+i \end{bmatrix}.$$

22. $P = \begin{bmatrix} 1+4i & 1-4i \\ 1 & 1 \end{bmatrix}, D = \begin{bmatrix} 8-12i & 0 \\ 0 & 8+12i \end{bmatrix}$

23. $P = \begin{bmatrix} -2i & 2i \\ 1 & 1 \end{bmatrix}, D = \begin{bmatrix} 1-3i & 0 \\ 0 & 1+i \end{bmatrix}$

24. $P = \begin{bmatrix} -1 & -1 \\ 2 & 3 \end{bmatrix}, D = \begin{bmatrix} 1-2i & 0 \\ 0 & 1+2i \end{bmatrix}$

25. $P = \begin{bmatrix} -1 & -1 & 1 \\ 2+i & 2-i & -1 \\ 1 & 1 & 1 \end{bmatrix},$
$D = \begin{bmatrix} 1+i & 0 & 0 \\ 0 & 1-i & 0 \\ 0 & 0 & 2 \end{bmatrix}$

26. $P = \begin{bmatrix} 1 & 1 & i \\ 0 & 1 & 1 \\ 0 & 0 & 1 \end{bmatrix}, D = \begin{bmatrix} 1 & 0 & 0 \\ 0 & i & 0 \\ 0 & 0 & 0 \end{bmatrix}$

27. $P = \begin{bmatrix} 1 & 0 & 0 \\ 0 & 1 & i \\ 0 & 0 & 1 \end{bmatrix}, D = \begin{bmatrix} 2i & 0 & 0 \\ 0 & 1 & 0 \\ 0 & 0 & 0 \end{bmatrix}$

28. $P = \begin{bmatrix} 2i & 1 & 1 \\ -1 & 0 & 0 \\ i & 0 & -2i \end{bmatrix}, D = \begin{bmatrix} 1 & 0 & 0 \\ 0 & 2i & 0 \\ 0 & 0 & 4 \end{bmatrix}$

29. False, see Example 1.

30. True   31. True   32. True

33. False, the eigenvalues of A may occur in any sequence as the diagonal entries of D.

34. False, if an $n \times n$ matrix has n *linearly independent* eigenvectors, then it is diagonalizable.

35. False, I_n is diagonalizable and has only one eigenvalue.

36. True

37. False, see Example 1.

38. False, for A to be diagonalizable, its characteristic polynomial must also factor as a product of linear factors.

39. True

40. False, the dimension of the eigenspace corresponding to λ is the *nullity* of $A - \lambda I_n$.

41. False, for example, I_n has only one eigenvalue, namely 1.

42. True   43. True

44. False, $P^{-1}AP$ is a diagonal matrix.

45. False, for example, any nonzero multiple of an eigenvector is an eigenvector.

46. True

47. False, this is true only if the multiplicity of each eigenvalue is equal to the dimension of the corresponding eigenspace.

48. False, this is true only if the sum of the multiplicities of the eigenvalues is equal to the size of A.

49. The first boxed statement on page 318 implies that the matrix is diagonalizable.

50. The first boxed statement on page 318 implies that the matrix is diagonalizable.

51. (a) By the test for a diagonalizable matrix on page 319, the matrix is diagonalizable if the eigenspace corresponding to -1 is 2-dimensional.
    (b) The matrix is not diagonalizable if the eigenspace corresponding to -1 is 1-dimensional.

52. The matrix is diagonalizable if and only if the eigenspace corresponding to the eigenvalue -4 is 3-dimensional.

53. The matrix is diagonalizable if and only if the eigenspace corresponding to the eigenvalue -3 is 4-dimensional.

54. (a) $(t-2)^3(t-7)$
    (b) Insufficient information is given because eigenvalue 2 might have multiplicity 2, or eigenvalue 7 might have multiplicity 3.
    (c) $(t-2)^2(t-7)^2$

55. (a) $-(t-4)^2(t-5)(t-8)^2$
    (b) There is insufficient information because the dimensions of W_1 and W_2 are not given. Therefore the multiplicities of the eigenvalues 4 and 8 are not determined.
    (c) $-(t-4)(t-5)^2(t-8)^2$

56. (a) Both AB and BA have 0 as an eigenvalue of multiplicity 2.
    (b) $AB = O$, the zero matrix, and so is diagonalizable.
    (c) $BA = \begin{bmatrix} 2 & -4 \\ 1 & -2 \end{bmatrix}$ is not diagonalizable since 0 is an eigenvalue of multiplicity 2, but its eigenspace has dimension 1.

57. We have $A = PDP^{-1}$, where
$$P = \begin{bmatrix} 1 & 2 \\ 1 & 1 \end{bmatrix} \quad \text{and} \quad D = \begin{bmatrix} 4 & 0 \\ 0 & 3 \end{bmatrix}.$$

Thus, as in the example on page 314, we have

$$A^k = PD^kP^{-1} = \begin{bmatrix} 1 & 2 \\ 1 & 1 \end{bmatrix} \begin{bmatrix} 4^k & 0 \\ 0 & 3^k \end{bmatrix} \begin{bmatrix} -1 & 2 \\ 1 & -1 \end{bmatrix}$$

$$= \begin{bmatrix} 4^k & 2 \cdot 3^k \\ 4^k & 3^k \end{bmatrix} \begin{bmatrix} -1 & 2 \\ 1 & -1 \end{bmatrix}$$

$$= \begin{bmatrix} 2 \cdot 3^k - 4^k & 2 \cdot 4^k - 2 \cdot 3^k \\ 3^k - 4^k & 2 \cdot 4^k - 3^k \end{bmatrix}.$$

58. $\begin{bmatrix} 2(-3)^k - (-2)^k & (-2)^k - (-3)^k \\ 2(-3)^k - 2(-2)^k & 2(-2)^k - (-3)^k \end{bmatrix}$

59. $\begin{bmatrix} -2 \cdot 2^k + 3 \cdot 3^k & -6 \cdot 2^k + 6 \cdot 3^k \\ 2^k - 3^k & 3 \cdot 2^k - 2 \cdot 3^k \end{bmatrix}$

60. $\begin{bmatrix} 2(2^k) - (-3)^k & 2^k - (-3)^k \\ 2(-3)^k - 2(2^k) & 2(-3)^k - 2^k \end{bmatrix}$

61. $\begin{bmatrix} -5^k + 2 & -2 \cdot 5^k + 2 & 0 \\ 5^k - 1 & 2 \cdot 5^k - 1 & 0 \\ 0 & 0 & 5^k \end{bmatrix}$

62. $\begin{bmatrix} 2 - 3^k & 0 & 3^k - 1 \\ 0 & 2^k & 0 \\ 2 - 2(3^k) & 0 & 2(3^k) - 1 \end{bmatrix}$

63. The first boxed statement on page 318 shows that the matrix is diagonalizable if $c \neq 2$ and $c \neq 3$. Checking the cases $c = 2$ and $c = 3$ separately, we see that the matrix is diagonalizable if $c = 2$ but not if $c = 3$.

64. $-3, -2$

65. Since there is only one eigenvalue (with multiplicity 1), this matrix is not diagonalizable for any scalar c.

66. 4   67. no c 68. all real numbers

69. It follows from the first boxed statement on page 318 that the given matrix is diagonalizable if $c \neq -2$ and $c \neq -1$. Thus we must check only the values of -2 and -1. For $c = -2$, we see that -2 is an eigenvalue of multiplicity 2, but the reduced row echelon form of $A + 2I_3$, which is

$$\begin{bmatrix} 1 & 0 & 0 \\ 0 & 0 & 1 \\ 0 & 0 & 0 \end{bmatrix},$$

has rank 2. Hence A is not diagonalizable if $c = -2$. Likewise, for $c = -1$, the eigenvalue -1 has multiplicity 2, but the reduced row echelon form of $A + I_3$, which is

$$\begin{bmatrix} 1 & 0 & 0 \\ 0 & 0 & 1 \\ 0 & 0 & 0 \end{bmatrix},$$

has rank 2. Thus A is also not diagonalizable if $c = -1$.

70. none **71.** 2 **72.** 1

73. The desired matrix A satisfies $A = PDP^{-1}$, where

$$P = \begin{bmatrix} 1 & 1 \\ 1 & 3 \end{bmatrix} \quad \text{and} \quad D = \begin{bmatrix} -3 & 0 \\ 0 & 5 \end{bmatrix}.$$

(Here the columns of P are the given eigenvectors of A, and the diagonal entries of D are the corresponding eigenvalues.) Thus

$$A = \begin{bmatrix} -7 & 4 \\ -12 & 9 \end{bmatrix}.$$

74. $\begin{bmatrix} -26 & -11 \\ 66 & 29 \end{bmatrix}$ **75.** $\begin{bmatrix} -1 & 5 & 4 \\ 0 & -2 & 0 \\ 2 & -5 & 5 \end{bmatrix}$

76. $\begin{bmatrix} 4 & 0 & -2 \\ 1 & 2 & -1 \\ 1 & 0 & 1 \end{bmatrix}$ **77.** $\begin{bmatrix} 0 & 0 \\ 0 & 1 \end{bmatrix}$ and $\begin{bmatrix} 0 & -1 \\ 0 & -1 \end{bmatrix}$

78. $A = \begin{bmatrix} 1 & 1 \\ 0 & 0 \end{bmatrix}, B = \begin{bmatrix} 0 & 0 \\ 0 & 1 \end{bmatrix}$

79. If D is a diagonal $n \times n$ matrix, then $D = I_n D I_n^{-1}$; so D is diagonalizable.

80. (a) Suppose A is diagonalizable. Then $A = PDP^{-1}$ for some diagonal matrix D and invertible matrix P. Since A has only one eigenvalue c, necessarily $D = cI_n$. Hence $A = P(cI_n)P^{-1} = cI_n$.
(b) The given 2×2 matrix has 2 as an eigenvalue of multiplicity two, but the matrix does not equal $2I_2$.

81. Let $A = PDP^{-1}$, where D is a diagonal matrix and P is an invertible matrix. Then for $Q = (P^T)^{-1}$, we have

$$A^T = (PDP^{-1})^T = (P^{-1})^T D^T P^T$$
$$= (P^T)^{-1} D P^T = QDQ^{-1},$$

and so A^T is also diagonalizable.

82. Let A be a matrix that is both invertible and diagonalizable. Then $A = PDP^{-1}$ for some diagonal matrix D and invertible matrix P. By Exercise 67 in Section 5.1, the diagonal entries of D, which are eigenvalues of A, are nonzero. Hence

$$A^{-1} = (PDP^{-1})^{-1} = PD^{-1}P^{-1}$$

is diagonalizable because

$$D^{-1} = \begin{bmatrix} \frac{1}{d_{11}} & 0 & \cdots & 0 \\ 0 & \frac{1}{d_{22}} & \cdots & 0 \\ \vdots & \vdots & & \vdots \\ 0 & 0 & \cdots & \frac{1}{d_{nn}} \end{bmatrix}$$

is a diagonal matrix.

83. Let $A = PDP^{-1}$, where D is a diagonal matrix and P is an invertible matrix. Then

$$A^2 = (PDP^{-1})(PDP^{-1}) = (PD^2P^{-1})$$

is also diagonalizable because D^2 is a diagonal matrix.

84. Using the notation in the solution to Exercise 83, we have $A^k = PD^kP^{-1}$, which is a diagonalizable matrix since D^k is a diagonal matrix.

85. (a) Suppose that A is diagonalizable. Then $A = QDQ^{-1}$ for some diagonal matrix D and invertible matrix Q. Since

$$B = PAP^{-1} = P(QDQ^{-1})P^{-1}$$
$$= (PQ)D(PQ)^{-1},$$

B is also diagonalizable. The proof of the converse is similar.
(b) See the box on page 307 in Section 5.2.
(c) We claim that $\mathbf{v}$ is an eigenvector of A if and only if $P\mathbf{v}$ is an eigenvector of B. For if $A\mathbf{v} = \lambda\mathbf{v}$, then

$$B(P\mathbf{v}) = (PAP^{-1})(P\mathbf{v})$$
$$= PA\mathbf{v} = P(\lambda\mathbf{v}) = \lambda(P\mathbf{v}).$$

Conversely, if $B(P\mathbf{v}) = \lambda(P\mathbf{v})$, then

$$A\mathbf{v} = (P^{-1}BP)\mathbf{v} = P^{-1}(\lambda P\mathbf{v}) = \lambda\mathbf{v}.$$

86. Let $A = PDP^{-1}$, where D is a diagonal matrix and P is an invertible matrix. Let C be the diagonal matrix such that $c_{ij} = \sqrt[3]{d_{ij}}$. For $B = PCP^{-1}$, we have

$$B^3 = (PCP^{-1})(PCP^{-1})(PCP^{-1})$$
$$= PC^3P^{-1} = PDP^{-1} = A.$$

Hence B is a cube root of A.

87. Let A be a nilpotent $n \times n$ matrix. If A is diagonalizable, then Exercise 72 of Section 5.1 shows that 0 is the only eigenvalue of A. Hence $A = 0I_n = O$ by Exercise 80(a).

88. If λ is an eigenvalue of A, then $f(\lambda) = 0$. Let $A = PDP^{-1}$, where D is a diagonal matrix and P is an invertible matrix. Then $f(D) = O$ because each diagonal entry of D is an eigenvalue of A. Hence
$$f(A) = f(PDP^{-1})$$
$$= Pf(D)P^{-1} = POP^{-1} = O.$$

89. (a) Let $A = PDP^{-1}$, where D is a diagonal matrix and P is an invertible matrix. By the hint, the trace of $A = PDP^{-1}$ equals the trace of $PP^{-1}D = D$, which is the sum of the eigenvalues of A.

(b) In $p(t)$, the characteristic polynomial of A, the coefficient of t^{n-1} is
$$(-1)^n(-\lambda_1 - \lambda_2 - \cdots - \lambda_n)$$
$$= (-1)^{n+1}(\lambda_1 + \lambda_2 + \cdots + \lambda_n),$$
which by (a) equals $(-1)^{n-1}$ times the trace of A.

(c) The constant term of $p(t)$ is
$$(-1)^n(-\lambda_1)(-\lambda_2)\cdots(-\lambda_n) = \lambda_1\lambda_2\cdots\lambda_n,$$
which equals $\det D = \det A$.

90. The eigenvalue 1 has multiplicity 3, but its eigenspace has dimension 1.

91. The characteristic polynomial of the given matrix is
$$t^4 + 6t^3 + 13t^2 + 12t + 4 = (t+2)^2(t+1)^2.$$
Since the reduced row echelon form of $A + I_4$ is
$$\begin{bmatrix} 1 & 0 & \frac{3}{2} & \frac{1}{2} \\ 0 & 1 & \frac{1}{2} & \frac{1}{2} \\ 0 & 0 & 0 & 0 \\ 0 & 0 & 0 & 0 \end{bmatrix}$$
and the reduced row echelon form of $A + 2I_4$ is
$$\begin{bmatrix} 1 & 0 & \frac{8}{3} & \frac{1}{3} \\ 0 & 1 & \frac{1}{3} & \frac{2}{3} \\ 0 & 0 & 0 & 0 \\ 0 & 0 & 0 & 0 \end{bmatrix}$$
the sets
$$\left\{ \begin{bmatrix} -3 \\ -1 \\ 2 \\ 0 \end{bmatrix}, \begin{bmatrix} -1 \\ -1 \\ 0 \\ 2 \end{bmatrix} \right\}$$
and
$$\left\{ \begin{bmatrix} -8 \\ -1 \\ 3 \\ 0 \end{bmatrix}, \begin{bmatrix} -1 \\ -2 \\ 0 \\ 3 \end{bmatrix} \right\}$$
are bases for the eigenspaces of A corresponding to eigenvalues -1 and -2, respectively. Thus $A = PDP^{-1}$, where
$$P = \begin{bmatrix} -3 & -1 & -8 & -1 \\ -1 & -1 & -1 & -2 \\ 2 & 0 & 3 & 0 \\ 0 & 2 & 0 & 3 \end{bmatrix},$$
and
$$D = \begin{bmatrix} -1 & 0 & 0 & 0 \\ 0 & -1 & 0 & 0 \\ 0 & 0 & -2 & 0 \\ 0 & 0 & 0 & -2 \end{bmatrix}.$$

92. $P = \begin{bmatrix} -3 & 1 & -1 & -1 & 0 \\ 0 & -1 & -1 & -3 & -3 \\ 1 & 0 & 0 & 1 & 0 \\ 0 & 1 & 0 & 0 & -2 \\ 0 & 0 & 1 & 0 & 1 \end{bmatrix},$

$D = \begin{bmatrix} -1 & 0 & 0 & 0 & 0 \\ 0 & -1 & 0 & 0 & 0 \\ 0 & 0 & -1 & 0 & 0 \\ 0 & 0 & 0 & 1 & 0 \\ 0 & 0 & 0 & 0 & 1 \end{bmatrix}$

93. The eigenvalue 1 has multiplicity 3, but its eigenspace is 2-dimensional.

94. (a) $A^m\mathbf{u}$ approaches $\mathbf{0}$.
(b) about 0.974, 0.829, and -0.803
(c) $A^m\mathbf{u}$ approaches $\dfrac{1}{3}\begin{bmatrix} -325 \\ -52 \\ 26 \end{bmatrix}$.
(d) $1, 0.8, -0.8$
(e) $A^m\mathbf{u}$ diverges.
(f) about 1.020, 0.777, and -0.797
(g) $B^m\mathbf{u}$ approaches $\mathbf{0}$.

5.4 DIAGONALIZATION OF LINEAR OPERATORS

1. $\begin{bmatrix} 0 & 0 & 2 \\ 0 & 3 & 0 \\ 4 & 0 & 0 \end{bmatrix}$, no

2. $\begin{bmatrix} 0 & 1 & -1 \\ 4 & 4 & 2 \\ 0 & 0 & 0 \end{bmatrix}$, no

3. $\begin{bmatrix} 2 & 1 & 0 \\ 0 & 2 & 0 \\ 0 & 0 & 1 \end{bmatrix}$, no

4. $\begin{bmatrix} -1 & 0 & 0 \\ 0 & 0 & 0 \\ 0 & 0 & 1 \end{bmatrix}$, yes

5. $\begin{bmatrix} 0 & 0 & 3 \\ 0 & -2 & 0 \\ -4 & 0 & 0 \end{bmatrix}$, no

6. $\begin{bmatrix} 1 & 0 & 0 \\ -1 & 2 & 0 \\ 1 & 1 & -1 \end{bmatrix}$, no

7. The standard matrix of T is
$$A = \begin{bmatrix} -3 & 5 & -5 \\ 2 & -3 & 2 \\ 2 & -5 & 4 \end{bmatrix}.$$

Chapter 5 Eigenvalues, Eigenvectors, and Diagonalization

If B is the matrix whose columns are the vectors in $\mathcal{B}$, then

$$[T]_\mathcal{B} = B^{-1}AB = \begin{bmatrix} 2 & 0 & 0 \\ 0 & -1 & 0 \\ 0 & 0 & -3 \end{bmatrix}.$$

Since $[T]_\mathcal{B}$ is a diagonal matrix, the basis $\mathcal{B}$ consists of eigenvectors of T.

8. $\begin{bmatrix} -2 & 0 & 0 \\ 0 & -2 & 1 \\ 0 & 0 & -2 \end{bmatrix}$, no

9. There are no real eigenvalues.

10. The eigenvalue -2 has multiplicity 2, but its eigenspace has dimension 1.

11. The standard matrix of T is

$$A = \begin{bmatrix} 7 & -5 \\ 10 & -8 \end{bmatrix}.$$

A basis for the eigenspace of T corresponding to the eigenvalue -3 can be obtained by solving $(A+3I_2)\mathbf{x} = \mathbf{0}$, and a basis for the eigenspace of T corresponding to eigenvalue 2 can be obtained by solving $(A - 2I_2)\mathbf{x} = \mathbf{0}$. The resulting bases are

$$\mathcal{B}_1 = \left\{ \begin{bmatrix} 1 \\ 2 \end{bmatrix} \right\} \quad \text{and} \quad \mathcal{B}_2 = \left\{ \begin{bmatrix} 1 \\ 1 \end{bmatrix} \right\}.$$

Combining these two sets produces a basis for $\mathcal{R}^2$ consisting of eigenvectors of T.

12. $\left\{ \begin{bmatrix} -1 \\ 1 \end{bmatrix}, \begin{bmatrix} -1 \\ 2 \end{bmatrix} \right\}$

13. $\left\{ \begin{bmatrix} 1 \\ -1 \\ 1 \end{bmatrix}, \begin{bmatrix} 0 \\ -1 \\ 1 \end{bmatrix}, \begin{bmatrix} 0 \\ 0 \\ 1 \end{bmatrix} \right\}$

14. $\left\{ \begin{bmatrix} -1 \\ 1 \\ 0 \end{bmatrix}, \begin{bmatrix} 0 \\ 1 \\ 0 \end{bmatrix}, \begin{bmatrix} 0 \\ 0 \\ 1 \end{bmatrix} \right\}$

15. The standard matrix of T is

$$A = \begin{bmatrix} -1 & -1 & 0 \\ 0 & -1 & 0 \\ 1 & 1 & 0 \end{bmatrix}.$$

Since the reduced row echelon form of $A + I_3$ is

$$\begin{bmatrix} 1 & 0 & 1 \\ 0 & 1 & 0 \\ 0 & 0 & 0 \end{bmatrix},$$

the dimension of the eigenspace of T corresponding to eigenvalue -1 is

$$3 - \text{rank}\,(A + I_3) = 1.$$

But the multiplicity of the eigenvalue -1 is 2, so that T is not diagonalizable. That is, there is no basis for $\mathcal{R}^3$ consisting of eigenvectors of T.

16. $\left\{ \begin{bmatrix} 0 \\ 0 \\ 1 \end{bmatrix}, \begin{bmatrix} 1 \\ -1 \\ 7 \end{bmatrix}, \begin{bmatrix} 3 \\ 0 \\ 4 \end{bmatrix} \right\}$

17. $\left\{ \begin{bmatrix} 1 \\ 1 \\ 0 \end{bmatrix}, \begin{bmatrix} 0 \\ 1 \\ 1 \end{bmatrix}, \begin{bmatrix} 1 \\ 0 \\ 0 \end{bmatrix} \right\}$

18. The eigenvalue -1 has multiplicity 3, but its eigenspace has dimension 2.

19. $\left\{ \begin{bmatrix} 1 \\ 0 \\ 1 \\ 0 \end{bmatrix}, \begin{bmatrix} -1 \\ 2 \\ 0 \\ 0 \end{bmatrix}, \begin{bmatrix} 1 \\ 0 \\ 2 \\ 0 \end{bmatrix}, \begin{bmatrix} -1 \\ 0 \\ 0 \\ 2 \end{bmatrix} \right\}$

20. $\left\{ \begin{bmatrix} -1 \\ -1 \\ -1 \\ 1 \end{bmatrix}, \begin{bmatrix} 1 \\ 0 \\ 0 \\ 0 \end{bmatrix}, \begin{bmatrix} 0 \\ 1 \\ 0 \\ 0 \end{bmatrix}, \begin{bmatrix} 0 \\ 0 \\ 0 \\ 1 \end{bmatrix} \right\}$

21. T has no real eigenvalues.

22. $\left\{ \begin{bmatrix} 1 \\ 1 \end{bmatrix}, \begin{bmatrix} 3 \\ 4 \end{bmatrix} \right\}$

23. The standard matrix for T is

$$\begin{bmatrix} -2 & 3 \\ 4 & -3 \end{bmatrix},$$

and its characteristic polynomial is $t^2 + 5t - 6 = (t-1)(t+6)$. As in Exercise 11, we find that

$$\mathcal{B}_1 = \left\{ \begin{bmatrix} 1 \\ 1 \end{bmatrix} \right\} \quad \text{and} \quad \mathcal{B}_2 = \left\{ \begin{bmatrix} -3 \\ 4 \end{bmatrix} \right\}$$

are bases for the eigenspaces of T corresponding to the eigenvalues -6 and 1, respectively. Combining these two sets produces a basis $\mathcal{B}$ for $\mathcal{R}^2$ consisting of eigenvectors of T, and so

$$[T]_\mathcal{B} = \begin{bmatrix} 1 & 0 \\ 0 & -6 \end{bmatrix}$$

is a diagonal matrix.

24. The eigenvalue -1 has multiplicity 2, but its eigenspace has dimension 1.

25. $\left\{ \begin{bmatrix} 0 \\ 1 \\ 0 \end{bmatrix}, \begin{bmatrix} -1 \\ 0 \\ 1 \end{bmatrix}, \begin{bmatrix} 0 \\ 1 \\ 1 \end{bmatrix} \right\}$

26. $\left\{ \begin{bmatrix} 1 \\ 0 \\ 0 \end{bmatrix}, \begin{bmatrix} 1 \\ 1 \\ 0 \end{bmatrix}, \begin{bmatrix} 0 \\ 0 \\ 1 \end{bmatrix} \right\}$

27. The standard matrix of T is

$$\begin{bmatrix} 1 & 0 & 0 \\ -1 & 1 & -1 \\ 0 & 0 & 1 \end{bmatrix},$$

and its characteristic polynomial is

$$-t^3 + 3t^2 - 3t + 1 = -(t-1)^3.$$

Since the reduced row echelon form of $A - I_3$ is

$$\begin{bmatrix} 1 & 0 & 1 \\ 0 & 0 & 0 \\ 0 & 0 & 0 \end{bmatrix},$$

the dimension of the eigenspace of T corresponding to eigenvalue 1 is

$$3 - \text{rank}\,(A - I_3) = 2.$$

Because this dimension does not equal the multiplicity of the eigenvalue 1, T is not diagonalizable. That is, there is no basis $\mathcal{B}$ for $\mathcal{R}^3$ such that $[T]_\mathcal{B}$ is a diagonal matrix.

28. The eigenvalue 3 has multiplicity 2, but its eigenspace has dimension 1.

29. False, its standard matrix is diagonalizable, that is, *similar* to a diagonal matrix.

30. False, the linear operator on $\mathcal{R}^2$ that rotates a vector by 90° is not diagonalizable.

31. True

32. False, $\mathcal{B}$ can be any basis for $\mathcal{R}^n$ consisting of eigenvectors of T.

33. False, the eigenvalues of T may occur in any sequence as the diagonal entries of D.

34. True 35. True 36. True 37. True

38. False, in addition, the multiplicity of each eigenvaue must equal the dimension of the corresponding eigenspace.

39. True

40. False, in addition, the sum of the multiplicities of the eigenvalues must equal n.

41. False, it is an eigenvector corresponding to the eigenvalue 1.

42. False, it is an eigenvector corresponding to the eigenvalue -1.

43. True

44. False, a linear operator on $\mathcal{R}^n$ may have no eigenvalues.

45. True 46. True 47. True

48. False, this statement is true only when T is diagonalizable.

49. If $c \neq 2$ and $c \neq 7$, then T has 3 distinct eigenvalues and so is diagonalizable. Separately checking the cases $c = 2$ and $c = 7$, we see that T is diagonalizable if $c = 2$ but not diagonalizable if $c = 7$.

50. 3, 4

51. The standard matrix of T is

$$A = \begin{bmatrix} c & 0 & 0 \\ -1 & -3 & -1 \\ -8 & 1 & -5 \end{bmatrix},$$

and so

$$A + 4I_3 = \begin{bmatrix} c+4 & 0 & 0 \\ -1 & 1 & -1 \\ -8 & 1 & -1 \end{bmatrix}.$$

Because the last two rows of $A + 4I_3$ are linearly independent, the rank of $A + 4I_3$ is at least 2. Hence the dimension of the eigenspace of T corresponding to eigenvalue -4 is 1. Since this dimension does not equal the multiplicity of the eigenvalue -4, T is not diagonalizable for any scalar c.

52. all real numbers

53. The standard matrix of T is

$$A = \begin{bmatrix} c & 0 & 0 \\ 2 & -3 & 2 \\ -3 & 0 & 1 \end{bmatrix}.$$

The characteristic polynomial of A is the same as that of T, namely $-(t-c)(t+3)(t+1)$. If $c \neq -3$, and $c \neq -1$, then the first boxed statement on page 318 shows that A (and hence T) is diagonalizable.

If $c = -3$, then -3 is an eigenvalue of A (and T) with multiplicity 2. But the reduced row echelon form of $A + 3I_3$ is

$$\begin{bmatrix} 1 & 0 & 0 \\ 0 & 1 & 0 \\ 0 & 0 & 0 \end{bmatrix},$$

so that the eigenspace of A (and T) corresponding to eigenvalue -3 has dimension $3 - 2 = 1$. Since the second condition in the test for a diagonalizable linear operator fails, T is not diagonalizable if $c = -3$.

If $c = -1$, then -1 is an eigenvalue of A (and T) with multiplicity 2. But the reduced row echelon form of $A + I_3$ is

$$\begin{bmatrix} 1 & 0 & 0 \\ 0 & 1 & -1 \\ 0 & 0 & 0 \end{bmatrix},$$

so that the eigenspace of A (and T) corresponding to eigenvalue -1 has dimension $3 - 2 = 1$. Since the second condition in the test for a diagonalizable linear operator fails, T is also not diagonalizable if $c = -1$.

54. -4 **55.** all scalars c **56.** 6, 7

57. no scalars c **58.** all scalars c

59. The vector form of the general solution of the equation $x + y + z = 0$ is

$$\begin{bmatrix} x \\ y \\ z \end{bmatrix} = y \begin{bmatrix} -1 \\ 1 \\ 0 \end{bmatrix} + z \begin{bmatrix} -1 \\ 0 \\ 1 \end{bmatrix}.$$

Hence

$$\left\{ \begin{bmatrix} -1 \\ 1 \\ 0 \end{bmatrix}, \begin{bmatrix} -1 \\ 0 \\ 1 \end{bmatrix} \right\}$$

is a basis for W, the eigenspace of T_W corresponding to eigenvalue 1. As on page 329, the vector

$$\begin{bmatrix} 1 \\ 1 \\ 1 \end{bmatrix}$$

whose components are the coefficients of the equation $x + y + z = 0$, is normal to W, and so is an eigenvector of T_W corresponding to eigenvalue -1. Thus

$$\mathcal{B} = \left\{ \begin{bmatrix} -1 \\ 1 \\ 0 \end{bmatrix}, \begin{bmatrix} -1 \\ 0 \\ 1 \end{bmatrix}, \begin{bmatrix} 1 \\ 1 \\ 1 \end{bmatrix} \right\}$$

is a basis for $\mathcal{R}^3$ consisting of eigenvectors of T_W. Hence

$$[T_W]_\mathcal{B} = \begin{bmatrix} 1 & 0 & 0 \\ 0 & 1 & 0 \\ 0 & 0 & -1 \end{bmatrix}.$$

Let B be the matrix whose columns are the vectors in $\mathcal{B}$. Then the standard matrix of T_W is

$$A = B[T_W]_\mathcal{B} B^{-1} = \frac{1}{3} \begin{bmatrix} 1 & -2 & -2 \\ -2 & 1 & -2 \\ -2 & -2 & 1 \end{bmatrix}.$$

Therefore

$$T_W\left(\begin{bmatrix} x_1 \\ x_2 \\ x_3 \end{bmatrix}\right) = A \begin{bmatrix} x_1 \\ x_2 \\ x_3 \end{bmatrix}$$

$$= \frac{1}{3} \begin{bmatrix} x_1 - 2x_2 - 2x_3 \\ -2x_1 + x_2 - 2x_3 \\ -2x_1 - 2x_2 + x_3 \end{bmatrix}.$$

60. $T_W\left(\begin{bmatrix} x_1 \\ x_2 \\ x_3 \end{bmatrix}\right) = \frac{1}{6} \begin{bmatrix} 2x_1 - 2x_2 - 2x_3 \\ -2x_1 + 5x_2 - x_3 \\ -2x_1 - x_2 + 5x_3 \end{bmatrix}$

61. $T_W\left(\begin{bmatrix} x_1 \\ x_2 \\ x_3 \end{bmatrix}\right) = \frac{1}{3} \begin{bmatrix} 2x_1 - 2x_2 + x_3 \\ -2x_1 - x_2 + 2x_3 \\ x_1 + 2x_2 + 2x_3 \end{bmatrix}$

62. $T_W\left(\begin{bmatrix} x_1 \\ x_2 \\ x_3 \end{bmatrix}\right) = \begin{bmatrix} -x_3 \\ x_2 \\ -x_1 \end{bmatrix}$

63. $T_W\left(\begin{bmatrix} x_1 \\ x_2 \\ x_3 \end{bmatrix}\right) = \frac{1}{90} \begin{bmatrix} 88x_1 - 16x_2 + 10x_3 \\ -16x_1 - 38x_2 + 80x_3 \\ 10x_1 + 80x_2 + 40x_3 \end{bmatrix}$

64. $T_W\left(\begin{bmatrix} x_1 \\ x_2 \\ x_3 \end{bmatrix}\right) = \frac{1}{25} \begin{bmatrix} 16x_1 + 12x_2 - 15x_3 \\ 12x_1 + 9x_2 + 20x_3 \\ -15x_1 + 20x_2 \end{bmatrix}$

65. $2x + 2y + z = 0$ **66.** $x + 2y + z = 0$

67. (a) Let $\{\mathbf{v}_1, \mathbf{v}_2\}$ be a basis for W and $\mathbf{v}_3$ be a nonzero vector orthogonal to W. Then $\mathcal{B} = \{\mathbf{v}_1, \mathbf{v}_2, \mathbf{v}_3\}$ is a basis for $\mathcal{R}^3$. Furthermore, $U_W(\mathbf{v}_1) = \mathbf{v}_1$, $U_W(\mathbf{v}_2) = \mathbf{v}_2$, and $U_W(\mathbf{v}_3) = \mathbf{0}$. The result now follows.

(b) Observe that $T_W(\mathbf{v}_1) = \mathbf{v}_1$, $T_W(\mathbf{v}_2) = \mathbf{v}_2$, and $T_W(\mathbf{v}_3) = -\mathbf{v}_3$. The result now follows.

(c) $\frac{1}{2}([T_W]_\mathcal{B} + I_3)$

$$= \frac{1}{2}\left(\begin{bmatrix} 1 & 0 & 0 \\ 0 & 1 & 0 \\ 0 & 0 & -1 \end{bmatrix} + \begin{bmatrix} 1 & 0 & 0 \\ 0 & 1 & 0 \\ 0 & 0 & 1 \end{bmatrix}\right)$$

$$= \begin{bmatrix} 1 & 0 & 0 \\ 0 & 1 & 0 \\ 0 & 0 & 0 \end{bmatrix} = [U_W]_\mathcal{B}.$$

(d) Let I denote the identity operator on $\mathcal{R}_3$. Then $[I]_\mathcal{B} = I_3$, and hence

$$[U_W]_\mathcal{B} = \frac{1}{2}([T_W]_\mathcal{B} + [I]_\mathcal{B}) = \frac{1}{2}([T_W + I]_\mathcal{B}).$$

So $U_W = \frac{1}{2}(T_W + I)$. Therefore

$$U_W\left(\begin{bmatrix} x_1 \\ x_2 \\ x_3 \end{bmatrix}\right)$$

$$= \frac{1}{2}\left(\frac{1}{3}\begin{bmatrix} x_1 - 2x_2 - 2x_3 \\ -2x_1 + x_2 - 2x_3 \\ -2x_1 - 2x_2 + x_3 \end{bmatrix} + \begin{bmatrix} x_1 \\ x_2 \\ x_3 \end{bmatrix}\right)$$

$$= \frac{1}{3}\begin{bmatrix} 2x_1 - x_2 - x_3 \\ -x_1 + 2x_2 - x_3 \\ -x_1 - x_2 + 2x_3 \end{bmatrix}.$$

68. $U_W\left(\begin{bmatrix} x_1 \\ x_2 \\ x_3 \end{bmatrix}\right) = \frac{1}{12}\begin{bmatrix} 8x_1 - 2x_2 - 2x_3 \\ -2x_1 + 11x_2 - x_3 \\ -2x_1 - x_2 + 11x_3 \end{bmatrix}$

69. $U_W\left(\begin{bmatrix} x_1 \\ x_2 \\ x_3 \end{bmatrix}\right) = \frac{1}{6}\begin{bmatrix} 5x_1 - 2x_2 + x_3 \\ -2x_1 + 2x_2 + 2x_3 \\ x_1 + 2x_2 + 5x_3 \end{bmatrix}$

70. $U_W\left(\begin{bmatrix} x_1 \\ x_2 \\ x_3 \end{bmatrix}\right) = \frac{1}{2}\begin{bmatrix} x_1 - x_3 \\ 2x_2 \\ -x_1 + x_3 \end{bmatrix}$

5.4 Diagonalization of Linear Operators

71. $U_W\left(\begin{bmatrix} x_1 \\ x_2 \\ x_3 \end{bmatrix}\right) = \dfrac{1}{90}\begin{bmatrix} 89x_1 - 8x_2 + 5x_3 \\ -8x_1 + 26x_2 + 40x_3 \\ 5x_1 + 40x_2 + 65x_3 \end{bmatrix}$

72. $U_W\left(\begin{bmatrix} x_1 \\ x_2 \\ x_3 \end{bmatrix}\right) = \dfrac{1}{50}\begin{bmatrix} 31x_1 + 12x_2 - 15x_3 \\ 12x_1 + 34x_2 + 20x_3 \\ -15x_1 + 20x_2 + 25x_3 \end{bmatrix}$

73. $U_W\left(\begin{bmatrix} x_1 \\ x_2 \\ x_3 \end{bmatrix}\right) = \dfrac{1}{9}\begin{bmatrix} 5x_1 - 4x_2 - 2x_3 \\ -4x_1 + 5x_2 - 2x_3 \\ -2x_1 - 2x_2 + 8x_3 \end{bmatrix}$

74. $U_W\left(\begin{bmatrix} x_1 \\ x_2 \\ x_3 \end{bmatrix}\right) = \dfrac{1}{6}\begin{bmatrix} 5x_1 - 2x_2 - x_3 \\ -2x_1 + 2x_2 - 2x_3 \\ -x_1 - 2x_2 + 5x_3 \end{bmatrix}$

75. We combine (a) and (b). Let $\mathcal{B} = \{\mathbf{u}, \mathbf{v}, \mathbf{w}\}$. Observe that $T(\mathbf{u}) = \mathbf{u}$, $T(\mathbf{v}) = \mathbf{v}$, and $T(\mathbf{w}) = \mathbf{0}$. Hence $\mathbf{u}$ and $\mathbf{v}$ are eigenvectors of T with corresponding eigenvalue 1, and $\mathbf{w}$ is an eigenvector of T with corresponding eigenvalue 0. Consequently, $\{\mathbf{u}, \mathbf{v}\}$ is a basis for the eigenspace of T corresponding to the eigenvalue 1, and $\{\mathbf{w}\}$ is a basis for the eigenspace of T corresponding to the eigenvalue 0. It also follows that T is diagonalizable because there is a basis for $\mathcal{R}^3$ consisting of eigenvectors of T.

76. This is similar to Exercise 75, except that the eigenvalue to which $\mathbf{w}$ corresponds is -1 instead of 0.

77. Let $[T]_\mathcal{B}$ be a diagonal matrix D, and let $\mathcal{B} = \{\mathbf{b}_1, \mathbf{b}_2, \ldots, \mathbf{b}_n\}$. Since the jth column of $[T]_\mathcal{B}$ is $[T(\mathbf{b}_j)]_\mathcal{B}$, we must have $T(\mathbf{b}_j) = d_{jj}\mathbf{b}_j$. Also, $\mathbf{b}_j \ne \mathbf{0}$ because $\mathcal{B}$ is a basis, and hence each $\mathbf{b}_j$ must be an eigenvector of T.

78. No, let T and U be the matrix transformations induced by the matrices in Exercise 47 of Section 5.3.

79. Yes, if $\mathcal{B}$ is a basis for $\mathcal{R}^n$ consisting of eigenvectors of T, then $\mathcal{B}$ is also a basis for $\mathcal{R}^n$ consisting of eigenvectors of cT for any scalar c.

80. No, let T and U be the matrix transformations induced by the matrices in Exercise 48 of Section 5.3.

81. Let $T(\mathbf{v}_i) = \lambda_i \mathbf{v}_i$ for $i = 1, 2, \ldots, k$. Suppose
$$a_1 T(\mathbf{v}_1) + a_2 T(\mathbf{v}_2) + \cdots + a_k T(\mathbf{v}_k) = \mathbf{0}$$
for some scalars $a_1, a_2, \ldots, a_k$. Then
$$a_1 \lambda_1 \mathbf{v}_1 + a_2 \lambda_2 \mathbf{v}_2 + \cdots + a_k \lambda_k \mathbf{v}_k = \mathbf{0}.$$
By Theorem 5.3, $\mathbf{v}_1, \mathbf{v}_2, \ldots, \mathbf{v}_k$ are linearly independent. So $a_i \lambda_i = 0$ for all i. Because every λ_i is nonzero, every a_i is zero. Hence the set $\{T(\mathbf{v}_1), T(\mathbf{v}_2), \ldots, T(\mathbf{v}_k)\}$ is linearly independent.

82. Let $\mathcal{B}$ be a basis for $\mathcal{R}^n$ consisting of eigenvectors of both T and U. Then $[T]_\mathcal{B} = C$ and $[U]_\mathcal{B} = D$ are both diagonal matrices. Hence
$$[TU]_\mathcal{B} = [T]_\mathcal{B}[U]_\mathcal{B} = CD = DC$$
$$= [U]_\mathcal{B}[T]_\mathcal{B} = [UT]_\mathcal{B},$$
Thus $TU = UT$.

83. Let U be a diagonalizable linear operator on $\mathcal{R}^n$ having only nonnegative eigenvalues, and let $\mathcal{B}$ be a basis for $\mathcal{R}^n$ consisting of eigenvectors of U. If C is the standard matrix of U and B is the matrix whose columns are the vectors in $\mathcal{B}$, then $[U]_\mathcal{B} = B^{-1}CB$. Let A be the diagonal matrix whose entries are the square roots of the entries of the diagonal matrix $[U]_\mathcal{B}$. Then $A^2 = [U]_\mathcal{B}$, and so
$$C = B[U]_\mathcal{B} B^{-1} = BA^2 B^{-1} = (BAB^{-1})^2.$$
So if T is the matrix transformation induced by BAB^{-1}, then T is a square root of U.

84. If T is diagonalizable, then so is its standard matrix. Thus the test for a diagonalizable matrix implies that $\mathcal{B}_1 \cup \mathcal{B}_2 \cup \cdots \cup \mathcal{B}_k$ is a set containing n vectors. Since this set is linearly independent, it must also be a generating set for $\mathcal{R}^n$ by Theorem 4.5.

    Conversely, if $\mathcal{B}_1 \cup \mathcal{B}_2 \cup \cdots \cup \mathcal{B}_k$ is a generating set for $\mathcal{R}^n$, then it must contain a basis for $\mathcal{R}^n$ by Theorem 4.3. Since this basis consists of eigenvectors of T, the operator T is diagonalizable.

85. $\left\{\begin{bmatrix} 2 \\ -2 \\ -4 \\ 3 \\ 0 \end{bmatrix}, \begin{bmatrix} -1 \\ 1 \\ 2 \\ 0 \\ 3 \end{bmatrix}, \begin{bmatrix} -1 \\ 1 \\ -3 \\ 2 \\ 0 \end{bmatrix}, \begin{bmatrix} 1 \\ 1 \\ 3 \\ 0 \\ 2 \end{bmatrix}, \begin{bmatrix} 1 \\ 0 \\ 1 \\ 0 \\ 0 \end{bmatrix}\right\}$

86. The eigenvalue -1 has multiplicity 2, but its eigenspace has dimension 1.

5.5 APPLICATIONS OF EIGENVALUES

1. False, the *column* sums of the transition matrix of a Markov chain are all 1.
2. False, see the matrix A on page 334.
3. True
4. False, consider $A = \begin{bmatrix} 0 & 1 \\ 1 & 0 \end{bmatrix}$ and $\mathbf{p} = \begin{bmatrix} .8 \\ .2 \end{bmatrix}$.
5. True 6. True 7. True
8. False, the general solution of $y' = ky$ is $y = ce^{kt}$.
9. False, the change of variable $\mathbf{y} = P\mathbf{z}$ transforms $\mathbf{y}' = A\mathbf{y}$ into $\mathbf{z}' = D\mathbf{z}$.
10. True
11. False, the solution is $\mathbf{y} = P\mathbf{z}$.
12. True
13. No, the $(1,2)$-entry of A^k is zero for every k.
14. Yes, A^2 has no zero entries.
15. Since there are no zero entries in
$$A^3 = \begin{bmatrix} .475 & .35 & .385 \\ .275 & .35 & .265 \\ .250 & .30 & .350 \end{bmatrix},$$
A is a regular transition matrix.
16. No, the $(2,1)$-entry of A^k is zero for every k.
17. No, the $(1,2)$-entry of A^k is zero for every k.
18. Yes, A^2 has no zero entries.
19. Yes, A^2 has no zero entries.
20. No, the $(1,2)$-entry of A^k is zero for every k.
21. $\begin{bmatrix} .75 \\ .25 \end{bmatrix}$ 22. $\begin{bmatrix} .2 \\ .8 \end{bmatrix}$
23. A steady-state vector is a probability vector that is also an eigenvector corresponding to eigenvalue 1. We begin by finding the eigenvectors corresponding to eigenvalue 1. Since the reduced row echelon form of $A - I_3$ is
$$\begin{bmatrix} 1 & 0 & -.5 \\ 0 & 1 & -.5 \\ 0 & 0 & 0 \end{bmatrix},$$
a basis for the eigenspace corresponding to eigenvalue 1 is
$$\left\{ \begin{bmatrix} 1 \\ 1 \\ 2 \end{bmatrix} \right\}.$$
Thus the eigenvectors corresponding to eigenvalue 1 have the form
$$c \begin{bmatrix} 1 \\ 1 \\ 2 \end{bmatrix} = \begin{bmatrix} c \\ c \\ 2c \end{bmatrix}.$$

We seek a vector of this form that is also a probability vector, that is, such that
$$c + c + 2c = 1.$$
So $c = .25$, and the steady-state vector is
$$\begin{bmatrix} .25 \\ .25 \\ .50 \end{bmatrix}.$$

24. $\dfrac{1}{3} \begin{bmatrix} 2 \\ 0 \\ 1 \end{bmatrix}$ 25. $\dfrac{1}{6} \begin{bmatrix} 1 \\ 3 \\ 2 \end{bmatrix}$ 26. $\dfrac{1}{9} \begin{bmatrix} 2 \\ 4 \\ 3 \end{bmatrix}$

27. $\dfrac{1}{29} \begin{bmatrix} 3 \\ 4 \\ 10 \\ 12 \end{bmatrix}$ 28. $\dfrac{1}{53} \begin{bmatrix} 5 \\ 8 \\ 20 \\ 20 \end{bmatrix}$

29. (a) The two states of this Markov chain are buying a root beer float (F) and buying a chocolate sundae (S). A transition matrix for this Markov chain is

$$\begin{array}{c} \text{Next visit} \end{array} \begin{array}{c} \text{Last visit} \\ \begin{array}{cc} \text{F} & \text{S} \end{array} \\ \begin{array}{c} \text{F} \\ \text{S} \end{array} \begin{bmatrix} .25 & .5 \\ .75 & .5 \end{bmatrix} = A. \end{array}$$

Note that the $(1,2)$-entry and the $(2,1)$-entry of A can be determined from the condition that each column sum in A must be 1.

(b) If Alison bought a sundae on her next-to-last visit, we can take
$$\mathbf{p} = \begin{bmatrix} 0 \\ 1 \end{bmatrix}.$$
Then the probabilities of each purchase on her last visit are
$$A\mathbf{p} = \begin{bmatrix} .5 \\ .5 \end{bmatrix},$$
and the probabilities of each purchase on her next visit are
$$A^2\mathbf{p} = A(A\mathbf{p}) = \begin{bmatrix} .375 \\ .625 \end{bmatrix}.$$
Thus the probability that she will buy a float on her next visit is .375.

(c) Over the long run, the proportion of purchases of each kind is given by the steady-state vector for A. As in Exercise 9, we first find a basis for the eigenspace corresponding to eigenvalue 1, which is

$$\left\{ \begin{bmatrix} 2 \\ 3 \end{bmatrix} \right\}.$$

The vector in this eigenspace that is also a probability vector is

$$.2 \begin{bmatrix} 2 \\ 3 \end{bmatrix} = \begin{bmatrix} .4 \\ .6 \end{bmatrix}.$$

Hence, over the long run, Alison buys a sundae on 60% of her trips to the ice cream store.

30. (a) $\begin{bmatrix} .75 & .35 \\ .25 & .65 \end{bmatrix}$ (b) 47%, 53.8%, 56.52%
    (c) $\frac{7}{12}$

31. (a) $\begin{bmatrix} .7 & .1 & .1 \\ .1 & .6 & .1 \\ .2 & .3 & .8 \end{bmatrix}$ (b) .6 (c) .33
    (d) .25 buy brand A, .20 buy brand B, and .55 buy brand C

32. (a) $\begin{bmatrix} .4 & .2 & .2 \\ .1 & .7 & .2 \\ .5 & .1 & .6 \end{bmatrix}$
    (b) After 1 year, 30% live in the city, 30% in the suburbs, and 40% in the country.
    After 2 years, 26% live in the city, 32% in the suburbs, and 42% in the country.
    After 3 years, 25.2% live in the city, 33.4% in the suburbs, and 41.4% in the country.
    (c) After 5 years, 25.008% live in the city, 34.586% in the suburbs, and 40.406% in the country.
    (d) After 8 years, about 25% live in the city, about 35% in the suburbs, and about 40% in the country.

33. For a fixed j, $1 \le j \le n$, the probability of moving from page j to page i, for $1 \le i \le n$, is a_{ij}. Since it is certain that the surfer will move from page j to *some* page, the sum of all these probabilities must be 1. That is, the sum of the entries of the jth column of A is 1.

34. We must show that a_{ij} is equal to the ij-entry of $pM + \dfrac{1-p}{n} W$ for all i,j. There are two cases to consider.

    Case 1: $s_j \neq 0$
    Here, the (i,j)-entry of $pM + \dfrac{1-p}{n}W$ is

    $$pm_{ij} + \frac{1-p}{n} w_{ij} = p\frac{1}{s_j} c_{ij} + \frac{1-p}{n} = a_{ij}.$$

    Case 2: $s_j = 0$
    Here, the (i,j)-entry of $pM + \dfrac{1-p}{n}W$ is

    $$pm_{ij} + \frac{1-p}{n} w_{ij} = \frac{p}{n} + \frac{1-p}{n} = \frac{1}{n} = a_{ij}.$$

    So the result is established in every case.

35. (a) This Markov chain has two states: wet (W) and dry (D). A transition matrix is

    $$\begin{array}{c} \\ \text{Next day} \end{array} \begin{array}{c} \\ \text{W} \\ \text{D} \end{array} \overset{\substack{\text{Current day} \\ \text{W} \quad\quad \text{D}}}{\begin{bmatrix} \frac{117}{195} & \frac{80}{615} \\ \frac{78}{195} & \frac{535}{615} \end{bmatrix}} = A.$$

    (b) The probability that the following day will be dry if the current day is dry is the $(2,2)$-entry of A, which is $\frac{535}{615} = \frac{107}{123}$.

    (c) A dry day corresponds to the probability vector

    $$\mathbf{p} = \begin{bmatrix} 0 \\ 1 \end{bmatrix}.$$

    If a Tuesday in November was dry, then the probabilities of each type of weather on the following Thursday are given by

    $$A^2 \mathbf{p} = A(A\mathbf{p}) = A \begin{bmatrix} \frac{16}{123} \\ \frac{107}{123} \end{bmatrix} = \begin{bmatrix} \frac{14,464}{75,645} \\ \frac{61,181}{75,645} \end{bmatrix} \approx \begin{bmatrix} .191 \\ .809 \end{bmatrix}.$$

    So the probability that the following Thursday will be dry is about .809.

    (d) As in (c), the probabilities of each type of weather on Saturday if the previous Wednesday was wet are given by

    $$A^3 \begin{bmatrix} 1 \\ 0 \end{bmatrix} = \begin{bmatrix} .324 \\ .676 \end{bmatrix}.$$

    So the probability that Saturday will be dry if the previous Wednesday was wet is about .676.

    (e) Over the long run, the probability of each type of weather is given by the steady-state vector, which can be computed as in Exercise 9. Since the reduced row echelon form of $A - I_2$ is

    $$\begin{bmatrix} 1 & -\frac{40}{123} \\ 0 & 0 \end{bmatrix},$$

the steady state vector is
$$\frac{1}{163}\begin{bmatrix} 40 \\ 123 \end{bmatrix} \approx \begin{bmatrix} .245 \\ .755 \end{bmatrix}.$$

Thus, over the long run, the probability of a wet day is about .245.

36. (a) $\begin{bmatrix} .6 & .1 & .1 \\ .2 & .8 & .2 \\ .2 & .1 & .7 \end{bmatrix}$ (b) .10 (c) .32

   (d) 20% at Midway, 50% at O'Hare, and 30% at the Loop

37. (a) .05 (b) .1 (c) .3

   (d) In general, suppose we have a transition matrix of the form
   $$M = \begin{bmatrix} 1-2a & b & c \\ a & 1-2b & c \\ a & b & 1-2c \end{bmatrix},$$
   where $0 < a, b, c < 1$. For example, in the given matrix A, we have $a = .05$, $b = .1$, and $c = .3$. Suppose furthermore that $\mathbf{p}$ is the steady state vector for M. Then $(M - I_3)\mathbf{p} = \mathbf{0}$, and hence
   $$\begin{bmatrix} -2a & b & c \\ a & -2b & c \\ a & b & -2c \end{bmatrix} \begin{bmatrix} p_1 \\ p_2 \\ p_3 \end{bmatrix}$$
   $$= \begin{bmatrix} -2 & 1 & 1 \\ 1 & -2 & 1 \\ 1 & 1 & -2 \end{bmatrix} \begin{bmatrix} ap_1 \\ bp_2 \\ cp_3 \end{bmatrix}$$
   $$= \begin{bmatrix} 0 \\ 0 \\ 0 \end{bmatrix}.$$

   Since a basis for the null space of the matrix
   $$\begin{bmatrix} -2 & 1 & 1 \\ 1 & -2 & 1 \\ 1 & 1 & -2 \end{bmatrix} \text{ is } \left\{ \begin{bmatrix} 1 \\ 1 \\ 1 \end{bmatrix} \right\},$$
   it follows that $ap_1 = bp_2 = cp_3$. So
   $$p_1 = \frac{cp_3}{a}, \quad p_2 = \frac{cp_3}{b}, \text{ and } p_3 = \frac{cp_3}{c}.$$
   It follows that
   $$\mathbf{p} = k \begin{bmatrix} \frac{1}{a} \\ \frac{1}{b} \\ \frac{1}{c} \end{bmatrix},$$
   where $k = \frac{1}{a} + \frac{1}{b} + \frac{1}{c}$. So, for the given matrix A,
   $$\mathbf{p} = k \begin{bmatrix} \frac{1}{.05} \\ \frac{1}{.1} \\ \frac{1}{.3} \end{bmatrix} = \begin{bmatrix} .6 \\ .3 \\ .1 \end{bmatrix}.$$

   (e) For the vector $\mathbf{p}$ in (d), we have $A\mathbf{p} = \mathbf{p}$.

38. $\begin{bmatrix} 0 & .5 & .2 \\ 1 & 0 & .8 \\ 0 & .5 & 0 \end{bmatrix}$

39. (a) The sum of the entries of each row of A^T is 1. Hence $A^T\mathbf{u} = \mathbf{u}$.
   (b) It follows from (a) that 1 is an eigenvalue of A^T.
   (c) Since 1 is an eigenvalue of A^T, we have $\det(A^T - I_n) = 0$. Hence
   $$\det(A - I_n) = \det(A - I_n)^T$$
   $$= \det(A^T - I_n) = 0.$$
   (d) It follows from (c) that 1 is an eigenvalue of A.

40. From Exercise 37, we see that the entries of the given steady-state vector $\mathbf{p} = \begin{bmatrix} .4 \\ .2 \\ .4 \end{bmatrix}$ are proportional to $\frac{1}{a}$, $\frac{1}{b}$, and $\frac{1}{c}$, respectively, where a, b, and c are as in Exercise 37. Since $p_1 = p_3$, it follows that $a = c$. Furthermore,
   $$2 = \frac{p_1}{p_2} = \frac{\frac{1}{a}}{\frac{1}{b}} = \frac{b}{a},$$
   and hence $b = 2a$. It follows that a regular stochastic matrix with steady-state vector $\mathcal{P}$ is of the form
   $$\begin{bmatrix} 1-2a & 2a & a \\ a & 1-4a & a \\ a & 2a & 1-2a \end{bmatrix}.$$
   For example, the stochastic matrices for $a = .1$ and $a = .15$ are
   $$\begin{bmatrix} .8 & .2 & .1 \\ .1 & .6 & .1 \\ .1 & .2 & .8 \end{bmatrix} \text{ and } \begin{bmatrix} .70 & .3 & .15 \\ .15 & .4 & .15 \\ .15 & .3 & .70 \end{bmatrix},$$
   respectively.

41. Let A be an $n \times n$ stochastic matrix and $\mathbf{p}$ be a probability vector in $\mathcal{R}^n$. Then each component of $A\mathbf{p}$ is nonnegative, and the sum of the components is
   $$(a_{11}p_1 + a_{12}p_2 + \cdots + a_{1n}p_n) + \cdots$$

$$+ (a_{n1}p_1 + a_{n2}p_2 + \cdots + a_{nn}p_n)$$
$$= (a_{11} + \cdots + a_{n1})p_1 + \cdots$$
$$+ (a_{1n} + \cdots + a_{nn})p_n$$
$$= p_1 + \cdots + p_n$$
$$= 1.$$

42. (a) $1, a+b-1$

   (b) $\left\{\begin{bmatrix}1-b\\1-a\end{bmatrix}\right\}, \left\{\begin{bmatrix}-1\\1\end{bmatrix}\right\}$ if $a \neq 1$ or $b \neq 1$

   $\{\mathbf{e}_1, \mathbf{e}_2\}$ if $a = b = 1$

   (c) A is always diagonalizable.

43. (a) The absolute value of the ith component of $A^T\mathbf{v}$ is
    $$|a_{1i}v_1 + a_{2i}v_2 + \cdots + a_{ni}v_n|$$
    $$\leq |a_{1i}||v_1| + |a_{2i}||v_2| + \cdots + |a_{ni}||v_n|$$
    $$\leq (|a_{1i}| + |a_{2i}| + \cdots + |a_{ni}|)|v_k|$$
    $$\leq |v_k|.$$

   (b) Let $\mathbf{v}$ be an eigenvector of A^T corresponding to eigenvalue λ. Then $A^T\mathbf{v} = \lambda\mathbf{v}$. It follows from (a) that the absolute value of the kth component of $A^T\mathbf{v}$ is $|\lambda v_k| \leq |v_k|$. Hence $|\lambda| \cdot |v_k| \leq |v_k|$.

   (c) Since $|v_k| \neq 0$, the preceding inequality implies that $|\lambda| \leq 1$.

44. If A and B are $n \times n$ stochastic matrices, then each column of B is a probability vector, and so each column of
    $$AB = [A\mathbf{b}_1 \ A\mathbf{b}_2 \ \cdots \ A\mathbf{b}_n]$$
    is a probability vector by Exercise 41.

45. $y_1 = -ae^{-3t} + 2be^{4t}$
    $y_2 = 3ae^{-3t} + be^{4t}$

46. $y_1 = 2ae^{2t} + be^{3t}$
    $y_2 = ae^{2t} + be^{3t}$

47. The given system of differential equations can be written in the form $\mathbf{y}' = A\mathbf{y}$, where
    $$A = \begin{bmatrix}2 & 4\\-6 & -8\end{bmatrix}.$$
    Now the characteristic polynomial of A is
    $$t^2 + 6t + 8 = (t+4)(t+2);$$
    so A has the eigenvalues -4 and -2. Bases for the corresponding eigenspaces of A are
    $$\left\{\begin{bmatrix}-2\\3\end{bmatrix}\right\} \text{ and } \left\{\begin{bmatrix}-1\\1\end{bmatrix}\right\}.$$

Hence $A = PDP^{-1}$, where
$$P = \begin{bmatrix}-2 & -1\\3 & 1\end{bmatrix} \text{ and } D = \begin{bmatrix}-4 & 0\\0 & -2\end{bmatrix}.$$

The solution of $\mathbf{z}' = D\mathbf{z}$ is
$$\mathbf{z}_1 = ae^{-4t}$$
$$\mathbf{z}_2 = be^{-2t}.$$

The algorithm on page 341 gives the solution of the original system to be
$$\begin{bmatrix}y_1\\y_2\end{bmatrix} = \mathbf{y} = P\mathbf{z} = \begin{bmatrix}-2 & -1\\3 & 1\end{bmatrix}\begin{bmatrix}ae^{-4t}\\be^{-2t}\end{bmatrix}$$
$$= \begin{bmatrix}-2ae^{-4t} - be^{-2t}\\3ae^{-4t} + be^{-2t}\end{bmatrix}.$$

48. $y_1 = 2ae^{4t} + 3be^{5t}$
    $y_2 = 3ae^{4t} + 5be^{5t}$

49. $y_1 = \phantom{-ae^{-t} +\ } -ce^{2t}$
    $y_2 = -ae^{-t} + be^{2t}$
    $y_3 = ae^{-t} + ce^{2t}$

50. $y_1 = -ae^t - 2be^{2t} - ce^{3t}$
    $y_2 = ae^t + be^{2t} + ce^{3t}$
    $y_3 = 2ae^t + 4be^{2t} + 4ce^{3t}$

51. $y_1 = ae^{-2t} + be^{-t}$
    $y_2 = -ae^{-2t} - ce^{2t}$
    $y_3 = 2ae^{-2t} + 2be^{-t} + ce^{2t}$

52. $y_1 = ae^{2t} + be^{2t} + 2ce^{-3t}$
    $y_2 = ae^{2t} \phantom{+ be^{2t}} + 2ce^{-3t}$
    $y_3 = \phantom{ae^{2t} +\ } be^{2t} + ce^{-3t}$

53. The given system of differential equations can be written in the form $\mathbf{y}' = A\mathbf{y}$, where
    $$A = \begin{bmatrix}1 & 1\\4 & 1\end{bmatrix}.$$
    Since the characteristic polynomial of A is
    $$t^2 - 2t - 3 = (t+1)(t-3),$$
    A has eigenvalues of -1 and 3. Bases for the corresponding eigenspaces of A are
    $$\left\{\begin{bmatrix}-1\\2\end{bmatrix}\right\} \text{ and } \left\{\begin{bmatrix}1\\2\end{bmatrix}\right\}.$$
    Hence $A = PDP^{-1}$, where
    $$P = \begin{bmatrix}-1 & 1\\2 & 2\end{bmatrix} \text{ and } D = \begin{bmatrix}-1 & 0\\0 & 3\end{bmatrix}.$$

The solution of $\mathbf{z}' = D\mathbf{z}$ is

$$z_1 = ae^{-t}$$
$$z_2 = be^{3t}.$$

Thus the general solution of the original system is

$$\begin{bmatrix} y_1 \\ y_2 \end{bmatrix} = \mathbf{y} = P\mathbf{z} = \begin{bmatrix} -1 & 1 \\ 2 & 2 \end{bmatrix} \begin{bmatrix} ae^{-t} \\ be^{3t} \end{bmatrix}$$
$$= \begin{bmatrix} -ae^{-t} + be^{3t} \\ 2ae^{-t} + 2be^{3t} \end{bmatrix}.$$

Taking $t = 0$, we obtain

$$15 = y_1(0) = -a + b$$

and

$$-10 = y_2(0) = 2a + 2b.$$

Solving this system, we obtain $a = -10$ and $b = 5$. Thus the solution of the original system of differential equations and initial conditions is

$$y_1 = 10e^{-t} + 5e^{3t}$$
$$y_2 = -20e^{-t} + 10e^{3t}.$$

54. $y_1 = 4e^{3t} + 3e^{4t}$
$y_2 = 2e^{3t} + 3e^{4t}$

55. $y_1 = -3e^{4t} + 5e^{6t}$
$y_2 = 6e^{4t} - 5e^{6t}$

56. $y_1 = -4e^{-t} + 5e^{3t}$
$y_2 = 2e^{-t} - 5e^{3t}$

57. $y_1 = 4e^{-t} + 5e^{t} - 9e^{2t}$
$y_2 = \phantom{4e^{-t} +}\; 5e^{t} - 3e^{2t}$
$y_3 = 4e^{-t} \phantom{+ 5e^{t}}\; - 3e^{2t}$

58. $y_1 = -3e^{t} + 2e^{3t}$
$y_2 = 3e^{t} - 2e^{3t}$
$y_3 = \phantom{-3e^{t} +}\; 2e^{3t}$

59. $y_1 = 6e^{-t} - 4e^{t} - 6e^{-2t}$
$y_2 = 6e^{-t} - 8e^{t} - 3e^{-2t}$
$y_3 = -6e^{-t} + 12e^{t} - 3e^{-2t}$

60. $y_1 = {-e^{t}} + 5e^{-t}$
$y_2 = -3e^{t} + 8e^{-t}$
$y_3 = e^{t} + 7e^{-t}$

61. Setting $y_1 = y$ and $y_2 = y_1' = y''$, we obtain the system

$$y_1' = y_2$$
$$y_2' = 3y_1 + 2y_2,$$

which in matrix form can be written $\mathbf{y}' = A\mathbf{y}$, where

$$A = \begin{bmatrix} 0 & 1 \\ 3 & 2 \end{bmatrix}.$$

Observe that $A = PDP^{-1}$, for

$$P = \begin{bmatrix} 1 & 1 \\ 3 & -1 \end{bmatrix} \quad \text{and} \quad D = \begin{bmatrix} 3 & 0 \\ 0 & -1 \end{bmatrix},$$

and the general solution of $\mathbf{z}' = D\mathbf{z}$ is $z_1 = ae^{3t}$, $z_2 = be^{-t}$. It follows that

$$\begin{bmatrix} y_1 \\ y_2 \end{bmatrix} = \mathbf{y} = P\mathbf{z} = \begin{bmatrix} 1 & 1 \\ 3 & -1 \end{bmatrix} \begin{bmatrix} z_1 \\ z_2 \end{bmatrix}$$
$$= \begin{bmatrix} ae^{3t} + be^{-t} \\ 3ae^{3t} - be^{-t} \end{bmatrix},$$

and hence $y = y_1 = ae^{3t} + be^{-t}$.

62. Let $y_1 = y$, $y_2 = y'$, and $y_3 = y''$. The given differential equation is equivalent to the system

$$y_1' = y_2$$
$$y_2' = y_3$$
$$y_3' = 8y_2 + 2y_3.$$

Its general solution is $y = a + be^{-2t} + ce^{4t}$.

63. Let $y_1 = y$, $y_2 = y'$, and $y_3 = y''$. The given equation can be written

$$y''' = -2y + y' + 2y''$$

or

$$y_3' = -2y_1 + y_2 + 2y_3.$$

So the given equation is equivalent to the system

$$y_1' = \; y_2$$
$$y_2' = \; y_3$$
$$y_3' = -2y_1 + y_2 + 2y_3.$$

Write this system in the matrix form $\mathbf{y}' = A\mathbf{y}$, where

$$A = \begin{bmatrix} 0 & 1 & 0 \\ 0 & 0 & 1 \\ -2 & 1 & 2 \end{bmatrix}.$$

The characteristic polynomial of A is

$$\det(A - tI_3) = -t^3 + 2t^2 + t - 2$$
$$= -(t+1)(t-1)(t-2),$$

and so the eigenvalues of A are -1, 1, and 2. Bases for the eigenspaces of A corresponding to the eigenvalues -1, 1, and 2 are

$$\left\{ \begin{bmatrix} 1 \\ -1 \\ 1 \end{bmatrix} \right\}, \quad \left\{ \begin{bmatrix} 1 \\ 1 \\ 1 \end{bmatrix} \right\}, \quad \text{and} \quad \left\{ \begin{bmatrix} 1 \\ 2 \\ 4 \end{bmatrix} \right\},$$

respectively. Hence $A = PDP^{-1}$, where

$$P = \begin{bmatrix} 1 & 1 & 1 \\ -1 & 1 & 2 \\ 1 & 1 & 4 \end{bmatrix} \quad \text{and} \quad D = \begin{bmatrix} -1 & 0 & 0 \\ 0 & 1 & 0 \\ 0 & 0 & 2 \end{bmatrix}.$$

The solution of $\mathbf{z}' = D\mathbf{z}$ is

$$\begin{aligned} z_1 &= ae^{-t} \\ z_2 &= be^t \\ z_3 &= ce^{2t} \end{aligned}.$$

Thus the general solution of the original system is

$$\begin{bmatrix} y_1 \\ y_2 \\ y_3 \end{bmatrix} = \mathbf{y} = P\mathbf{z} = \begin{bmatrix} 1 & 1 & 1 \\ -1 & 1 & 2 \\ 1 & 1 & 4 \end{bmatrix} \begin{bmatrix} ae^{-t} \\ be^t \\ ce^{2t} \end{bmatrix}$$

$$= \begin{bmatrix} ae^{-t} + be^t + ce^{2t} \\ -ae^{-t} + be^t + 2ce^{2t} \\ ae^{-t} + be^t + 4ce^{2t} \end{bmatrix}.$$

Taking $t = 0$, we have

$$\begin{aligned} 2 &= y(0) = y_1(0) = a + b + c \\ -3 &= y'(0) = y_2(0) = -a + b + 2c \\ 5 &= y''(0) = y_3(0) = a + b + 4x. \end{aligned}$$

Solving this system yields $a = 3$, $b = -2$, and $c = 1$. Hence the particular solution of the original system of differential equations and initial conditions is

$$\begin{aligned} y_1 &= 3e^{-t} - 2e^t + e^{2t} \\ y_2 &= -3e^{-t} - 2e^t + 2e^{2t} \\ y_3 &= 3e^{-t} - 2e^t + 4e^{2t}. \end{aligned}$$

Therefore the particular solution of the given third-order differential equation and initial conditions is

$$y(t) = y_1(t) = 3e^{-t} - 2e^t + e^{2t}.$$

64. $y = ce^{-2t}\cos 2\sqrt{2}\,t + de^{-2t}\sin 2\sqrt{2}\,t$

65. Take $w = 10$ lbs, $b = 0.625$, and $k = 1.25$ lbs/foot in the equation

$$\frac{w}{g}y''(t) + by'(t) + ky(t) = 0.$$

Then the equation simplifies to the form

$$y'' + 2y' + 4y = 0.$$

We transform this differential equation into a system of differential equations by letting $y_1 = y$ and $y_2 = y'$. These substitutions produce

$$\begin{aligned} y_1' &= y_2 \\ y_2' &= -4y_1 - 2y_2, \end{aligned}$$

which can be written as $\mathbf{y}' = A\mathbf{y}$, where

$$A = \begin{bmatrix} 0 & 1 \\ -4 & -2 \end{bmatrix}.$$

The characteristic polynomial of this matrix is $t^2 + 2t + 4$, which has the roots $-1 + \sqrt{3}\,i$ and $-1 - \sqrt{3}\,i$. The general solution of the original differential equation can be written using Euler's formula as

$$\begin{aligned} y &= ae^{(-1-\sqrt{3}\,i)t} + be^{(-1+\sqrt{3}\,i)t} \\ &= ae^{-t}(\cos\sqrt{3}\,t + i\sin\sqrt{3}\,t) \\ &\quad + be^{-t}(\cos\sqrt{3}\,t - i\sin\sqrt{3}\,t) \end{aligned}$$

or, equivalently, as

$$y = ce^{-t}\cos\sqrt{3}\,t + de^{-t}\sin\sqrt{3}\,t.$$

66. Since $\mathbf{y} = P\mathbf{z}$, we have $y_i = p_{i1}z_1 + p_{i2}z_2 + p_{i3}z_3$ for $i = 1, 2, 3$. Hence $y_i' = p_{i1}z_1' + p_{i2}z_2' + p_{i3}z_3'$ for $i = 1, 2, 3$. So $\mathbf{y}' = P\mathbf{z}'$.

67. (a) $y_1 = 100e^{-2t} + 800e^t$, $y_2 = 100e^{-2t} + 200e^t$
    (b) 2188 and 557 at time 1, 5913 and 1480 at time 2, and 16069 and 4017 at time 3
    (c) .25, no

68. The characteristic polynomial is

$$\det \begin{bmatrix} -t & 1 & 0 \\ 0 & -t & 1 \\ -c & -b & -a-t \end{bmatrix}$$

$$= \det \begin{bmatrix} 0 & 1 & 0 \\ -t^2 & -t & 1 \\ -c-bt & -b & -a-t \end{bmatrix}$$

$$= (-1)\cdot\det \begin{bmatrix} -t^2 & 1 \\ -c-bt & -a-t \end{bmatrix}$$

$$= -\left[-t^2(-a-t) - (-c-bt)\right]$$

$$= -(t^3 + at^2 + bt + c).$$

69. The differential equation $y''' + ay'' + by' + cy = 0$ can be written as $\mathbf{y}' = A\mathbf{y}$, where

$$A = \begin{bmatrix} 0 & 1 & 0 \\ 0 & 0 & 1 \\ -c & -b & -a \end{bmatrix}.$$

By Exercise 68, the characteristic polynomial of A is $-t^3 - at^2 - bt - c$; so $\lambda_i^3 = -a\lambda_i^2 - b\lambda_i - c$ for $i = 1, 2, 3$. Now

$$A\begin{bmatrix} 1 \\ \lambda_i \\ \lambda_i^2 \end{bmatrix} = \begin{bmatrix} \lambda_i \\ \lambda_i^2 \\ -c - b\lambda_i - a\lambda_i^2 \end{bmatrix}$$

128 Chapter 5 Eigenvalues, Eigenvectors, and Diagonalization

$$= \begin{bmatrix} \lambda_i \\ \lambda_i^2 \\ \lambda_i^3 \end{bmatrix} = \lambda_i \begin{bmatrix} 1 \\ \lambda_i \\ \lambda_i^2 \end{bmatrix}.$$

Thus

$$\mathbf{v}_i = \begin{bmatrix} 1 \\ \lambda_i \\ \lambda_i^2 \end{bmatrix}$$

is an eigenvector of A with λ_i as its corresponding eigenvalue. So $\{\mathbf{v}_1, \mathbf{v}_2, \mathbf{v}_3\}$ is a basis for $\mathcal{R}^3$ consisting of eigenvectors of A. Thus the solution of the given equation is given by the boxed result of $\mathbf{y}' = A\mathbf{y}$ on page 341 with

$$P = [\mathbf{v}_1 \ \mathbf{v}_2 \ \mathbf{v}_3] \quad \text{and} \quad D = \begin{bmatrix} \lambda_1 & 0 & 0 \\ 0 & \lambda_2 & 0 \\ 0 & 0 & \lambda_3 \end{bmatrix}.$$

70. $r_n = 5 \cdot 2^n$ for $n \geq 0$, $r_6 = 320$
71. $r_n = 8(-3)^n$, $r_6 = 5832$
72. $3 \cdot 2^n + 4(-1)^n$, $r_6 = 196$
73. The given difference equation can be written as $\mathbf{s}_n = A\mathbf{s}_{n-1}$, where

$$\mathbf{s}_n = \begin{bmatrix} r_n \\ r_{n+1} \end{bmatrix} \quad \text{and} \quad A = \begin{bmatrix} 0 & 1 \\ 4 & 3 \end{bmatrix}.$$

Taking

$$P = \begin{bmatrix} 1 & 1 \\ -1 & 4 \end{bmatrix} \quad \text{and} \quad D = \begin{bmatrix} -1 & 0 \\ 0 & 4 \end{bmatrix},$$

we have $A = PDP^{-1}$. Hence $A^n = PD^nP^{-1}$, and so

$$\mathbf{s}_n = A^n \mathbf{s}_0 = PD^n P^{-1} \mathbf{s}_0$$

$$= \begin{bmatrix} 1 & 1 \\ -1 & 4 \end{bmatrix} \begin{bmatrix} (-1)^n & 0 \\ 0 & 4^n \end{bmatrix} \begin{bmatrix} .8 & -.2 \\ .2 & .2 \end{bmatrix} \begin{bmatrix} 1 \\ 1 \end{bmatrix}$$

$$= \begin{bmatrix} (-1)^n & 4^n \\ (-1)^{n+1} & 4^{n+1}n \end{bmatrix} \begin{bmatrix} .6 \\ .4 \end{bmatrix}$$

$$= \begin{bmatrix} .6(-1)^n + .4(4^n) \\ .6(-1)^{n+1} + .4(4^{n+1}) \end{bmatrix}.$$

Equating the first components of $\mathbf{s}_n$ and the preceding vector, we have

$$r_n = .6(-1)^n + .4(4^n) \text{ for } n \geq 0.$$

Hence

$$r_6 = .6(-1)^6 + .4(4^6) = 1639.$$

74. $r_n = -1 + 2^{n+1}$, $r_6 = 127$
75. $r_n = 3(-3)^n + 5(2^n)$, $r_6 = 2507$
76. $r_n = 9(-1)^n - 6(-4)^n$, $r_6 = -24567$
77. $r_n = 6(-1)^n 3(2^n)$, $r_6 = 198$
78. $r_n = 2 + (-1)^n$ for $n \geq 0$, $r_6 = 3$
79. (a) $r_0 = 1, r_1 = 2, r_2 = 7, r_3 = 20$
    (b) $r_n = 2r_{n-1} + 3r_{n-2}$
    (c) $r_n = \left(\frac{3}{4}\right)3^n + \left(\frac{1}{4}\right)(-1)^n$
80. $\$1000(1.08)^n$ after n years, $\$1469.33$ after 5 years, $\$2158.92$ after 10 years, $\$3172.17$ after 15 years.
81. Since the given difference equation is of the third order, $\mathbf{s}_n$ is a vector in $\mathcal{R}^3$ and A is a 3×3 matrix. Taking

$$\mathbf{s}_n = \begin{bmatrix} r_n \\ r_{n+1} \\ r_{n+2} \end{bmatrix} \quad \text{and} \quad A = \begin{bmatrix} 0 & 1 & 0 \\ 0 & 0 & 1 \\ 5 & -2 & 4 \end{bmatrix},$$

we see that the matrix form of the given equation is $\mathbf{s}_n = A\mathbf{s}_{n-1}$.

82. The last component of $\mathbf{s}_n = A\mathbf{s}_{n-1}$ yields

$$r_{n+k-1} = a_k r_{n-1} + a_{k-1} r_n + \cdots + a_1 r_{n+k-2}$$
$$= a_1 r_{n+k-2} + a_2 r_{n+k-3} + \cdots + a_k r_{n-1}.$$

Replacing n by $n - k + 1$ for each n, we obtain (4).

83. (a) As a consequence of Theorem 5.3, $\{\mathbf{v}_1, \mathbf{v}_2, \ldots, \mathbf{v}_k\}$ is a linearly independent subset of $\mathcal{R}^n$, and hence it is a basis by Theorem 4.5.
    (b) Apply A^n to both sides of the equation in (a) to obtain

$$\mathbf{s}_n = \lambda_1^n t_1 \mathbf{v}_1 + \lambda_2^n t_2 \mathbf{v}_2 + \cdots + \lambda_k^n t_k \mathbf{v}_k.$$

84. If λ is a solution of equation (6), then

$$A\mathbf{w}_\lambda = \begin{bmatrix} \lambda \\ \lambda^2 \\ \vdots \\ a_k + a_{k-1}\lambda + \cdots + a_1 \lambda^{k-1} \end{bmatrix} = \begin{bmatrix} \lambda \\ \lambda^2 \\ \vdots \\ \lambda^k \end{bmatrix}$$

$$= \lambda \mathbf{w}_\lambda.$$

Conversely, if $\mathbf{w}_\lambda$ is an eigenvector of A, then the first component of $A\mathbf{w}_\lambda$ shows that λ is the eigenvalue that corresponds to $\mathbf{w}_\lambda$. The last components of the vector equation $A\mathbf{w}_\lambda = \lambda \mathbf{w}_\lambda$ show that λ is a solution of equation (6). Hence λ is a solution of equation (6) if and only if $\mathbf{w}_\lambda$ is an eigenvector of A.

5.5 Applications of Eigenvalues

85. We have
$$A\mathbf{s}_0 = \begin{bmatrix} 1 & 0 \\ c & a \end{bmatrix} \begin{bmatrix} 1 \\ r_0 \end{bmatrix} = \begin{bmatrix} 1 \\ c + ar_0 \end{bmatrix} = \begin{bmatrix} 1 \\ r_1 \end{bmatrix} = \mathbf{s}_1,$$

$$A\mathbf{s}_1 = \begin{bmatrix} 1 & 0 \\ c & a \end{bmatrix} \begin{bmatrix} 1 \\ r_1 \end{bmatrix} = \begin{bmatrix} 1 \\ c + ar_1 \end{bmatrix} = \begin{bmatrix} 1 \\ r_2 \end{bmatrix} = \mathbf{s}_2,$$

and, in general,
$$A\mathbf{s}_{n-1} = \begin{bmatrix} 1 & 0 \\ c & a \end{bmatrix} \begin{bmatrix} 1 \\ r_{n-1} \end{bmatrix} = \begin{bmatrix} 1 \\ c + ar_{n-1} \end{bmatrix} = \begin{bmatrix} 1 \\ r_n \end{bmatrix}$$
$$= \mathbf{s}_n.$$

Hence
$$\mathbf{s}_n = A\mathbf{s}_{n-1} = A(A\mathbf{s}_{n-2}) = A^2 \mathbf{s}_{n-2}$$
$$= A^2(A\mathbf{s}_{n-3}) = A^3 \mathbf{s}_{n-3} = \cdots = A^n \mathbf{s}_0.$$

(For those familiar with mathematical induction, this proof can be given using induction.)

86. (a) First, observe that for any real numbers c_1 and c_2, we have
$$\begin{bmatrix} 1 & 0 \\ c_1 & 1 \end{bmatrix} \begin{bmatrix} 1 & 0 \\ c_2 & 1 \end{bmatrix} = \begin{bmatrix} 1 & 0 \\ c_1 + c_2 & 1 \end{bmatrix},$$

and hence, in general,
$$A^n = \begin{bmatrix} 1 & 0 \\ c & 1 \end{bmatrix} \begin{bmatrix} 1 & 0 \\ c & 1 \end{bmatrix} \cdots \begin{bmatrix} 1 & 0 \\ c & 1 \end{bmatrix}$$
$$= \begin{bmatrix} 1 & 0 \\ c + c + \cdots + c & 1 \end{bmatrix} = \begin{bmatrix} 1 & 0 \\ nc & 1 \end{bmatrix}.$$

(b) By Exercise 85 and (a), we have
$$\begin{bmatrix} 1 \\ r_n \end{bmatrix} = \mathbf{s}_n = A^n \mathbf{s}_0$$
$$= \begin{bmatrix} 1 & 0 \\ nc & 1 \end{bmatrix} \begin{bmatrix} 1 \\ r_0 \end{bmatrix} = \begin{bmatrix} 1 \\ nc + r_0 \end{bmatrix}.$$

Hence $r_n = r_0 + nc$.

87. (a) The characteristic polynomial of A is
$$\det(A - tI_2) = (1 - t)(a - t).$$

(b) The vectors
$$\mathbf{v}_1 = \begin{bmatrix} 1 - a \\ c \end{bmatrix} \quad \text{and} \quad \mathbf{v}_2 = \begin{bmatrix} 0 \\ 1 \end{bmatrix}$$

are eigenvectors of A corresponding to eigenvalues 1 and a, respectively. Thus $\{\mathbf{v}_1, \mathbf{v}_2\}$ is a basis for $\mathcal{R}^2$ consisting of eigenvectors of A. So $\mathbf{s}_0 = b_1 \mathbf{v}_1 + b_2 \mathbf{v}_2$ for some scalars b_1 and b_2.

(c) Solving the equation $\mathbf{s}_0 = b_1 \mathbf{v}_1 + b_2 \mathbf{v}_2$ for b_1 and b_2 gives
$$b_1 = \frac{-1}{a-1} \quad \text{and} \quad b_2 = r_0 + \frac{c}{a-1}.$$

For these values, we have
$$\begin{bmatrix} 1 \\ r_n \end{bmatrix} = \mathbf{s}_n = A^n \mathbf{s}_0 = b_1(A^n \mathbf{v}_1) + b_2(A^n \mathbf{v}_2)$$
$$= b_1 \mathbf{v}_1 + b_2 a^n \mathbf{v}_2$$
$$= \left(\frac{-1}{a-1}\right) \begin{bmatrix} 1 - a \\ c \end{bmatrix}$$
$$+ \left(r_0 + \frac{c}{a-1}\right) a^n \begin{bmatrix} 0 \\ 1 \end{bmatrix}$$
$$= \begin{bmatrix} 1 \\ \left(r_0 + \dfrac{c}{a-1}\right) a^n - \dfrac{c}{a-1} \end{bmatrix}$$
$$= \begin{bmatrix} 1 \\ b_2 a^n + cb_1 \end{bmatrix}.$$

88. $\$38{,}333.33(1.06)^n - \$33{,}333.33$

89. The given system of differential equations has the form $\mathbf{y}' = A\mathbf{y}$, where
$$\mathbf{y} = \begin{bmatrix} y_1 \\ y_2 \\ y_3 \\ y_4 \end{bmatrix}$$

and
$$A = \begin{bmatrix} 3.2 & 4.1 & 7.7 & 3.7 \\ -0.3 & 1.2 & 0.2 & 0.5 \\ -1.8 & -1.8 & -4.4 & -1.8 \\ 1.7 & -0.7 & 2.9 & 0.4 \end{bmatrix}.$$

The characteristic polynomial of A is
$$t^4 - 0.4t^3 - 0.79t^2 + .166t + 0.24$$
$$= (t + 0.8)(t + 0.1)(t - 0.3)(t - 1).$$

Since A has four distinct eigenvalues, it is diagonalizable; in fact, $A = PDP^{-1}$, where
$$P = \begin{bmatrix} 1 & -1 & -1 & 2 \\ 0 & -1 & -2 & -1 \\ -1 & 0 & 0 & -1 \\ 1 & 2 & 3 & 2 \end{bmatrix}$$

and
$$D = \begin{bmatrix} -0.8 & 0.0 & 0.0 & 0 \\ 0.0 & -0.1 & 0.0 & 0 \\ 0.0 & 0.0 & 0.3 & 0 \\ 0.0 & 0.0 & 0.0 & 1 \end{bmatrix}.$$

The solution of $\mathbf{z}' = D\mathbf{z}$ is $\begin{bmatrix} z_1 \\ z_2 \\ z_3 \\ z_4 \end{bmatrix} = \begin{bmatrix} ae^{-0.8t} \\ be^{-0.1t} \\ ce^{0.3t} \\ de^t \end{bmatrix}.$

Hence the general solution of the original equation is

$\begin{bmatrix} y_1 \\ y_2 \\ y_3 \\ y_4 \end{bmatrix} =$

$\begin{bmatrix} ae^{-0.8t} - be^{-0.1t} - ce^{0.3t} + 2de^t \\ -be^{-0.1t} - 2ce^{0.3t} - de^t \\ -ae^{-0.8t} \qquad\qquad\qquad - de^t \\ ae^{-0.8t} + 2be^{-0.1t} + 3ce^{0.3t} + 2de^t \end{bmatrix}.$

When $t = 0$, the preceding equation takes the form

$\begin{bmatrix} 1 \\ -4 \\ 2 \\ 3 \end{bmatrix} = P \begin{bmatrix} a \\ b \\ c \\ d \end{bmatrix},$

and so $a = -6$, $b = 2$, $c = -1$, and $d = 4$. Thus the particular solution of the original system that satisfies the given initial condition is

$y_1 = -6e^{-0.8t} - 2e^{-0.1t} + e^{0.3t} + 8e^t$
$y_2 = \qquad\qquad -2e^{-0.1t} + 2e^{0.3t} - 4e^t$
$y_3 = \quad 6e^{-0.8t} \qquad\qquad\qquad\qquad - 4e^t$
$y_4 = -6e^{-0.8t} + 4e^{-0.1t} - 3e^{0.3t} + 8e^t.$

90. (a) about 5.75%, 11.85%, 12.35%, 7.25%, 6.77%, 19.40%, 12.82%, 11.22%, 0.45%, and 12.14%

(b) about 5.02%, 8.39%, 11.86%, 8.16%, 15.07%, 18.99%, 11.42%, 11.54%, 0.42%, 9.14%

91. (a) First obtain the 10×10 matrix C, where $c_{ij} = 1$ if there is a link from page j to page i, and $c_{ij} = 0$, otherwise. Then use C to obtain the matrix M, as defined in the subsection on Google searches. Finally, obtain $A = (.85)M + (.015)W$, where W is the 10×10 matrix whose entries are all equal to 1. The result is

$\begin{bmatrix} b & b & d & .1 & b & b & c & b & b & .1 \\ c & b & b & .1 & a & b & b & b & b & .1 \\ b & b & b & .1 & b & b & b & b & b & .1 \\ b & a & d & .1 & b & b & b & a & b & .1 \\ b & b & d & .1 & b & a & c & b & a & .1 \\ c & b & b & .1 & b & b & c & b & b & .1 \\ b & a & b & .1 & b & b & b & b & b & .1 \\ c & b & d & .1 & b & b & b & a & b & .1 \\ c & b & b & .1 & b & a & c & b & b & .1 \\ b & b & d & .1 & a & b & b & b & a & .1 \end{bmatrix}$

where $a = .4400$, $b = .0150$, $c = .2275$, and $d = .1850$.

(b) To obtain the steady state vector for this Markov chain, choose a vector corresponding to the eigenvalue 1. Then multiply this vector by the reciprocal of the sum the entries to obtain the the probability vector

$\begin{bmatrix} 0.0643 \\ 0.1114 \\ 0.0392 \\ 0.1372 \\ 0.1377 \\ 0.0712 \\ 0.0865 \\ 0.1035 \\ 0.1015 \\ 0.1475 \end{bmatrix},$

which results in the ranking: 10, 5, 4, 2, 8, 9, 6, 1, 3.

CHAPTER 5 REVIEW EXERCISES

1. True
2. False, there are infinitely many eigenvectors that correspond to a particular eigenvalue.
3. True   4. True   5. True   6. True
7. False, the linear operator on $\mathcal{R}^2$ that rotates a vector by 90° has no real eigenvalues.
8. False, the rotation matrix $A_{90°}$ has no (real) eigenvalues.
9. False, I_n has only one eigenvalue, namely 1.
10. False, if two $n \times n$ matrices have the same characteristic polynomial, they have the same *eigenvalues*.
11. True   12. True
13. False, if $A = PDP^{-1}$, where P is an invertible matrix and D is a diagonal matrix, then the columns of P are a basis for $\mathcal{R}^n$ consisting of eigenvectors of A.
14. True   15. True   16. True   17. True
18. The characteristic polynomial is $t^2 + t + 4$, which has no (real) roots.
19. $1, \left\{ \begin{bmatrix} -3 \\ 2 \end{bmatrix} \right\}$ and $2, \left\{ \begin{bmatrix} -2 \\ 1 \end{bmatrix} \right\}$
20. $-2, \left\{ \begin{bmatrix} 3 \\ 1 \end{bmatrix} \right\}$

21. The characteristic polynomial of the given matrix A is
$$-t^3 - 5t^2 - 8t - 4 = -(t+1)(t+2)^2.$$
So the eigenvalues of A are -1 and -2 (with multiplicity 2). A basis for the eigenspace of A corresponding to eigenvalue -1 is obtained from the vector form of the general solution of $(A + I_3)\mathbf{x} = \mathbf{0}$. Such a basis is
$$\left\{ \begin{bmatrix} 0 \\ 1 \\ 0 \end{bmatrix} \right\}.$$
In a similar manner, a basis for the eigenspace of A corresponding to eigenvalue -2 is obtained from the vector form of the general solution of $(A + 2I_3)\mathbf{x} = \mathbf{0}$. Such a basis is
$$\left\{ \begin{bmatrix} -1 \\ 1 \\ 0 \end{bmatrix} \right\}.$$

22. -1, $\left\{ \begin{bmatrix} -1 \\ 1 \\ 0 \end{bmatrix}, \begin{bmatrix} -1 \\ 0 \\ 1 \end{bmatrix} \right\}$

23. $P = \begin{bmatrix} 2 & 1 \\ 1 & 3 \end{bmatrix}$, $D = \begin{bmatrix} 2 & 0 \\ 0 & 7 \end{bmatrix}$

24. No such matrices exist because eigenvalue -2 has multiplicity 2, but its eigenspace has dimension 1.

25. The characteristic polynomial of the given matrix A is $-t^3 - t^2 + t + 1 = -(t-1)(t+1)^2$, and so the eigenvalues of A are 1 and -1 (with multiplicity 2). Since the reduced row echelon form of $A + I_3$ is
$$\begin{bmatrix} 1 & 0 & 0 \\ 0 & 1 & 1 \\ 0 & 0 & 0 \end{bmatrix},$$
the eigenspace of A corresponding to eigenvalue -1 has dimension $3 - \text{rank}(A + 3I_3) = 3 - 2 = 1$. Since this eigenvalue has multiplicity 2, A is not diagonalizable.

26. $P = \begin{bmatrix} -1 & 0 & 0 \\ -1 & 0 & -1 \\ 1 & 1 & 1 \end{bmatrix}$, $D = \begin{bmatrix} -2 & 0 & 0 \\ 0 & -1 & 0 \\ 0 & 0 & 2 \end{bmatrix}$

27. $\left\{ \begin{bmatrix} -2 \\ 1 \end{bmatrix}, \begin{bmatrix} -1 \\ 4 \end{bmatrix} \right\}$

28. The characteristic polynomial of T is $t^2 + 7$, which has no (real) roots.

29. The standard matrix of T is
$$A = \begin{bmatrix} 2 & 0 & 0 \\ 0 & 2 & 0 \\ -3 & 3 & -1 \end{bmatrix}.$$
The characteristic polynomial of A is
$$-t^3 + 3t^2 - 4 = -(t+1)(t-2)^2,$$
and so A has eigenvalues of -1 and 2 (with multiplicity 2). Bases for the corresponding eigenspaces of A (and T) are
$$\left\{ \begin{bmatrix} 0 \\ 0 \\ 1 \end{bmatrix} \right\} \text{ and } \left\{ \begin{bmatrix} 1 \\ 1 \\ 0 \end{bmatrix}, \begin{bmatrix} -1 \\ 0 \\ 1 \end{bmatrix} \right\}.$$
Combining these two sets produces a basis for $\mathcal{R}^3$ consisting of eigenvectors of T.

30. $\left\{ \begin{bmatrix} 0 \\ 1 \\ 0 \end{bmatrix}, \begin{bmatrix} 1 \\ 0 \\ 1 \end{bmatrix}, \begin{bmatrix} 0 \\ 1 \\ 1 \end{bmatrix} \right\}$ 31. none 32. -1

33. The eigenvalues of the given matrix are c, 2, and -2. Since an $n \times n$ matrix having n distinct eigenvalues is diagonalizable, A is diagonalizable if $c \neq 2$ and $c \neq -2$. For $c = 2$, the reduced row echelon form of $A - 2I_3$ is
$$\begin{bmatrix} 0 & 1 & 0 \\ 0 & 0 & 1 \\ 0 & 0 & 0 \end{bmatrix}.$$
So the eigenspace corresponding to eigenvalue 2 has dimension 1. Hence A is not diagonalizable if $c = 2$. Likewise, for $c = -2$, the reduced row echelon form of $A + 2I_3$ is
$$\begin{bmatrix} 0 & 1 & 0 \\ 0 & 0 & 1 \\ 0 & 0 & 0 \end{bmatrix},$$
and so A is not diagonalizable for $c = -2$.

34. no real numbers

35. The characteristic polynomial of A is
$$\det(A - tI_2) = t^2 - t - 2 = (t+1)(t-2).$$
So the eigenvalues of A are -1 and 2. Bases for the eigenspaces of A corresponding to the eigenvalues -1 and 2 are
$$\left\{ \begin{bmatrix} 1 \\ 1 \end{bmatrix} \right\} \text{ and } \left\{ \begin{bmatrix} 2 \\ 1 \end{bmatrix} \right\},$$
respectively. Hence $A = PDP^{-1}$, where
$$P = \begin{bmatrix} 1 & 2 \\ 1 & 1 \end{bmatrix} \text{ and } D = \begin{bmatrix} -1 & 0 \\ 0 & 2 \end{bmatrix}.$$

So, for any positive integer k,

$$A^k = PD^kP^{-1}$$
$$= \begin{bmatrix} 1 & 2 \\ 1 & 1 \end{bmatrix} \begin{bmatrix} (-1)^k & 0 \\ 0 & 2^k \end{bmatrix} \begin{bmatrix} -1 & 2 \\ 1 & -1 \end{bmatrix}$$
$$= \begin{bmatrix} 1 & 2 \\ 1 & 1 \end{bmatrix} \begin{bmatrix} -(-1)^k & 2(-1)^k \\ 2^k & -2^k \end{bmatrix}$$
$$= \begin{bmatrix} (-1)^{k+1} + 2^{k+1} & 2(-1)^k - 2^{k+1} \\ (-1)^{k+1} + 2^k & 2(-1)^k - 2^k \end{bmatrix}.$$

36. $\begin{bmatrix} 3^{k+1} + 2(-1)^{k+1} & 2 \cdot 3^k + 2(-1)^{k+1} \\ 3(-1)^k - 3^{k+1} & 3(-1)^k - 2(3^k) \end{bmatrix}$

37. A basis $\mathcal{B}$ such that $[T]_\mathcal{B}$ is a diagonal matrix is a basis for $\mathcal{R}^3$ consisting of eigenvectors of T; so we proceed as in Exercise 13. The standard matrix of T is

$$A = \begin{bmatrix} -4 & -3 & -3 \\ 0 & -1 & 0 \\ 6 & 6 & 5 \end{bmatrix},$$

and its characteristic polynomial is

$$-t^3 + 3t + 2 = -(t-2)(t+1)^2.$$

Thus the eigenvalues of T are 2 and -1 (with multiplicity 2). Solving $(A - 2I_3)\mathbf{x} = \mathbf{0}$, we obtain the following basis for the eigenspace of T corresponding to eigenvalue 2:

$$\left\{ \begin{bmatrix} -1 \\ 0 \\ 2 \end{bmatrix} \right\}.$$

Solving $(A + I_3)\mathbf{x} = \mathbf{0}$, we obtain the following basis for the eigenspace of T corresponding to eigenvalue -1:

$$\left\{ \begin{bmatrix} -1 \\ 1 \\ 0 \end{bmatrix}, \begin{bmatrix} -1 \\ 0 \\ 1 \end{bmatrix} \right\}.$$

Combining these two sets produces a basis for $\mathcal{R}^3$ consisting of eigenvectors of T, and hence $[T]_\mathcal{B}$ is a diagonal matrix. In fact,

$$[T]_\mathcal{B} = \begin{bmatrix} 2 & 0 & 0 \\ 0 & -1 & 0 \\ 0 & 0 & -1 \end{bmatrix}.$$

38. $\begin{bmatrix} -1 & -2 & -1 \\ 1 & 1 & 1 \\ 1 & 2 & 2 \end{bmatrix} \begin{bmatrix} -1 & 0 & 0 \\ 0 & 2 & 0 \\ 0 & 0 & 3 \end{bmatrix} \begin{bmatrix} -1 & -2 & -1 \\ 1 & 1 & 1 \\ 1 & 2 & 2 \end{bmatrix}^{-1}$
$= \begin{bmatrix} 1 & 6 & -4 \\ 1 & -4 & 4 \\ 2 & -6 & 7 \end{bmatrix}$

39. If $a = b$, then the eigenvalue a has multiplicity 3, but its eigenspace has dimension 2. If $a \neq b$, then the eigenvalue a has multiplicity 2, but its eigenspace has dimension 1. In either case, A is not diagonalizable.

40. It follows from the test for a diagonalizable matrix that A is diagonalizable if and only if the eigenspace corresponding to λ_2 has dimension $n-1$.

41. If $I_n - A$ is invertible, then $\det(I_n - A) \neq 0$. Hence

$$\det(A - I_n) = (-1)^n \det(I_n - A) \neq 0,$$

and so 1 is not a root of $\det(A - tI_n)$. Conversely, if $I_n - A$ is not invertible, then

$$\det(A - I_n) = (-1)^n \det(I_n - A) = 0.$$

So 1 is a root of $\det(A - tI_n)$.

42. Suppose that A and B are simultaneously diagonalizable. Then there is an invertible matrix P such that $P^{-1}AP = C$ and $P^{-1}BP = D$ are diagonal matrices. Hence

$$AB = (PCP^{-1})(PDP^{-1}) = P(CD)P^{-1}$$
$$= P(DC)P^{-1} = (PDP^{-1})(PCP^{-1}) = BA.$$

43. If B is the matrix whose columns are the vectors in $\mathcal{B}$, then $[T]_\mathcal{B} = B^{-1}AB$. So the characteristic polynomial of $[T]_\mathcal{B}$ is

$$\det(B^{-1}AB - tI_n)$$
$$= \det(B^{-1}(A - tI_n)B)$$
$$= (\det B)^{-1}(\det(A - tI_n))(\det B)$$
$$= \det(A - tI_n).$$

44. Let V be the eigenspace of T corresponding to eigenvalue λ. Then $T(\mathbf{v}) = \lambda \mathbf{v}$ for every $\mathbf{v}$ in V. But $\lambda \mathbf{v}$ is in V because V is closed under scalar multiplication. Hence V is T-invariant.

CHAPTER 5 MATLAB EXERCISES

1. (a) $P = \begin{bmatrix} 1.0 & 0.8 & 0.75 & 1 & 1.0 \\ -0.5 & -0.4 & -0.50 & 1 & -1.0 \\ 0.0 & -0.2 & -0.25 & 0 & -0.5 \\ 0.5 & 0.4 & 0.50 & 0 & 0.0 \\ 1.0 & 1.0 & 1.00 & 1 & 1.0 \end{bmatrix},$

$D = \begin{bmatrix} 3 & 0 & 0 & 0 & 0 \\ 0 & 1 & 0 & 0 & 0 \\ 0 & 0 & 0 & 0 & 0 \\ 0 & 0 & 0 & -1 & 0 \\ 0 & 0 & 0 & 0 & 2 \end{bmatrix}$

(b) Eigenvalue $\frac{1}{2}$ has multiplicity 2, but rank $(A - \frac{1}{2}I_4) = 3$.

(c) $P = \begin{bmatrix} -1.25 & -1.00 & -0.50 & -1.00 \\ -0.25 & -0.50 & 0.50 & 0.00 \\ 0.75 & 0.50 & 1.00 & 0.00 \\ 1.00 & 1.00 & 0.00 & 1.00 \end{bmatrix}$,

$D = \begin{bmatrix} -1 & 0 & 0 & 0 \\ 0 & 2 & 0 & 0 \\ 0 & 0 & 1 & 0 \\ 0 & 0 & 0 & 1 \end{bmatrix}$

(d) Eigenvalue 0 has multiplicity 2, but rank $(A - 0I_5) = 4$.

2. Let J be the matrix obtained from I_n by interchanging columns i and j. Then $AJ = B$ and $JA = C$. Hence $A = BJ^{-1} = J^{-1}C$. Thus $C = JBJ^{-1}$, and therefore B and C are similar matrices. It follows that B and C have the same eigenvalues. Suppose that $\mathbf{v}$ is an eigenvector of B with corresponding eigenvalue λ. Then

$$C(J\mathbf{v}) = JBJ^{-1}J\mathbf{v} = JB\mathbf{v} = J\lambda\mathbf{v} = \lambda(J\mathbf{v}).$$

Hence $J\mathbf{v}$ is an eigenvector of C with corresponding eigenvalue λ. Similarly, if $\mathbf{w}$ is an eigenvector of C with corresponding eigenvalue λ, then $J^{-1}\mathbf{w}$ is an eigenvector of B with corresponding eigenvalue λ.

3. $\begin{bmatrix} -9 & 20 & -5 & -8 & 2 \\ -5 & 11 & -2 & -4 & 1 \\ 0 & 0 & 2 & 0 & 0 \\ -10 & 16 & -4 & -5 & 2 \\ -27 & 60 & -15 & -24 & 6 \end{bmatrix}$

4. (a) The transition matrix is

$\begin{bmatrix} b & c & b & b & d & b & d & .1 & b & a \\ d & b & b & b & b & b & b & .1 & a & b \\ b & b & b & b & b & b & d & .1 & b & a \\ b & b & b & b & d & c & b & .1 & b & b \\ d & b & b & a & b & b & d & .1 & a & b \\ b & b & b & a & b & b & d & .1 & b & b \\ d & b & a & b & d & c & b & .1 & b & b \\ b & c & b & b & b & c & b & .1 & b & b \\ b & c & b & b & d & b & b & .1 & b & b \\ d & b & a & b & b & b & .1 & b & b \end{bmatrix}$

where $a = .4400$, $b = .0150$, $c = .2983$, and $d = .2275$.

(b) The steady-state vector is $\begin{bmatrix} 0.1442 \\ 0.0835 \\ 0.0895 \\ 0.0756 \\ 0.1463 \\ 0.0835 \\ 0.1442 \\ 0.0681 \\ 0.0756 \\ 0.0895 \end{bmatrix}$,

which results in the rankings: 5, 1, 7, 3, 10, 2, 6, 4, 9, 8

5. (a) A basis does not exist because the sum of the multiplicities of the eigenvalues of the standard matrix of T is not 4.

(b) $\left\{ \begin{bmatrix} -1 \\ -1 \\ 0 \\ 1 \\ 0 \end{bmatrix}, \begin{bmatrix} 0 \\ -1 \\ -1 \\ 0 \\ 1 \end{bmatrix}, \begin{bmatrix} 11 \\ 10 \\ -3 \\ -13 \\ 3 \end{bmatrix}, \begin{bmatrix} 15 \\ 8 \\ -4 \\ -15 \\ 1 \end{bmatrix}, \begin{bmatrix} 5 \\ 10 \\ 0 \\ -7 \\ 1 \end{bmatrix} \right\}$

6. (a) Let

$V = [\mathbf{v}_1 \; \mathbf{v}_2 \; \mathbf{v}_3 \; \mathbf{v}_4]$

$= \begin{bmatrix} 1 & -1 & 2 & 3 \\ 1 & 0 & 1 & 2 \\ 3 & 0 & -2 & 1 \\ 2 & -3 & 1 & 1 \end{bmatrix}.$

By computing the reduced row echelon form of V, we see that V has rank 4. Thus the columns of V form a basis for $\mathcal{R}^4$.

(b) Let A be the standard matrix of T and

$W = [2\mathbf{v}_1 \; 3\mathbf{v}_2 \; -\mathbf{v}_3 \; \mathbf{v}_3 - \mathbf{v}_4]$

$= \begin{bmatrix} 2 & -3 & -2 & -1 \\ 2 & 0 & -1 & -1 \\ 6 & 0 & 2 & -3 \\ 4 & -9 & -1 & 0 \end{bmatrix}.$

Then $AV = [A\mathbf{v}_1 \; A\mathbf{v}_2 \; A\mathbf{v}_3 \; A\mathbf{v}_4] = W$, and hence

$A = WV^{-1}$

$= \begin{bmatrix} 1.5 & -3.5 & 1.0 & 0.5 \\ -3.0 & 3.6 & -0.2 & 1.0 \\ -16.5 & 22.3 & -3.6 & 5.5 \\ 4.5 & -8.3 & 1.6 & 1.5 \end{bmatrix}.$

Therefore $T\left(\begin{bmatrix} x_1 \\ x_2 \\ x_3 \\ x_4 \end{bmatrix}\right) =$

$\begin{bmatrix} 1.5x_1 - 3.5x_2 + 1.0x_3 + 0.5x_4 \\ -3.0x_1 + 3.6x_2 - 0.2x_3 + 1.0x_4 \\ -16.5x_1 + 22.3x_2 - 3.6x_3 + 5.5x_4 \\ 4.5x_1 - 8.3x_2 + 1.6x_3 + 1.5x_4 \end{bmatrix}.$

(c) The linear operator T is not diagonalizable because the eigenspace of T corresponding to the eigenvalue $\lambda = 1$ has dimension 2 but the multiplicity of λ is 1.

7. (a) $T\left(\begin{bmatrix} x_1 \\ x_2 \\ x_3 \\ x_4 \end{bmatrix}\right) = \begin{bmatrix} 11.5x_1 - 13.7x_2 + 3.4x_3 - 4.5x_4 \\ 5.5x_1 - 5.9x_2 + 1.8x_3 - 2.5x_4 \\ -6.0x_1 + 10.8x_2 - 1.6x_3 \\ 5.0x_1 - 5.6x_2 + 1.2x_3 - 3.0x_4 \end{bmatrix}$

(b) The vectors listed below, obtained using the MATLAB `eig` function, form a basis of eigenvectors of T. (Answers are correct to 4 places after the decimal point.)

$\begin{bmatrix} 0.7746 \\ 0.5164 \\ 0.2582 \\ 0.2582 \end{bmatrix}, \begin{bmatrix} 0.0922 \\ 0.3147 \\ 0.9440 \\ -0.0382 \end{bmatrix},$

$\begin{bmatrix} 0.6325 \\ 0.3162 \\ -0.6325 \\ 0.3162 \end{bmatrix},$ and $\begin{bmatrix} 0.3122 \\ 0.1829 \\ 0.5486 \\ 0.7537 \end{bmatrix}.$

8. Answers are correct to 4 places after the decimal point.

(a) Clearly A is a transition matrix since its entries are nonnegative and each of its columns sums to 1. To show that A is regular, it suffices to show that all of the entries of A^2 are positive. This can be done by computing A^2 directly or by noting that the sum of products of the corresponding entries in any row and column of A is positive because at least one such product is positive.

(b) $\begin{bmatrix} .2344 \\ .1934 \\ .1732 \\ .2325 \\ .1665 \end{bmatrix}$ (c) $\begin{bmatrix} 5.3 \\ 5.2 \\ 5.1 \\ 6.1 \\ 3.3 \end{bmatrix}, \begin{bmatrix} 5.8611 \\ 4.8351 \\ 4.3299 \\ 5.8114 \\ 4.1626 \end{bmatrix}, \begin{bmatrix} 5.8610 \\ 4.8351 \\ 4.3299 \\ 5.8114 \\ 4.1625 \end{bmatrix}$

(d) $A^{100}\mathbf{p} \approx 25\mathbf{v}$

9. $r_n = (0.2)3^n - 2^n - (0.2)(-2)^n + 4 + 2(-1)^n$

Chapter 6

Orthogonality

6.1 THE GEOMETRY OF VECTORS

1. $\|\mathbf{u}\| = \sqrt{34}$, $\|\mathbf{v}\| = \sqrt{20}$, and $d = \sqrt{58}$
2. $\|\mathbf{u}\| = \sqrt{5}$, $\|\mathbf{v}\| = \sqrt{58}$, and $d = \sqrt{29}$
3. $\|\mathbf{u}\| = \sqrt{2}$, $\|\mathbf{v}\| = \sqrt{5}$, and $d = \sqrt{5}$
4. $\|\mathbf{u}\| = \sqrt{11}$, $\|\mathbf{v}\| = \sqrt{21}$, and $d = \sqrt{6}$
5. $\|\mathbf{u}\| = \sqrt{1^2 + (-1)^2 + 3^2} = \sqrt{11}$,
   $\|\mathbf{v}\| = \sqrt{2^2 + 1^2 + 0^2} = \sqrt{5}$,
   and
   $$d = \|\mathbf{u} - \mathbf{v}\|$$
   $$= \sqrt{(1-2)^2 + (-1-1)^2 + (3-0)^2}$$
   $$= \sqrt{14}$$
6. $\|\mathbf{u}\| = \sqrt{7}$, $\|\mathbf{v}\| = \sqrt{17}$, and $d = 2$
7. $\|\mathbf{u}\| = \sqrt{7}$, $\|\mathbf{v}\| = \sqrt{15}$, and $d = \sqrt{26}$
8. $\|\mathbf{u}\| = \sqrt{6}$, $\|\mathbf{v}\| = \sqrt{15}$, and $d = \sqrt{21}$
9. 0, yes   10. 17, no
11. $\mathbf{u} \cdot \mathbf{v} = (1)(2) + (-1)(1) = 1$, and hence $\mathbf{u}$ and $\mathbf{v}$ are not orthogonal because $\mathbf{u} \cdot \mathbf{v} \neq 0$.
12. 13, no   13. 0, yes   14. 2, no
15. -2, no   16. 0, yes
17. $\|\mathbf{u}\|^2 = 20$, $\|\mathbf{v}\|^2 = 45$, $\|\mathbf{u} + \mathbf{v}\|^2 = 65$
18. $\|\mathbf{u}\|^2 = 10$, $\|\mathbf{v}\|^2 = 10$, $\|\mathbf{u} + \mathbf{v}\|^2 = 20$
19. $\|\mathbf{u}\|^2 = 13$, $\|\mathbf{v}\|^2 = 0$, $\|\mathbf{u} + \mathbf{v}\|^2 = 13$
20. $\|\mathbf{u}\|^2 = 40$, $\|\mathbf{v}\|^2 = 90$, $\|\mathbf{u} + \mathbf{v}\|^2 = 130$
21. $\|\mathbf{u}\|^2 = 14$, $\|\mathbf{v}\|^2 = 3$, $\|\mathbf{u} + \mathbf{v}\|^2 = 17$
22. $\|\mathbf{u}\|^2 = 6$, $\|\mathbf{v}\|^2 = 5$, $\|\mathbf{u} + \mathbf{v}\|^2 = 11$
23. We have
    $$\|\mathbf{u}\|^2 = 1^2 + 2^2 + 3^2 = 14,$$
    $$\|\mathbf{v}\|^2 = (-11)^2 + 4^2 + 1^2 = 138, \text{ and}$$
    $$\|\mathbf{u} + \mathbf{v}\|^2 = (1 - 11)^2 + (2 + 4)^2 + (3 + 1)^2$$
    $$= 152.$$
    So $\|\mathbf{u}\|^2 + \|\mathbf{v}\|^2 = 14 + 138 = \|\mathbf{u} + \mathbf{v}\|^2$.

24. $\|\mathbf{u}\|^2 = 21$, $\|\mathbf{v}\|^2 = 17$, $\|\mathbf{u} + \mathbf{v}\|^2 = 38$
25. $\|\mathbf{u}\| = \sqrt{13}$, $\|\mathbf{v}\| = \sqrt{44}$, $\|\mathbf{u} + \mathbf{v}\| = \sqrt{13}$
26. $\|\mathbf{u}\| = \sqrt{5}$, $\|\mathbf{v}\| = \sqrt{13}$, $\|\mathbf{u} + \mathbf{v}\| = \sqrt{26}$
27. $\|\mathbf{u}\| = \sqrt{20}$, $\|\mathbf{v}\| = \sqrt{10}$, $\|\mathbf{u} + \mathbf{v}\| = \sqrt{50}$
28. $\|\mathbf{u}\| = \sqrt{29}$, $\|\mathbf{v}\| = \sqrt{10}$, $\|\mathbf{u} + \mathbf{v}\| = \sqrt{37}$
29. $\|\mathbf{u}\| = \sqrt{21}$, $\|\mathbf{v}\| = \sqrt{11}$, $\|\mathbf{u} + \mathbf{v}\| = \sqrt{34}$
30. $\|\mathbf{u}\| = \sqrt{14}$, $\|\mathbf{v}\| = \sqrt{6}$, $\|\mathbf{u} + \mathbf{v}\| = \sqrt{22}$
31. We have
    $$\|\mathbf{u}\| = \sqrt{2^2 + (-1)^2 + 3^2} = \sqrt{14},$$
    $$\|\mathbf{v}\| = \sqrt{4^2 + 0^2 + 1^2} = \sqrt{17}, \text{ and}$$
    $$\|\mathbf{u} + \mathbf{v}\| = \sqrt{(2+4)^2 + (-1+0)^2 + (3+1)^2}$$
    $$= \sqrt{53}.$$
    So
    $$\|\mathbf{u} + \mathbf{v}\| = \sqrt{53} \leq \sqrt{14} + \sqrt{17} = \|\mathbf{u}\| + \|\mathbf{v}\|.$$
32. $\|\mathbf{u}\| = \sqrt{14}$, $\|\mathbf{v}\| = \sqrt{56}$, $\|\mathbf{u} + \mathbf{v}\| = \sqrt{14}$
33. $\|\mathbf{u}\| = \sqrt{13}$, $\|\mathbf{v}\| = \sqrt{34}$, $\mathbf{u} \cdot \mathbf{v} = -1$
34. $\|\mathbf{u}\| = \sqrt{29}$, $\|\mathbf{v}\| = 5$, $\mathbf{u} \cdot \mathbf{v} = 26$
35. $\|\mathbf{u}\| = \sqrt{17}$, $\|\mathbf{v}\| = 2$, $\mathbf{u} \cdot \mathbf{v} = -2$
36. $\|\mathbf{u}\| = 5$, $\|\mathbf{v}\| = \sqrt{5}$, $\mathbf{u} \cdot \mathbf{v} = 5$
37. $\|\mathbf{u}\| = \sqrt{41}$, $\|\mathbf{v}\| = \sqrt{18}$, $\mathbf{u} \cdot \mathbf{v} = 0$
38. $\|\mathbf{u}\| = \sqrt{2}$, $\|\mathbf{v}\| = \sqrt{14}$, $\mathbf{u} \cdot \mathbf{v} = 4$
39. We have
    $$\|\mathbf{u}\| = \sqrt{4^2 + 2^2 + 1^2} = \sqrt{21},$$
    $$\|\mathbf{v}\| = \sqrt{2^2 + (-1)^2 + (-1)^2} = \sqrt{6}, \text{ and}$$
    $$\mathbf{u} \cdot \mathbf{v} = (4)(2) + 2(-1) + (1)(-1) = -5.$$
    So $|\mathbf{u} \cdot \mathbf{v}| = 5 \leq \sqrt{21}\sqrt{6} = \|\mathbf{u}\|\|\mathbf{v}\|$.
40. $\|\mathbf{u}\| = \sqrt{14}$, $\|\mathbf{v}\| = \sqrt{11}$, $\mathbf{u} \cdot \mathbf{v} = -2$
41. $\mathbf{w} = \begin{bmatrix} 5 \\ 0 \end{bmatrix}$ and $d = 0$

136   Chapter 6   Orthogonality

42. Let $\mathbf{u} = \begin{bmatrix} 1 \\ 2 \end{bmatrix}$, a nonzero vector that lies along the line $y = 2x$. Then $\mathbf{w} = c\mathbf{u}$, where
$$c = \frac{\mathbf{u} \cdot \mathbf{v}}{\mathbf{u} \cdot \mathbf{u}} = \frac{8}{5},$$
and hence $\mathbf{w} = \begin{bmatrix} 1.6 \\ 3.2 \end{bmatrix}$. Therefore
$$d = \|\mathbf{v} - \mathbf{w}\| = \left\| \begin{bmatrix} 2 \\ 3 \end{bmatrix} - \begin{bmatrix} 1.6 \\ 3.2 \end{bmatrix} \right\| = \sqrt{0.2}.$$

43. Use the same method as in Exercise 42 to obtain $\mathbf{w} = \frac{1}{2}\begin{bmatrix} -1 \\ 1 \end{bmatrix}$ and $d = \frac{7\sqrt{2}}{2}$.

44. Use the same method as in Exercise 42 to obtain $\mathbf{w} = \begin{bmatrix} -1 \\ 2 \end{bmatrix}$ and $d = \sqrt{20}$.

45. Use the same method as in Exercise 42 to obtain $\mathbf{w} = \begin{bmatrix} 0.7 \\ 2.1 \end{bmatrix}$ and $d = 1.1\sqrt{10}$.

46. $\mathbf{w} = -\begin{bmatrix} 0.5 \\ 0.5 \end{bmatrix}$ and $d = \sqrt{12.5}$

47. $\mathbf{w} = \begin{bmatrix} -1.3 \\ 3.9 \end{bmatrix}$ and $d = \sqrt{12.1}$

48. $\mathbf{w} = \begin{bmatrix} -0.823 \\ 3.294 \end{bmatrix}$ and $d = \sqrt{49.471}$

49. $(\mathbf{u} + \mathbf{v}) \cdot \mathbf{w} = \mathbf{u} \cdot \mathbf{w} + \mathbf{v} \cdot \mathbf{w} = 1 - 4 = -3$

50. $\|4\mathbf{w}\| = 4\|\mathbf{w}\| = 4 \cdot 5 = 20$

51. $\|\mathbf{u} + \mathbf{v}\|^2 = \|\mathbf{u}\|^2 + 2\mathbf{u} \cdot \mathbf{v} + \|\mathbf{v}\|^2 = 4 - 2 + 9 = 11$

52. $(\mathbf{u} + \mathbf{w}) \cdot \mathbf{v} = \mathbf{u} \cdot \mathbf{v} + \mathbf{w} \cdot \mathbf{v} = -1 - 4 = -5$

53.
$$\|\mathbf{v} - 4\mathbf{w}\|^2 = \|\mathbf{v}\|^2 - 8\mathbf{v} \cdot \mathbf{w} + 16\|\mathbf{w}\|^2$$
$$= 9 + 32 + 400 = 441$$

54.
$$\|2\mathbf{u} + 3\mathbf{v}\|^2 = 4\|\mathbf{u}\|^2 + 12\mathbf{u} \cdot \mathbf{v} + 9\|\mathbf{v}\|^2$$
$$= 16 - 12 + 81 = 85$$

55. 21   56. $3\sqrt{14}$ 57. 7   58. -25
59. 49   60. $\sqrt{246}$

61. True
62. False, the dot product of two vectors is a scalar.
63. False, the norm of a vector equals the square root of the dot product of the vector with itself.
64. False, the norm is the product of the *absolute value* of the multiple and the norm of the vector.

65. False, for example, if $\mathbf{v}$ is a nonzero vector, then
$$\|\mathbf{v} + (-\mathbf{v})\| = 0 \neq \|\mathbf{v}\| + \|-\mathbf{v}\|.$$

66. True   67. True   68. True   69. True
70. False, consider nonzero orthogonal vectors.
71. False, we need to replace $=$ by $\leq$.
72. True   73. True   74. True   75. True
76. False, $A\mathbf{u} \cdot \mathbf{v} = \mathbf{u} \cdot A^T\mathbf{v}$.
77. True
78. False, we need to replace $=$ by $\leq$.
79. True   80. True

81.
$$\mathbf{u} \cdot \mathbf{u} = u_1(u_1) + u_2(u_2) + \cdots + u_n(u_n)$$
$$= \left(\sqrt{u_1^2 + u_2^2 + \cdots + u_n^2}\right)^2 = \|\mathbf{u}\|^2$$

82. Suppose that $\mathbf{u}$ is in $\mathcal{R}^n$ and $\mathbf{u} \cdot \mathbf{u} = 0$. Then $u_1^2 + u_2^2 + \cdots + u_n^2 = 0$. Since $u_i^2 \geq 0$ for all i, it follows that $u_i^2 = 0$ for all i. Thus $u_i = 0$ for each i, and hence $\mathbf{u} = \mathbf{0}$.
Conversely, suppose that $\mathbf{u} = \mathbf{0}$. Then $\mathbf{u} \cdot \mathbf{u} = 0^2 + 0^2 + \cdots + 0^2 = 0$.

83. Suppose that $\mathbf{u}$ and $\mathbf{v}$ are in $\mathcal{R}^n$. Then
$$\mathbf{u} \cdot \mathbf{v} = u_1v_1 + u_2v_2 + \cdots + u_nv_n$$
$$= v_1u_1 + v_2u_2 + \cdots + v_nu_n$$
$$= \mathbf{v} \cdot \mathbf{u}.$$

84. Suppose that $\mathbf{u}$, $\mathbf{v}$, and $\mathbf{w}$ are in $\mathcal{R}^n$. Then
$$(\mathbf{v} + \mathbf{w}) \cdot \mathbf{u} = (v_1 + w_1)u_1 + \cdots + (v_n + w_n)u_n$$
$$= (v_1u_1 + w_1u_1) + \cdots$$
$$\quad + (v_nu_n + w_nu_n)$$
$$= (v_1u_1 + v_2u_2 + \cdots + v_nu_n)$$
$$\quad + (w_1u_1 + w_2u_2 + \cdots + w_nu_n)$$
$$= \mathbf{v} \cdot \mathbf{u} + \mathbf{w} \cdot \mathbf{u}.$$

85. Suppose that $\mathbf{u}$ and $\mathbf{v}$ are in $\mathcal{R}^n$. Then
$$(c\mathbf{u}) \cdot \mathbf{v} = (cu_1)v_1 + (cu_2)v_2 + \cdots + (cu_n)v_n$$
$$= c(u_1v_1 + u_2v_2 + \cdots + u_nv_n)$$
$$= c(\mathbf{u} \cdot \mathbf{v}).$$

The proof that $(c\mathbf{u}) \cdot \mathbf{v} = \mathbf{u} \cdot (c\mathbf{v})$ is similar.

86. Let $a\mathbf{v} + b\mathbf{w}$ be a linear combination of $\mathbf{v}$ and $\mathbf{w}$, where a and b are scalars. Then
$$\mathbf{u} \cdot (a\mathbf{v} + b\mathbf{w}) = a(\mathbf{u} \cdot \mathbf{v}) + b(\mathbf{u} \cdot \mathbf{w})$$
$$= a(0) + b(0) = 0.$$

So $\mathbf{u}$ is orthogonal to $a\mathbf{v} + b\mathbf{w}$.

87. Since $\mathbf{v}$ is in a basis for W and $\mathbf{z}$ is a nonzero linear combination of basis vectors for W, both $\mathbf{v}$ and $\mathbf{z}$ are nonzero vectors. Also, by Theorem 6.1,
$$\mathbf{v} \cdot \mathbf{z} = \mathbf{v} \cdot \left(\mathbf{w} - \frac{\mathbf{v} \cdot \mathbf{w}}{\mathbf{v} \cdot \mathbf{v}}\mathbf{v}\right)$$
$$= \mathbf{v} \cdot \mathbf{w} - \left(\frac{\mathbf{v} \cdot \mathbf{w}}{\mathbf{v} \cdot \mathbf{v}}\right)(\mathbf{v} \cdot \mathbf{v})$$
$$= \mathbf{v} \cdot \mathbf{w} - \mathbf{v} \cdot \mathbf{w} = 0.$$

So $\mathbf{v}$ and $\mathbf{z}$ are orthogonal.

By Exercise 96, $\{\mathbf{v}, \mathbf{z}\}$ is linearly independent. Since this set lies in the 2-dimensional subspace W, the set is a basis for W.

88. Let $\mathbf{u}$ and $\mathbf{v}$ be vectors in $\mathcal{R}^n$. Then $|\mathbf{u} \cdot \mathbf{v}| = \|\mathbf{u}\| \cdot \|\mathbf{v}\|$ if and only if $\{\mathbf{u}, \mathbf{v}\}$ is linearly dependent.

Proof Suppose that $\{\mathbf{u}, \mathbf{v}\}$ is linearly dependent. If $\mathbf{u} = \mathbf{0}$, then the equality $|\mathbf{u} \cdot \mathbf{v}| = \|\mathbf{u}\| \cdot \|\mathbf{v}\|$ is immediate. Otherwise, if $\mathbf{u} \neq \mathbf{0}$, then $\mathbf{v} = c\mathbf{u}$ for some scalar c. Hence
$$|\mathbf{u} \cdot \mathbf{v}| = |\mathbf{u} \cdot (c\mathbf{u})| = |c| \cdot |\mathbf{u} \cdot \mathbf{u}| = |c| \cdot \|\mathbf{u}\|^2$$
$$= \|\mathbf{u}\|(|c|\|\mathbf{u}\|) = \|\mathbf{u}\| \cdot \|c\mathbf{u}\|$$
$$= \|\mathbf{u}\| \cdot \|\mathbf{v}\|.$$

Conversely, suppose that $|\mathbf{u} \cdot \mathbf{v}| = \|\mathbf{u}\| \cdot \|\mathbf{v}\|$. If $\mathbf{v} = \mathbf{0}$, then $\{\mathbf{u}, \mathbf{v}\}$ is linearly dependent. If $\mathbf{u} \cdot \mathbf{v} \geq 0$, then $\mathbf{u} \cdot \mathbf{v} = \|\mathbf{u}\| \cdot \|\mathbf{v}\|$. Let $c = \frac{\|\mathbf{u}\|}{\|\mathbf{v}\|}$. Then
$$\|\mathbf{u} - c\mathbf{v}\|^2 = \|\mathbf{u}\|^2 - 2c\mathbf{u} \cdot \mathbf{v} + c^2\|\mathbf{v}\|^2$$
$$= \|\mathbf{u}\|^2 - 2c\|\mathbf{u}\| \cdot \|\mathbf{v}\| + c^2\|\mathbf{v}\|^2$$
$$= \left(\|\mathbf{u}\| - \frac{\|\mathbf{u}\|}{\|\mathbf{v}\|}\|\mathbf{v}\|\right)^2$$
$$= (\|\mathbf{u}\| - \|\mathbf{u}\|)^2$$
$$= 0,$$

and hence $\mathbf{u} - c\mathbf{v} = \mathbf{0}$. Therefore $\mathbf{u} = c\mathbf{v}$, and $\{\mathbf{u}, \mathbf{v}\}$ is linearly dependent. If $\mathbf{u} \cdot \mathbf{v} < 0$, repeat the computation above with $c = -\frac{\|\mathbf{u}\|}{\|\mathbf{v}\|}$.

89. Let $\mathbf{u}$ and $\mathbf{v}$ be vectors in $\mathcal{R}^n$. Then
$$\|\mathbf{u} + \mathbf{v}\| = \|\mathbf{u}\| + \|\mathbf{v}\|$$

if and only if $\mathbf{u}$ is a nonnegative multiple of $\mathbf{v}$ or $\mathbf{v}$ is a nonnegative multiple of $\mathbf{u}$.

Proof Suppose that $\|\mathbf{u} + \mathbf{v}\| = \|\mathbf{u}\| + \|\mathbf{v}\|$. If $\mathbf{u} = \mathbf{0}$ or $\mathbf{v} = \mathbf{0}$, the result is immediate; so assume that the vectors are nonzero. Then
$$\|\mathbf{u}\|^2 + 2\mathbf{u} \cdot \mathbf{v} + \|\mathbf{v}\|^2$$
$$= \|\mathbf{u} + \mathbf{v}\|^2$$
$$= (\|\mathbf{u}\| + \|\mathbf{v}\|)^2$$
$$= \|\mathbf{u}\|^2 + 2\|\mathbf{u}\| \cdot \|\mathbf{v}\| + \|\mathbf{v}\|^2,$$

and therefore $\mathbf{u} \cdot \mathbf{v} = \|\mathbf{u}\| \cdot \|\mathbf{v}\|$. Thus $\mathbf{u} \cdot \mathbf{v}$ is nonnegative. By Exercise 88, it follows that $\mathbf{u}$ or $\mathbf{v}$ is a multiple of the other. Suppose $\mathbf{u} = k\mathbf{v}$ for some scalar k. Then
$$0 \leq \mathbf{u} \cdot \mathbf{v} = k\mathbf{v} \cdot \mathbf{v} = k(\mathbf{v} \cdot \mathbf{v}) = k\|\mathbf{v}\|^2.$$

So $k \geq 0$. A similar argument may be used if $\mathbf{v} = k\mathbf{u}$.

Conversely, suppose one of the vectors is a nonnegative multiple of the other. Since both of these vectors are nonzero, we may assume that $\mathbf{v} = c\mathbf{u}$, where c is a nonnegative scalar. Then
$$\|\mathbf{u} + \mathbf{v}\| = \|\mathbf{u} + c\mathbf{u}\| = \|(1 + c)\mathbf{u}\|$$
$$= (1 + c)\|\mathbf{u}\|,$$

and
$$\|\mathbf{u}\| + \|\mathbf{v}\| = \|\mathbf{u}\| + \|c\mathbf{u}\| = \|\mathbf{u}\| + c\|\mathbf{u}\|$$
$$= (1 + c)\|\mathbf{u}\|.$$

Therefore $\|\mathbf{u} + \mathbf{v}\| = \|\mathbf{u}\| + \|\mathbf{v}\|$.

90. By the triangle inequality,
$$\|\mathbf{v}\| = \|(\mathbf{v} - \mathbf{w}) + \mathbf{w}\| \leq \|\mathbf{v} - \mathbf{w}\| + \|\mathbf{w}\|,$$

and hence $\|\mathbf{v}\| - \|\mathbf{w}\| \leq \|\mathbf{v} - \mathbf{w}\|$. Similarly, $\|\mathbf{w}\| - \|\mathbf{v}\| \leq \|\mathbf{v} - \mathbf{w}\|$. Combining these two inequalities, we obtain the result.

91. $(\mathbf{u} + \mathbf{v}) \cdot \mathbf{w} = \mathbf{w} \cdot (\mathbf{u} + \mathbf{v}) = \mathbf{w} \cdot \mathbf{u} + \mathbf{w} \cdot \mathbf{v} = \mathbf{u} \cdot \mathbf{w} + \mathbf{v} \cdot \mathbf{w}$

92. Since $\mathbf{0} \cdot \mathbf{z} = 0$, $\mathbf{0}$ is in W. Now suppose that $\mathbf{u}$ and $\mathbf{v}$ are V. Then
$$(\mathbf{u} + \mathbf{v}) \cdot \mathbf{z} = \mathbf{u} \cdot \mathbf{z} + \mathbf{v} \cdot \mathbf{z} = 0 + 0 = 0,$$

and hence $\mathbf{u} + \mathbf{v}$ is in W. Finally, let c be any scalar. Then
$$(c\mathbf{u}) \cdot \mathbf{z} = c(\mathbf{u} \cdot \mathbf{z}) = c \cdot 0 = 0,$$

and hence $c\mathbf{u}$ is in W. We conclude that W is a subspace.

93. The proof of this exercise is similar to the proof of Exercise 92 except that we must state that the various equalities hold for all $\mathbf{z}$ in $\mathcal{S}$.

94. Let $\mathbf{z} = \begin{bmatrix} -2 \\ 1 \end{bmatrix}$ and $V = \{\mathbf{u} \in \mathcal{R}^2 : \mathbf{u} \cdot \mathbf{z} = 0\}$. To show that $W = V$, suppose $\mathbf{w}$ is in W. Then $\mathbf{w} = \begin{bmatrix} a \\ 2a \end{bmatrix}$ for some scalar a. Thus
$$\mathbf{w} \cdot \mathbf{z} = \begin{bmatrix} a \\ 2a \end{bmatrix} \cdot \begin{bmatrix} -2 \\ 1 \end{bmatrix} = -2a + 2a = 0,$$
and hence $\mathbf{w}$ is in V. Therefore W is contained in V.

Conversely, consider any vector $\mathbf{u} = \begin{bmatrix} a \\ b \end{bmatrix}$ in V. Then
$$\mathbf{u} \cdot \mathbf{z} = -2a + b = 0,$$
and hence $b = 2a$. Thus $\mathbf{u} = \begin{bmatrix} a \\ 2a \end{bmatrix}$, which lies in W. Therefore V is contained in W, and we conclude that $W = V$.

95. We have
$$\|\mathbf{u}+\mathbf{v}\|^2 + \|\mathbf{u}-\mathbf{v}\|^2 = (\|\mathbf{u}\|^2 + 2\mathbf{u}\cdot\mathbf{v} + \|\mathbf{v}\|^2)$$
$$+ (\|\mathbf{u}\|^2 - 2\mathbf{u}\cdot\mathbf{v} + \|\mathbf{v}\|^2)$$
$$= 2\|\mathbf{u}\|^2 + 2\|\mathbf{v}\|^2.$$

96. Assume $\mathbf{u}$ and $\mathbf{v}$ are nonzero orthogonal vectors, and suppose that $a\mathbf{u} + b\mathbf{v} = \mathbf{0}$. Then
$$0 = (a\mathbf{u}+b\mathbf{v})\cdot\mathbf{u} = a(\mathbf{u}\cdot\mathbf{u}) + b(\mathbf{v}\cdot\mathbf{u})$$
$$= a\|\mathbf{u}\|^2 + b\cdot 0 = a\|\mathbf{u}\|^2.$$
Since $\|\mathbf{u}\| \neq 0$, it follows that $a = 0$. So $b\mathbf{v} = \mathbf{0}$, which implies that $b = 0$. Therefore the vectors are linearly independent.

97. (a) Suppose that $\mathbf{v}$ is in Null A. Then
$$(A^T A)\mathbf{v} = A^T(A\mathbf{v}) = A^T\mathbf{0} = \mathbf{0},$$
and hence $\mathbf{v}$ is in Null $A^T A$. Thus Null A is contained in Null $A^T A$.

Conversely, suppose that $\mathbf{v}$ is in Null $A^T A$. Then
$$0 = \mathbf{0} \cdot \mathbf{v}$$
$$= (A^T A)\mathbf{v}\cdot\mathbf{v}$$
$$= (A^T A\mathbf{v})^T \mathbf{v}$$
$$= \mathbf{v}^T A^T A\mathbf{v}$$
$$= (A\mathbf{v})^T(A\mathbf{v})$$
$$= (A\mathbf{v})\cdot(A\mathbf{v})$$
$$= \|A\mathbf{v}\|^2,$$
and hence $\|A\mathbf{v}\| = 0$. Thus $A\mathbf{v} = \mathbf{0}$, and it follows that Null $A^T A$ is contained in Null A. Therefore
$$\text{Null } A^T A = \text{Null } A.$$

(b) Since
$$\text{Null } A^T A = \text{Null } A,$$
we have that
$$\text{nullity } A^T A = \text{nullity } A.$$
Notice that $A^T A$ and A have the same number of columns, say n, and hence
$$\text{rank } A^T A = n - \text{nullity } A^T A$$
$$= n - \text{nullity } A = \text{rank } A.$$

98. As a consequence of the law of cosines, we have
$$\|\mathbf{v}\|^2 - 2\mathbf{v}\cdot\mathbf{u} + \|\mathbf{u}\|^2 =$$
$$\|\mathbf{u}\|^2 + \|\mathbf{v}\|^2 - 2\|\mathbf{u}\|\,\|\mathbf{v}\|\cos\theta.$$
This equation simplifies to
$$-2\mathbf{u}\cdot\mathbf{v} = -2\|\mathbf{u}\|\,\|\mathbf{v}\|\cos\theta,$$
and hence $\mathbf{u}\cdot\mathbf{v} = \|\mathbf{u}\|\,\|\mathbf{v}\|\cos\theta$.

99. $135°$ 100. $45°$ 101. $180°$
102. $116.56°$ 103. $60°$ 104. $120°$
105. $150°$ 106. $30°$
107. For $\mathbf{u}$ in $\mathcal{R}^3$,
$$\mathbf{u}\times\mathbf{u} = \begin{bmatrix} u_2 u_3 - u_3 u_2 \\ u_3 u_1 - u_1 u_3 \\ u_1 u_2 - u_2 u_1 \end{bmatrix} = \begin{bmatrix} 0 \\ 0 \\ 0 \end{bmatrix} = \mathbf{0}.$$

108. For $\mathbf{u}$ and $\mathbf{v}$ in $\mathcal{R}^3$,
$$\mathbf{u}\times\mathbf{v} = \begin{bmatrix} u_2 v_3 - u_3 v_2 \\ u_3 v_1 - u_1 v_3 \\ u_1 v_2 - u_2 v_1 \end{bmatrix} = \begin{bmatrix} -(v_2 u_3 - v_3 u_2) \\ -(v_3 u_1 - v_1 u_3) \\ -(v_1 u_2 - v_2 u_1) \end{bmatrix}$$
$$= -\begin{bmatrix} v_2 u_3 - v_3 u_2 \\ v_3 u_1 - v_1 u_3 \\ v_1 u_2 - v_2 u_1 \end{bmatrix} = -(\mathbf{v}\times\mathbf{u}).$$

109. For $\mathbf{u}$ in $\mathcal{R}^3$,
$$\mathbf{u}\times\mathbf{0} = \begin{bmatrix} u_2 0 - u_3 0 \\ u_3 0 - u_1 0 \\ u_1 0 - u_2 0 \end{bmatrix} = \begin{bmatrix} 0 \\ 0 \\ 0 \end{bmatrix} = \mathbf{0}.$$
By Exercise 108, $\mathbf{0}\times\mathbf{u} = -(\mathbf{u}\times\mathbf{0}) = -\mathbf{0} = \mathbf{0}$.

110. Suppose that $\mathbf{u}$ and $\mathbf{v}$ are parallel. Then $\mathbf{u} = c\mathbf{v}$ for some scalar c, and hence,

$$\mathbf{u} \times \mathbf{v} = (c\mathbf{v}) \times \mathbf{v} = \begin{bmatrix} cv_2v_3 - cv_3v_2 \\ cv_3v_1 - cv_1v_3 \\ cv_1v_2 - cv_2v_1 \end{bmatrix}$$

$$= \begin{bmatrix} 0 \\ 0 \\ 0 \end{bmatrix} = \mathbf{0}.$$

Now suppose that $\mathbf{u} \times \mathbf{v} = \mathbf{0}$. If $\mathbf{u}$ or $\mathbf{v}$ is zero, the result follows immediately. So assume, without loss of generality, that $\mathbf{v} \neq \mathbf{0}$. Then one of the components of $\mathbf{v}$ is not zero, say $v_1 \neq 0$. Since

$$u_2v_3 = u_3v_2, \quad u_3v_1 = u_1v_3,$$

and

$$u_1v_2 = u_2v_1,$$

it follows that

$$u_1 = \left(\frac{u_1}{v_1}\right)v_1, \quad u_2 = \left(\frac{u_1}{v_1}\right)v_2,$$

and

$$u_3 = \left(\frac{u_1}{v_1}\right)v_3.$$

Therefore $\mathbf{u} = c\mathbf{v}$ for $c = u_1/v_1$.

111. For $\mathbf{u}$ and $\mathbf{v}$ in $\mathcal{R}^3$, and any scalar c,

$$c(\mathbf{u} \times \mathbf{v}) = c\begin{bmatrix} u_2v_3 - u_3v_2 \\ u_3v_1 - u_1v_3 \\ u_1v_2 - u_2v_1 \end{bmatrix}$$

$$= \begin{bmatrix} c(u_2v_3 - u_3v_2) \\ c(u_3v_1 - u_1v_3) \\ c(u_1v_2 - u_2v_1) \end{bmatrix}$$

$$= \begin{bmatrix} (cu_2)v_3 - (cu_3)v_2 \\ (cu_3)v_1 - (cu_1)v_3 \\ (cu_1)v_2 - (cu_2)v_1 \end{bmatrix}$$

$$= c\mathbf{u} \times \mathbf{v}.$$

To establish the second equality, apply Exercise 108 and the first equality to obtain

$$\mathbf{u} \times c\mathbf{v} = -(c\mathbf{v} \times \mathbf{u}) = -c(\mathbf{v} \times \mathbf{u})$$
$$= c(-(\mathbf{v} \times \mathbf{u})) = c(\mathbf{u} \times \mathbf{v}).$$

112. For $\mathbf{u}$, $\mathbf{v}$, and $\mathbf{w}$ in $\mathcal{R}^3$,

$$\mathbf{u} \times (\mathbf{v} + \mathbf{w}) = \begin{bmatrix} u_2(v_3 + w_3) - u_3(v_2 + w_2) \\ u_3(v_1 + w_1) - u_1(v_3 + w_3) \\ u_1(v_2 + w_2) - u_2(v_1 + w_1) \end{bmatrix}$$

$$= \begin{bmatrix} u_2v_3 - u_3v_2 \\ u_3v_1 - u_1v_3 \\ u_1v_2 - u_2v_1 \end{bmatrix} + \begin{bmatrix} u_2w_3 - u_3w_2 \\ u_3w_1 - u_1w_3 \\ u_1w_2 - u_2w_1 \end{bmatrix}$$

$$= \mathbf{u} \times \mathbf{v} + \mathbf{u} \times \mathbf{w}.$$

113. Let $\mathbf{u}$, $\mathbf{v}$, and $\mathbf{w}$ be in $\mathcal{R}^3$. By Exercises 108 and 112,

$$(\mathbf{u} + \mathbf{v}) \times \mathbf{w} = -(\mathbf{w} \times (\mathbf{u} + \mathbf{v}))$$
$$= -(\mathbf{w} \times \mathbf{u} + \mathbf{w} \times \mathbf{v})$$
$$= -(\mathbf{w} \times \mathbf{u}) + -(\mathbf{w} \times \mathbf{v})$$
$$= \mathbf{u} \times \mathbf{w} + \mathbf{v} \times \mathbf{w}.$$

114. By Exercises 115 and 107,

$$(\mathbf{u} \times \mathbf{v}) \cdot \mathbf{v} = \mathbf{u} \cdot (\mathbf{v} \times \mathbf{v}) = \mathbf{u} \cdot \mathbf{0} = 0.$$

So $\mathbf{u} \times \mathbf{v}$ is orthogonal to $\mathbf{v}$. Similarly, $\mathbf{u} \times \mathbf{v}$ is orthogonal to $\mathbf{u}$.

115. For $\mathbf{u}$, $\mathbf{v}$, and $\mathbf{w}$ in $\mathcal{R}^3$,

$$(\mathbf{u} \times \mathbf{v}) \cdot \mathbf{w} = \begin{bmatrix} u_2v_3 - u_3v_2 \\ u_3v_1 - u_1v_3 \\ u_1v_2 - u_2v_1 \end{bmatrix} \cdot \begin{bmatrix} w_1 \\ w_2 \\ w_3 \end{bmatrix}$$

$$= (u_2v_3 - u_3v_2)w_1$$
$$+ (u_3v_1 - u_1v_3)w_2$$
$$+ (u_1v_2 - u_2v_1)w_3$$

and

$$\mathbf{u} \cdot (\mathbf{v} \times \mathbf{w}) = \begin{bmatrix} u_1 \\ u_2 \\ u_3 \end{bmatrix} \begin{bmatrix} v_2w_3 - v_3w_2 \\ v_3w_1 - v_1w_3 \\ v_1w_2 - v_2w_1 \end{bmatrix}$$

$$= u_1(v_2w_3 - v_3w_2)$$
$$+ u_2(v_3w_1 - v_1w_3)$$
$$+ u_3(v_1w_2 - v_2w_1).$$

Since the third expressions in the two preceding equations are equal, the result follows.

116. Let $\mathbf{u}$, $\mathbf{v}$, and $\mathbf{w}$ be in $\mathcal{R}^3$. If we compute the first component of $\mathbf{u} \times (\mathbf{v} \times \mathbf{w})$, we obtain

$$u_2(v_1w_2 - v_2w_1) - u_3(v_3w_1 - v_1w_3).$$

If we compute the first component of

$$(\mathbf{u} \cdot \mathbf{w})\mathbf{v} - (\mathbf{u} \cdot \mathbf{v})\mathbf{w},$$

we obtain

$$(u_1w_1 + u_2w_2 + u_3w_3)v_1 - (u_1v_1 + u_2v_2 + u_3v_3)w_1.$$

Expanding both expressions shows they are equal. The other components can be handled similarly.

117. Let **u**, **v**, and **w** be in $\mathcal{R}^3$. By Exercises 108 and 116, we have
$$(\mathbf{u} \times \mathbf{v}) \times \mathbf{w} = -(\mathbf{w} \times (\mathbf{u} \times \mathbf{v}))$$
$$= -((\mathbf{w} \cdot \mathbf{v})\mathbf{u} - (\mathbf{w} \cdot \mathbf{u})\mathbf{v})$$
$$= (\mathbf{w} \cdot \mathbf{u})\mathbf{v} - (\mathbf{w} \cdot \mathbf{v})\mathbf{u}.$$

118. Let **u** and **v** in $\mathcal{R}^3$. By Exercises 115 and 116, we have
$$\|\mathbf{u} \times \mathbf{v}\|^2 = (\mathbf{u} \times \mathbf{v}) \cdot (\mathbf{u} \times \mathbf{v})$$
$$= \mathbf{u} \cdot (\mathbf{v} \times (\mathbf{u} \times \mathbf{v}))$$
$$= \mathbf{u} \cdot ((\mathbf{v} \cdot \mathbf{v})\mathbf{u} - (\mathbf{v} \cdot \mathbf{u})\mathbf{v})$$
$$= (\mathbf{v} \cdot \mathbf{v})(\mathbf{u} \cdot \mathbf{u}) - (\mathbf{v} \cdot \mathbf{u})(\mathbf{u} \cdot \mathbf{v})$$
$$= \|\mathbf{u}\|^2 \|\mathbf{v}\|^2 - (\mathbf{u} \cdot \mathbf{v})^2.$$

119. Let **u** and **v** in $\mathcal{R}^3$. By Exercises 118 and 98, we have
$$\|\mathbf{u} \times \mathbf{v}\|^2 = \|\mathbf{u}\|^2 \|\mathbf{v}\|^2 - (\mathbf{u} \cdot \mathbf{v})^2$$
$$= \|\mathbf{u}\|^2 \|\mathbf{v}\|^2 - \|\mathbf{u}\|^2 \|\mathbf{v}\|^2 \cos^2 \theta$$
$$= \|\mathbf{u}\|^2 \|\mathbf{v}\|^2 (1 - \cos^2 \theta)$$
$$= \|\mathbf{u}\|^2 \|\mathbf{v}\|^2 \sin^2 \theta.$$

Since $0° \leq \theta \leq 180°$, it follows that $\sin \theta \geq 0$, and hence the result follows.

120. Let **u**, **v**, and **w** be in $\mathcal{R}^3$. By Exercise 117, we have
$$(\mathbf{u} \times \mathbf{v}) \times \mathbf{w} + (\mathbf{v} \times \mathbf{w}) \times \mathbf{u} + (\mathbf{w} \times \mathbf{u}) \times \mathbf{v}$$
$$= [(\mathbf{w} \cdot \mathbf{u})\mathbf{v} - (\mathbf{w} \cdot \mathbf{v})\mathbf{u}] + [(\mathbf{u} \cdot \mathbf{v})\mathbf{w} - (\mathbf{v} \cdot \mathbf{w})\mathbf{v}]$$
$$\quad + [(\mathbf{v} \cdot \mathbf{w})\mathbf{u} - (\mathbf{v} \cdot \mathbf{u})\mathbf{w}]$$
$$= \mathbf{0}.$$

121. The supervisor polls all 20 students and finds that the students are divided among three sections. The first has 8 students, the second has 12 students, and the class has 6 students. She divides the total number of students by the number of sections and computes
$$\bar{v} = \frac{8 + 12 + 6}{3} = \frac{26}{3} = 8.6667.$$

When the investigator polls 8 students in Section 1, they all report that their class sizes are 8. Likewise, for the other two sections, 12 students report that their class sizes are 12, and 6 students report that their class sizes are 6. Adding these sizes and dividing by the total number of polled students, the investigator obtains
$$v^* = \frac{8 \cdot 8 + 12 \cdot 12 + 6 \cdot 6}{8 + 12 + 6} = \frac{244}{26} = 9.3846.$$

122. $\bar{v} = 25$ and $v^* = 27$. The method is similar to that of Exercise 121.

123. $\bar{v} = v^* = 22$. The method is similar to that of Exercise 121.

124. The Cauchy-Schwarz inequality was applied within this application to the vectors **u** and **v**, where
$$\mathbf{u} = \frac{1}{n}\begin{bmatrix} 1 \\ 1 \\ \vdots \\ 1 \end{bmatrix} \quad \text{and} \quad \mathbf{v} = \begin{bmatrix} v_1 \\ v_2 \\ \vdots \\ v_n \end{bmatrix}.$$

By Exercise 88, equality in the Cauchy-Schwarz inequality occurs if and only if **u** is a multiple of **v** or **v** is a multiple of **u**. Suppose that **v** is a multiple of **u**, that is, there exists a scalar k such that $\mathbf{v} = k\mathbf{u}$. In this case, we have $v_i = k/n$ for all i, so that all of the class sizes are equal. The rest is proved similarly.

125. In (a), (b), and (c), we describe the use of MATLAB in the default (*short*) format.

    (a) Entering norm(**u** + **v**) produces the output 13.964, and entering norm(**u**)+norm(**v**) yields 17.3516. We conclude that
    $$\|\mathbf{u} + \mathbf{v}\| < \|\mathbf{u}\| + \|\mathbf{v}\|.$$

    (b) As in (a), we obtain norm(**u**+**v**$_1$) = 16.449 and norm(**u**) + norm(**v**$_1$) = 16.449. However, these outputs are rounded to 3 places after the decimal, so we need an additional test for equality. Entering the difference
    norm(**u**) + norm(**v**$_1$)
    − norm(**u** + **v**$_1$)
    yields the output 1.0114×10^{-6}, which indicates that the two are unequal but the difference is small. Thus
    $$\|\mathbf{u} + \mathbf{v}_1\| < \|\mathbf{u}\| + \|\mathbf{v}_1\|.$$

    (c) As in (b), we obtain a strict inequality. In this case, the difference, given by the MATLAB output, is 5.0617×10^{-7}.

    (d) Notice that in (b) and (c), **v**$_1$ and **v**$_2$ are "nearly" positive multiples of **u** and the triangle inequality is almost an equality. Thus we conjecture that the equality $\|\mathbf{u} + \mathbf{v}\| = \|\mathbf{u}\| + \|\mathbf{v}\|$ holds if and only if **u** is a nonnegative multiple of **v** or **v** is a nonnegative multiple of **u**. (For a proof of this conjecture, see Exercise 89.)

(e) If two sides of a triangle are parallel, then the triangle is "degenerate," that is, the third side coincides with the union of the other two sides.

6.2 ORTHOGONAL VECTORS

1. no   **2.** yes   **3.** no   **4.** yes
5. no   **6.** no
7. We have
$$\begin{bmatrix}1\\2\\3\\-3\end{bmatrix} \cdot \begin{bmatrix}1\\1\\-1\\0\end{bmatrix}$$
$$= (1)(1) + (2)(1) + (3)(-1) + (-3)(0) = 0,$$
$$\begin{bmatrix}1\\2\\3\\-3\end{bmatrix} \cdot \begin{bmatrix}3\\-3\\0\\-1\end{bmatrix}$$
$$= (1)(3) + (2)(-3) + (3)(0) + (-3)(-1) = 0,$$
and
$$\begin{bmatrix}1\\1\\-1\\0\end{bmatrix} \cdot \begin{bmatrix}3\\-3\\0\\-1\end{bmatrix}$$
$$= (1)(3) + (1)(-3) + (-1)(0) + (0)(-1) = 0.$$
Therefore the set is orthogonal.

8. no

9. (a) $\left\{ \begin{bmatrix}1\\1\\1\end{bmatrix}, \begin{bmatrix}3\\-3\\0\end{bmatrix} \right\}$

(b) $\left\{ \dfrac{1}{\sqrt{3}}\begin{bmatrix}1\\1\\1\end{bmatrix}, \dfrac{1}{\sqrt{2}}\begin{bmatrix}1\\-1\\0\end{bmatrix} \right\}$

10. (a) $\left\{ \begin{bmatrix}1\\-2\\1\end{bmatrix}, \dfrac{1}{2}\begin{bmatrix}1\\0\\-1\end{bmatrix} \right\}$

(b) $\left\{ \dfrac{1}{\sqrt{6}}\begin{bmatrix}1\\-2\\1\end{bmatrix}, \dfrac{1}{\sqrt{2}}\begin{bmatrix}1\\0\\-1\end{bmatrix} \right\}$

11. (a) $\left\{ \begin{bmatrix}1\\-2\\-1\end{bmatrix}, \begin{bmatrix}9\\3\\3\end{bmatrix} \right\}$

(b) $\left\{ \dfrac{1}{\sqrt{6}}\begin{bmatrix}1\\-2\\-1\end{bmatrix}, \dfrac{1}{\sqrt{11}}\begin{bmatrix}3\\1\\1\end{bmatrix} \right\}$

12. (a) Let
$$\mathbf{u}_1 = \begin{bmatrix}-1\\3\\4\end{bmatrix} \text{ and } \mathbf{u}_2 = \begin{bmatrix}-7\\1\\3\end{bmatrix}.$$
Then
$$\mathbf{v}_1 = \mathbf{u}_1 = \begin{bmatrix}-1\\3\\4\end{bmatrix}$$
and
$$\mathbf{v}_2 = \mathbf{u}_2 - \dfrac{\mathbf{u}_2 \cdot \mathbf{v}_1}{\|\mathbf{v}_1\|^2}\mathbf{v}_1$$
$$= \begin{bmatrix}-7\\11\\3\end{bmatrix} - \dfrac{\begin{bmatrix}-7\\11\\3\end{bmatrix} \cdot \begin{bmatrix}-1\\3\\4\end{bmatrix}}{\left\|\begin{bmatrix}-1\\3\\4\end{bmatrix}\right\|^2}\begin{bmatrix}-1\\3\\4\end{bmatrix}$$
$$= \begin{bmatrix}-7\\11\\3\end{bmatrix} - \dfrac{52}{26}\begin{bmatrix}-1\\3\\4\end{bmatrix}$$
$$= \begin{bmatrix}-5\\5\\-5\end{bmatrix}.$$
So the desired orthogonal set is
$$\left\{ \begin{bmatrix}-1\\3\\4\end{bmatrix}, \begin{bmatrix}-5\\5\\-5\end{bmatrix} \right\}.$$

(b) Normalizing the vectors above, we obtain the orthonormal set
$$\left\{ \dfrac{1}{\sqrt{26}}\begin{bmatrix}-1\\3\\4\end{bmatrix}, \dfrac{1}{\sqrt{3}}\begin{bmatrix}-1\\1\\-1\end{bmatrix} \right\}.$$

13. (a) Let
$$\mathbf{u}_1 = \begin{bmatrix}0\\1\\1\\1\end{bmatrix}, \mathbf{u}_2 = \begin{bmatrix}1\\0\\1\\1\end{bmatrix}, \text{ and } \mathbf{u}_3 = \begin{bmatrix}1\\1\\0\\1\end{bmatrix}.$$
Set
$$\mathbf{v}_1 = \mathbf{u}_1,$$
$$\mathbf{v}_2 = \mathbf{u}_2 - \dfrac{\mathbf{u}_2 \cdot \mathbf{v}_1}{\|\mathbf{v}_1\|^2}\mathbf{v}_1$$

$$= \begin{bmatrix} 1 \\ 0 \\ 1 \\ 1 \end{bmatrix} - \frac{2}{3}\begin{bmatrix} 0 \\ 1 \\ 1 \\ 1 \end{bmatrix} = \frac{1}{3}\begin{bmatrix} 3 \\ -2 \\ 1 \\ 1 \end{bmatrix},$$

and

$$\mathbf{v}_3 = \mathbf{u}_3 - \frac{\mathbf{u}_3 \cdot \mathbf{v}_1}{\|\mathbf{v}_1\|^2}\mathbf{v}_1 - \frac{\mathbf{u}_3 \cdot \mathbf{v}_2}{\|\mathbf{v}_2\|^2}\mathbf{v}_2$$

$$= \begin{bmatrix} 1 \\ 1 \\ 0 \\ 1 \end{bmatrix} - \frac{2}{3}\begin{bmatrix} 0 \\ 1 \\ 1 \\ 1 \end{bmatrix} - \left(\frac{\frac{2}{3}}{\frac{5}{3}}\right)\left(\frac{1}{3}\right)\begin{bmatrix} 3 \\ -2 \\ 1 \\ 1 \end{bmatrix}$$

$$= \frac{1}{5}\begin{bmatrix} 3 \\ 3 \\ -4 \\ 1 \end{bmatrix}.$$

Thus

$$\{\mathbf{v}_1, \mathbf{v}_2, \mathbf{v}_2\} = \left\{\begin{bmatrix} 0 \\ 1 \\ 1 \\ 1 \end{bmatrix}, \frac{1}{3}\begin{bmatrix} 3 \\ -2 \\ 1 \\ 1 \end{bmatrix}, \frac{1}{5}\begin{bmatrix} 3 \\ 3 \\ -4 \\ 1 \end{bmatrix}\right\}$$

is the corresponding orthogonal set.

(b) Using the notation in (a), we have

$$\left\{\frac{1}{\|\mathbf{v}_1\|^2}\mathbf{v}_1, \frac{1}{\|\mathbf{v}_2\|^2}\mathbf{v}_2, \frac{1}{\|\mathbf{v}_3\|^2}\mathbf{v}_3\right\} =$$

$$\left\{\frac{1}{\sqrt{3}}\begin{bmatrix} 0 \\ 1 \\ 1 \\ 1 \end{bmatrix}, \frac{1}{\sqrt{15}}\begin{bmatrix} 3 \\ -2 \\ 1 \\ 1 \end{bmatrix}, \frac{1}{\sqrt{35}}\begin{bmatrix} 3 \\ 3 \\ -4 \\ 1 \end{bmatrix}\right\}.$$

14. (a) $\left\{\begin{bmatrix} 1 \\ -1 \\ 0 \\ 2 \end{bmatrix}, \begin{bmatrix} 0 \\ 2 \\ 1 \\ 1 \end{bmatrix}, \frac{1}{3}\begin{bmatrix} 3 \\ 1 \\ -1 \\ -1 \end{bmatrix}\right\}$

(b) $\left\{\frac{1}{\sqrt{6}}\begin{bmatrix} 1 \\ -1 \\ 0 \\ 2 \end{bmatrix}, \frac{1}{\sqrt{6}}\begin{bmatrix} 0 \\ 2 \\ 1 \\ 1 \end{bmatrix}, \frac{1}{\sqrt{12}}\begin{bmatrix} 3 \\ 1 \\ -1 \\ -1 \end{bmatrix}\right\}$

15. (a) $\left\{\begin{bmatrix} 1 \\ 0 \\ -1 \\ 1 \end{bmatrix}, \begin{bmatrix} 1 \\ 1 \\ 0 \\ -1 \end{bmatrix}, \begin{bmatrix} 2 \\ -1 \\ 3 \\ 1 \end{bmatrix}\right\}$

(b) $\left\{\frac{1}{\sqrt{3}}\begin{bmatrix} 1 \\ 0 \\ -1 \\ 1 \end{bmatrix}, \frac{1}{\sqrt{3}}\begin{bmatrix} 1 \\ 1 \\ 0 \\ -1 \end{bmatrix}, \frac{1}{\sqrt{15}}\begin{bmatrix} 2 \\ -1 \\ 3 \\ 1 \end{bmatrix}\right\}$

16. (a) $\left\{\begin{bmatrix} 1 \\ -1 \\ 0 \\ 1 \\ 1 \end{bmatrix}, \begin{bmatrix} 0 \\ 1 \\ 0 \\ 1 \\ 0 \end{bmatrix}, \begin{bmatrix} 0 \\ 0 \\ 1 \\ 0 \\ 0 \end{bmatrix}, \begin{bmatrix} 2 \\ 1 \\ 0 \\ -1 \\ 0 \end{bmatrix}\right\}$

(b) $\left\{\frac{1}{2}\begin{bmatrix} 1 \\ -1 \\ 0 \\ 1 \\ 1 \end{bmatrix}, \frac{1}{\sqrt{2}}\begin{bmatrix} 0 \\ 1 \\ 0 \\ 1 \\ 0 \end{bmatrix}, \begin{bmatrix} 0 \\ 0 \\ 1 \\ 0 \\ 0 \end{bmatrix}, \frac{1}{\sqrt{6}}\begin{bmatrix} 2 \\ 1 \\ 0 \\ -1 \\ 0 \end{bmatrix}\right\}$

17. We have

$$\mathbf{v} = \frac{\begin{bmatrix} 1 \\ 8 \end{bmatrix} \cdot \begin{bmatrix} 2 \\ 1 \end{bmatrix}}{\left\|\begin{bmatrix} 2 \\ 1 \end{bmatrix}\right\|^2}\begin{bmatrix} 2 \\ 1 \end{bmatrix} + \frac{\begin{bmatrix} 1 \\ 8 \end{bmatrix} \cdot \begin{bmatrix} -1 \\ 2 \end{bmatrix}}{\left\|\begin{bmatrix} -1 \\ 2 \end{bmatrix}\right\|^2}\begin{bmatrix} -1 \\ 2 \end{bmatrix}$$

$$= \frac{10}{5}\begin{bmatrix} 2 \\ 1 \end{bmatrix} + \frac{15}{5}\begin{bmatrix} -1 \\ 2 \end{bmatrix}$$

$$= 2\begin{bmatrix} 2 \\ 1 \end{bmatrix} + 3\begin{bmatrix} -1 \\ 2 \end{bmatrix}.$$

18. $\mathbf{v} = 2\begin{bmatrix} 2 \\ 1 \end{bmatrix} + 3\begin{bmatrix} -1 \\ 2 \end{bmatrix}$

19. $\mathbf{v} = (-1)\begin{bmatrix} -1 \\ 3 \\ -2 \end{bmatrix} + (-2)\begin{bmatrix} -1 \\ 1 \\ 2 \end{bmatrix} + 4\begin{bmatrix} 1 \\ 1 \\ 1 \end{bmatrix}$

20. $\begin{bmatrix} 2 \\ 1 \\ 6 \end{bmatrix} = 3\begin{bmatrix} 1 \\ 1 \\ 1 \end{bmatrix} + (-1)\begin{bmatrix} 1 \\ 2 \\ -3 \end{bmatrix}$

21. $\mathbf{v} = \frac{5}{2}\begin{bmatrix} 1 \\ 0 \\ 1 \end{bmatrix} + \frac{3}{6}\begin{bmatrix} 1 \\ 2 \\ -1 \end{bmatrix} + 0\begin{bmatrix} 1 \\ -1 \\ -1 \end{bmatrix}$

22. $\mathbf{v} = -2\begin{bmatrix} 1 \\ -2 \\ 0 \\ -1 \end{bmatrix} + 5\begin{bmatrix} 2 \\ 1 \\ 1 \\ 0 \end{bmatrix} - 2\begin{bmatrix} 1 \\ 0 \\ -2 \\ 1 \end{bmatrix}$

23. $\mathbf{v} = (-4)\begin{bmatrix} 1 \\ -1 \\ -1 \\ 1 \end{bmatrix} + 2\begin{bmatrix} 2 \\ 1 \\ 1 \\ 0 \end{bmatrix} + (-1)\begin{bmatrix} -1 \\ 1 \\ 1 \\ 3 \end{bmatrix}$

24. $\mathbf{v} = \begin{bmatrix} 1 \\ 1 \\ 1 \\ 1 \end{bmatrix} - \frac{1}{2}\begin{bmatrix} 1 \\ -1 \\ 1 \\ -1 \end{bmatrix} + \begin{bmatrix} 1 \\ -1 \\ -1 \\ 1 \end{bmatrix} + \frac{1}{2}\begin{bmatrix} 1 \\ 1 \\ -1 \\ -1 \end{bmatrix}$

25. $Q = \begin{bmatrix} \frac{1}{\sqrt{3}} & \frac{1}{\sqrt{2}} \\ \frac{1}{\sqrt{3}} & -\frac{1}{\sqrt{2}} \\ \frac{1}{\sqrt{3}} & 0 \end{bmatrix}$ and $R = \begin{bmatrix} \sqrt{3} & 2\sqrt{3} \\ 0 & 3\sqrt{2} \end{bmatrix}$

26. (a) We use the solution to Exercise 10 and model the approach with the same notation used in Example 4. Let

$$\mathbf{w}_1 = \frac{1}{\sqrt{6}} \begin{bmatrix} 1 \\ -2 \\ 1 \end{bmatrix} \text{ and } \mathbf{w}_2 = \frac{1}{\sqrt{2}} \begin{bmatrix} 1 \\ 0 \\ -1 \end{bmatrix}.$$

So

$$Q = \begin{bmatrix} \frac{1}{\sqrt{6}} & \frac{1}{\sqrt{2}} \\ -\frac{2}{\sqrt{6}} & 0 \\ \frac{1}{\sqrt{6}} & -\frac{1}{\sqrt{2}} \end{bmatrix}.$$

The entries of R are computed as

$$r_{11} = \mathbf{a}_1 \cdot \mathbf{w}_1 = \sqrt{6}$$
$$r_{12} = \mathbf{a}_2 \cdot \mathbf{w}_1 = \frac{3}{\sqrt{6}}$$
$$r_{22} = \mathbf{a}_2 \cdot \mathbf{w}_2 = \frac{1}{\sqrt{2}}.$$

Hence, $R = \begin{bmatrix} \sqrt{6} & \frac{3}{\sqrt{6}} \\ 0 & \frac{1}{\sqrt{2}} \end{bmatrix}.$

27. $Q = \begin{bmatrix} \frac{1}{\sqrt{6}} & \frac{3}{\sqrt{11}} \\ -\frac{2}{\sqrt{6}} & \frac{1}{\sqrt{11}} \\ -\frac{1}{\sqrt{6}} & \frac{1}{\sqrt{11}} \end{bmatrix}, R = \begin{bmatrix} \sqrt{6} & -2\sqrt{6} \\ 0 & 3\sqrt{11} \end{bmatrix}$

28. $Q = \begin{bmatrix} -\frac{1}{\sqrt{26}} & -\frac{1}{\sqrt{3}} \\ \frac{3}{\sqrt{26}} & \frac{1}{\sqrt{3}} \\ \frac{4}{\sqrt{26}} & -\frac{1}{\sqrt{3}} \end{bmatrix}, R = \begin{bmatrix} \sqrt{26} & 2\sqrt{26} \\ 0 & 5\sqrt{3} \end{bmatrix}$

29.
$$Q = \begin{bmatrix} 0 & \frac{3}{\sqrt{15}} & \frac{3}{\sqrt{35}} \\ \frac{1}{\sqrt{3}} & -\frac{2}{\sqrt{15}} & \frac{3}{\sqrt{35}} \\ \frac{1}{\sqrt{3}} & \frac{1}{\sqrt{15}} & -\frac{4}{\sqrt{35}} \\ \frac{1}{\sqrt{3}} & \frac{1}{\sqrt{15}} & \frac{1}{\sqrt{35}} \end{bmatrix},$$

$$R = \begin{bmatrix} \sqrt{3} & \frac{2}{\sqrt{3}} & \frac{2}{\sqrt{3}} \\ 0 & \frac{\sqrt{15}}{3} & \frac{2}{\sqrt{15}} \\ 0 & 0 & \frac{7}{\sqrt{35}} \end{bmatrix}$$

30.
$$Q = \begin{bmatrix} \frac{1}{\sqrt{6}} & 0 & \frac{3}{\sqrt{12}} \\ -\frac{1}{\sqrt{6}} & \frac{2}{\sqrt{6}} & \frac{1}{\sqrt{12}} \\ 0 & \frac{1}{\sqrt{6}} & -\frac{1}{\sqrt{12}} \\ \frac{2}{\sqrt{6}} & \frac{1}{\sqrt{6}} & -\frac{1}{\sqrt{12}} \end{bmatrix},$$

$$R = \begin{bmatrix} \sqrt{6} & \sqrt{6} & 2\sqrt{6} \\ 0 & \sqrt{6} & \frac{8}{\sqrt{6}} \\ 0 & 0 & \frac{4}{\sqrt{12}} \end{bmatrix}$$

31.
$$Q = \begin{bmatrix} \frac{1}{\sqrt{3}} & \frac{1}{\sqrt{3}} & \frac{2}{\sqrt{15}} \\ 0 & \frac{1}{\sqrt{3}} & -\frac{1}{\sqrt{15}} \\ -\frac{1}{\sqrt{3}} & 0 & \frac{3}{\sqrt{15}} \\ \frac{1}{\sqrt{3}} & -\frac{1}{\sqrt{3}} & \frac{1}{\sqrt{15}} \end{bmatrix},$$

$$R = \begin{bmatrix} \sqrt{3} & \sqrt{3} & 2\sqrt{3} \\ 0 & \sqrt{3} & -\frac{2}{\sqrt{3}} \\ 0 & 0 & \frac{5}{\sqrt{15}} \end{bmatrix}$$

32.
$$Q = \begin{bmatrix} \frac{1}{2} & 0 & 0 & \frac{2}{\sqrt{6}} \\ -\frac{1}{2} & \frac{1}{\sqrt{2}} & 0 & \frac{1}{\sqrt{6}} \\ 0 & 0 & 1 & 0 \\ \frac{1}{2} & \frac{1}{\sqrt{2}} & 0 & -\frac{1}{\sqrt{6}} \\ \frac{1}{2} & 0 & 0 & 0 \end{bmatrix},$$

$$R = \begin{bmatrix} 2 & 4 & 2 & 2 \\ 0 & \sqrt{2} & 0 & \sqrt{2} \\ 0 & 0 & 1 & 0 \\ 0 & 0 & 0 & \sqrt{6} \end{bmatrix}$$

33. $\begin{bmatrix} 2 \\ -1 \end{bmatrix}$

34. We use the technique in Example 5 to solve $A\mathbf{x} = \mathbf{b}$, where

$$A = \begin{bmatrix} 1 & 1 \\ -2 & -1 \\ 1 & 0 \end{bmatrix} \quad \text{and} \quad \mathbf{b} = \begin{bmatrix} 6 \\ -8 \\ 2 \end{bmatrix}.$$

In Exercise 26, we found the matrices Q and R in a QR factorization of A.

$$Q = \begin{bmatrix} \frac{1}{\sqrt{6}} & \frac{1}{\sqrt{2}} \\ -\frac{2}{\sqrt{6}} & 0 \\ \frac{1}{\sqrt{6}} & -\frac{1}{\sqrt{2}} \end{bmatrix} \text{ and } R = \begin{bmatrix} \sqrt{6} & \frac{3}{\sqrt{6}} \\ 0 & \frac{1}{\sqrt{2}} \end{bmatrix}.$$

We must solve the equivalent system $R\mathbf{x} = Q^T\mathbf{b}$, which has the form

$$\begin{array}{ll} \sqrt{6}x_1 + \frac{\sqrt{6}}{2}x_2 = 4\sqrt{6} & x_1 + \frac{1}{2}x_2 = 4 \\ \frac{1}{\sqrt{2}}x_2 = 2\sqrt{2} & \text{or} \quad \frac{1}{2}x_2 = 2 \end{array}$$

The second system is easily solved: $x_2 = 4$ and $x_1 = 4 - 2 = 2$. In vector form, the solution is

$$\mathbf{x} = \begin{bmatrix} 2 \\ 4 \end{bmatrix}.$$

144 Chapter 6 Orthogonality

35. $\begin{bmatrix} 3 \\ -2 \end{bmatrix}$ 36. $\begin{bmatrix} 1 \\ -2 \end{bmatrix}$ 37. $\begin{bmatrix} -2 \\ 1 \\ 3 \end{bmatrix}$

38. $\begin{bmatrix} 1 \\ -2 \\ 3 \end{bmatrix}$ 39. $\begin{bmatrix} 2 \\ -4 \\ 3 \end{bmatrix}$ 40. $\begin{bmatrix} 2 \\ 0 \\ 3 \\ 1 \end{bmatrix}$

41. False, if **0** lies in the set, then the set is linearly dependent.
42. True   43. True   44. True   45. True
46. True   47. True   48. True
49. False, consider the sets $\{\mathbf{e}_1\}$ and $\{-\mathbf{e}_1\}$. The combined set is $\{\mathbf{e}_1, -\mathbf{e}_1\}$, which is not orthonormal.
50. False, consider $\mathbf{x} = \mathbf{e}_1$, $\mathbf{y} = \mathbf{0}$, and $\mathbf{z} = \mathbf{e}_1$.
51. True
52. False, in Example 4, Q is not upper triangular.
53. For any $i \ne j$,
$$(c_i\mathbf{v}_i) \cdot (c_j\mathbf{v}_j) = (c_ic_j)(\mathbf{v}_i \cdot \mathbf{v}_j) = (c_ic_j) \cdot 0 = 0.$$
Hence $c_i\mathbf{v}_i$ and $c_j\mathbf{v}_j$ are orthogonal.
54. Suppose that $\mathcal{S} = \{\mathbf{u}_1, \mathbf{u}_2, \ldots, \mathbf{u}_k\}$ and $\mathcal{S}' = \{\mathbf{v}_1, \mathbf{v}_2, \ldots, \mathbf{v}_k\}$. Clearly $\mathbf{v}_1 = \mathbf{u}_1$. Now assume that $\mathbf{v}_j = \mathbf{u}_j$, for $j = 1, \ldots, i-1$, where i is a fixed integer, $1 < i \le k$. Then
$$\mathbf{v}_i = \mathbf{u}_i - \frac{\mathbf{u}_i \cdot \mathbf{v}_1}{\|\mathbf{v}_1\|^2}\mathbf{v}_1 - \frac{\mathbf{u}_i \cdot \mathbf{v}_2}{\|\mathbf{v}_2\|^2}\mathbf{v}_2 -$$
$$\cdots - \frac{\mathbf{u}_i \cdot \mathbf{v}_{i-1}}{\|\mathbf{v}_{i-1}\|^2}\mathbf{v}_{i-1}$$
$$= \mathbf{u}_i - 0 - 0 - \cdots - 0$$
$$= \mathbf{u}_i.$$
It follows that $\mathbf{v}_i = \mathbf{u}_i$ for all i, and so $\mathcal{S}' = \mathcal{S}$.
55. (a) We have
$$\mathbf{u} + \mathbf{v} = (\mathbf{u} \cdot \mathbf{w}_1)\mathbf{w}_1 + \cdots + (\mathbf{u} \cdot \mathbf{w}_n)\mathbf{w}_n$$
$$+ (\mathbf{v} \cdot \mathbf{w}_1)\mathbf{w}_1 + \cdots + (\mathbf{v} \cdot \mathbf{w}_n)\mathbf{w}_n$$
$$= (\mathbf{u} \cdot \mathbf{w}_1 + \mathbf{v} \cdot \mathbf{w}_1)\mathbf{w}_1 + \cdots$$
$$+ (\mathbf{u} \cdot \mathbf{w}_n + \mathbf{v} \cdot \mathbf{w}_n)\mathbf{w}_n.$$

(b) We have
$$\mathbf{u} \cdot \mathbf{v} = [(\mathbf{u} \cdot \mathbf{w}_1)\mathbf{w}_1 + \cdots + (\mathbf{u} \cdot \mathbf{w}_n)\mathbf{w}_n] \cdot$$
$$[(\mathbf{v} \cdot \mathbf{w}_1)\mathbf{w}_1 + \cdots + (\mathbf{v} \cdot \mathbf{w}_n)\mathbf{w}_n]$$
$$= (\mathbf{u} \cdot \mathbf{w}_1)(\mathbf{v} \cdot \mathbf{w}_1)\mathbf{w}_1 \cdot \mathbf{w}_1 + \cdots$$
$$+ (\mathbf{u} \cdot \mathbf{w}_n)(\mathbf{v} \cdot \mathbf{w}_n)\mathbf{w}_n \cdot \mathbf{w}_n$$
$$= (\mathbf{u} \cdot \mathbf{w}_1)(\mathbf{v} \cdot \mathbf{w}_1) + \cdots$$
$$+ (\mathbf{u} \cdot \mathbf{w}_n)(\mathbf{v} \cdot \mathbf{w}_n).$$

(c) This follows from (b) by setting $\mathbf{v}$ equal $\mathbf{u}$.

56. By the Extension Theorem, $\{\mathbf{v}_1, \mathbf{v}_2, \ldots, \mathbf{v}_k\}$ can be extended to a basis
$$\{\mathbf{v}_1, \mathbf{v}_2, \ldots, \mathbf{v}_k, \mathbf{u}_{k+1}, \ldots, \mathbf{u}_n\}$$
for $\mathcal{R}^n$. Applying the Gram-Schmidt process to this basis gives an orthogonal basis
$$\{\mathbf{v}_1, \mathbf{v}_2, \ldots, \mathbf{v}_k, \mathbf{w}_{k+1}, \ldots, \mathbf{w}_n\}.$$
(Note that by Exercise 54, the first k vectors in this new basis remain unchanged when applying the Gram-Schmidt process.) Finally, we replace each $\mathbf{w}_i$, $k+1 \le i \le n$, by $\mathbf{u}_i = \frac{1}{\|\mathbf{w}_i\|}\mathbf{w}_i$ to obtain the desired orthonormal basis for $\mathcal{R}^n$.

57. By Exercise 56, $\mathcal{S}$ can be extended to an orthonormal basis
$$\{\mathbf{v}_1, \mathbf{v}_2, \ldots, \mathbf{v}_k, \mathbf{v}_{k+1}, \ldots, \mathbf{v}_n\}$$
for $\mathcal{R}^n$. By Exercise 55(c),
$$\|\mathbf{u}\|^2 = (\mathbf{u} \cdot \mathbf{v}_1)^2 + (\mathbf{u} \cdot \mathbf{v}_2)^2 + \cdots + (\mathbf{u} \cdot \mathbf{v}_k)^2$$
$$+ (\mathbf{u} \cdot \mathbf{v}_{k+1})^2 + \cdots + (\mathbf{u} \cdot \mathbf{v}_n)^2.$$

(a) The desired inequality follows immediately from the equation above since
$$(\mathbf{u} \cdot \mathbf{v}_{k+1})^2 + \cdots + (\mathbf{u} \cdot \mathbf{v}_n)^2 \ge 0.$$

(b) The inequality in (a) is an equality if and only if
$$(\mathbf{u} \cdot \mathbf{v}_{k+1})^2 + \cdots + (\mathbf{u} \cdot \mathbf{v}_n)^2 = 0,$$
which is true if and only if $\mathbf{u} \cdot \mathbf{v}_i = 0$ for $i > k$. In this case,
$$\mathbf{u} = (\mathbf{u} \cdot \mathbf{v}_1)\mathbf{v}_1 + (\mathbf{u} \cdot \mathbf{v}_2)\mathbf{v}_2 + \cdots + (\mathbf{u} \cdot \mathbf{v}_k)\mathbf{v}_k,$$
which is true if and only if $\mathbf{u}$ is in Span $\mathcal{S}$.

58. Write $Q = \begin{bmatrix} a & b & c \\ 0 & d & e \\ 0 & 0 & f \end{bmatrix}$. Now $1 = \|\mathbf{q}_1\| = |a|$ implies that $a \ne 0$. So $0 = \mathbf{q}_1 \cdot \mathbf{q}_2 = ab$ implies $b = 0$. Also, $0 = \mathbf{q}_1 \cdot \mathbf{q}_3 = ac$ implies $c = 0$. Thus $Q = \begin{bmatrix} a & 0 & 0 \\ 0 & d & e \\ 0 & 0 & f \end{bmatrix}$. Using arguments similar to the above, $1 = \|\mathbf{q}_2\| = |d|$ implies $d \ne 0$. Therefore $\mathbf{q}_2 \cdot \mathbf{q}_3 = de$ implies $e = 0$. So $Q = \begin{bmatrix} a & 0 & 0 \\ 0 & d & 0 \\ 0 & 0 & f \end{bmatrix}$, a diagonal matrix.

59. Let A by an $m \times n$ matrix with linearly independent columns and a QR factorization $A = QR$. By Exercise 62 of Section 2.3,

$$n = \operatorname{rank} A = \operatorname{rank} QR \leq \operatorname{rank} R \leq n.$$

So $\operatorname{rank} A = n$, and therefore R is invertible by parts (a) and (c) of Theorem 2.6.

60. By Exercise 59, R is invertible; so $\det R \neq 0$. But, by Theorem 3.2, $\det R$ is a product of the diagonal entries of R, and therefore all the diagonal entries of R are nonzero.

61. From Exercise 60, we have $r_{ii} \neq 0$ for every i. If $r_{ii} < 0$, then replacing $\mathbf{q}_i$ by $-\mathbf{q}_i$ changes the corresponding entry of R to $-r_{ii}$, which is positive.

62. By Exercise 54, applying the Gram-Schmidt process to an orthonormal set leaves it unchanged. So $Q = A$, and hence $A = QR = AR$. Thus $A(I_n - R) = O$. For each j, the jth column of $A(I_n - R)$ is a linear combination of the columns of A, where the coefficients are the entries of the jth column of $I_n - R$. Since the columns of A are linearly independent, it follows that the jth column of $I_n - R$ is the zero vector, and hence $I_n - R = O$, that is, $R = I_n$.

63. First observe that for all i, j, the (i, j)-entry of $Q^T Q$ is $\mathbf{q}_i \cdot \mathbf{q}_j$. Therefore the columns of Q form an orthonormal set (and hence, an orthonormal basis for $\mathcal{R}^n$) if and only if $Q^T Q = I_n$.

64. By Exercise 63, we have

$$(PQ)^T (PQ) = (Q^T P^T)(PQ)$$
$$= Q^T (P^T P) Q = Q^T I_n Q$$
$$= Q^T Q = I_n.$$

Using Exercise 63 again, we conclude that the columns of PQ also form an orthonormal basis for $\mathcal{R}^n$.

65. Suppose $QR = Q'R'$, where both R and R' have positive diagonal entries. Multiplying both sides on the left by Q^T and on the right by R'^{-1}, we obtain

$$Q^T Q R R'^{-1} = Q^T Q' R' R'^{-1},$$

which by Exercise 63, reduces to $RR'^{-1} = Q^T Q'$. By Exercises 42 and 43 of Section 2.6, RR'^{-1}, and hence $Q^T Q'$, is an upper triangular matrix with positive diagonal entries. By Exercise 64, the columns of $Q^T Q'$ form an orthonormal basis. Hence, by Exercise 58, $Q^T Q'$ is a diagonal matrix. But a diagonal matrix with positive diagonal entries and whose columns are unit vectors must equal I_3. So $RR'^{-1} = Q^T Q' = I_3$, and therefore $Q = Q'$ and $R = R'$.

66. Suppose that A has n columns. Given a QR factorization $A = QR$, observe that for each j, the jth column of A is a linear combination of the columns of Q in which the coefficients are the entries of the jth column of R. Hence the columns of A are contained in $\operatorname{Col} Q$. It follows that $\operatorname{Col} A$ is a subspace of $\operatorname{Col} Q$. Since the dimensions of each of these spaces is n, it follows that $\operatorname{Col} A = \operatorname{Col} Q$. Finally, observe that the columns of Q form an orthonormal basis for $\operatorname{Col} Q$, and hence they form an orthonormal basis for $\operatorname{Col} A$.

67. (a) $\operatorname{rank} A = 3$
(b)

$$Q = \begin{bmatrix} -0.3172 & -0.4413 & -0.5587 \\ 0.2633 & -0.4490 & -0.2951 \\ 0.7182 & -0.4040 & -0.0570 \\ -0.5386 & -0.5875 & 0.3130 \\ -0.1556 & 0.3087 & -0.7068 \end{bmatrix},$$

$$R = \begin{bmatrix} -16.7096 & 6.4460 & 6.3700 \\ 0.0000 & -20.7198 & -3.7958 \\ 0.0000 & 0.0000 & -15.6523 \end{bmatrix}$$

68. (a) $\operatorname{rank} A = 4$
(b) $Q =$

$$\begin{bmatrix} -0.1191 & 0.2063 & 0.8244 & 0.1292 \\ 0.0000 & -0.4622 & 0.5221 & 0.1226 \\ -0.5359 & -0.7369 & -0.1320 & 0.0273 \\ 0.3156 & -0.2295 & 0.1707 & -0.8967 \\ 0.7740 & -0.3849 & -0.0341 & 0.4044 \end{bmatrix},$$

$R =$

$$\begin{bmatrix} -16.7955 & 1.1908 & 9.7824 & -0.9824 \\ 0.0000 & -15.7934 & -1.8825 & -11.5991 \\ 0.0000 & 0.0000 & 8.6765 & 6.8708 \\ 0.0000 & 0.0000 & 0.0000 & -6.1063 \end{bmatrix}$$

6.3 ORTHOGONAL PROJECTIONS

1. $\mathbf{v}$ is in $\mathcal{S}^\perp$ if and only if

$$\mathbf{v} \cdot \begin{bmatrix} 1 \\ -1 \\ 2 \end{bmatrix} = v_1 - v_2 + 2v_3 = 0.$$

A basis for the solution set of this system is

$$\left\{ \begin{bmatrix} 1 \\ 1 \\ 0 \end{bmatrix}, \begin{bmatrix} -2 \\ 0 \\ 1 \end{bmatrix} \right\}.$$

2. $\left\{ \begin{bmatrix} 0 \\ 1 \\ 0 \end{bmatrix}, \begin{bmatrix} -2 \\ 0 \\ 1 \end{bmatrix} \right\}$
3. $\left\{ \begin{bmatrix} 1 \\ 1 \\ -1 \end{bmatrix} \right\}$
4. $\left\{ \begin{bmatrix} 0 \\ -1 \\ 1 \end{bmatrix} \right\}$

5. $\left\{ \begin{bmatrix} -5 \\ -2 \\ 1 \\ 0 \end{bmatrix}, \begin{bmatrix} -3 \\ -1 \\ 0 \\ 1 \end{bmatrix} \right\}$
6. $\left\{ \begin{bmatrix} 2 \\ -3 \\ 1 \\ 0 \end{bmatrix}, \begin{bmatrix} -1 \\ -2 \\ 0 \\ 1 \end{bmatrix} \right\}$

7. $\left\{ \begin{bmatrix} 1 \\ -2 \\ 1 \\ 0 \end{bmatrix}, \begin{bmatrix} -3 \\ 1 \\ 0 \\ 1 \end{bmatrix} \right\}$
8. $\left\{ \begin{bmatrix} -1 \\ 1 \\ 1 \\ 1 \end{bmatrix} \right\}$

9. Use the method of Exercise 10.
   (a) $\mathbf{w} = \begin{bmatrix} -1 \\ 1 \end{bmatrix}$ and $\mathbf{z} = \begin{bmatrix} 2 \\ 2 \end{bmatrix}$
   (b) $\begin{bmatrix} -1 \\ 1 \end{bmatrix}$ (c) $\sqrt{8}$

10. (a) Using the method and notation of Example 3, we have

$$\mathbf{w} = (\mathbf{u} \cdot \mathbf{v}_1)\mathbf{v}_1 + (\mathbf{u} \cdot \mathbf{v}_2)\mathbf{v}_2$$

$$= \left(\begin{bmatrix} 2 \\ 3 \\ -1 \end{bmatrix} \cdot \frac{1}{\sqrt{2}} \begin{bmatrix} 1 \\ 1 \\ 0 \end{bmatrix} \right) \frac{1}{\sqrt{2}} \begin{bmatrix} 1 \\ 1 \\ 0 \end{bmatrix} +$$

$$\left(\begin{bmatrix} 2 \\ 3 \\ -1 \end{bmatrix} \cdot \frac{1}{\sqrt{3}} \begin{bmatrix} 1 \\ -1 \\ 1 \end{bmatrix} \right) \frac{1}{\sqrt{3}} \begin{bmatrix} 1 \\ -1 \\ 1 \end{bmatrix}$$

$$= \frac{5}{2} \begin{bmatrix} 1 \\ 1 \\ 0 \end{bmatrix} - \frac{2}{3} \begin{bmatrix} 1 \\ -1 \\ 1 \end{bmatrix} = \frac{1}{6} \begin{bmatrix} 11 \\ 19 \\ -4 \end{bmatrix}.$$

So

$$\mathbf{z} = \mathbf{u} - \mathbf{w} = \begin{bmatrix} 2 \\ 3 \\ -1 \end{bmatrix} - \frac{1}{6} \begin{bmatrix} 11 \\ 19 \\ -4 \end{bmatrix} = \frac{1}{6} \begin{bmatrix} 1 \\ -1 \\ -2 \end{bmatrix}.$$

   (b) $\frac{1}{6} \begin{bmatrix} 11 \\ 19 \\ -4 \end{bmatrix}$
   (c) The distance is

$$d = \|\mathbf{z}\| = \left\| \frac{1}{6} \begin{bmatrix} 1 \\ -1 \\ -2 \end{bmatrix} \right\| = \sqrt{\frac{1}{6}} = \frac{\sqrt{6}}{6}.$$

11. Use the method of Exercise 10.
    (a) $\mathbf{w} = \mathbf{u}$ and $\mathbf{z} = \mathbf{0}$
    (b) $\begin{bmatrix} 1 \\ 4 \\ -1 \end{bmatrix}$ (c) 0

12. Use the method of Exercise 10.
    (a) $\mathbf{w} = \begin{bmatrix} 1 \\ -1 \\ 0 \end{bmatrix}$ and $\mathbf{z} = \begin{bmatrix} 2 \\ 2 \\ 1 \end{bmatrix}$
    (b) $\begin{bmatrix} 1 \\ -1 \\ 0 \end{bmatrix}$ (c) 3

13. Use the method of Exercise 10.
    (a) $\mathbf{w} = \begin{bmatrix} 2 \\ 2 \\ 3 \\ 1 \end{bmatrix}$ and $\mathbf{z} = \begin{bmatrix} 0 \\ 2 \\ -2 \\ 2 \end{bmatrix}$
    (b) $\begin{bmatrix} 2 \\ 2 \\ 3 \\ 1 \end{bmatrix}$ (c) $\sqrt{12}$

14. Use the method of Exercise 10.
    (a) $\mathbf{w} = \begin{bmatrix} 1 \\ -3 \\ 3 \\ -1 \end{bmatrix}$ and $\mathbf{z} = \begin{bmatrix} 2 \\ 1 \\ 1 \\ 2 \end{bmatrix}$
    (b) $\begin{bmatrix} 1 \\ -3 \\ 3 \\ -1 \end{bmatrix}$ (c) $\sqrt{10}$

15. Use the method of Exercise 10.
    (a) $\mathbf{w} = \begin{bmatrix} 3 \\ 4 \\ -2 \\ 3 \end{bmatrix}$ and $\mathbf{z} = \begin{bmatrix} -3 \\ 1 \\ -1 \\ 1 \end{bmatrix}$
    (b) $\begin{bmatrix} 3 \\ 4 \\ -2 \\ 3 \end{bmatrix}$ (c) $\sqrt{12}$

16. Use the method of Exercise 10.
    (a) $\mathbf{w} = \begin{bmatrix} 1 \\ -3 \\ 3 \\ -5 \end{bmatrix}$ and $\mathbf{z} = \begin{bmatrix} 2 \\ 2 \\ 2 \\ 2 \end{bmatrix}$
    (b) $\begin{bmatrix} 1 \\ -3 \\ 3 \\ -5 \end{bmatrix}$ (c) 4

17. The method is similar to that of Exercise 19.
    (a) $P_W = \frac{1}{25} \begin{bmatrix} 9 & -12 \\ -12 & 16 \end{bmatrix}$
    (b) $\mathbf{w} = \begin{bmatrix} -6 \\ 8 \end{bmatrix}, \mathbf{z} = \begin{bmatrix} -4 \\ -3 \end{bmatrix}$ (c) 5

18. The method is similar to that of Exercise 19.

(a) $P_W = \dfrac{1}{14}\begin{bmatrix} 13 & 2 & -3 \\ 2 & 10 & 6 \\ -3 & 6 & 5 \end{bmatrix}$

(b) $\mathbf{w} = \dfrac{1}{7}\begin{bmatrix} -1 \\ 37 \\ 25 \end{bmatrix}$, $\mathbf{z} = \dfrac{1}{7}\begin{bmatrix} 8 \\ -16 \\ 24 \end{bmatrix}$

(c) $\dfrac{8\sqrt{14}}{7}$

19. (a) By solving the system, we obtain a basis $\left\{\begin{bmatrix} -1 \\ 2 \\ 1 \end{bmatrix}\right\}$ for the solution set. Let $C = \begin{bmatrix} -1 \\ 2 \\ 1 \end{bmatrix}$. The desired matrix is

$$P_W = C(C^T C)^{-1} C^T$$
$$= \dfrac{1}{6}\begin{bmatrix} 1 & -2 & -1 \\ -2 & 4 & 2 \\ -1 & 2 & 1 \end{bmatrix}.$$

(b) $\mathbf{w} = P_W \mathbf{u} = \dfrac{1}{3}\begin{bmatrix} -1 \\ 2 \\ 1 \end{bmatrix}$,

$\mathbf{z} = \mathbf{u} - \mathbf{w} = \dfrac{4}{3}\begin{bmatrix} 1 \\ 1 \\ -1 \end{bmatrix}$.

(c) The distance equals $\|\mathbf{z}\| = \dfrac{4}{3}\sqrt{3}$.

20. The method is similar to that of Exercise 19.

(a) $P_W = \dfrac{1}{66}\begin{bmatrix} 17 & 7 & 28 \\ 7 & 65 & -4 \\ 28 & -4 & 50 \end{bmatrix}$

(b) $\mathbf{w} = \begin{bmatrix} 1 \\ 3 \\ 1 \end{bmatrix}$, $\mathbf{z} = \begin{bmatrix} -7 \\ 1 \\ 4 \end{bmatrix}$ (c) $\sqrt{66}$

21. The method is similar to that of Exercise 19. Note that the given 4×3 matrix has rank 2.

(a) $P_W = \dfrac{1}{33}\begin{bmatrix} 22 & 11 & 0 & 11 \\ 11 & 19 & 9 & -8 \\ 0 & 9 & 6 & -9 \\ 11 & -8 & -9 & 19 \end{bmatrix}$

(b) $\mathbf{w} = \begin{bmatrix} 3 \\ 0 \\ -1 \\ 3 \end{bmatrix}$, $\mathbf{z} = \begin{bmatrix} -2 \\ 1 \\ 3 \\ 3 \end{bmatrix}$ (c) $\sqrt{23}$

22. The method is similar to that of Exercise 19.

(a) $P_W = \dfrac{1}{7}\begin{bmatrix} 6 & 1 & -2 & -1 \\ 1 & 6 & 2 & 1 \\ -2 & 2 & 3 & -2 \\ -1 & 1 & -2 & 6 \end{bmatrix}$

(b) $\mathbf{w} = \begin{bmatrix} -2 \\ 6 \\ 1 \\ 6 \end{bmatrix}$, $\mathbf{z} = \begin{bmatrix} -1 \\ 1 \\ -2 \\ -1 \end{bmatrix}$ (c) $\sqrt{7}$

23. The method is similar to that of Exercise 19.

(a) $P_W = \dfrac{1}{12}\begin{bmatrix} 11 & 1 & -3 & 1 \\ 1 & 11 & 3 & -1 \\ -3 & 3 & 3 & 3 \\ 1 & -1 & 3 & 11 \end{bmatrix}$

(b) $\mathbf{w} = \begin{bmatrix} 3 \\ -1 \\ 0 \\ 4 \end{bmatrix}$, $\mathbf{z} = \begin{bmatrix} -1 \\ 1 \\ -3 \\ 1 \end{bmatrix}$ (c) $\sqrt{12}$

24. The method is similar to that of Exercise 19.

(a) $P_W = \dfrac{1}{77}\begin{bmatrix} 11 & 22 & 11 & -11 \\ 22 & 51 & 29 & -1 \\ 11 & 29 & 18 & 10 \\ -11 & -1 & 10 & 74 \end{bmatrix}$

(b) $\mathbf{w} = \begin{bmatrix} 2 \\ 5 \\ 3 \\ 1 \end{bmatrix}$, $\mathbf{z} = \begin{bmatrix} 5 \\ -1 \\ -2 \\ 1 \end{bmatrix}$ (c) $\sqrt{31}$

25. The method is similar to that of Exercise 19.

(a) $P_W = \dfrac{1}{6}\begin{bmatrix} 5 & -2 & 1 \\ -2 & 2 & 2 \\ 1 & 2 & 5 \end{bmatrix}$

(b) $\mathbf{w} = \begin{bmatrix} 2 \\ -1 \\ 0 \end{bmatrix}$, $\mathbf{z} = \begin{bmatrix} 1 \\ 2 \\ -1 \end{bmatrix}$ (c) $\sqrt{6}$

26. The method is similar to that of Exercise 19.

(a) $P_W = \dfrac{1}{10}\begin{bmatrix} 9 & 0 & 3 \\ 0 & 0 & 0 \\ 3 & 0 & 1 \end{bmatrix}$

(b) $\mathbf{w} = \dfrac{1}{10}\begin{bmatrix} 3 \\ 0 \\ 1 \end{bmatrix}$, $\mathbf{z} = \dfrac{1}{10}\begin{bmatrix} 7 \\ 30 \\ -21 \end{bmatrix}$

(c) $\dfrac{\sqrt{1390}}{10}$

27. The method is similar to that of Exercise 19.

(a) $P_W = \dfrac{1}{42}\begin{bmatrix} 25 & -20 & 5 \\ -20 & 16 & -4 \\ 5 & -4 & 1 \end{bmatrix}$

(b) $\mathbf{w} = \begin{bmatrix} 5 \\ -4 \\ 1 \end{bmatrix}$, $\mathbf{z} = \begin{bmatrix} 3 \\ 4 \\ 1 \end{bmatrix}$ (c) $\sqrt{26}$

28. The method is similar to that of Exercise 19.

(a) $P_W = \dfrac{1}{4}\begin{bmatrix} 3 & -1 & 1 & 1 \\ -1 & 3 & 1 & 1 \\ 1 & 1 & 3 & -1 \\ 1 & 1 & -1 & 3 \end{bmatrix}$

(b) $\mathbf{w} = \dfrac{1}{2}\begin{bmatrix} -1 \\ 3 \\ -3 \\ 5 \end{bmatrix}$, $\mathbf{z} = \dfrac{1}{2}\begin{bmatrix} 3 \\ 3 \\ -3 \\ -3 \end{bmatrix}$ (c) 3

29. The method is similar to that of Exercise 19.

(a) $P_W = \dfrac{1}{11}\begin{bmatrix} 6 & -2 & -1 & -5 \\ -2 & 8 & 4 & -2 \\ -1 & 4 & 2 & -1 \\ -5 & -2 & -1 & 6 \end{bmatrix}$

(b) $\mathbf{w} = \begin{bmatrix} 0 \\ 4 \\ 2 \\ -2 \end{bmatrix}$, $\mathbf{z} = \begin{bmatrix} 1 \\ 1 \\ -1 \\ 1 \end{bmatrix}$ (c) 2

30. The method is similar to that of Exercise 19.

(a) $P_W = \dfrac{1}{11}\begin{bmatrix} 6 & 5 & -2 & 1 \\ 5 & 6 & 2 & -1 \\ -2 & 2 & 8 & -4 \\ 1 & -1 & -4 & 2 \end{bmatrix}$

(b) $\mathbf{w} = \begin{bmatrix} 2 \\ 0 \\ -4 \\ 2 \end{bmatrix}$, $\mathbf{z} = \begin{bmatrix} -1 \\ 1 \\ -1 \\ -1 \end{bmatrix}$ (c) 2

31. The method is similar to that of Exercise 19.

(a) $P_W = \begin{bmatrix} 1 & 0 \\ 0 & 1 \end{bmatrix}$

(b) $\mathbf{w} = \begin{bmatrix} 2 \\ 3 \end{bmatrix}$, $\mathbf{z} = \begin{bmatrix} 0 \\ 0 \end{bmatrix}$ (c) 0

32. The method is similar to that of Exercise 19.

(a) $P_W = \dfrac{1}{11}\begin{bmatrix} 5 & -5 & 2 & -1 \\ -5 & 5 & -2 & 1 \\ 2 & -2 & 3 & 4 \\ -1 & 1 & 4 & 9 \end{bmatrix}$

(b) $\mathbf{w} = \begin{bmatrix} 2 \\ -2 \\ 1 \\ 0 \end{bmatrix}$, $\mathbf{z} = \begin{bmatrix} 2 \\ 3 \\ 2 \\ -1 \end{bmatrix}$ (c) $3\sqrt{2}$

33. False, $(S^\perp)^\perp$ is a subspace for any set S. So, if S is not a subspace, then $S \neq (S^\perp)^\perp$.

34. False, in $\mathcal{R}^2$, let $F = \{\mathbf{e}_1, \mathbf{e}_2\}$ and $G = \{\mathbf{e}_1, 2\mathbf{e}_2\}$. Then $F^\perp = \{\mathbf{0}\} = G^\perp$, but $F \neq G$.

35. True

36. False, $(\text{Row } A)^\perp = \text{Null } A$.

37. True 38. True 39. True 40. True

41. True

42. False, $\dim W = n - \dim W^\perp$.

43. False, we need the given basis to be orthonormal.

44. True 45. True

46. False, the only invertible orthogonal projection matrix is the identity matrix.

47. True

48. False, the columns of C can form *any* basis for W.

49. False, we need the columns of C to form a basis for W.

50. True

51. False, see Example 4.

52. True 53. True

54. False, the distance is given by $\|\mathbf{u} - P_W \mathbf{u}\|$.

55. True 56. True

57. Suppose that $\mathbf{v}$ is in $W^\perp$. Because every vector in $\mathcal{S}$ is also in W, it follows that $\mathbf{v}$ is orthogonal to each vector in $\mathcal{S}$. Therefore $W^\perp$ is contained in $\mathcal{S}^\perp$.

Conversely, let $\mathbf{u}$ be in $\mathcal{S}^\perp$ and $\mathbf{w}$ be in W. There exist scalars $a_1, a_2, \ldots, a_k$ and vectors $\mathbf{v}_1, \mathbf{v}_2, \ldots, \mathbf{v}_k$ in $\mathcal{S}$ such that

$$\mathbf{w} = a_1\mathbf{v}_1 + \cdots + a_k\mathbf{v}_k.$$

Thus

$$\begin{aligned}\mathbf{w} \cdot \mathbf{u} &= (a_1\mathbf{v}_1 + \cdots + a_k\mathbf{v}_k) \cdot \mathbf{u} \\ &= a_1(\mathbf{v}_1 \cdot \mathbf{u}) + \cdots + a_k(\mathbf{v}_k \cdot \mathbf{u}) \\ &= a_1 \cdot 0 + \cdots + a_k \cdot 0 \\ &= 0,\end{aligned}$$

and therefore $\mathbf{u}$ is in $W^\perp$. Thus $\mathcal{S}^\perp$ is contained in $W^\perp$, and we conclude that $W^\perp = \mathcal{S}^\perp$.

58. (a) First, observe that $\mathcal{B}_1 \cup \mathcal{B}_2$ is a generating set for $\mathcal{R}^n$. Let $\mathbf{v}$ be in $\mathcal{R}^n$. By Theorem 6.7, there are vectors $\mathbf{w}$ in W and $\mathbf{z}$ in $W^\perp$ such that $\mathbf{v} = \mathbf{w} + \mathbf{z}$. Since $\mathcal{B}_1$ is a basis for W and $\mathcal{B}_2$ is a basis for $W^\perp$, $\mathbf{w}$ is a linear combination of the vectors in $\mathcal{B}_1$ and $\mathbf{z}$ is a linear combination of the vectors in $\mathcal{B}_2$. It follows that $\mathbf{v} = \mathbf{w} + \mathbf{z}$ is a linear combination of the vectors in $\mathcal{B}_1 \cup \mathcal{B}_2$, which therefore is a generating set for $\mathcal{R}^n$.

Finally, we show that $\mathcal{B}_1 \cup \mathcal{B}_2$ is a linearly independent set. Let

$$\mathcal{B}_1 = \{\mathbf{w}_1, \mathbf{w}_2, \ldots, \mathbf{w}_p\}$$

and

$$\mathcal{B}_2 = \{\mathbf{z}_1, \mathbf{z}_2, \ldots, \mathbf{z}_q\},$$

and suppose that
$$a_1\mathbf{w}_1 + a_2\mathbf{w}_2 + \cdots + a_p\mathbf{w}_p \\ + b_1\mathbf{z}_1 + b_2\mathbf{z}_2 + \cdots + b_q\mathbf{z}_q = \mathbf{0}.$$

Let
$$\mathbf{w} = a_1\mathbf{w}_1 + a_2\mathbf{w}_2 + \cdots + a_p\mathbf{w}_p$$
and
$$\mathbf{z} = b_1\mathbf{z}_1 + b_2\mathbf{z}_2 + \cdots + b_q\mathbf{z}_q.$$

Then $\mathbf{w} + \mathbf{z} = \mathbf{0}$, and by uniqueness it follows that $\mathbf{w} = \mathbf{0}$ and $\mathbf{z} = \mathbf{0}$. But $\mathcal{B}_1$ and $\mathcal{B}_2$ are linearly independent, and hence all of the coefficients a_i and b_i equal 0. So $\mathcal{B}_1 \cup \mathcal{B}_2$ is linearly independent.

(b) Using the notation of (a), $p = \dim W$, $q = \dim W^\perp$, and since $\mathcal{B}_1 \cup \mathcal{B}_2$ is a basis for $\mathcal{R}^n$, it follows that $p + q = n$. Combining these equations, we obtain the result.

59. First, note that $\dim W = k$, and hence by Exercise 58,
$$\dim W^\perp = n - \dim W = n - k.$$

Next, observe that. for $i > k$. each $\mathbf{v}_i$ is orthogonal to a generating set for W and hence lies in $W^\perp$ by Exercise 57. Since $\{\mathbf{v}_{k+1}, \mathbf{v}_{k+2}, \ldots, \mathbf{v}_n\}$ is orthogonal and consists of nonzero vectors, it is a linearly independent subset of $W^\perp$. Also, it contains $n - k$ vectors. Therefore this set is a basis for $W^\perp$.

60. Suppose that $\mathbf{w}$ is in W. Then $\mathbf{w}$ is orthogonal to every vector in $W^\perp$, and hence $\mathbf{w}$ is in $(W^\perp)^\perp = W^{\perp\perp}$. Therefore W is a subspace of $W^{\perp\perp}$.

To obtain the reverse containment, we apply Exercise 58(b) to both W and $W^\perp$ to obtain
$$\dim W + \dim W^\perp = n$$
and
$$\dim W^\perp + \dim W^{\perp\perp} = n.$$

It follows that $\dim W = \dim W^{\perp\perp}$. Since W is a subspace of $W^{\perp\perp}$ and the two subspaces have the same dimension, they must be equal.

61. (a) Suppose that $\mathbf{v}$ is in $(\text{Row } A)^\perp$. Then $\mathbf{v}$ is orthogonal to every row in A. But each component of $A\mathbf{v}$ is the dot product of a row of A with $\mathbf{v}$, and hence every component of $A\mathbf{v}$ is zero. So $A\mathbf{v} = \mathbf{0}$, and hence $\mathbf{v}$ is in Null A. Thus $(\text{Row } A)^\perp$ is contained in Null A.

Now suppose that $\mathbf{v}$ is in Null A. Then $A\mathbf{v} = \mathbf{0}$, and hence $\mathbf{v}$ is orthogonal to every row of A. So $\mathbf{v}$ is in $(\text{Row } A)^\perp$ by Exercise 57. Thus Null A is contained in $(\text{Row } A)^\perp$, and the result follows.

(b) By (a), $(\text{Row } A^T)^\perp = \text{Null } A^T$. But the rows of A^T are the columns of A, and hence $(\text{Col } A)^\perp = \text{Null } A^T$.

62. Suppose that $\mathbf{v}$ is in $\mathcal{S}_2^\perp$. Then $\mathbf{v}$ is orthogonal to every vector in $\mathcal{S}_2$. Since $\mathcal{S}_1$ is a subset of $\mathcal{S}_2$, it follows that $\mathbf{v}$ is orthogonal to every vector in $\mathcal{S}_1$. Therefore $\mathbf{v}$ is in $\mathcal{S}_1^\perp$, and we conclude that $\mathcal{S}_2^\perp$ is contained in $\mathcal{S}_1^\perp$.

63. Let $W = \text{Span}\,\mathcal{S}$. Then $W^\perp = \mathcal{S}^\perp$ by Exercise 57. Thus $\mathcal{S}^{\perp\perp} = W^{\perp\perp} = W$ by Exercise 60.

64. Let A be a $k \times n$ matrix whose rows form a basis for W. Then
$$\text{rank } A = \dim (\text{Row } A)$$
by the first boxed result on page 258, and
$$(\text{Row } A)^\perp = \text{Null } A$$
by Exercise 61(a). Therefore
$$\dim W + \dim W^\perp$$
$$= \dim (\text{Row } A) + \dim (\text{Row } A)^\perp$$
$$= \dim (\text{Row } A) + \dim (\text{Null } A)$$
$$= \text{rank } A + \text{nullity } A = n.$$

65. Suppose $\mathbf{v}$ is in both Row A and Null A. Because Null $A = (\text{Row } A)^\perp$ by Exercise 61(a), $\mathbf{v}$ is orthogonal to itself, that is, $\mathbf{v} \cdot \mathbf{v} = 0$. So $\mathbf{v} = \mathbf{0}$.

66. By assumption, V is contained in $W^\perp$. So
$$\dim V + \dim W \leq \dim V + \dim W^\perp = n$$
by Exercise 58(b).

67. (a) Let C be an $n \times k$ matrix whose columns form a basis for W. Then
$$(P_W)^2 = C(C^TC)^{-1}C^TC(C^TC)^{-1}C^T$$
$$= CI_k(C^TC)^{-1}C^T = P_W.$$

(b) Let C be an $n \times k$ matrix whose columns form a basis for W. Then
$$(P_W)^T = [C(C^TC)^{-1}C^T]^T$$
$$= C(C^TC)^{-1}C^T = P_W.$$

68. Let C be an $n \times k$ matrix whose columns form a basis for W, and suppose that $P_W \mathbf{u} = \mathbf{u}$. Then $\mathbf{u} = C[(C^TC)^{-1}C^T \mathbf{u}]$, and hence $\mathbf{u}$ is a linear combination of the columns of C, and therefore $\mathbf{u}$ is in W.

Conversely, suppose that $\mathbf{u}$ is in W. Since $\mathbf{u}$ is a linear combination of the columns of C, there exists a vector $\mathbf{v}$ in $\mathcal{R}^k$ such that $\mathbf{u} = C\mathbf{v}$. Thus
$$P_W \mathbf{u} = C(C^TC)^{-1}C^T C\mathbf{v} = C\mathbf{v} = \mathbf{u}.$$

ALTERNATE PROOF: By Theorem 6.7, there are unique vectors $\mathbf{w}$ in W and $\mathbf{z}$ in $W^\perp$ such that $\mathbf{u} = \mathbf{w} + \mathbf{z}$. It easily follows that $\mathbf{u}$ is in W if and only if $\mathbf{u} = \mathbf{w}$, and hence if and only if $\mathbf{z} = \mathbf{0}$. By Theorem 6.8, $P_W \mathbf{u} = \mathbf{w}$, and hence $P_W \mathbf{u} = \mathbf{u}$ if and only if $\mathbf{u}$ is in W.

69. By Theorem 6.7, there are unique vectors $\mathbf{w}$ in W and $\mathbf{z}$ in $W^\perp$ such that $\mathbf{u} = \mathbf{w} + \mathbf{z}$. It follows that $\mathbf{u}$ is in $W^\perp$ if and only if $\mathbf{u} = \mathbf{z}$, and hence if and only if $\mathbf{w} = \mathbf{0}$. By Theorem 6.8, $P_W \mathbf{u} = \mathbf{w}$, and hence $P_W \mathbf{u} = \mathbf{0}$ if and only if $\mathbf{u}$ is in $W^\perp$.

ALTERNATE PROOF: By Theorem 6.8, $P_W = C(C^TC)^{-1}C^T$, where C is a matrix whose columns form a basis for W. Now suppose that $\mathbf{u}$ is in $W^\perp$. Then $\mathbf{u}$ is orthogonal to each column of C, and hence $C^T \mathbf{u} = \mathbf{0}$. Therefore
$$P_W \mathbf{u} = C(C^TC)^{-1}C^T \mathbf{u}$$
$$= C(C^TC)^{-1}(C^T \mathbf{u})$$
$$= C(C^TC)^{-1}\mathbf{0} = \mathbf{0}.$$

Conversely, suppose that $P_W \mathbf{u} = \mathbf{0}$. Then
$$C(C^TC)^{-1}C^T \mathbf{u} = \mathbf{0}$$
$$C^TC(C^TC)^{-1}C^T \mathbf{u} = C^T \mathbf{0} = \mathbf{0}$$
$$C^T \mathbf{u} = \mathbf{0}.$$

This last equation asserts that $\mathbf{u}$ is orthogonal to the columns of C, a generating set for W. Therefore $\mathbf{u}$ is in $W^\perp$ by Exercise 57.

70. Suppose that $\mathbf{u}$ is an eigenvector of P_W. Then $P_W \mathbf{u} = \lambda \mathbf{u}$ for some scalar λ. Let $\mathbf{w}$ and $\mathbf{z}$ be the unique vectors in W and $W^\perp$, respectively, such that $\mathbf{u} = \mathbf{w} + \mathbf{z}$. Then $\mathbf{w} = P_W \mathbf{u} = \lambda \mathbf{u}$ by Theorem 6.8. There are two cases to consider.
Case 1: $\lambda = 0$. Then $\mathbf{w} = \mathbf{0}$, and hence $\mathbf{u} = \mathbf{z}$. Thus $P_{W^\perp} \mathbf{u} = \mathbf{z} = \mathbf{u}$, and so $\mathbf{u}$ is an eigenvector of $P_{W^\perp}$ with corresponding eigenvalue 1.
Case 2: $\lambda \neq 0$. Then $\mathbf{w} = P_W \mathbf{u} = \lambda \mathbf{u}$ implies that $\mathbf{u}$ is in W, and hence $\mathbf{z} = \mathbf{0}$. Thus $P_{W^\perp} \mathbf{u} = \mathbf{z} = \mathbf{0}$, and so $\mathbf{u}$ is an eigenvector of $P_{W^\perp}$ with corresponding eigenvalue 0.
The proof of the converse is similar.

71. Applying Exercise 67(b), we have
$$P_W \mathbf{u} \cdot \mathbf{v} = (P_W \mathbf{u})^T \mathbf{v} = \mathbf{u}^T P_W^T \mathbf{v}$$
$$= \mathbf{u}^T P_W \mathbf{v} = \mathbf{u} \cdot P_W \mathbf{v}.$$

72. Let $\mathbf{u}$ be a vector in $\mathcal{R}^n$, and let $\mathbf{w}$ and $\mathbf{z}$ be the unique vectors in W and $W^\perp$, respectively, such that $\mathbf{u} = \mathbf{w} + \mathbf{z}$. Then
$$P_W P_{W^\perp} \mathbf{u} = P_W P_{W^\perp}(\mathbf{w} + \mathbf{z})$$
$$= P_W P_{W^\perp} \mathbf{w} + P_W P_{W^\perp} \mathbf{z}$$
$$= \mathbf{0} + P_W \mathbf{z} = \mathbf{0}.$$

Similarly, $P_{W^\perp} P_W \mathbf{u} = \mathbf{0}$. Since $\mathbf{u}$ is an arbitrary vector in $\mathcal{R}^n$, $P_W P_{W^\perp} = P_{W^\perp} P_W = O$.

73. Let $\mathbf{u}$ be a vector in $\mathcal{R}^n$, and let $\mathbf{w}$ and $\mathbf{z}$ be the unique vectors in W and $W^\perp$, respectively, such that $\mathbf{u} = \mathbf{w} + \mathbf{z}$. Then
$$(P_W + P_{W^\perp})\mathbf{u} = P_W \mathbf{u} + P_{W^\perp} \mathbf{u}$$
$$= \mathbf{w} + \mathbf{z} = \mathbf{u} = I_n \mathbf{u}.$$

Thus $P_W + P_{W^\perp} = I_n$.

74. Let $Z = \{\mathbf{v} + \mathbf{w} : \mathbf{v} \text{ is in } V \text{ and } \mathbf{w} \text{ is in } W\}$. By Exercise 52 of the Chapter 4 Review Exercises, Z is a subspace of $\mathcal{R}^n$. We will show that $P_Z = P_V + P_W$.

Since every vector in V is orthogonal to every vector in W, it follows that V is contained in $W^\perp$. Similarly, W is contained in $V^\perp$. Let $\mathbf{z}$ be any vector in Z. Then there exist vectors $\mathbf{v}$ in V and $\mathbf{w}$ in W such that $\mathbf{z} = \mathbf{v} + \mathbf{w}$. Since $\mathbf{v}$ is in V, we have $P_V \mathbf{v} = \mathbf{v}$ and $P_{V^\perp} \mathbf{v} = \mathbf{0}$. Furthermore, since W is contained in $V^\perp$, we have $P_V \mathbf{w} = \mathbf{0}$ and $P_{V^\perp} \mathbf{w} = \mathbf{w}$. Similarly,
$$P_W \mathbf{v} = \mathbf{0}, \quad P_{W^\perp} \mathbf{v} = \mathbf{v}, \quad P_W \mathbf{w} = \mathbf{w},$$
and
$$P_{W^\perp} \mathbf{w} = \mathbf{0}.$$
Thus
$$(P_V + P_W)\mathbf{z} = P_V(\mathbf{v} + \mathbf{w}) + P_W(\mathbf{v} + \mathbf{w})$$
$$= P_V \mathbf{v} + P_V \mathbf{w} + P_W \mathbf{v} + P_W \mathbf{w}$$
$$= \mathbf{v} + \mathbf{0} + \mathbf{0} + \mathbf{w} = \mathbf{z}.$$

Now suppose that $\mathbf{u}$ is a vector in $Z^\perp$. Since V is a subspace of Z, it follows that $Z^\perp$ is a subspace of $V^\perp$ by Exercise 37 of Section 6.2. Hence $\mathbf{u}$ is in $V^\perp$, and so $P_V \mathbf{u} = \mathbf{0}$. Similarly, $P_W \mathbf{u} = \mathbf{0}$. Therefore
$$(P_V + P_W)\mathbf{u} = P_V \mathbf{u} + P_W \mathbf{u} = \mathbf{0} + \mathbf{0} = \mathbf{0}.$$

Finally, let **y** be any vector in $\mathcal{R}^n$. Then there exist vectors **z** in Z and **u** in $Z^\perp$ such that $\mathbf{y} = \mathbf{z}+\mathbf{u}$. Applying what we have learned about the various orthogonal projection matrices, we have

$$P_Z \mathbf{y} = P_Z(\mathbf{z}+\mathbf{u}) = P_Z \mathbf{z} + P_Z \mathbf{u}$$
$$= \mathbf{z} + \mathbf{0}$$
$$= (P_V + P_W)\mathbf{z} + (P_V + P_W)\mathbf{u}$$
$$= (P_V + P_W)(\mathbf{z}+\mathbf{u})$$
$$= (P_V + P_W)\mathbf{y}.$$

Since **y** is an arbitrarily chosen vector in $\mathcal{R}^n$, we conclude that $P_Z = P_V + P_W$.

75. (a) Since C^T is a $k \times n$ matrix and C is an $n \times k$ matrix, $C^T C$ is a $k \times k$ matrix. The (i,j)-entry of $C^T C$ is equal to $\mathbf{v}_i \cdot \mathbf{v}_j$. This entry is equal to 1 if $i = j$ and equal to 0 if $i \neq j$. Therefore $C^T C = I_k$.

(b) By Theorem 6.8 and (a), we have

$$P_W = C(C^T C)^{-1} C^T$$
$$= C I_k C^T = C C^T.$$

76. Let $W = \{\mathbf{0}\}$. For any vector **u** in $\mathcal{R}^n$, the product $P_W \mathbf{u}$ lies in W. Since **0** is the only vector in W, we have that $P_W \mathbf{u} = \mathbf{0}$.

77. (a) We first show that 1 and 0 are eigenvalues of P_W. Since $k \neq 0$, we can choose a nonzero vector **w** in W. Then $P_W \mathbf{w} = \mathbf{w}$, and hence **w** is an eigenvector with corresponding eigenvalue 1. Since $k \neq n$, we can choose a nonzero vector **z** in $W^\perp$. Then $P_W \mathbf{z} = \mathbf{0}$, and hence **z** is an eigenvector with corresponding eigenvalue 0.

Next we show that 1 and 0 are the only eigenvalues of P_W. Suppose that λ is a nonzero eigenvalue of P_W, and let **u** be an eigenvector corresponding to λ. Then $P_W \mathbf{u} = \lambda \mathbf{u}$, and hence $P_W(\frac{1}{\lambda}\mathbf{u}) = \mathbf{u}$. Thus **u** is an image of U_W, and so **u** is in W. Therefore

$$\mathbf{u} = P_W \left(\frac{1}{\lambda}\mathbf{u}\right) = \frac{1}{\lambda} P_W \mathbf{u} = \frac{1}{\lambda} \mathbf{u}.$$

Hence $1 = \frac{1}{\lambda}$, that is, $\lambda = 1$.

(b) Since $P_W \mathbf{u} = \mathbf{u}$ if and only **u** is in W, we see that W is the eigenspace of P_W corresponding to eigenvalue 1. Similarly, since $P_W \mathbf{u} = \mathbf{0}$ if and only if **u** is in $W^\perp$, we have that $W^\perp$ is the eigenspace of P_W corresponding to eigenvalue 0.

(c) Let T be the matrix transformation induced by P_W. Since $T(\mathbf{u}) = 1 \cdot \mathbf{u}$ for all **u** in $\mathcal{B}_1$ and $T(\mathbf{u}) = 0 \cdot \mathbf{v}$ for all **v** in $\mathcal{B}_2$, we have $[T]_\mathcal{B} = D$, and hence $P_W = BDB^{-1}$ by Theorem 4.12.

78. Since nonzero rows of the reduced row echelon form

$$R = \begin{bmatrix} 1 & 0 & -2 & 2 \\ 0 & 1 & -2 & 1 \\ 0 & 0 & 0 & 0 \\ 0 & 0 & 0 & 0 \end{bmatrix}$$

of A form a basis for Row A, we may take the first two rows of R

$$\mathcal{B}_1 = \left\{ \begin{bmatrix} 1 \\ 0 \\ -2 \\ 2 \end{bmatrix}, \begin{bmatrix} 0 \\ 1 \\ -2 \\ 1 \end{bmatrix} \right\}$$

as the required basis. The subspace $(\text{Row } A)^\perp$ is the solution space of the system $A\mathbf{x} = \mathbf{0}$, which is equivalent to $R\mathbf{x} = \mathbf{0}$. Thus a basis for this subspace is

$$\mathcal{B}_2 = \left\{ \begin{bmatrix} 2 \\ 2 \\ 1 \\ 0 \end{bmatrix}, \begin{bmatrix} -2 \\ -1 \\ 0 \\ 1 \end{bmatrix} \right\}.$$

Let

$$B = \begin{bmatrix} 1 & 0 & 2 & -2 \\ 0 & 1 & 2 & -1 \\ -2 & -2 & 1 & 0 \\ 2 & 1 & 0 & 1 \end{bmatrix}$$

and

$$D = \begin{bmatrix} 1 & 0 & 0 & 0 \\ 0 & 1 & 0 & 0 \\ 0 & 0 & 0 & 0 \\ 0 & 0 & 0 & 0 \end{bmatrix}.$$

Notice that B is the matrix whose columns are the vectors in $\mathcal{B}_1 \cup \mathcal{B}_2$. Then, by Exercise 77,

$$P_V = BDB^{-1} = \frac{1}{6} \begin{bmatrix} 2 & -2 & 0 & 2 \\ -2 & 3 & -2 & -1 \\ 0 & -2 & 4 & -2 \\ 2 & -1 & -2 & 3 \end{bmatrix}.$$

79. (a) Because $W^\perp = \text{Row } C$ and rank $C = m$, the rows of C, and hence the columns of C^T, form a basis for $W^\perp$. Therefore

$$P_{W^\perp} = C^T \left((C^T)^T C^T\right)^{-1} (C^T)^T$$
$$= C^T (CC^T)^{-1} C.$$

So, by Exercise 73,
$$P_W = I_n - P_{W^\perp}$$
$$= I_n - C^T(CC^T)^{-1}C.$$

(b) Using the hint, we let
$$C = \begin{bmatrix} 1 & 0 & -2 & 2 \\ 0 & 1 & -2 & 1 \end{bmatrix}.$$

Observe that for any $\mathbf{v}$ in $\mathcal{R}^4$, $A\mathbf{v} = \mathbf{0}$ if and only if $C\mathbf{v} = \mathbf{0}$. (This is true because elementary row operations do not affect the null space of a matrix.) So
$$P_W = I_4 - P_W = I_4 - C^T(CC^T)^{-1}C$$
$$= \frac{1}{6}\begin{bmatrix} 4 & 2 & 0 & -2 \\ 2 & 3 & 2 & 1 \\ 0 & 2 & 2 & 2 \\ -2 & 1 & 2 & 3 \end{bmatrix}.$$

(c) Let
$$B = \begin{bmatrix} 2 & -2 & -1 & 0 \\ 2 & -1 & 1 & 1 \\ 1 & 0 & 0 & -2 \\ 0 & 1 & -1 & 1 \end{bmatrix},$$
where the first two columns form a basis for W and the last two columns form a basis for $W^\perp$. Then $BDB^{-1} = P_W$.

80. Since $V^\perp = \operatorname{Row} A^\perp = \operatorname{Null} A = W$, we have $P_V + P_W = I_4$ by Exercise 73.

81. The first two columns of A are its pivot columns, and so the rank of A is 2 and a basis for $\operatorname{Col} A$ is
$$\mathcal{B}_1 = \left\{ \begin{bmatrix} 1 \\ 0 \\ -1 \\ 1 \end{bmatrix}, \begin{bmatrix} 0 \\ 1 \\ -2 \\ 1 \end{bmatrix} \right\}.$$

As in Example 2 on pages 390–391, we can find a basis
$$\mathcal{B}_2 = \left\{ \begin{bmatrix} 1 \\ 2 \\ 1 \\ 0 \end{bmatrix}, \begin{bmatrix} -1 \\ -1 \\ 0 \\ 1 \end{bmatrix} \right\}.$$
for $(\operatorname{Col} A)^\perp$. Let
$$B = \begin{bmatrix} 1 & 0 & 1 & -1 \\ 0 & 1 & 2 & -1 \\ -1 & -2 & 1 & 0 \\ 1 & 1 & 0 & 1 \end{bmatrix}$$

and
$$D = \begin{bmatrix} 1 & 0 & 0 & 0 \\ 0 & 1 & 0 & 0 \\ 0 & 0 & 0 & 0 \\ 0 & 0 & 0 & 0 \end{bmatrix}.$$

Notice that B is the matrix whose columns are the vectors in $\mathcal{B}_1 \cup \mathcal{B}_2$. Then by Exercise 77(c),
$$P_W = BDB^{-1}$$
$$= \frac{1}{3}\begin{bmatrix} 2 & -1 & 0 & 1 \\ -1 & 1 & -1 & 0 \\ 0 & -1 & 2 & -1 \\ 1 & 0 & -1 & 1 \end{bmatrix}.$$

82. Clearly, $\mathbf{u} = P\mathbf{u} + (I_n - P)\mathbf{u}$ for every vector $\mathbf{u}$ in $\mathcal{R}^n$. Furthermore, $P\mathbf{u}$ is in W because it is a linear combination of the columns of P. To show that $(I_n - P)\mathbf{u}$ is in $W^\perp$, consider any vector $\mathbf{w}$ in W. Then $\mathbf{w} = P\mathbf{v}$ for some vector $\mathbf{v}$ in $\mathcal{R}^n$. So
$$\mathbf{w} \cdot (I_n - P)\mathbf{u} = (P\mathbf{v}) \cdot (I_n - P)\mathbf{u}$$
$$= (P\mathbf{v})^T(I_n - P)\mathbf{u}$$
$$= \mathbf{v}^T P^T(I_n - P)\mathbf{u}$$
$$= \mathbf{v}^T P(I_n - P)\mathbf{u}$$
$$= \mathbf{v}^T(P - P^2)\mathbf{u}$$
$$= \mathbf{v}^T(P - P)\mathbf{u}$$
$$= \mathbf{0}.$$

Hence $(I_n - P)\mathbf{u}$ is in $W^\perp$. Since, for every $\mathbf{u}$ in $\mathcal{R}^n$, $\mathbf{u} = P\mathbf{u} + (I_n - P)\mathbf{u}$ is a decomposition of $\mathbf{u}$ into a sum of a vector in W and a vector in $W^\perp$, it follows that $P\mathbf{u} = P_W\mathbf{u}$ for every vector $\mathbf{u}$ in $\mathcal{R}^n$. Therefore $P = P_W$.

83. By definition, $A = \mathbf{v}\mathbf{v}^T$. Because $\dim W = 1$ and $\mathbf{v} \neq \mathbf{0}$, the set $\{\mathbf{v}\}$ is a basis for W. Let C be the $n \times 1$ matrix whose only column is $\mathbf{v}$. By Theorem 6.8, we have
$$P_W = C(C^T C)^{-1}C^T = \mathbf{v}(\mathbf{v}^T\mathbf{v})^{-1}\mathbf{v}^T$$
$$= \mathbf{v}(\|\mathbf{v}\|^2)^{-1}\mathbf{v}^T = \frac{1}{\|\mathbf{v}\|^2}\mathbf{v}\mathbf{v}^T = \frac{1}{\|\mathbf{v}\|^2}A.$$

84. (a) Let $\mathbf{u}$ be a vector in $\mathcal{R}^3$. Then there exist vectors $\mathbf{w}$ in W and $\mathbf{z}$ in $W^\perp$ such that $\mathbf{u} = \mathbf{w} + \mathbf{z}$. Based on the description of T_W given in Figure 5.5 in Section 5.4, we have
$$T_W(\mathbf{u}) = \mathbf{w} - \mathbf{z}$$
$$= U_W(\mathbf{u}) - (\mathbf{u} - U_W(\mathbf{u}))$$
$$= 2U_W(\mathbf{u}) - \mathbf{u}.$$

(b) Observe that, for any vector $\mathbf{u}$ in $\mathcal{R}^3$,

$$2U_W(\mathbf{u}) - \mathbf{u} = 2P_W\mathbf{u} - I_3\mathbf{u}$$
$$= (2P_W - I_3)\mathbf{u},$$

and hence $T_W(\mathbf{u}) = (2P_W - I_3)\mathbf{u}$. It follows that T_W is the matrix transformation induced by the matrix $2P_W - I_3$. Matrix transformations are linear, and hence T_W is a linear transformation.

85. (a) There is no unique answer. Using Q in the MATLAB function $[Q\ R] = \text{qr}(A, 0)$ (see Table D.3 in Appendix D), where A is the matrix whose columns are the vectors in $\mathcal{S}$, we obtain

$$\left\{\begin{bmatrix} 0 \\ .2914 \\ -.8742 \\ 0 \\ .3885 \end{bmatrix}, \begin{bmatrix} .7808 \\ -.5828 \\ -.1059 \\ 0 \\ .1989 \end{bmatrix}, \begin{bmatrix} -.0994 \\ -.3243 \\ -.4677 \\ .1082 \\ -.8090 \end{bmatrix},\right.$$

$$\left.\begin{bmatrix} -.1017 \\ -.1360 \\ -.0589 \\ -.9832 \\ -.0304 \end{bmatrix}\right\}$$

(b) $\mathbf{w} = \begin{bmatrix} -6.3817 \\ 6.8925 \\ 7.2135 \\ 1.3687 \\ 2.3111 \end{bmatrix}$

(c) $\|\mathbf{u} - \mathbf{w}\| = 4.3033$

86. For

$$A = \begin{bmatrix} 0 & -3 & 9 & 0 & -4 \\ -8 & 9 & -8 & 0 & 2 \\ -4 & 8 & 1 & -1 & 8 \\ -9 & 5 & 5 & 6 & -7 \end{bmatrix},$$

$\text{null}(A, \text{'r'})$ produces the basis

$$\left\{\begin{bmatrix} -1.5503 \\ -1.7126 \\ -0.1264 \\ 0.3738 \\ 1.0000 \end{bmatrix}\right\}.$$

87. $P_W =$
$$\begin{bmatrix} 0.6298 & -0.4090 & -0.0302 & 0.0893 & 0.2388 \\ -0.4090 & 0.5482 & -0.0334 & 0.0986 & 0.2638 \\ -0.0302 & -0.0334 & 0.9975 & 0.0073 & 0.0195 \\ 0.0893 & 0.0986 & 0.0073 & 0.9785 & -0.0576 \\ 0.2388 & 0.2638 & 0.0195 & -0.0576 & 0.8460 \end{bmatrix}$$

and $P_W\mathbf{u} = \begin{bmatrix} -6.3817 \\ 6.8925 \\ 7.2135 \\ 1.3687 \\ 2.3111 \end{bmatrix}$

88. $P_W =$
$$\begin{bmatrix} 0.7201 & 0.0001 & -0.1845 & -0.3943 & -0.1098 \\ 0.0001 & 0.4915 & 0.4391 & -0.1547 & -0.1823 \\ -0.1845 & 0.4391 & 0.4993 & -0.1263 & 0.0850 \\ -0.3943 & -0.1547 & -0.1263 & 0.3975 & -0.2102 \\ -0.1098 & -0.1823 & 0.0850 & -0.2102 & 0.8915 \end{bmatrix}$$

and $P_W\mathbf{u} = \begin{bmatrix} 4.4048 \\ -2.8160 \\ -4.0734 \\ -0.8866 \\ -1.2014 \end{bmatrix}$

The distance is $\|\mathbf{u} - P_W\mathbf{u}\| = 3.4418$.

6.4 LEAST-SQUARES APPROXIMATIONS AND ORTHOGONAL PROJECTION MATRICES

1. Using the data, let

$$C = \begin{bmatrix} 1 & 1 \\ 1 & 3 \\ 1 & 5 \\ 1 & 7 \end{bmatrix} \quad \text{and} \quad \mathbf{y} = \begin{bmatrix} 14 \\ 17 \\ 19 \\ 20 \end{bmatrix}.$$

Then the equation of the least squares line for the given data is $y = a_0 + a_1 x$, where

$$\begin{bmatrix} a_0 \\ a_1 \end{bmatrix} = (C^T C)^{-1} C^T \mathbf{y}$$

$$= \begin{bmatrix} 4 & 16 \\ 16 & 84 \end{bmatrix}^{-1} \begin{bmatrix} 1 & 1 & 1 & 1 \\ 1 & 3 & 5 & 7 \end{bmatrix} \begin{bmatrix} 14 \\ 17 \\ 19 \\ 20 \end{bmatrix}$$

$$= \frac{1}{20} \begin{bmatrix} 21 & -4 \\ -4 & 1 \end{bmatrix} \begin{bmatrix} 70 \\ 300 \end{bmatrix} = \begin{bmatrix} 13.5 \\ 1.0 \end{bmatrix}.$$

Therefore $y = 13.5 + x$.

2. $y = \frac{97}{3} - \frac{8}{3}x$ 3. $y = 3.2 + 1.6x$
4. $y = \frac{1}{5} + 2x$ 5. $y = 44 - 3x$
6. $y = \frac{207}{10} - \frac{17}{10}x$ 7. $y = 9.6 + 2.6x$
8. $y = 22.2 + 1.8x$
9. The equation of the least squares line is $y = -6.35 + 2.1x$, and so the estimates of k and L are 2.1 and

$$L = -\frac{a}{k} = -\frac{(-6.35)}{2.1} \approx 3.02,$$

respectively.

10. $y = 2 - x + x^2$ 11. $y = 3 - x + x^2$

12. $y = 1 + x$ 13. $2 + 0.5x + 0.5x^2$

14. $y = -1 + x^2 + x^3$ 15. $-1 + x^2 + x^3$

16. $\dfrac{1}{30}\begin{bmatrix}18\\67\end{bmatrix}$ 17. $\dfrac{1}{3}\begin{bmatrix}4\\-2\\0\end{bmatrix} + x_3\begin{bmatrix}-1\\1\\1\end{bmatrix}$

18. $\begin{bmatrix}-1\\2\\4\\0\end{bmatrix} + x_4\begin{bmatrix}-2\\-1\\-1\\1\end{bmatrix}$ 19. $\dfrac{1}{19}\begin{bmatrix}-35\\-50\\31\\0\end{bmatrix} + x_4\begin{bmatrix}0\\1\\-1\\1\end{bmatrix}$

20. $\begin{bmatrix}1.0\\0.4\\0.8\end{bmatrix}$ 21. $\dfrac{1}{3}\begin{bmatrix}7\\-1\\8\end{bmatrix}$ 22. $\dfrac{1}{14}\begin{bmatrix}5\\-15\\10\end{bmatrix}$

23. $\begin{bmatrix}1\\-1\\1\\0\end{bmatrix}$ 24. $\dfrac{1}{30}\begin{bmatrix}18\\67\end{bmatrix}$ 25. $\dfrac{1}{3}\begin{bmatrix}2\\0\\2\end{bmatrix}$

26. $\dfrac{1}{7}\begin{bmatrix}-15\\10\\24\\4\end{bmatrix}$ 27. $\dfrac{1}{19}\begin{bmatrix}-35\\-23\\4\\27\end{bmatrix}$

28. False, the least-squares line is the line that minimizes the sum of the *squares* of the vertical distances from the data points to the line.

29. True

30. False, in Example 2, the method is used to approximate data with a polynomial of degree 2.

31. False, the inconsistent system in Example 3 has infinitely many vectors that minimize this distance.

32. True

33. We have
$$\|\mathbf{y} - (a_0\mathbf{v}_1 + a_1\mathbf{v}_2)\|^2$$
$$= \left\| \begin{bmatrix}y_1\\y_2\\\vdots\\y_n\end{bmatrix} - a_0\begin{bmatrix}1\\1\\\vdots\\1\end{bmatrix} - a_1\begin{bmatrix}x_1\\x_2\\\vdots\\x_n\end{bmatrix} \right\|^2$$
$$= \left\| \begin{bmatrix}y_1 - (a_0 + a_1x_1)\\y_2 - (a_0 + a_1x_2)\\\vdots\\y_n - (a_0 + a_1x_n)\end{bmatrix} \right\|^2$$
$$= [y_1 - (a_0 + a_1x_1)]^2 + \cdots + [y_n - (a_0 + a_1x_n)]^2$$
$$= E.$$

34. Let A be the 3×3 matrix consisting of the first three rows of $[\mathbf{v}_1\ \mathbf{v}_2\ \mathbf{v}_3]$. Observe that the reduced row echelon form of A is I_3, and hence the columns of A are linearly independent. So the vectors $\mathbf{v}_1$, $\mathbf{v}_2$, and $\mathbf{v}_3$ are linearly independent.

35. Suppose that $\mathbf{v} = \mathbf{v}_0 + \mathbf{z}$ for some vector $\mathbf{z}$ in Z. Then
$$A\mathbf{v} = A(\mathbf{v}_0 + \mathbf{z}) = A\mathbf{v}_0 + A\mathbf{z} = \mathbf{c} + \mathbf{0} = \mathbf{c},$$
and hence $\mathbf{v}$ is a solution of the system. Conversely, suppose that $\mathbf{v}$ is a solution of the system, and let $\mathbf{z} = \mathbf{v} - \mathbf{v}_0$. Then
$$A\mathbf{z} = A(\mathbf{v} - \mathbf{v}_0) = A\mathbf{v} - A\mathbf{v}_0 = \mathbf{c} - \mathbf{c} = \mathbf{0}.$$
Thus $\mathbf{z}$ is in Z. Furthermore, $\mathbf{v} = \mathbf{v}_0 + \mathbf{z}$.

36. Let A be an $m \times n$ matrix. Since the columns of A are linearly independent, rank $A = n$, and hence, by Exercise 62 of Section 2.3,
$$n = \operatorname{rank} A = \operatorname{rank} QR \leq \operatorname{rank} R \leq n,$$
It follows that rank $R = n$, and so R is invertible. Furthermore, since the columns of Q are orthonormal, it follows that $Q^TQ = I_n$. Thus
$$P_W = A(A^TA)^{-1}A^T$$
$$= (QR)((QR)^T(QR))^{-1}(QR)^T$$
$$= QR(R^TQ^TQR)^{-1}R^TQ^T$$
$$= QR(R^TI_nR)^{-1}R^TQ^T$$
$$= QRR^{-1}(R^T)^{-1}R^TQ^T$$
$$= QQ^T.$$

37. Let $A = QR$ be a QR-factorization of A, and let W denote the column space of A. The solutions of $A\mathbf{x} = P_W\mathbf{b}$ minimize $\|A\mathbf{x} - \mathbf{b}\|$. It follows from Exercise 36 that $A\mathbf{x} = P_W\mathbf{b}$ if and only if $QR\mathbf{x} = QQ^T\mathbf{b}$. If we multiply both sides of this equation by Q^T and simplify, we obtain $R\mathbf{x} = Q^T\mathbf{b}$.

38. Form the matrices
$$C = \begin{bmatrix}1 & 5 & 25\\1 & 10 & 100\\1 & 15 & 225\\1 & 20 & 400\\1 & 25 & 625\\1 & 30 & 900\end{bmatrix} \quad \text{and} \quad \mathbf{y} = \begin{bmatrix}140\\290\\560\\910\\1400\\2000\end{bmatrix}.$$
Then
$$\begin{bmatrix}a_0\\a_1\\a_2\end{bmatrix} = (C^TC)^{-1}C^T\mathbf{v} \approx \begin{bmatrix}107.0000\\-4.0786\\2.2357\end{bmatrix},$$
and therefore the best quadratic fit is given by $y = 107 - 4.0786t + 2.2357t^2$.

39. Using the data, let

$$C = \begin{bmatrix} 1 & -2 & 4 & -8 \\ 1 & -1 & 1 & -1 \\ 1 & 0 & 0 & 0 \\ 1 & 2 & 4 & 8 \\ 1 & 3 & 9 & 27 \end{bmatrix} \quad \text{and} \quad \mathbf{y} = \begin{bmatrix} -4 \\ 1 \\ 1 \\ 10 \\ 26 \end{bmatrix}.$$

Then the best cubic fit for the data is given by $y = a_0 + a_1 x + a_2 x^2 + a_3 x^3$, where

$$\begin{bmatrix} a_0 \\ a_1 \\ a_2 \\ a_3 \end{bmatrix} = (C^T C)^{-1} C^T \mathbf{y}$$

$$= \begin{bmatrix} 5 & 2 & 18 & 26 \\ 2 & 18 & 26 & 114 \\ 18 & 26 & 114 & 242 \\ 26 & 114 & 242 & 858 \end{bmatrix}^{-1} \begin{bmatrix} 34 \\ 105 \\ 259 \\ 813 \end{bmatrix}$$

$$\approx \begin{bmatrix} 1.42 \\ 0.49 \\ 0.38 \\ 0.73 \end{bmatrix}.$$

Thus the best cubic fit is given by

$$y = 1.42 + 0.49x + 0.38x^2 + 0.73x^3.$$

40. (rounded to 4 places after the decimal)
- (a) $y = 6.9947 - 2.2265x$
- (b) 284.7263
- (d) $y = -0.6351 + 17.0683x - 8.0529x^2 + 0.8544x^3$
- (e) 7.2383

In the following figures, the darker curves are the graphs of the least squares line (Exercise 40(c)) and the best cubic fit (Exercise 40(f)).

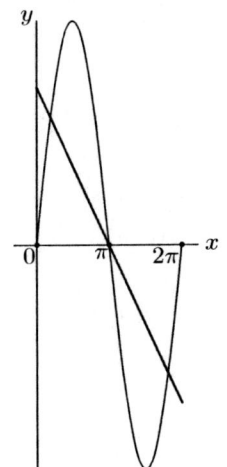

Figure for Exercise 40(c)

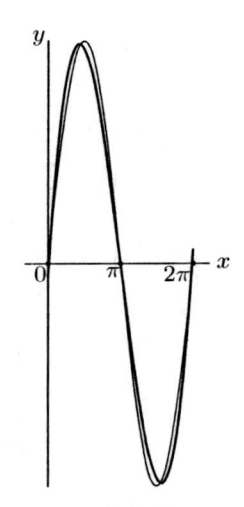

Figure for Exercise 40(f)

41. Following the hint, let W denote the column space of A. Then we have (with entries rounded to 4 places after the decimal)

$$A = \begin{bmatrix} 0.9962 & 0.0872 \\ 0.9848 & 0.1736 \\ 0.9659 & 0.2588 \\ 0.9397 & 0.3420 \\ 0.9063 & 0.4226 \\ 0.8660 & 0.5000 \end{bmatrix}$$

and

$$P_W \mathbf{y} = A(A^T A)^{-1} A^T \mathbf{y} = \begin{bmatrix} 2.8039 \\ 2.6003 \\ 2.3769 \\ 2.1355 \\ 1.8777 \\ 1.6057 \end{bmatrix}.$$

The system of linear equations $A\mathbf{x} = P_W \mathbf{y}$ has a solution (rounded to 4 places after the decimal) of $a = 2.9862$ and $b = -1.9607$, which, when rounded to 2 significant figures, gives $a = 3.0$ and $b = -2.0$.

6.5 ORTHOGONAL MATRICES AND OPERATORS

1. no **2.** no **3.** no **4.** yes

5. Since

$$\begin{bmatrix} 0 & 1 & 0 \\ 0 & 0 & 1 \\ 1 & 0 & 0 \end{bmatrix}^T \begin{bmatrix} 0 & 1 & 0 \\ 0 & 0 & 1 \\ 1 & 0 & 0 \end{bmatrix} = \begin{bmatrix} 0 & 0 & 1 \\ 1 & 0 & 0 \\ 0 & 1 & 0 \end{bmatrix} \begin{bmatrix} 0 & 1 & 0 \\ 0 & 0 & 1 \\ 1 & 0 & 0 \end{bmatrix}$$

156 Chapter 6 Orthogonality

$= I_3,$

the matrix is orthogonal by Theorem 6.9(b).

6. no **7.** No, the matrix is not a square matrix.

8. yes

9. Since $\det\left(\frac{1}{\sqrt{2}}\begin{bmatrix} 1 & 1 \\ 1 & -1 \end{bmatrix}\right) = \frac{1}{2}(-1-1) = -1$, the operator is a reflection. The line of reflection is the same as the eigenspace of the matrix corresponding to eigenvalue 1. This is the solution set of the system

$$\left(\frac{1}{\sqrt{2}}\begin{bmatrix} 1 & 1 \\ 1 & -1 \end{bmatrix} - \begin{bmatrix} 1 & 0 \\ 0 & 1 \end{bmatrix}\right)\begin{bmatrix} x \\ y \end{bmatrix} = \begin{bmatrix} 0 \\ 0 \end{bmatrix},$$

or

$(\frac{1}{\sqrt{2}}-1)x + \frac{1}{\sqrt{2}}y = 0$
$\frac{1}{\sqrt{2}}x - \frac{1}{\sqrt{2}}y = 0.$

Treating x as a free variable, the general solution of this system is $y = (\sqrt{2}-1)x$, which is therefore the equation of the line of reflection.

10. a rotation, $\theta = 45°$

11. Since $\det\left(\frac{1}{2}\begin{bmatrix} \sqrt{3} & -1 \\ 1 & \sqrt{3} \end{bmatrix}\right) = \frac{1}{4}(3+1) = 1$, the operator is a rotation. Comparing this matrix with the general form of a rotation matrix,

$$A_\theta = \begin{bmatrix} \cos\theta & -\sin\theta \\ \sin\theta & \cos\theta \end{bmatrix} = \begin{bmatrix} \frac{\sqrt{3}}{2} & -\frac{1}{2} \\ \frac{1}{2} & \frac{\sqrt{3}}{2} \end{bmatrix},$$

we see that $\cos\theta = \frac{\sqrt{3}}{2}$ and $\sin\theta = \frac{1}{2}$. Therefore $\theta = 30°$.

12. a reflection, $y = \dfrac{1}{2-\sqrt{3}}x$

13. Since

$\det\left(\frac{1}{13}\begin{bmatrix} 5 & 12 \\ 12 & -5 \end{bmatrix}\right) = \frac{1}{13^2}(-25-144) = -1,$

the operator is a reflection. As in Exercise 9, the line of reflection is the eigenspace of the given matrix corresponding to eigenvalue 1, which is the solution set of the system

$$\left(\frac{1}{13}\begin{bmatrix} 5 & 12 \\ 12 & -5 \end{bmatrix} - \begin{bmatrix} 1 & 0 \\ 0 & 1 \end{bmatrix}\right)\begin{bmatrix} x \\ y \end{bmatrix} = \begin{bmatrix} 0 \\ 0 \end{bmatrix}.$$

Treating x as a free variable, the general solution of this system is given by $y = \frac{2}{3}x$, which is the equation of the line of reflection.

14. a reflection, $y = x$ **15.** a rotation, $\theta = 270°$

16. a reflection, $y = \sqrt{3}\,x$

17. True

18. False; for example, if T is a translation by a nonzero vector, then T preserves distances, but T is not linear.

19. False, only orthogonal linear operators preserve dot products.

20. True   **21.** True   **22.** True

23. False, for example, let $P = I_n$ and $Q = -I_n$.

24. False, for example, let $P = \begin{bmatrix} 1 & 1 \\ 1 & 2 \end{bmatrix}$.

25. True

26. False, consider $\begin{bmatrix} 1 & 1 \\ 1 & -1 \end{bmatrix}$.

27. False, consider P_W from Example 4 in Section 6.3.

28. True   **29.** True   **30.** True

31. False, we need $\det Q = 1$.

32. False, for example, if T is a translation by a nonzero vector, then T is a rigid motion, but T is not linear, and hence is not orthogonal.

33. False, for example, if T is a translation by a nonzero vector, then T is a rigid motion, but T is not linear.

34. True   **35.** True   **36.** True

37. Using the method of Example 3, one possibility is to let $T = T_A$, for

$$A = \frac{1}{35}\begin{bmatrix} 10 & 33 & 6 \\ -30 & 6 & 17 \\ 15 & -10 & 30 \end{bmatrix}.$$

38. We begin by applying the method in Example 3 to obtain orthogonal matrices A and B such that $A\mathbf{e}_1 = \mathbf{v}$ and $B\mathbf{e}_1 = \mathbf{w}$. For this purpose, choose the columns of A and B to be orthonormal bases of $\mathcal{R}^3$ so that $\mathbf{a}_1 = \mathbf{v}$ and $\mathbf{b}_1 = \mathbf{w}$. For example, let

$$A = \frac{1}{\sqrt{10}}\begin{bmatrix} 3 & 1 & 0 \\ 1 & -3 & 0 \\ 0 & 0 & \sqrt{10} \end{bmatrix}$$

and

$$B = \frac{1}{\sqrt{5}}\begin{bmatrix} 0 & \sqrt{5} & 0 \\ -2 & 0 & 1 \\ 1 & 0 & 2 \end{bmatrix}.$$

Then $\mathbf{e}_1 = A^T\mathbf{v} = B^T\mathbf{w}$, and therefore

$$BA^T\mathbf{v} = \frac{1}{5\sqrt{2}}\begin{bmatrix} \sqrt{5} & -3\sqrt{5} & 0 \\ -6 & -2 & \sqrt{10} \\ 3 & 1 & 2\sqrt{10} \end{bmatrix}\mathbf{v} = \mathbf{w}.$$

Finally, let $C = BA^T$ and $T = T_C$.

39. (a) Let Q be the standard matrix of T. Then
$$Q = \begin{bmatrix} \cos\theta & -\sin\theta & 0 \\ \sin\theta & \cos\theta & 0 \\ 0 & 0 & 1 \end{bmatrix}.$$

Since $QQ^T = I_3$, matrix Q is orthogonal. It follows that T is an orthogonal operator.

(b) First, we compute the characteristic polynomial of Q in (a):

$\det(Q - tI_3)$
$= \det \begin{bmatrix} \cos\theta - t & -\sin\theta & 0 \\ \sin\theta & \cos\theta - t & 0 \\ 0 & 0 & 1-t \end{bmatrix}$
$= (1-t)(t^2 - 2t\cos\theta + 1).$

Since the discriminant of $t^2 - 2t\cos\theta + 1$ is $4\cos^2\theta - 4 < 0$, the only (real) eigenvalue of Q is 1, which has multiplicity 1. It follows that the eigenspace corresponding to 1 is 1-dimensional. Since $T(\mathbf{e}_3) = \mathbf{e}_3$, we see that $\mathbf{e}_3$ is an eigenvector for the eigenvalue 1. Therefore the eigenspace corresponding to 1 is the span of $\{\mathbf{e}_3\}$, the z-axis.

(c) For any vector $\mathbf{v}$ in $\mathcal{R}^3$, the image $T(\mathbf{v})$ is obtained by rotating $\mathbf{v}$ about the z-axis by the angle θ. Looking down from the positive direction of the z-axis, this rotation is counter-clockwise.

40. Both B and C are orthogonal matrices since their columns are orthonormal bases for $\mathcal{R}^n$. Furthermore, $B\mathbf{e}_i = \mathbf{v}_i$ and $C\mathbf{e}_i = \mathbf{w}_i$ for $1 \le i \le k$. Let $A = CB^T$. Then A is an orthogonal matrix because it is a product of orthogonal matrices. Thus, for $1 \le i \le k$, we have
$$A\mathbf{v}_i = CB^T\mathbf{v}_i = CB^T B\mathbf{e}_i = C\mathbf{e}_i = \mathbf{w}_i.$$

Finally, let $T = T_A$. Then T is an orthogonal operator on $\mathcal{R}^n$ and $T(\mathbf{v}_i) = \mathbf{w}_i$ for $1 \le i \le k$.

41. We extend $\{\mathbf{v}_1, \mathbf{v}_2\}$ and $\{\mathbf{w}_1, \mathbf{w}_2\}$ to orthonormal bases for $\mathcal{R}^3$ by including
$$\mathbf{v}_3 = \frac{1}{3}\begin{bmatrix} 2 \\ -2 \\ 1 \end{bmatrix} \quad \text{and} \quad \mathbf{w}_3 = \frac{1}{7}\begin{bmatrix} 3 \\ -6 \\ 2 \end{bmatrix},$$
respectively. Let
$$B = \begin{bmatrix} \mathbf{v}_1 & \mathbf{v}_2 & \mathbf{v}_3 \end{bmatrix} = \frac{1}{3}\begin{bmatrix} 1 & 2 & 2 \\ 2 & 1 & -2 \\ 2 & -2 & 1 \end{bmatrix}$$

and
$$C = \begin{bmatrix} \mathbf{w}_1 & \mathbf{w}_2 & \mathbf{w}_3 \end{bmatrix} = \frac{1}{7}\begin{bmatrix} 2 & 6 & 3 \\ 3 & 2 & -6 \\ 6 & -3 & 2 \end{bmatrix},$$

which are orthogonal matrices. Take
$$A = CB^T = \frac{1}{21}\begin{bmatrix} 20 & 4 & -5 \\ -5 & 20 & -4 \\ 4 & 5 & 20 \end{bmatrix}$$

and $T = T_A$ to obtain an orthogonal operator that meets the given requirements.

42. Let Q be the standard matrix of T. Then
$$Q = \begin{bmatrix} -1 & 0 & 0 \\ 0 & 1 & 0 \\ 0 & 0 & 1 \end{bmatrix}.$$

Since $QQ^T = I_3$, matrix Q is orthogonal, and hence T is an orthogonal operator.

43. By Theorem 6.11, Q is a reflection or a rotation. If Q is a reflection, then it is diagonalizable by Example 3. So suppose that Q is a rotation by the angle θ. Then $Q = \begin{bmatrix} \cos\theta & -\sin\theta \\ \sin\theta & \cos\theta \end{bmatrix}$, which has the characteristic polynomial
$$\det Q = \begin{bmatrix} \cos\theta - t & -\sin\theta \\ \sin\theta & \cos\theta - t \end{bmatrix} = t^2 - 2t\cos\theta + 1.$$

The discriminant of this polynomial, $4\cos^2\theta - 4$, is nonnegative if and only if $\cos\theta = \pm 1$, that is, if and only if $Q = \pm I_2$. So if $Q \ne \pm I_2$, Q has no eigenvalues and hence is not diagonalizable.

44. (a) Let $\mathcal{B}_1$ and $\mathcal{B}_2$ be orthonormal bases for W and $W^\perp$, respectively, and let $\mathcal{B} = \mathcal{B}_1 \cup \mathcal{B}_2$, which is an orthonormal basis for $\mathcal{R}^n$. If $Q = [T]_\mathcal{B}$, then
$$Q = \left[\begin{array}{c|c} I_k & O \\ \hline O & -I_{n-k} \end{array}\right].$$

Since $QQ^T = I_n$, we see that Q is an orthogonal matrix. The matrix B whose columns are the vectors in $\mathcal{B}$ is an orthogonal matrix. Let A be the standard matrix of T. Then $A = BQB^{-1}$, which is an orthogonal matrix by Theorem 6.10. Therefore T is an orthogonal operator.

(b) Consider any vector $\mathbf{v}$ in $\mathcal{R}^n$, and let $\mathbf{w}$ and $\mathbf{z}$ be the unique vectors in W and $W^\perp$, respectively, such that $\mathbf{v} = \mathbf{w} + \mathbf{z}$. Then
$$U(\mathbf{v}) = \frac{1}{2}(\mathbf{v} + T(\mathbf{v})) = \frac{1}{2}(\mathbf{w} + \mathbf{z} + \mathbf{w} - \mathbf{z})$$

158   Chapter 6   Orthogonality

$= \mathbf{w} = P_W \mathbf{v}$.

It follows that P_W is the standard matrix of U.

45. Let $\mathcal{B} = \{\mathbf{v}, \mathbf{w}\}$. Observe that

$$T(\mathbf{v}) = (\mathbf{v} \cdot \mathbf{v} \cos\theta + \mathbf{v} \cdot \mathbf{w} \sin\theta)\mathbf{v}$$
$$+ (-\mathbf{v} \cdot \mathbf{v} \sin\theta + \mathbf{v} \cdot \mathbf{w} \cos\theta)\mathbf{w}$$
$$= \cos\theta \mathbf{v} - \sin\theta \mathbf{w}.$$

Similarly, $T(\mathbf{w}) = \sin\theta \mathbf{v} + \cos\theta \mathbf{w}$, and hence

$$[T]_\mathcal{B} = \begin{bmatrix} \cos\theta & \sin\theta \\ -\sin\theta & \cos\theta \end{bmatrix}.$$

Since $[T]_\mathcal{B}$ is an orthogonal matrix, T is an orthogonal operator.

46. For any i and j, the (i,j)-entry of QQ^T is the dot product of the ith row of Q with the jth column of Q^T, that is, the jth row of Q. Thus the (i,j)-entry of QQ^T is 1 if $i = j$ and 0 if $i \neq j$. So $QQ^T = I_n$, and therefore Q is an orthogonal matrix.

47. Let T and U be orthogonal operators on $\mathcal{R}^n$, and let P and Q be the standard matrices of T and U, respectively. By Theorem 6.1(b), PQ is an orthogonal matrix. But PQ is the standard matrix of TU, and hence TU is an orthogonal operator. Similarly, T^{-1} is an orthogonal operator by Theorem 6.10(c).

48. Suppose T is a reflection and U is a rotation. Then $\det P = -1$ and $\det Q = 1$. Hence

$$\det PQ = (\det P)(\det Q) = (-1)(1) = -1,$$

and therefore PQ is a reflection by Theorem 6.11(b). The proof for the case that T is a rotation and Q is a reflection is similar.

49. Suppose that λ is an eigenvalue for Q, and let $\mathbf{v}$ be a corresponding eigenvector. Then

$$\|\mathbf{v}\| = \|Q\mathbf{v}\| = \|\lambda\mathbf{v}\| = |\lambda|\|\mathbf{v}\|.$$

Since $\|\mathbf{v}\| \neq 0$, it follows that $|\lambda| = 1$. Therefore $\lambda = \pm 1$.

50. (a) Let Q be the standard matrix of U. By the results of Section 4.5, there is an invertible matrix P such that $Q = PDP^{-1}$, where $D = \begin{bmatrix} 1 & 0 \\ 0 & -1 \end{bmatrix}$. Observe that $D^2 = I_2$. Hence

$$Q^2 = PDP^{-1}PDP^{-1} = PD^2P^{-1}$$
$$= PI_2P^{-1} = I_2,$$

and thus $U^2 = I$. Therefore $U^{-1} = U$.

(b) Since T is a rotation and U is a reflection, TU is a reflection by Theorem 6.12. So by (a) we have $TUTU = I$. Since $U^2 = I$, we have

$$TUT = TUTUU = (TUTU)U = IU = U.$$

(c) As in (b), we have $UTUT = I$, and hence

$$UTU = UTUI = UTUTT^{-1}$$
$$= IT^{-1} = T^{-1}.$$

51. (a) Let Q be the standard matrix of T. Then Q^{-1} is the standard matrix of T^{-1}. Furthermore, since T is an orthogonal operator, both Q and Q^{-1} are orthogonal matrices. Also, $\det Q = 1$ because T is a rotation. So

$$\det Q^{-1} = (\det Q)^{-1} = (1)^{-1} = 1.$$

It follows that T^{-1} is a rotation.
Suppose T rotates any nonzero vector $\mathbf{v}$ in $\mathcal{R}^2$ by the angle θ. Then $T^{-1}(T(\mathbf{v})) = \mathbf{v}$, and hence T^{-1} rotates a nonzero vector by the angle $-\theta$.

(b) Since T is a reflection, $T = T^{-1}$ by Exercise 50(a). It follows that T^{-1} is a reflection, and the line of reflection of T^{-1} is the same as the line of reflection of T.

52. (a) Notice that $S^2 = T$, and since S is a rotation, $SUS = U$ by Exercise 50(b). Thus

$$TU(S(\mathbf{b})) = S(SUS)(\mathbf{b})$$
$$= S(U(\mathbf{b})) = S(\mathbf{b}),$$

and hence $S(\mathbf{b})$ is an eigenvector of TU corresponding to eigenvalue 1.

(b) Because T is a rotation and U is a reflection, TU is a reflection. By (a), $S(\mathbf{b})$ is an eigenvector of TU corresponding to eigenvalue 1, and hence $S(\mathbf{b})$ is in the direction of $\mathcal{L}'$. It follows that TU is the reflection about $\mathcal{L}'$.

53. (a)

$$Q_W^T = (2P_W - I_2)^T = 2P_W^T - I_2^T$$
$$= 2P_W - I_2 = Q_W$$

(b)

$$Q_W^2 = (2P_W - I_2)^2 = 4P_W^2 - 4P_W I_2 + I_2$$
$$= 4P_W - 4P_W + I_2 = I_2$$

(c) By (a) and (b), we have
$$Q_W^T Q_W = Q_W Q_W = I_2,$$
and hence Q_W is an orthogonal matrix.

(d)
$$Q_W \mathbf{w} = (2P_W - I_2)\mathbf{w} = 2P_W \mathbf{w} - I_2 \mathbf{w}$$
$$= 2\mathbf{w} - \mathbf{w} = \mathbf{w}$$

(e)
$$Q_W \mathbf{v} = (2P_W - I_2)\mathbf{v} = 2P_W \mathbf{v} - I_2 \mathbf{v}$$
$$= \mathbf{0} - \mathbf{v} = -\mathbf{v}$$

(f) Select nonzero vectors $\mathbf{w}$ in W and $\mathbf{v}$ in $W^\perp$. Then $\{\mathbf{w}, \mathbf{v}\}$ is a basis for $\mathcal{R}^2$ since it is an orthogonal set of nonzero vectors. Let $P = [\mathbf{w} \ \mathbf{v}]$ and T be the matrix transformation induced by Q_W. Then Q_W is the standard matrix of T, and T is an orthogonal operator because Q_W is an orthogonal matrix. Also,
$$Q_W = PDP^{-1}, \quad \text{where} \quad D = \begin{bmatrix} 1 & 0 \\ 0 & -1 \end{bmatrix}.$$
Thus
$$\det Q_W = \det(PDP^{-1})$$
$$= (\det P)(\det D)(\det P^{-1})$$
$$= (\det P)(-1)(\det P)^{-1} = -1.$$

It follows that T is a reflection. Furthermore, since $T(\mathbf{w}) = \mathbf{w}$, T is the reflection of $\mathcal{R}^2$ about W.

54. Suppose that T is an orthogonal operator. Then for any i and j,
$$T(\mathbf{v}_i) \cdot T(\mathbf{v}_j) = \mathbf{v}_i \cdot \mathbf{v}_j = \begin{cases} 1 & \text{if } i = j \\ 0 & \text{if } i \neq j. \end{cases}$$

Hence $\{T(\mathbf{v}_1), T(\mathbf{v}_2), \ldots, T(\mathbf{v}_n)\}$ is an orthonormal set of n vectors. Therefore it is an orthonormal basis for $\mathcal{R}^n$.

Conversely, suppose $\{T(\mathbf{v}_1), T(\mathbf{v}_2), \ldots, T(\mathbf{v}_n)\}$ is an orthonormal basis for $\mathcal{R}^n$. Let $P = [\mathbf{v}_1 \ \mathbf{v}_2 \ \cdots \ \mathbf{v}_n]$, $Q = [T(\mathbf{v}_1) \ T(\mathbf{v}_2) \ \ldots \ T(\mathbf{v}_n)]$, and A be the standard matrix of T. Then P and Q are orthogonal matrices because their columns form orthonormal sets. Furthermore
$$Q = [T(\mathbf{v}_1) \ T(\mathbf{v}_2) \ \ldots \ T(\mathbf{v}_n)]$$
$$= [A\mathbf{v}_1 \ A\mathbf{v}_2 \ \cdots \ A\mathbf{v}_n]$$
$$= A[\mathbf{v}_1 \ \mathbf{v}_2 \ \cdots \ \mathbf{v}_n]$$
$$= AP.$$

Therefore $A = QP^{-1}$ is an orthogonal matrix, and it follows that T is an orthogonal operator.

55. Let $P = [\mathbf{v}_1 \ \mathbf{v}_2 \ \cdots \ \mathbf{v}_n]$ and $Q = [\mathbf{w}_1 \ \mathbf{w}_2 \ \cdots \ \mathbf{w}_n]$. Both P and Q are orthogonal matrices since their columns form orthonormal bases. Let $A = QP^{-1}$, and let T be the matrix operator induced by A. Then A is the standard matrix of T, and A is an orthogonal matrix. So T is an orthogonal operator. Furthermore
$$[T(\mathbf{v}_1) \ T(\mathbf{v}_2) \ \cdots \ T(\mathbf{v}_n)] = [A\mathbf{v}_1 \ A\mathbf{v}_2 \ \cdots \ A\mathbf{v}_n]$$
$$= AP = Q$$
$$= [\mathbf{w}_1 \ \mathbf{w}_2 \ \cdots \ \mathbf{w}_n].$$

Hence $T(\mathbf{v}_i) = \mathbf{w}_i$ for all i.

56. Let F and G be rigid motions on $\mathcal{R}^n$. For any $\mathbf{u}$ and $\mathbf{v}$ in $\mathcal{R}^n$, we have
$$\|G(F(\mathbf{u})) - G(F(\mathbf{v}))\| = \|F(\mathbf{u}) - F(\mathbf{v})\|$$
$$= \|\mathbf{u} - \mathbf{v}\|,$$
and hence GF is a rigid motion.

57.
$$\|T(\mathbf{u})\| = \|T(\mathbf{u}) - \mathbf{0}\| = \|T(\mathbf{u}) - T(\mathbf{0})\|$$
$$= \|\mathbf{u} - \mathbf{0}\| = \|\mathbf{u}\|$$

58. Let $\mathbf{u}$ be a vector in $\mathcal{R}^n$ and c be a scalar. Then
$$\|T(c\mathbf{u}) - cT(\mathbf{u})\|^2$$
$$= [T(c\mathbf{u}) - cT(\mathbf{u})] \cdot [T(c\mathbf{u}) - cT(\mathbf{u})]$$
$$= T(c\mathbf{u}) \cdot T(c\mathbf{u}) - 2cT(\mathbf{u}) \cdot T(c\mathbf{u}) + c^2 T(\mathbf{u}) \cdot T(\mathbf{u})$$
$$= (c\mathbf{u}) \cdot (c\mathbf{u}) - 2c(\mathbf{u} \cdot c\mathbf{u}) + c^2 \mathbf{u} \cdot \mathbf{u}$$
$$= c^2 \|\mathbf{u}\|^2 - 2c^2 \|\mathbf{u}\|^2 + c^2 \|\mathbf{u}\|^2 = 0.$$

Thus $T(c\mathbf{u}) - cT(\mathbf{u}) = \mathbf{0}$, and therefore $T(c\mathbf{u}) = cT(\mathbf{u})$.

59. Suppose that
$$F(\mathbf{v}) = Q\mathbf{v} + \mathbf{b} = R\mathbf{v} + \mathbf{c}$$
for every vector $\mathbf{v}$ in $\mathcal{R}^n$. Then $F(\mathbf{0}) = \mathbf{b} = \mathbf{c}$, and therefore $Q\mathbf{v} = R\mathbf{v}$ for every vector $\mathbf{v}$ in $\mathcal{R}^n$. It follows that $Q = R$.

60. For any vector $\mathbf{v}$ in $\mathcal{R}^n$, we have
$$F(G(\mathbf{v})) = F(P\mathbf{v} + \mathbf{a}) = Q(P\mathbf{v} + \mathbf{a}) + \mathbf{b}$$
$$= QP\mathbf{v} + (Q\mathbf{a} + \mathbf{b}).$$

Therefore $R = QP$ and $\mathbf{c} = Q\mathbf{a} + \mathbf{b}$.

61. Since

$$F\left(\begin{bmatrix}1\\0\end{bmatrix}\right) + F\left(\begin{bmatrix}0\\1\end{bmatrix}\right) = Q\begin{bmatrix}1\\0\end{bmatrix} + \mathbf{b} + Q\begin{bmatrix}0\\1\end{bmatrix} + \mathbf{b}$$
$$= \mathbf{q}_1 + \mathbf{q}_2 + 2\mathbf{b}$$

and

$$F\left(\begin{bmatrix}1\\1\end{bmatrix}\right) = Q\left(\begin{bmatrix}1\\0\end{bmatrix} + \begin{bmatrix}0\\1\end{bmatrix}\right) + \mathbf{b}$$
$$= \mathbf{q}_1 + \mathbf{q}_2 + \mathbf{b},$$

it follows that

$$\mathbf{b} = F\left(\begin{bmatrix}1\\0\end{bmatrix}\right) + F\left(\begin{bmatrix}0\\1\end{bmatrix}\right) - F\left(\begin{bmatrix}1\\1\end{bmatrix}\right)$$
$$= \begin{bmatrix}2\\4\end{bmatrix} + \begin{bmatrix}1\\3\end{bmatrix} - \begin{bmatrix}2\\3\end{bmatrix} = \begin{bmatrix}1\\4\end{bmatrix}.$$

Thus

$$\mathbf{q}_1 = Q\begin{bmatrix}1\\0\end{bmatrix} = F\left(\begin{bmatrix}1\\0\end{bmatrix}\right) - \mathbf{b} = \begin{bmatrix}2\\4\end{bmatrix} - \begin{bmatrix}1\\4\end{bmatrix} = \begin{bmatrix}1\\0\end{bmatrix}$$

and

$$\mathbf{q}_2 = Q\begin{bmatrix}0\\1\end{bmatrix} = F\left(\begin{bmatrix}0\\1\end{bmatrix}\right) - \mathbf{b} = \begin{bmatrix}1\\3\end{bmatrix} - \begin{bmatrix}1\\4\end{bmatrix} = \begin{bmatrix}0\\-1\end{bmatrix}.$$

Therefore $Q = [\mathbf{q}_1 \; \mathbf{q}_2] = \begin{bmatrix}1 & 0\\0 & -1\end{bmatrix}$.

62. $Q = \frac{1}{5}\begin{bmatrix}3 & 4\\4 & -3\end{bmatrix}$ and $\mathbf{b} = \begin{bmatrix}-1\\1\end{bmatrix}$

63. $Q = \begin{bmatrix}.8 & -.6\\.6 & .8\end{bmatrix}$ and $\mathbf{b} = \begin{bmatrix}0\\3\end{bmatrix}$

64. $Q = \begin{bmatrix}-.6 & .8\\.8 & .6\end{bmatrix}$ and $\mathbf{b} = \begin{bmatrix}4\\1\end{bmatrix}$

65. Let $\mathbf{u}$ and $\mathbf{v}$ be in $\mathcal{R}^n$. Then

$$\|T(\mathbf{u}) - T(\mathbf{v})\|^2$$
$$= [T(\mathbf{u}) - T(\mathbf{v})] \cdot [T(\mathbf{u}) - T(\mathbf{v})]$$
$$= T(\mathbf{u}) \cdot T(\mathbf{u}) - 2T(\mathbf{u}) \cdot T(\mathbf{v}) + T(\mathbf{v}) \cdot T(\mathbf{v})$$
$$= \mathbf{u} \cdot \mathbf{u} - 2\mathbf{u} \cdot \mathbf{v} + \mathbf{v} \cdot \mathbf{v} \|\mathbf{u} - \mathbf{v}\|^2.$$

Hence $\|T(\mathbf{u}) - T(\mathbf{v})\| = \|\mathbf{u} - \mathbf{v}\|$. It follows that T is a rigid motion. Furthermore,

$$\|T(\mathbf{0})\|^2 = T(\mathbf{0}) \cdot T(\mathbf{0}) = \mathbf{0} \cdot \mathbf{0} = 0,$$

and hence $T(\mathbf{0}) = \mathbf{0}$. Therefore T is an orthogonal operator by Theorem 6.13.

66. Let $\mathbf{u}$ and $\mathbf{v}$ be nonzero vectors in $\mathcal{R}^2$, and let $0 \le \alpha \le 180°$ and $0 \le \beta \le 180°$ be the angles between $\mathbf{u}$ and $\mathbf{v}$ and between $T(\mathbf{u})$ and $T(\mathbf{v})$, respectively. By Exercise 98 of Section 6.1, we have

$$\mathbf{u} \cdot \mathbf{v} = \|\mathbf{u}\| \cdot \|\mathbf{u}\| \cos \alpha$$

and

$$T(\mathbf{u}) \cdot T(\mathbf{v}) = \|T(\mathbf{u})\| \cdot \|T(\mathbf{u})\| \cos \beta.$$

But $T(\mathbf{u}) \cdot T(\mathbf{v}) = \mathbf{u} \cdot \mathbf{v}$, $\|T(\mathbf{u})\| = \|\mathbf{u}\|$, and $\|T(\mathbf{v})\| = \|\mathbf{v}\|$. It follows that $\cos \alpha = \cos \beta$, and hence $\alpha = \beta$.

67. (a) $A_2 = \begin{bmatrix}0 & -1\\-1 & 0\end{bmatrix}$,

$$A_3 = \frac{1}{3}\begin{bmatrix}1 & -2 & -2\\-2 & 1 & -2\\-2 & -2 & 1\end{bmatrix},$$

and

$$A_6 = \frac{1}{3}\begin{bmatrix}2 & -1 & -1 & -1 & -1 & -1\\-1 & 2 & -1 & -1 & -1 & -1\\-1 & -1 & 2 & -1 & -1 & -1\\-1 & -1 & -1 & 2 & -1 & -1\\-1 & -1 & -1 & -1 & 2 & -1\\-1 & -1 & -1 & -1 & -1 & 2\end{bmatrix}$$

(b) For $n = 3$, we have

$$A_3^T A_3 = \frac{1}{3}\begin{bmatrix}1 & -2 & -2\\-2 & 1 & -2\\-2 & -2 & 1\end{bmatrix}\frac{1}{3}\begin{bmatrix}1 & -2 & -2\\-2 & 1 & -2\\-2 & -2 & 1\end{bmatrix}$$

$$= \frac{1}{9}\begin{bmatrix}9 & 0 & 0\\0 & 9 & 0\\0 & 0 & 9\end{bmatrix} = I_3.$$

So A_3 is invertible, and $(A_3)^{-1} = A_3^T$. Therefore A_3 is orthogonal by Theorem 6.2(b).

(c) Since the entries of E_n are all 1s, the same is true for the entries of E_n^T, and hence $E_n^T = E_n$. Thus

$$A_n^T = (I_n - \frac{2}{n}E_n)^T = I_n^T - \frac{2}{n}E_n^T$$
$$= I_n - \frac{2}{n}E_n = A_n,$$

proving that A_n is symmetric.

(d) First observe that the (i,j)-entry of E_n^2 is $1^2 + 1^2 + \cdots + 1^2 = n$, and therefore $E_n^2 = nE_n$. Hence

$$A_n^2 = (I_n - \frac{2}{n}E_n)(I_n - \frac{2}{n}E_n)$$
$$= I_n - \frac{4}{n}I_n E_n + \frac{4}{n^2}e_n^2$$

$$= I_n - \frac{4}{n}E_n + \frac{4}{n^2}nE_n$$
$$= I_n.$$

Thus, by (c), we have
$$A_n A_n^T = A_n A_n = I_n.$$

It follows that A_n is invertible and $(A_n)^{-1} = A_n^T$. Therefore A_n is orthogonal by Theorem 6.2(b).

68. Let Q be the standard matrix of U, $\mathbf{w} = \begin{bmatrix} \cos\theta \\ \sin\theta \end{bmatrix}$, $\mathbf{z} = \begin{bmatrix} -\sin\theta \\ \cos\theta \end{bmatrix}$, and $\mathcal{B} = \{\mathbf{w}, \mathbf{z}\}$. Then $\mathcal{B}$ is a basis for $\mathcal{R}^2$, $U(\mathbf{w}) = \mathbf{w}$, and $U(\mathbf{z}) = -\mathbf{z}$. Let $P = [\mathbf{w} \ \mathbf{z}]$. Then
$$[U]_{\mathcal{B}} = \begin{bmatrix} 1 & 0 \\ 0 & -1 \end{bmatrix} \quad \text{and} \quad Q = P[U]_{\mathcal{B}} P^{-1}.$$

Thus
$$Q = P[U]_{\mathcal{B}} P^{-1}$$
$$= \begin{bmatrix} \cos\theta & -\sin\theta \\ \sin\theta & \cos\theta \end{bmatrix} \begin{bmatrix} 1 & 0 \\ 0 & -1 \end{bmatrix} \begin{bmatrix} \cos\theta & -\sin\theta \\ \sin\theta & \cos\theta \end{bmatrix}^{-1}$$
$$= \begin{bmatrix} \cos\theta & -\sin\theta \\ \sin\theta & \cos\theta \end{bmatrix} \begin{bmatrix} 1 & 0 \\ 0 & -1 \end{bmatrix} \begin{bmatrix} \cos\theta & \sin\theta \\ -\sin\theta & \cos\theta \end{bmatrix}$$
$$= \begin{bmatrix} \cos\theta & \sin\theta \\ \sin\theta & -\cos\theta \end{bmatrix} \begin{bmatrix} \cos\theta & \sin\theta \\ -\sin\theta & \cos\theta \end{bmatrix}$$
$$= \begin{bmatrix} \cos^2\theta - \sin^2\theta & 2\cos\theta\cos\theta \\ 2\cos\theta\cos\theta & \sin^2\theta - \cos^2\theta \end{bmatrix}$$
$$= \begin{bmatrix} \cos 2\theta & \sin 2\theta \\ \sin 2\theta & -\cos 2\theta \end{bmatrix}.$$

69. Let Q be the standard matrix of U, $\mathbf{w} = \begin{bmatrix} 1 \\ m \end{bmatrix}$, and $\mathbf{z} = \begin{bmatrix} -m \\ 1 \end{bmatrix}$. Since $\mathbf{w}$ lies along $\mathcal{L}$ and $\mathbf{z}$ is perpendicular to $\mathcal{L}$, we have $U(\mathbf{w}) = \mathbf{w}$ and $U(\mathbf{z}) = -\mathbf{z}$. Let $\mathcal{B} = \{\mathbf{w}, \mathbf{z}\}$ and $P = [\mathbf{w} \ \mathbf{z}]$. Then $\mathcal{B}$ is a basis for $\mathcal{R}^2$, and hence
$$[U]_{\mathcal{B}} = \begin{bmatrix} 1 & 0 \\ 0 & -1 \end{bmatrix} \quad \text{and} \quad Q = P[U]_{\mathcal{B}} P^{-1}.$$

Thus
$$Q = P[U]_{\mathcal{B}} P^{-1}$$
$$= \begin{bmatrix} 1 & -m \\ -m & 1 \end{bmatrix} \begin{bmatrix} 1 & 0 \\ 0 & -1 \end{bmatrix} \begin{bmatrix} 1 & m \\ -m & 1 \end{bmatrix}^{-1}$$

$$= \begin{bmatrix} 1 & m \\ -m & 1 \end{bmatrix} \begin{bmatrix} 1 & 0 \\ 0 & -1 \end{bmatrix} \begin{bmatrix} 1 & m \\ -m & 1 \end{bmatrix}^{-1}$$
$$= \begin{bmatrix} 1 & m \\ -m & 1 \end{bmatrix} \begin{bmatrix} 1 & 0 \\ 0 & -1 \end{bmatrix} \left(\frac{1}{1+m^2} \begin{bmatrix} 1 & m \\ -m & 1 \end{bmatrix} \right)$$
$$= \frac{1}{1+m^2} \begin{bmatrix} 1-m^2 & 2m \\ 2m & m^2-1 \end{bmatrix}.$$

70. $\begin{bmatrix} 0.5496 & 0.8354 \\ 0.8354 & -0.5496 \end{bmatrix}$ (rounded to 4 places after the decimal)

71. $\begin{bmatrix} 0.7833 & 0.6217 \\ 0.6217 & -0.7833 \end{bmatrix}$ (rounded to 4 places after the decimal)

72. $-31.4002°$ 73. $231°$

6.6 SYMMETRIC MATRICES

1. (a) $A = \begin{bmatrix} 2 & -7 \\ -7 & 50 \end{bmatrix}$ (b) about $8.1°$

   (c) $x = \frac{7}{\sqrt{50}}x' - \frac{1}{\sqrt{50}}y'$
   $y = \frac{1}{\sqrt{50}}x' + \frac{7}{\sqrt{50}}y'$

   (d) $(x')^2 + 51(y')^2 = 255$ (e) an ellipse

2. (a) $A = \begin{bmatrix} 2 & 1 \\ 1 & 2 \end{bmatrix}$ (b) $45°$

   (c) $x = \frac{1}{\sqrt{2}}x' - \frac{1}{\sqrt{2}}y'$
   $y = \frac{1}{\sqrt{2}}x' + \frac{1}{\sqrt{2}}y'$

   (d) $3(x')^2 + (y')^2 = 1$ (e) an ellipse

3. (a) $\begin{bmatrix} 1 & -6 \\ -6 & -4 \end{bmatrix}$ (b) about $56.3°$

   (c) $x = \frac{2}{\sqrt{13}}x' - \frac{3}{\sqrt{13}}y'$
   $y = \frac{3}{\sqrt{13}}x' + \frac{2}{\sqrt{13}}y'$

   (d) $-8(x')^2 + 5(y')^2 = 40$ (e) a hyperbola

4. (a) $A = \begin{bmatrix} 3 & -2 \\ -2 & 3 \end{bmatrix}$ (b) $45°$

   (c) $x = \frac{1}{\sqrt{2}}x' - \frac{1}{\sqrt{2}}y'$
   $y = \frac{1}{\sqrt{2}}x' + \frac{1}{\sqrt{2}}y'$

   (d) $(x')^2 + 5(y')^2 = 5$ (e) an ellipse

5. (a) Let $a_{11} = 5$, $a_{22} = 5$, and $a_{12} = a_{21} = \frac{4}{2} = 2$. Thus
$$A = \begin{bmatrix} 5 & 2 \\ 2 & 5 \end{bmatrix}.$$

A has eigenvalues 7 and 3, and
$$\left\{ \begin{bmatrix} \frac{1}{\sqrt{2}} \\ \frac{1}{\sqrt{2}} \end{bmatrix}, \begin{bmatrix} -\frac{1}{\sqrt{2}} \\ \frac{1}{\sqrt{2}} \end{bmatrix} \right\}$$

is an orthonormal basis for $\mathcal{R}^2$ consisting of corresponding eigenvectors. Choose the x'-axis to be in the direction of the first vector in this basis.

(b) Since the x'-axis is in the direction of $\begin{bmatrix} \frac{1}{\sqrt{2}} \\ \frac{1}{\sqrt{2}} \end{bmatrix}$, it has a slope of 1, and hence the angle of rotation of the coordinate system is 45°.

(c) Let P be the 45 rotation matrix. Then

$$P = \begin{bmatrix} \frac{1}{\sqrt{2}} & -\frac{1}{\sqrt{2}} \\ \frac{1}{\sqrt{2}} & \frac{1}{\sqrt{2}} \end{bmatrix}, \text{ and } \begin{bmatrix} x \\ y \end{bmatrix} = P \begin{bmatrix} x' \\ y' \end{bmatrix}.$$

Therefore

$$x = \tfrac{1}{\sqrt{2}}x' - \tfrac{1}{\sqrt{2}}y'$$
$$y = \tfrac{1}{\sqrt{2}}x' + \tfrac{1}{\sqrt{2}}y'.$$

(d) Since the eigenvalues are 7 and 3, the equation becomes $7(x')^2 + 3(y')^2 = 9$.

(e) The conic section is an ellipse.

6. (a) $A = \begin{bmatrix} 11 & 12 \\ 12 & 4 \end{bmatrix}$ (b) about 36.9°

   (c) $x = \tfrac{4}{5}x' - \tfrac{3}{5}y'$
   $y = \tfrac{3}{5}x' + \tfrac{4}{5}y'$

   (d) $20(x')^2 - 5(y')^2 = 15$ (e) a hyperbola

7. (a) $\begin{bmatrix} 1 & 2 \\ 2 & 1 \end{bmatrix}$ (b) 45°

   (c) $x = \tfrac{1}{\sqrt{2}}x' - \tfrac{1}{\sqrt{2}}y'$
   $y = \tfrac{1}{\sqrt{2}}x' + \tfrac{1}{\sqrt{2}}y'$

   (d) $3(x')^2 - (y')^2 = 7$ (e) a hyperbola

8. (a) $A = \begin{bmatrix} 4 & 3 \\ 3 & -4 \end{bmatrix}$ (b) about 18.4°

   (c) $x = \tfrac{3}{\sqrt{10}}x' - \tfrac{1}{\sqrt{10}}y'$
   $y = \tfrac{1}{\sqrt{10}}x' + \tfrac{3}{\sqrt{10}}y'$

   (d) $(x')^2 - (y')^2 = 36$ (e) a hyperbola

9. (a) Let $a_{11} = 2$, $a_{22} = -7$, and $a_{12} = a_{21} = \tfrac{-12}{2} = -6$. Thus

$$A = \begin{bmatrix} 2 & -6 \\ -6 & -7 \end{bmatrix}.$$

A has the eigenvalues -10 and 5, and

$$\left\{ \begin{bmatrix} \frac{1}{\sqrt{5}} \\ \frac{2}{\sqrt{5}} \end{bmatrix}, \begin{bmatrix} -\frac{2}{\sqrt{5}} \\ \frac{1}{\sqrt{5}} \end{bmatrix} \right\}$$

is an orthonormal basis for $\mathcal{R}^2$ consisting of corresponding eigenvectors. Choose the x'-axis to be in the direction of the first vector in this basis.

(b) Since the x'-axis is in the direction of $\begin{bmatrix} \frac{1}{\sqrt{5}} \\ \frac{2}{\sqrt{5}} \end{bmatrix}$, it has a slope of 2, and hence the angle of rotation of the coordinate system is $\tan^{-1} 2 \approx 63.4°$.

(c) Let P be the rotation matrix corresponding to this angle of rotation. Then

$$P = \begin{bmatrix} \frac{1}{\sqrt{5}} & -\frac{2}{\sqrt{5}} \\ \frac{2}{\sqrt{5}} & \frac{1}{\sqrt{5}} \end{bmatrix} \text{ and } \begin{bmatrix} x \\ y \end{bmatrix} = P \begin{bmatrix} x' \\ y' \end{bmatrix}.$$

Therefore

$$x = \tfrac{1}{\sqrt{5}}x' - \tfrac{2}{\sqrt{5}}y'$$
$$y = \tfrac{2}{\sqrt{5}}x' + \tfrac{1}{\sqrt{5}}y'.$$

(d) Since the eigenvalues are -10 and 5, the equation becomes $-10(x')^2 + 5(y')^2 = 200$.

(e) The conic section is a hyperbola.

10. (a) $A = \begin{bmatrix} 6.0 & 2.5 \\ 2.5 & 6.0 \end{bmatrix}$ (b) about 11.3°

    (c) $x = \tfrac{5}{\sqrt{26}}x' - \tfrac{1}{\sqrt{26}}y'$
    $y = \tfrac{1}{\sqrt{26}}x' + \tfrac{5}{\sqrt{26}}y'$

    (d) $(x')^2 - (y')^2 = 4$ (e) a hyperbola

11. (a) $\begin{bmatrix} 1 & 1 \\ 1 & 1 \end{bmatrix}$ (b) 45°

    (c) $x = \tfrac{1}{\sqrt{2}}x' - \tfrac{1}{\sqrt{2}}y'$
    $y = \tfrac{1}{\sqrt{2}}x' + \tfrac{1}{\sqrt{2}}y'$

    (d) $2\sqrt{2}(x')^2 + 9x' - 7y' = 0$ (e) a parabola

12. (a) $A = \begin{bmatrix} 52 & 36 \\ 36 & 73 \end{bmatrix}$ (b) about 53.1°

    (c) $x = \tfrac{3}{5}x' - \tfrac{4}{5}y'$
    $y = \tfrac{4}{5}x' + \tfrac{3}{5}y'$

    (d) $100(x')^2 + 25(y')^2 - 200x' + 50y' + 25 = 0$
    (e) an ellipse

13. The characteristic polynomial of A is

$$\det(A - tI_2) = \det \begin{bmatrix} t-3 & 1 \\ 1 & t-3 \end{bmatrix} = (t-3)^2 - 1$$
$$= (t-2)(t-4),$$

and hence A has the eigenvalues $\lambda_1 = 2$ and $\lambda_2 = 4$. The vectors $\begin{bmatrix} 1 \\ -1 \end{bmatrix}$ and $\begin{bmatrix} 1 \\ 1 \end{bmatrix}$ are eigenvectors that correspond to these eigenvalues. Normalizing these vectors, we obtain unit vectors

$$\mathbf{u}_1 = \begin{bmatrix} \frac{1}{\sqrt{2}} \\ -\frac{1}{\sqrt{2}} \end{bmatrix} \quad \text{and} \quad \mathbf{u}_2 = \begin{bmatrix} \frac{1}{\sqrt{2}} \\ \frac{1}{\sqrt{2}} \end{bmatrix},$$

which form an orthonormal basis $\{\mathbf{u}_1, \mathbf{u}_2\}$ for $\mathcal{R}^2$. We use these eigenvectors and corresponding eigenvalues to obtain the spectral decomposition

$$A = \lambda_1 \mathbf{u}_1 \mathbf{u}_1^T + \lambda_2 \mathbf{u}_2 \mathbf{u}_2^T$$

$$= 2 \begin{bmatrix} \frac{1}{\sqrt{2}} \\ -\frac{1}{\sqrt{2}} \end{bmatrix} \begin{bmatrix} \frac{1}{\sqrt{2}} & -\frac{1}{\sqrt{2}} \end{bmatrix} + 4 \begin{bmatrix} \frac{1}{\sqrt{2}} \\ \frac{1}{\sqrt{2}} \end{bmatrix} \begin{bmatrix} \frac{1}{\sqrt{2}} & \frac{1}{\sqrt{2}} \end{bmatrix}$$

$$= 2 \begin{bmatrix} 0.5 & -0.5 \\ -0.5 & 0.5 \end{bmatrix} + 4 \begin{bmatrix} 0.5 & 0.5 \\ 0.5 & 0.5 \end{bmatrix}.$$

14. $\left\{ \dfrac{1}{\sqrt{5}} \begin{bmatrix} 1 \\ -2 \end{bmatrix}, \dfrac{1}{\sqrt{5}} \begin{bmatrix} 2 \\ 1 \end{bmatrix} \right\}$, -5 and 10,

$$A = (-5) \begin{bmatrix} .2 & -.4 \\ -.4 & .8 \end{bmatrix} + (10) \begin{bmatrix} .8 & .4 \\ .4 & .2 \end{bmatrix}$$

15. $\left\{ \dfrac{1}{\sqrt{2}} \begin{bmatrix} 1 \\ 1 \end{bmatrix}, \dfrac{1}{\sqrt{2}} \begin{bmatrix} 1 \\ -1 \end{bmatrix} \right\}$, 3 and -1,

$$A = 3 \begin{bmatrix} 0.5 & 0.5 \\ 0.5 & 0.5 \end{bmatrix} + (-1) \begin{bmatrix} 0.5 & -0.5 \\ -0.5 & 0.5 \end{bmatrix}$$

16. $\left\{ \dfrac{1}{\sqrt{2}} \begin{bmatrix} 1 \\ -1 \end{bmatrix}, \dfrac{1}{\sqrt{2}} \begin{bmatrix} 1 \\ 1 \end{bmatrix} \right\}$, 2 and 0,

$$A = 2 \begin{bmatrix} \frac{1}{2} & -\frac{1}{2} \\ -\frac{1}{2} & \frac{1}{2} \end{bmatrix} + 0 \begin{bmatrix} \frac{1}{2} & \frac{1}{2} \\ \frac{1}{2} & \frac{1}{2} \end{bmatrix}$$

17. The characteristic polynomial of A is

$$\det(A - tI_3) = \det \begin{bmatrix} 3-t & 2 & 2 \\ 2 & 2-7 & 0 \\ 2 & 0 & 4-t \end{bmatrix}$$

$$= -18t + 9t^2 - t^3$$

$$= -t(t-3)(t-6),$$

and hence A has the eigenvalues $\lambda_1 = 3$, $\lambda_2 = 6$, and $\lambda_3 = 0$. For each λ_i, select a nonzero solution of $(A - \lambda_i I_3)\mathbf{x} = \mathbf{0}$ to obtain an eigenvector corresponding to each eigenvalue. Since these eigenvalues are distinct, the eigenvectors are orthogonal, and hence normalizing these eigenvectors produces an orthonormal basis for $\mathcal{R}^3$ consisting of eigenvectors of A. Hence if

$$\mathbf{u}_1 = \frac{1}{3} \begin{bmatrix} -1 \\ -2 \\ 2 \end{bmatrix}, \quad \mathbf{u}_2 = \frac{1}{3} \begin{bmatrix} 2 \\ 1 \\ 2 \end{bmatrix},$$

and

$$\mathbf{u}_3 = \frac{1}{3} \begin{bmatrix} -2 \\ 2 \\ 1 \end{bmatrix},$$

then $\{\mathbf{u}_1, \mathbf{u}_2, \mathbf{u}_3\}$ is an orthonormal basis for $\mathcal{R}^3$ consisting of eigenvectors of A. We use these eigenvectors and corresponding eigenvalues to obtain the spectral decomposition

$$A = \lambda_1 \mathbf{u}_1 \mathbf{u}_1^T + \lambda_2 \mathbf{u}_2 \mathbf{u}_2^T + \lambda_3 \mathbf{u}_3 \mathbf{u}_3^T$$

$$= 3 \left(\frac{1}{3}\right) \begin{bmatrix} -1 \\ -2 \\ 2 \end{bmatrix} \frac{1}{3}[-1 \ -2 \ 2]$$

$$+ 6 \left(\frac{1}{3}\right) \begin{bmatrix} 2 \\ 1 \\ 2 \end{bmatrix} \frac{1}{3}[2 \ 1 \ 2]$$

$$+ 0 \left(\frac{1}{3}\right) \begin{bmatrix} -2 \\ 2 \\ 1 \end{bmatrix} \frac{1}{3}[-2 \ 2 \ 1]$$

$$= 3 \begin{bmatrix} \frac{1}{9} & \frac{2}{9} & -\frac{2}{9} \\ \frac{2}{9} & \frac{4}{9} & -\frac{4}{9} \\ -\frac{2}{9} & -\frac{4}{9} & \frac{2}{9} \end{bmatrix} + 6 \begin{bmatrix} \frac{4}{9} & \frac{2}{9} & \frac{4}{9} \\ \frac{2}{9} & \frac{1}{9} & \frac{2}{9} \\ \frac{4}{9} & \frac{2}{9} & \frac{4}{9} \end{bmatrix}$$

$$+ 0 \begin{bmatrix} \frac{4}{9} & -\frac{4}{9} & -\frac{2}{9} \\ -\frac{4}{9} & \frac{4}{9} & \frac{2}{9} \\ -\frac{2}{9} & \frac{2}{9} & \frac{1}{9} \end{bmatrix}.$$

18. $\left\{ \dfrac{1}{\sqrt{2}} \begin{bmatrix} 1 \\ -1 \\ 0 \end{bmatrix}, \dfrac{1}{\sqrt{6}} \begin{bmatrix} 1 \\ 1 \\ -2 \end{bmatrix}, \dfrac{1}{\sqrt{3}} \begin{bmatrix} 1 \\ 1 \\ 1 \end{bmatrix} \right\}$,

-2, -2, and 4

$$A = (-2) \begin{bmatrix} \frac{1}{2} & -\frac{1}{2} & 0 \\ -\frac{1}{2} & \frac{1}{2} & 0 \\ 0 & 0 & 0 \end{bmatrix}$$

$$+ (-2) \begin{bmatrix} \frac{1}{6} & \frac{1}{6} & -\frac{2}{6} \\ \frac{1}{6} & \frac{1}{6} & -\frac{2}{6} \\ -\frac{2}{6} & -\frac{2}{6} & \frac{4}{6} \end{bmatrix}$$

$$+ 4 \begin{bmatrix} \frac{1}{3} & \frac{1}{3} & \frac{1}{3} \\ \frac{1}{3} & \frac{1}{3} & \frac{1}{3} \\ \frac{1}{3} & \frac{1}{3} & \frac{1}{3} \end{bmatrix}$$

19. $\left\{ \begin{bmatrix} 1 \\ 0 \\ 0 \end{bmatrix}, \frac{1}{\sqrt{5}} \begin{bmatrix} 0 \\ -2 \\ 1 \end{bmatrix}, \frac{1}{\sqrt{5}} \begin{bmatrix} 0 \\ 1 \\ 2 \end{bmatrix} \right\}$ $-1, -1,$ and $4,$

$$A = (-1) \begin{bmatrix} 1 & 0 & 0 \\ 0 & 0 & 0 \\ 0 & 0 & 0 \end{bmatrix} + (-1) \begin{bmatrix} 0 & 0 & 0 \\ 0 & .8 & -.4 \\ 0 & -.4 & .2 \end{bmatrix}$$

$$+ 4 \begin{bmatrix} 0 & 0 & 0 \\ 0 & .2 & .4 \\ 0 & .4 & .8 \end{bmatrix}$$

20. $\left\{ \begin{bmatrix} 0 \\ 1 \\ 0 \end{bmatrix}, \frac{1}{5} \begin{bmatrix} -4 \\ 0 \\ 3 \end{bmatrix}, \frac{1}{5} \begin{bmatrix} 3 \\ 0 \\ 4 \end{bmatrix} \right\},$ $-3, 25,$ and $-50,$

$$A = (-3) \begin{bmatrix} 0 & 0 & 0 \\ 0 & 1 & 0 \\ 0 & 0 & 0 \end{bmatrix} + 25 \begin{bmatrix} \frac{16}{25} & 0 & -\frac{12}{25} \\ 0 & 0 & 0 \\ -\frac{12}{25} & 0 & \frac{9}{25} \end{bmatrix}$$

$$+ (-50) \begin{bmatrix} \frac{9}{25} & 0 & \frac{12}{25} \\ 0 & 0 & 0 \\ \frac{12}{25} & 0 & \frac{16}{25} \end{bmatrix}$$

21. True

22. False, any nonzero vector in $\mathcal{R}^2$ is an eigenvector of the symmetric matrix I_2, but not every 2×2 matrix with nonzero columns is an orthogonal matrix.

23. True

24. False, let $A = \begin{bmatrix} 1 & 4 \\ 1 & 1 \end{bmatrix}$. Then $\begin{bmatrix} 2 \\ 1 \end{bmatrix}$ and $\begin{bmatrix} 2 \\ -1 \end{bmatrix}$ are eigenvectors of A that correspond to the eigenvalues 3 and -1, respectively. But these two eigenvectors are not orthogonal.

25. False, if $\mathbf{v}$ is an eigenvector, then so is $2\mathbf{v}$. But these two eigenvectors are not orthogonal.

26. True   27. True

28. False, $\begin{bmatrix} 1 & 0 \\ 0 & -1 \end{bmatrix}$ is not the sum of orthogonal projection matrices.

29. True   30. True   31. True

32. False, if θ is an acceptable angle of rotation, then so is $\theta \pm \frac{\pi}{2}$.

33. True

34. False, see Exercise 41.

35. False, the matrix must be symmetric.

36. False, the correct matrix is $\begin{bmatrix} a & b \\ b & c \end{bmatrix}$.

37. True   38. True

39. False, we also require that $\det P = 1$.

40. False, we need the coefficient of xy to be $2b$.

41. $2 \begin{bmatrix} 1 & 0 \\ 0 & 0 \end{bmatrix} + 2 \begin{bmatrix} 0 & 0 \\ 0 & 1 \end{bmatrix}$ and

$2 \begin{bmatrix} .5 & .5 \\ .5 & .5 \end{bmatrix} + 2 \begin{bmatrix} .5 & -.5 \\ -.5 & .5 \end{bmatrix}$

42. In addition to the answer given for Exercise 19, another possibility is

$$A = (-1) \frac{1}{6} \begin{bmatrix} 1 & 2 & -1 \\ 2 & 4 & -2 \\ -1 & -2 & 1 \end{bmatrix}$$

$$+ (-1) \frac{1}{30} \begin{bmatrix} 25 & -10 & 5 \\ -10 & 4 & -2 \\ 5 & -2 & 1 \end{bmatrix}$$

$$+ (4) \frac{1}{5} \begin{bmatrix} 0 & 0 & 0 \\ 0 & 1 & 2 \\ 0 & 2 & 4 \end{bmatrix}$$

43. Let $W = \text{Span } \{\mathbf{u}\}$. Using $C = [\mathbf{u}]$ in Theorem 6.8, we obtain

$P_W = \mathbf{u}(\mathbf{u}^T\mathbf{u})^{-1}\mathbf{u}^T = \mathbf{u}(\mathbf{u}^T\mathbf{u})^{-1}\mathbf{u}^T$
$= \mathbf{u}(1)\mathbf{u}^T = \mathbf{u}\mathbf{u}^T = P.$

44. Applying Exercise 97(b) of Section 6.1, we have

$\text{rank } P_i = \text{rank } \mathbf{u}_i\mathbf{u}_i^T = \text{rank } \mathbf{u}_i^T = 1 = \text{rank } \mathbf{u}_i.$

45. For $i = j$, we have

$P_iP_i = \mathbf{u}_i\mathbf{u}_i^T\mathbf{u}_i\mathbf{u}_i^T = \mathbf{u}_i(\mathbf{u}_i^T\mathbf{u}_i)\mathbf{u}_i^T = \mathbf{u}_i(1)\mathbf{u}^T$
$= \mathbf{u}_i\mathbf{u}_i^T = P_i;$

and for $i \neq j$, we have

$P_iP_j = \mathbf{u}_i\mathbf{u}_i^T\mathbf{u}_j\mathbf{u}_j^T = \mathbf{u}_i(\mathbf{u}_i^T\mathbf{u}_j)\mathbf{u}_j^T$
$= \mathbf{u}_i[0]\mathbf{u}_j^T = O.$

46. $P_i\mathbf{u}_i = \mathbf{u}_i\mathbf{u}_i^T\mathbf{u}_i = \mathbf{u}_i[1] = \mathbf{u}_i$

47. The sum $\mu_1Q_1 + \mu_2Q_2 + \cdots + \mu_kQ_k$ is simply a rearrangement of the terms of the spectral decomposition $A = \lambda_1P_1 + \lambda_2P_2 + \cdots + \lambda_nP_n,$ where the terms λ_iP_i with $\lambda_i = \mu_j$ are grouped together and the scalar μ_j is factored out.

48. Suppose that

$Q_i = P_r + P_{r+1} + \cdots + P_s$ and
$Q_j = P_t + P_{t+1} + \cdots + P_y.$

Then Q_iQ_j is the sum of terms of the form P_aP_b, where $r \leq a \leq s$ and $t \leq b \leq y$, and $P_aP_b = P_a$

if $a = b$ and $P_a P_b = O$ if $a \neq b$. If $i = j$, therefore, these terms simplify to the sum

$$P_r + P_{r+1} + \cdots + P_s = Q_i.$$

If $i \neq j$, then $a \neq b$ for all a and b, and hence all of the terms in the product are zeros; that is, $Q_i Q_j = O$.

49. Suppose that $Q_j = P_r + P_{r+1} + \cdots + P_s$. Then
$$\begin{aligned}Q_j^T &= (P_r + P_{r+1} + \cdots + P_s)^T \\ &= P_r^T + P_{r+1}^T + \cdots + P_s^T \\ &= P_r + P_{r+1} + \cdots + P_s = Q_j.\end{aligned}$$

50. By repeatedly applying Exercise 74 of Section 6.3 and using that Q_j is the sum of orthogonal projections of mutually orthogonal subspaces of $\mathcal{R}^n$, we see that Q_j is the orthogonal projection on the span of the set of vectors that form the bases for the range of each P_i in the sum. This is the eigenspace corresponding to μ_j.

51. We have
$$Q_j = \mathbf{w}_1 \mathbf{w}_1^t + \mathbf{w}_2 \mathbf{w}_2^T + \cdots + \mathbf{w}_s \mathbf{w}_s^T.$$

52. By Exercise 50, Q_j is the orthogonal projection matrix for the eigenspace associated with μ_j. It follows that the columns of Q_j form a generating set foe this eigenspace, and hence the rank of Q_j equals the dimension of this eigenspace.

53. Let $A = \mu_1 Q_1 + \mu_2 Q_2 + \cdots + \mu_k Q_k$ be the spectral decomposition, as in Exercise 47. Then A^s is the sum of all products of s terms (with possible duplication) from the sum above. Any such term containing factors Q_i and Q_j with $i \neq j$ equals O. Otherwise, each factor of the term is of the form $\mu_i Q_i$, and hence the nonzero terms are of the form $\mu_i^s Q_i^s = \mu_i^s Q_i$. Therefore
$$A^s = \mu_1^s Q_1 + \mu_2^s Q_2 + \cdots + \mu_k^s Q_k.$$

54. Let $A = \mu_1 Q_1 + \mu_2 Q_2 + \cdots + \mu_k Q_k$ be the spectral decomposition, as in Exercise 47. Define
$$C = \sqrt[3]{\mu_1} Q_1 + \sqrt[3]{\mu_2} Q_2 + \cdots + \sqrt[3]{\mu_k} Q_k.$$

This gives a spectral decomposition of C, and $C^3 = A$ by Exercise 53.

55. First observe that $Q_1 + Q_2 + \cdots + Q_k$ is a spectral decomposition for I_n. Next, we apply Exercise 53 to obtain
$$\begin{aligned}g(A) &= a_n A^n + \cdots + a_1 A + a_0 I_n \\ &= a_n(\mu_1^n Q_1 + \cdots + \mu_k^n Q_k) + \cdots\end{aligned}$$
$$\begin{aligned}&+ a_1(\mu_1 Q_1 + \cdots + \mu_k Q_k) \\ &+ a_0(Q_1 + \cdots + Q_k) \\ &= (a_n \mu_1^n + \cdots + a_1 \mu_1 + a_0) Q_1 + \cdots \\ &\quad + (a_n \mu_k^n + \cdots + a_1 \mu_k + a_0) Q_k \\ &= g(\mu_1) Q_1 + \cdots + g(\mu_k) Q_k.\end{aligned}$$

56. Suppose that $A = \mu_1 Q_1 + \mu_2 Q_2 + \cdots + \mu_k Q_k$, as in Exercise 47. Since each μ_i is an eigenvalue of A, we have that $f(\mu_i) = 0$ for all i. Hence, by Exercise 55,
$$\begin{aligned}f(A) &= f(\mu_1) Q_1 + f(\mu_2) Q_2 + \cdots + f(\mu_k) Q_k \\ &= 0 Q_1 + 0 Q_2 + \cdots + 0 Q_k = O.\end{aligned}$$

57. Let $A = \mu_1 Q_1 + \mu_2 Q_2 + \cdots + \mu_k Q_k$ be the spectral decomposition, as in Exercise 47. By Exercise 55, we have
$$\begin{aligned}&f_j(A) \\ &= f_j(\mu_1) Q_1 + \cdots + f_j(\mu_j) Q_j + \cdots + f_k(\mu_j) Q_k \\ &= 0 Q_1 + \cdots + 1 Q_j + \cdots + 0 Q_k = Q_j.\end{aligned}$$

58. Let $A = \mu_1 Q_1 + \mu_2 Q_2 + \cdots + \mu_k Q_k$ be the spectral decomposition, as in Exercise 47. Suppose that B commutes with A. For each j, let f_j be the polynomial such that $f_j(A) = Q_j$, as in Exercise 57. Since B commutes with A, it commutes with $f_j(A) = Q_j$ for each j.

Conversely, suppose that B commutes with each Q_j. Then
$$\begin{aligned}AB &= (\mu_1 Q_1 + \mu_2 Q_2 + \cdots + \mu_k Q_k) B \\ &= \mu_1 Q_1 B + \mu_2 Q_2 B + \cdots + \mu_k Q_k B \\ &= B \mu_1 Q_1 + B \mu_2 Q_2 + \cdots + B \mu_k Q_k \\ &= B(\mu_1 Q_1 + \mu_2 Q_2 + \cdots + \mu_k Q_k) \\ &= BA.\end{aligned}$$

59. Suppose that A is positive definite, and let λ be an eigenvalue of A and $\mathbf{v}$ be a corresponding eigenvector. Then
$$0 < \mathbf{v}^T A \mathbf{v} = \mathbf{v} \cdot (\lambda \mathbf{v}) = \lambda \|\mathbf{v}\|^2,$$

and hence $\lambda > 0$.

Conversely, suppose all the eigenvalues of A are positive. Let $\{\mathbf{v}_1, \mathbf{v}_2, \ldots, \mathbf{v}_n\}$ be an orthonormal basis for $\mathcal{R}^n$ consisting of eigenvectors of A, and suppose that λ_i is the eigenvalue corresponding to $\mathbf{v}_i$ for all i. Let $\mathbf{v}$ be a nonzero vector in $\mathcal{R}^n$. Then there exist scalars $a_1, a_2, \ldots, a_n$, not all zero, such that $\mathbf{v} = a_1 \mathbf{v}_1 + a_2 \mathbf{v}_2 + \cdots + a_n \mathbf{v}_n$. Hence
$$\mathbf{v}^T A \mathbf{v} = \mathbf{v} \cdot (A\mathbf{v})$$

$$= (a_1\mathbf{v}_1 + a_2\mathbf{v}_2 + \cdots + a_n\mathbf{v}_n) \cdot$$
$$(a_1 A\mathbf{v}_1 + a_2 A\mathbf{v}_2 + \cdots + a_n A\mathbf{v}_n)$$
$$= (a_1\mathbf{v}_1 + a_2\mathbf{v}_2 + \cdots + a_n\mathbf{v}_n) \cdot$$
$$(a_1\lambda_1\mathbf{v}_1 + a_2\lambda_2\mathbf{v}_2 + \cdots + a_n\lambda_n\mathbf{v}_n)$$
$$= \lambda_1 a_1^2 + \lambda_2 a_2^2 + \cdots + \lambda_n a_n^2$$
$$> 0.$$

60. A symmetric matrix is positive semidefinite if and only if all its eigenvalues are nonnegative. The proof is nearly identical to the proof given in Exercise 59, except that strict inequalities of the form $x < y$ are replaced by the weaker inequalities of the form $x \leq y$.

61. Suppose that A is positive definite. Then A is symmetric, and hence A^{-1} is also symmetric. Furthermore, the eigenvalues of A^{-1} are the reciprocals of the eigenvalues of A. Therefore, since the eigenvalues of A are positive, the eigenvalues of A^{-1} are also positive. It follows that A^{-1} is positive definite by Exercise 59.

62. Suppose that A is positive definite. Then A is symmetric, and hence cA is also symmetric. Furthermore the eigenvalues of cA are the products of c and the eigenvalues of A. It follows that the eigenvalues of cA are positive, and hence cA is positive definite by Exercise 59.

63. Suppose A is positive semidefinite and $c > 0$. Then cA is positive semidefinite. The proof is nearly identical to the proof given in Exercise 62, except that we are concerned with *nonnegative* rather than *positive* products.

64. Note that $A + B$ is symmetric because both A and B are symmetric. Let $\mathbf{v}$ be a nonzero vector in $\mathcal{R}^n$. Then $\mathbf{v}^T A\mathbf{v} > 0$ and $\mathbf{v}^T B\mathbf{v} > 0$ because A and B are positive definite. Therefore
$$\mathbf{v}^T(A+B)\mathbf{v} = \mathbf{v}^T A\mathbf{v} + \mathbf{v}^T B\mathbf{v} > 0,$$
and hence $A + B$ is positive definite.

65. The sum of two positive semidefinite $n \times n$ matrices is also positive semidefinite. The proof is nearly identical to the proof given in Exercise 64, except that strict inequalities of the form $x < y$ are replaced by the weaker inequalities of the form $x \leq y$.

66. Let $\mathbf{v}$ be any nonzero vector in $\mathcal{R}^n$, and let B be an $n \times n$ matrix. Then $Q^T\mathbf{v}$ is nonzero, and hence
$$\mathbf{v}^T A\mathbf{v} = \mathbf{v}^T QBQ^T\mathbf{v} = (Q^T\mathbf{v})^T B(Q^T\mathbf{v}) > 0$$
because B is positive definite. Therefore A is positive definite.

67. If $A = QBQ^T$, where Q is orthogonal and B is positive semidefinite, then A is also positive semidefinite. The proof is nearly identical to the proof given in Exercise 66, except that strict inequalities of the form $x < y$ are replaced by the weaker inequalities of the form $x \leq y$.

68. Suppose that A is positive definite. Then the eigenvalues of A are positive, and hence each such eigenvalue has a positive square root. The rest of the proof is nearly identical to the proof given in Exercise 54, except that we use terms of the form $\sqrt{\mu_i}Q_i$ in place of $\sqrt[3]{\mu_i}Q_i$.

69. If A is a positive semidefinite matrix, then there exists a positive semidefinite matrix B such that $B^2 = A$. The proof is nearly identical to the proof given in Exercise 68, except that each eigenvalue has a *nonnegative* square root.

70. Observe that $\mathbf{e}_i^T A \mathbf{e}_j = a_{ij}$ for $1 \leq i, j \leq n$, where $\mathbf{e}_i$ and $\mathbf{e}_j$ are standard vectors in $\mathcal{R}^n$. Consider any vector $\mathbf{u}$ in $\mathcal{R}^n$. Then
$$\mathbf{u}^T A\mathbf{u} = \left(\sum_i u_i \mathbf{e}_i^T\right) A \left(\sum_j u_j \mathbf{e}_j\right)$$
$$= \sum_{i,j} u_i u_j \mathbf{e}_i^T A\mathbf{e}_j = \sum_{i,j} a_{ij} u_i u_j.$$

Now suppose that A is positive definite, and let $u_1, u_2, \ldots, u_n$ be scalars not all zero. Let $\mathbf{u}$ be the vector in $\mathcal{R}^n$ whose jth component is u_j. Then $\mathbf{u} \neq \mathbf{0}$, and hence
$$\sum_{i,j} a_{ij} u_i u_j = \mathbf{u}^T A\mathbf{u} > 0.$$

Conversely, assume this condition. Let $\mathbf{u}$ be any nonzero vector in $\mathcal{R}^n$. Then at least one component u_i is not zero, and hence
$$\mathbf{u}^T A\mathbf{u} = \sum_{i,j} a_{ij} u_i u_j > 0.$$

Therefore A is positive definite.

71. An $n \times n$ symmetric matrix A is positive semidefinite if and only if
$$\sum_{i,j} a_{ij} u_i u_j \geq 0$$
for all scalars $u_1, u_2, \ldots, u_n$. The proof is nearly identical to the proof given in Exercise 70, except that strict inequalities of the form $x < y$ are replaced by the weaker inequalities of the form $x \leq y$.

72. Let $\mathbf{v}$ be any vector in $\mathcal{R}^n$ and A be an $m \times n$ matrix. Then
$$\mathbf{v}^T A^T A \mathbf{v} = (A\mathbf{v})^T(A\mathbf{v}) = (A\mathbf{v}) \cdot (A\mathbf{v}) \geq 0,$$
and hence $A^T A$ is positive semidefinite. Similarly, AA^T is positive semidefinite.

73. Suppose that A is invertible. Let $\mathbf{v}$ be any nonzero vector in $\mathcal{R}^n$. Since nullity $A = 0$, it follows that $A\mathbf{v} \neq \mathbf{0}$. Thus
$$\mathbf{v}^T A^T A \mathbf{v} = (A\mathbf{v})^T(A\mathbf{v}) = (A\mathbf{v}) \cdot (A\mathbf{v}) > 0,$$
and hence $A^T A$ is positive definite. Similarly, AA^T is positive definite.

74. (a) $A^T = A$
(b) 2, 2, 2, 6, 8
(c) (rounded to 4 places after the decimal)
$$\left\{ \begin{bmatrix} -0.7071 \\ 0 \\ 0.7071 \\ 0 \\ 0 \end{bmatrix}, \begin{bmatrix} 0 \\ -0.7071 \\ 0 \\ 0.7071 \\ 0 \end{bmatrix}, \begin{bmatrix} -0.4082 \\ 0 \\ -0.4082 \\ 0 \\ 0.8165 \end{bmatrix}, \right.$$
$$\left. \begin{bmatrix} 0 \\ 0.7071 \\ 0 \\ 0.7071 \\ 0 \end{bmatrix}, \begin{bmatrix} 0.5774 \\ 0 \\ 0.5774 \\ 0 \\ 0.5774 \end{bmatrix} \right\}$$

(d) Using the notation in Section 6.6, we have
$$P_1 = \begin{bmatrix} 0.5000 & 0 & -0.5000 & 0 & 0 \\ 0 & 0 & 0 & 0 & 0 \\ -0.5000 & 0 & 0.5000 & 0 & 0 \\ 0 & 0 & 0 & 0 & 0 \\ 0 & 0 & 0 & 0 & 0 \end{bmatrix},$$

$$P_2 = \begin{bmatrix} 0 & 0 & 0 & 0 & 0 \\ 0 & 0.5000 & 0 & -0.5000 & 0 \\ 0 & 0 & 0 & 0 & 0 \\ 0 & -0.5000 & 0 & 0.5000 & 0 \\ 0 & 0 & 0 & 0 & 0 \end{bmatrix},$$

$$P_3 = \begin{bmatrix} 0.1667 & 0 & 0.1667 & 0 & -0.3333 \\ 0 & 0 & 0 & 0 & 0 \\ 0.1667 & 0 & 0.1667 & 0 & -0.3333 \\ 0 & 0 & 0 & 0 & 0 \\ -0.3333 & 0 & -0.3333 & 0 & 0.6667 \end{bmatrix},$$

(rounded to 4 places after the decimal)
$$P_4 = \begin{bmatrix} 0 & 0 & 0 & 0 & 0 \\ 0 & 0.5000 & 0 & 0.5000 & 0 \\ 0 & 0 & 0 & 0 & 0 \\ 0 & 0.5000 & 0 & 0.5000 & 0 \\ 0 & 0 & 0 & 0 & 0 \end{bmatrix},$$

$$P_5 = \begin{bmatrix} 0.3333 & 0 & 0.3333 & 0 & 0.3333 \\ 0 & 0 & 0 & 0 & 0 \\ 0.3333 & 0 & 0.3333 & 0 & 0.3333 \\ 0 & 0 & 0 & 0 & 0 \\ 0.3333 & 0 & 0.3333 & 0 & 0.3333 \end{bmatrix},$$

(rounded to 4 places after the decimal)
$$A = 2P_1 + 2P_2 + 2P_3 + 6P_4 + 8P_5$$

(e) $A^6 = \begin{bmatrix} 87424 & 0 & 87360 & 0 & 87360 \\ 0 & 23360 & 0 & 23296 & 0 \\ 87360 & 0 & 87424 & 0 & 87360 \\ 0 & 23296 & 0 & 23360 & 0 \\ 87360 & 0 & 87360 & 0 & 87424 \end{bmatrix}$

(f) $A^6 = 64P_1 + 64P_2 + 64P_3 + 46656P_4 + 262114P_5$

75. (a) $P =$
$$\begin{bmatrix} 0.8205 & 0.0874 & 0.1548 & 0.3058 & 0.4491 \\ -0.4572 & 0.4753 & -0.2423 & 0.5495 & 0.4522 \\ -0.1762 & -0.1468 & 0.0515 & -0.6158 & 0.7520 \\ -0.1601 & -0.8600 & -0.0463 & 0.4521 & 0.1680 \\ -0.2472 & 0.0726 & 0.9553 & 0.1449 & 0.0095 \end{bmatrix}$$

and $D =$
$$\begin{bmatrix} -4.7485 & 0 & 0 & 0 & 0 \\ 0 & -1.7083 & 0 & 0 & 0 \\ 0 & 0 & 2.2380 & 0 & 0 \\ 0 & 0 & 0 & 10.9901 & 0 \\ 0 & 0 & 0 & 0 & 288.2286 \end{bmatrix}$$

(b) The orthogonal projection matrices are

$P_1 =$
$$\begin{bmatrix} 0.2017 & 0.2031 & 0.3377 & 0.0754 & 0.0043 \\ 0.2031 & 0.2045 & 0.3400 & 0.0760 & 0.0043 \\ 0.3377 & 0.3400 & 0.5656 & 0.1263 & 0.0071 \\ 0.0754 & 0.0760 & 0.1263 & 0.0282 & 0.0016 \\ 0.0043 & 0.0043 & 0.0071 & 0.0016 & 0.0001 \end{bmatrix},$$

$P_2 =$
$$\begin{bmatrix} 0.0935 & 0.1680 & -0.1883 & 0.1382 & 0.0443 \\ 0.1680 & 0.3020 & -0.3384 & 0.2484 & 0.0796 \\ -0.1883 & -0.3384 & 0.3792 & -0.2784 & -0.0893 \\ 0.1382 & 0.2484 & -0.2784 & 0.2044 & 0.0655 \\ 0.0443 & 0.0796 & -0.0893 & 0.0655 & 0.0210 \end{bmatrix},$$

$P_3 =$
$$\begin{bmatrix} 0.6732 & -0.3751 & -0.1446 & -0.1313 & -0.2028 \\ -0.3751 & 0.2090 & 0.0806 & 0.0732 & 0.1130 \\ -0.1446 & 0.0806 & 0.0311 & 0.0282 & 0.0436 \\ -0.1313 & 0.0732 & 0.0282 & 0.0256 & 0.0396 \\ -0.2028 & 0.1130 & 0.0436 & 0.0396 & 0.0611 \end{bmatrix},$$

$P_4 =$

$$\begin{bmatrix} 0.0240 & -0.0375 & 0.0080 & -0.0072 & 0.1479 \\ -0.0375 & 0.0587 & -0.0125 & 0.0112 & -0.2315 \\ 0.0080 & -0.0125 & 0.0027 & -0.0024 & 0.0492 \\ -0.0072 & 0.0112 & -0.0024 & 0.0021 & -0.0442 \\ 0.1479 & -0.2315 & 0.0492 & -0.0442 & 0.9125 \end{bmatrix},$$

and

$$P_5 =$$

$$\begin{bmatrix} 0.0076 & 0.0415 & -0.0128 & -0.0752 & 0.0063 \\ 0.0415 & 0.2259 & -0.0698 & -0.4088 & 0.0345 \\ -0.0128 & -0.0698 & 0.0215 & 0.1262 & -0.0107 \\ -0.0752 & -0.4088 & 0.1262 & 0.7397 & -0.0625 \\ 0.0063 & 0.0345 & -0.0107 & -0.0625 & 0.0053 \end{bmatrix}.$$

So the spectral decomposition of A is

$$A = 288.2286 P_1 + 10.9901 P_2 + (-4.7485) P_3 + 2.2380 P_4 + (-1.7083) P_5.$$

76. (a) $A_1 = 288.2286 P_1 =$

$$\begin{bmatrix} 58.1288 & 58.5273 & 97.3428 & 21.7450 & 1.2298 \\ 58.5273 & 58.9284 & 98.0101 & 21.8940 & 1.2382 \\ 97.3428 & 98.0101 & 163.0109 & 36.4143 & 2.0594 \\ 21.7450 & 21.8940 & 36.4143 & 8.1344 & 0.4600 \\ 1.2298 & 1.2382 & 2.0594 & 0.4600 & 0.0260 \end{bmatrix}$$

(b) The norms of E_1 and A are 12.2987 and 288.4909, respectively.

(c) So the information lost is

$$\frac{12.2987}{288.4909} \approx .04263,$$

that is, about 4.263%.

6.7 SINGULAR VALUE DECOMPOSITION

1. $\begin{bmatrix} \frac{1}{\sqrt{2}} & -\frac{1}{\sqrt{2}} \\ \frac{1}{\sqrt{2}} & \frac{1}{\sqrt{2}} \end{bmatrix} \begin{bmatrix} \sqrt{2} & 0 \\ 0 & 0 \end{bmatrix} \begin{bmatrix} 1 & 0 \\ 0 & 1 \end{bmatrix}^T$

2. $\begin{bmatrix} 1 & 0 \\ 0 & 1 \end{bmatrix} \begin{bmatrix} \sqrt{2} & 0 \\ 0 & 0 \end{bmatrix} \begin{bmatrix} \frac{1}{\sqrt{2}} & \frac{1}{\sqrt{2}} \\ \frac{1}{\sqrt{2}} & \frac{-1}{\sqrt{2}} \end{bmatrix}^T$

3. $\begin{bmatrix} \frac{1}{3} & \frac{2}{\sqrt{5}} & \frac{2}{3\sqrt{5}} \\ \frac{2}{3} & \frac{-1}{\sqrt{5}} & \frac{4}{3\sqrt{5}} \\ \frac{2}{3} & 0 & \frac{-5}{3\sqrt{5}} \end{bmatrix} \begin{bmatrix} 3 \\ 0 \\ 0 \end{bmatrix} [1]^T$

4. $\begin{bmatrix} \frac{1}{2} & \frac{1}{\sqrt{2}} & \frac{1}{\sqrt{6}} & \frac{1}{\sqrt{12}} \\ \frac{1}{2} & \frac{-1}{\sqrt{2}} & \frac{1}{\sqrt{6}} & \frac{1}{\sqrt{12}} \\ \frac{-1}{2} & 0 & \frac{2}{\sqrt{6}} & \frac{-1}{\sqrt{12}} \\ \frac{1}{2} & 0 & 0 & \frac{-3}{\sqrt{12}} \end{bmatrix} \begin{bmatrix} 2 \\ 0 \\ 0 \\ 0 \end{bmatrix} [1]^T$

5. We begin by computing

$$A^T A = \begin{bmatrix} 1 & 1 & 1 \\ 1 & -1 & 2 \end{bmatrix} \begin{bmatrix} 1 & 1 \\ 1 & -1 \\ 1 & 2 \end{bmatrix} = \begin{bmatrix} 3 & 2 \\ 2 & 6 \end{bmatrix},$$

and determine the eigenvalues of $A^T A$ from its characteristic polynomial

$$\det(A^T A - tI_2) = \det \begin{bmatrix} 3-t & 2 \\ 2 & 6-t \end{bmatrix}$$
$$= (t-7)(t-2).$$

Thus $\lambda_1 = 7$ and $\lambda_2 = 2$ are the eigenvalues of $A^T A$. The singular values of A are therefore $\sigma_1 = \sqrt{7}$ and $\sigma_2 = \sqrt{2}$. For each eigenvalue λ_i, find an eigenvector corresponding to λ_i by choosing a nonzero solution of $(A - \lambda_i I_2)\mathbf{x} = \mathbf{0}$. For example, choose

$$\mathbf{w}_1 = \begin{bmatrix} 1 \\ 2 \end{bmatrix} \quad \text{and} \quad \mathbf{w}_2 = \begin{bmatrix} 2 \\ -1 \end{bmatrix}$$

as eigenvectors corresponding to λ_1 and λ_2, respectively. Now normalize each of these to obtain an orthonormal basis $\{\mathbf{v}_1, \mathbf{v}_2\}$ for $\mathcal{R}^2$ consisting of right singular vectors of A:

$$\mathbf{v}_1 = \frac{1}{\sqrt{5}} \begin{bmatrix} 1 \\ 2 \end{bmatrix} \quad \text{and} \quad \mathbf{v}_2 = \frac{1}{\sqrt{5}} \begin{bmatrix} 2 \\ -1 \end{bmatrix}.$$

Next, obtain an orthonormal basis for $\mathcal{R}^3$ of left singular vectors of A. The first two left singular vectors can be obtained from the right singular vectors as follows:

$$\mathbf{u}_1 = \frac{1}{\sigma_1} A \mathbf{v}_1 = \frac{1}{\sqrt{7}} \begin{bmatrix} 1 & 1 \\ 1 & -1 \\ 1 & 2 \end{bmatrix} \frac{1}{\sqrt{5}} \begin{bmatrix} 1 \\ 2 \end{bmatrix}$$

$$= \frac{1}{\sqrt{35}} \begin{bmatrix} 3 \\ -1 \\ 5 \end{bmatrix}$$

and

$$\mathbf{u}_2 = \frac{1}{\sigma_1} A \mathbf{v}_2 = \frac{1}{\sqrt{2}} \begin{bmatrix} 1 & 1 \\ 1 & -1 \\ 1 & 2 \end{bmatrix} \frac{1}{\sqrt{5}} \begin{bmatrix} 2 \\ -1 \end{bmatrix}$$

$$= \frac{1}{\sqrt{10}} \begin{bmatrix} 1 \\ 3 \\ 0 \end{bmatrix}.$$

For the third left singular vector, we can choose any unit vector that is orthogonal to both $\mathbf{u}_1$ and $\mathbf{u}_2$. This can be done in two steps. First find a nonzero solution of the system of linear equations

$$\begin{aligned} 3x_1 - x_2 + 5x_3 &= 0 \\ x_1 + 3x_2 &= 0, \end{aligned}$$

for example, $\begin{bmatrix} -3 \\ 1 \\ 2 \end{bmatrix}$. Then normalize this vector to obtain the third left singular vector:

$$\mathbf{u}_3 = \frac{1}{\sqrt{14}} \begin{bmatrix} -3 \\ 1 \\ 2 \end{bmatrix}.$$

Now let

$$U = [\mathbf{u}_1 \ \mathbf{u}_2 \ \mathbf{u}_3] = \begin{bmatrix} \frac{3}{\sqrt{35}} & \frac{1}{\sqrt{10}} & \frac{-3}{\sqrt{14}} \\ \frac{-1}{\sqrt{35}} & \frac{3}{\sqrt{10}} & \frac{1}{\sqrt{14}} \\ \frac{5}{\sqrt{35}} & 0 & \frac{2}{\sqrt{14}} \end{bmatrix},$$

$$\Sigma = \begin{bmatrix} \sigma_1 & 0 \\ 0 & \sigma_2 \\ 0 & 0 \end{bmatrix} = \begin{bmatrix} \sqrt{7} & 0 \\ 0 & \sqrt{2} \\ 0 & 0 \end{bmatrix},$$

and

$$V = [\mathbf{v}_1 \ \mathbf{v}_2] = \begin{bmatrix} \frac{1}{\sqrt{5}} & \frac{2}{\sqrt{5}} \\ \frac{2}{\sqrt{5}} & \frac{-1}{\sqrt{5}} \end{bmatrix}$$

to obtain the singular value decomposition

$$U\Sigma V^T =$$

$$\begin{bmatrix} \frac{3}{\sqrt{35}} & \frac{1}{\sqrt{10}} & \frac{-3}{\sqrt{14}} \\ \frac{-1}{\sqrt{35}} & \frac{3}{\sqrt{10}} & \frac{1}{\sqrt{14}} \\ \frac{5}{\sqrt{35}} & 0 & \frac{2}{\sqrt{14}} \end{bmatrix} \begin{bmatrix} \sqrt{7} & 0 \\ 0 & \sqrt{2} \\ 0 & 0 \end{bmatrix} \begin{bmatrix} \frac{1}{\sqrt{5}} & \frac{2}{\sqrt{5}} \\ \frac{2}{\sqrt{5}} & \frac{-1}{\sqrt{5}} \end{bmatrix}^T.$$

6. $\begin{bmatrix} \frac{1}{\sqrt{12}} & \frac{2}{\sqrt{6}} & \frac{1}{\sqrt{54}} & \frac{5}{\sqrt{108}} \\ \frac{3}{\sqrt{12}} & \frac{-1}{\sqrt{6}} & \frac{2}{\sqrt{54}} & \frac{1}{\sqrt{108}} \\ \frac{1}{\sqrt{12}} & 0 & \frac{-7}{\sqrt{54}} & \frac{1}{\sqrt{108}} \\ \frac{1}{\sqrt{12}} & \frac{1}{\sqrt{6}} & 0 & \frac{-9}{\sqrt{108}} \end{bmatrix} \begin{bmatrix} \sqrt{12} & 0 \\ 0 & \sqrt{6} \\ 0 & 0 \\ 0 & 0 \end{bmatrix} \begin{bmatrix} 1 & 0 \\ 0 & 1 \end{bmatrix}^T$

7. $\begin{bmatrix} \frac{1}{\sqrt{2}} & \frac{1}{\sqrt{2}} \\ \frac{-1}{\sqrt{2}} & \frac{1}{\sqrt{2}} \end{bmatrix} \begin{bmatrix} 2 & 0 & 0 \\ 0 & \sqrt{2} & 0 \end{bmatrix} \begin{bmatrix} 0 & 1 & 0 \\ \frac{1}{\sqrt{2}} & 0 & \frac{1}{\sqrt{2}} \\ \frac{1}{\sqrt{2}} & 0 & \frac{-1}{\sqrt{2}} \end{bmatrix}^T$

8. $\begin{bmatrix} 0 & 1 \\ 1 & 0 \end{bmatrix} \begin{bmatrix} 2 & 0 & 0 & 0 \\ 0 & 1 & 0 & 0 \end{bmatrix} \begin{bmatrix} 0 & 1 & 0 & 0 \\ 1 & 0 & 0 & 0 \\ 0 & 0 & 1 & 0 \\ 0 & 0 & 0 & 1 \end{bmatrix}^T$

9. We begin by computing

$$A^T A = \begin{bmatrix} 1 & 2 & 1 \\ 1 & 0 & -1 \\ 2 & -1 & 0 \end{bmatrix} \begin{bmatrix} 1 & 1 & 2 \\ 2 & 0 & -1 \\ 1 & -1 & 0 \end{bmatrix}$$

$$= \begin{bmatrix} 6 & 0 & 0 \\ 0 & 2 & 2 \\ 0 & 2 & 5 \end{bmatrix},$$

and determine the eigenvalues of $A^T A$ from its characteristic polynomial

$$\det(A^T A - I_3) = \det \begin{bmatrix} 6-t & 0 & 0 \\ 0 & 2-t & 2 \\ 0 & 2 & 5-t \end{bmatrix}$$

$$= -(t-6)^2(t-1).$$

Thus $\lambda_1 = 6$, $\lambda_2 = 6$, and $\lambda_3 = 1$ are the eigenvalues of $A^T A$. The singular values of A are therefore $\sigma_1 = \sqrt{6}$, $\sigma_2 = \sqrt{6}$, and $\sigma_3 = 1$. It can be shown that

$$\left\{ \begin{bmatrix} 1 \\ 0 \\ 0 \end{bmatrix}, \frac{1}{\sqrt{5}} \begin{bmatrix} 0 \\ 1 \\ 2 \end{bmatrix}, \frac{1}{\sqrt{5}} \begin{bmatrix} 0 \\ 2 \\ -1 \end{bmatrix} \right\}$$

is an orthonormal basis for $\mathcal{R}^3$ consisting of eigenvectors corresponding to the eigenvalues λ_1, λ_2, and λ_3, respectively. Let $\mathbf{v}_1$, $\mathbf{v}_2$, and $\mathbf{v}_3$ denote the vectors in this basis. These vectors are right singular vectors of A.

To obtain an orthonormal basis of eigenvectors of $\mathcal{R}^3$ consisting of left singular vectors, we compute

$$\mathbf{u}_1 = \frac{1}{\sigma_1} A \mathbf{v}_1 = \frac{1}{\sqrt{6}} \begin{bmatrix} 1 & 1 & 2 \\ 2 & 0 & -1 \\ 1 & -1 & 0 \end{bmatrix} \begin{bmatrix} 1 \\ 0 \\ 0 \end{bmatrix}$$

$$= \frac{1}{\sqrt{6}} \begin{bmatrix} 1 \\ 2 \\ 1 \end{bmatrix},$$

$$\mathbf{u}_2 = \frac{1}{\sigma_1} A \mathbf{v}_2 = \frac{1}{\sqrt{6}} \begin{bmatrix} 1 & 1 & 2 \\ 2 & 0 & -1 \\ 1 & -1 & 0 \end{bmatrix} \frac{1}{\sqrt{5}} \begin{bmatrix} 0 \\ 1 \\ 2 \end{bmatrix}$$

$$= \frac{1}{\sqrt{30}} \begin{bmatrix} 5 \\ -2 \\ -1 \end{bmatrix},$$

and

$$\mathbf{u}_3 = \frac{1}{\sigma_1} A \mathbf{v}_3 = \frac{1}{1} \begin{bmatrix} 1 & 1 & 2 \\ 2 & 0 & -1 \\ 1 & -1 & 0 \end{bmatrix} \frac{1}{\sqrt{5}} \begin{bmatrix} 0 \\ 2 \\ -1 \end{bmatrix}$$

$$= \frac{1}{\sqrt{5}} \begin{bmatrix} 0 \\ 1 \\ -2 \end{bmatrix}.$$

170   Chapter 6   Orthogonality

Now let
$$U = [\mathbf{u}_1 \ \mathbf{u}_2 \ \mathbf{u}_3] = \begin{bmatrix} \frac{1}{\sqrt{6}} & \frac{5}{\sqrt{30}} & 0 \\ \frac{2}{\sqrt{6}} & \frac{-2}{\sqrt{30}} & \frac{1}{\sqrt{5}} \\ \frac{1}{\sqrt{6}} & \frac{-1}{\sqrt{30}} & \frac{-2}{\sqrt{5}} \end{bmatrix},$$

$$\Sigma = \begin{bmatrix} \sigma_1 & 0 & 0 \\ 0 & \sigma_2 & 0 \\ 0 & 0 & \sigma_3 \end{bmatrix} = \begin{bmatrix} \sqrt{6} & 0 & 0 \\ 0 & \sqrt{6} & 0 \\ 0 & 0 & 1 \end{bmatrix},$$

and
$$V = [\mathbf{v}_1 \ \mathbf{v}_2 \ \mathbf{v}_3] = \begin{bmatrix} 1 & 0 & 0 \\ 0 & \frac{1}{\sqrt{5}} & \frac{2}{\sqrt{5}} \\ 0 & \frac{2}{\sqrt{5}} & \frac{-1}{\sqrt{5}} \end{bmatrix}$$

to obtain the singular value decomposition

$$\begin{bmatrix} \frac{1}{\sqrt{6}} & \frac{5}{\sqrt{30}} & 0 \\ \frac{2}{\sqrt{6}} & \frac{-2}{\sqrt{30}} & \frac{1}{\sqrt{5}} \\ \frac{1}{\sqrt{6}} & \frac{-1}{\sqrt{30}} & \frac{-2}{\sqrt{5}} \end{bmatrix} \begin{bmatrix} \sqrt{6} & 0 & 0 \\ 0 & \sqrt{6} & 0 \\ 0 & 0 & 1 \end{bmatrix} \begin{bmatrix} 1 & 0 & 0 \\ 0 & \frac{1}{\sqrt{5}} & \frac{2}{\sqrt{5}} \\ 0 & \frac{2}{\sqrt{5}} & \frac{-1}{\sqrt{5}} \end{bmatrix}^T.$$

10. $\begin{bmatrix} \frac{5}{\sqrt{30}} & \frac{1}{\sqrt{6}} & 0 \\ \frac{-1}{\sqrt{30}} & \frac{1}{\sqrt{6}} & \frac{-2}{\sqrt{5}} \\ \frac{-2}{\sqrt{30}} & \frac{2}{\sqrt{6}} & \frac{1}{\sqrt{5}} \end{bmatrix} \begin{bmatrix} \sqrt{12} & 0 & 0 \\ 0 & \sqrt{6} & 0 \\ 0 & 0 & 0 \end{bmatrix} \begin{bmatrix} 0 & 1 & 0 \\ \frac{-1}{\sqrt{10}} & 0 & \frac{3}{\sqrt{10}} \\ \frac{3}{\sqrt{10}} & 0 & \frac{1}{\sqrt{10}} \end{bmatrix}^T$

11. $\begin{bmatrix} \frac{1}{\sqrt{5}} & \frac{2}{\sqrt{5}} \\ \frac{-2}{\sqrt{5}} & \frac{1}{\sqrt{5}} \end{bmatrix} \begin{bmatrix} \sqrt{30} & 0 & 0 \\ 0 & 0 & 0 \end{bmatrix} \begin{bmatrix} \frac{1}{\sqrt{6}} & \frac{1}{\sqrt{3}} & \frac{1}{\sqrt{2}} \\ \frac{-1}{\sqrt{6}} & \frac{-1}{\sqrt{3}} & \frac{1}{\sqrt{2}} \\ \frac{2}{\sqrt{6}} & \frac{-1}{\sqrt{3}} & 0 \end{bmatrix}^T$

12. $\begin{bmatrix} \frac{1}{\sqrt{5}} & \frac{-2}{\sqrt{5}} \\ \frac{2}{\sqrt{5}} & \frac{1}{\sqrt{5}} \end{bmatrix} \begin{bmatrix} 5 & 0 & 0 \\ 0 & 0 & 0 \end{bmatrix} \begin{bmatrix} \frac{1}{\sqrt{5}} & 0 & \frac{2}{\sqrt{5}} \\ 0 & 1 & 0 \\ \frac{-2}{\sqrt{5}} & 0 & \frac{1}{\sqrt{5}} \end{bmatrix}^T$

13. $\begin{bmatrix} \frac{1}{\sqrt{5}} & \frac{2}{\sqrt{5}} \\ \frac{-2}{\sqrt{5}} & \frac{1}{\sqrt{5}} \end{bmatrix} \begin{bmatrix} \sqrt{7} & 0 & 0 \\ 0 & \sqrt{2} & 0 \end{bmatrix} \begin{bmatrix} \frac{3}{\sqrt{35}} & \frac{1}{\sqrt{10}} & \frac{3}{\sqrt{14}} \\ \frac{-5}{\sqrt{35}} & 0 & \frac{2}{\sqrt{14}} \\ \frac{-1}{\sqrt{35}} & \frac{3}{\sqrt{10}} & \frac{-1}{\sqrt{14}} \end{bmatrix}^T$

14. $\begin{bmatrix} \frac{1}{\sqrt{2}} & \frac{1}{\sqrt{2}} \\ \frac{-1}{\sqrt{2}} & \frac{1}{\sqrt{2}} \end{bmatrix} \begin{bmatrix} \sqrt{3} & 0 & 0 \\ 0 & 1 & 0 \end{bmatrix} \begin{bmatrix} \frac{2}{\sqrt{6}} & 0 & \frac{1}{\sqrt{3}} \\ \frac{-1}{\sqrt{6}} & \frac{1}{\sqrt{2}} & \frac{1}{\sqrt{3}} \\ \frac{-1}{\sqrt{6}} & \frac{-1}{\sqrt{2}} & \frac{1}{\sqrt{3}} \end{bmatrix}^T$

15. $\begin{bmatrix} \frac{2}{\sqrt{5}} & \frac{1}{\sqrt{5}} & 0 \\ \frac{1}{\sqrt{5}} & \frac{-2}{\sqrt{5}} & 0 \\ 0 & 0 & 1 \end{bmatrix} \begin{bmatrix} \sqrt{60} & 0 & 0 & 0 \\ 0 & \sqrt{15} & 0 & 0 \\ 0 & 0 & 0 & 0 \end{bmatrix} \begin{bmatrix} \frac{1}{\sqrt{3}} & \frac{-1}{\sqrt{3}} & \frac{1}{\sqrt{6}} & \frac{1}{\sqrt{6}} \\ \frac{1}{\sqrt{3}} & \frac{1}{\sqrt{3}} & \frac{1}{\sqrt{6}} & \frac{-1}{\sqrt{6}} \\ \frac{1}{\sqrt{3}} & 0 & \frac{-2}{\sqrt{6}} & 0 \\ 0 & \frac{1}{\sqrt{3}} & 0 & \frac{2}{\sqrt{6}} \end{bmatrix}^T$

16. $\begin{bmatrix} \frac{1}{\sqrt{10}} & \frac{-3}{\sqrt{10}} & 0 \\ \frac{3}{\sqrt{10}} & \frac{1}{\sqrt{10}} & 0 \\ 0 & 0 & 1 \end{bmatrix} \begin{bmatrix} \sqrt{80} & 0 & 0 & 0 \\ 0 & \sqrt{20} & 0 & 0 \\ 0 & 0 & 0 & 0 \end{bmatrix} \begin{bmatrix} \frac{1}{\sqrt{2}} & 0 & \frac{1}{\sqrt{2}} & 0 \\ 0 & \frac{1}{\sqrt{2}} & 0 & \frac{-1}{\sqrt{2}} \\ \frac{1}{\sqrt{2}} & 0 & \frac{-1}{\sqrt{2}} & 0 \\ 0 & \frac{1}{\sqrt{2}} & 0 & \frac{1}{\sqrt{2}} \end{bmatrix}^T$

17. From the given characteristic polynomial, we see that the eigenvalues of $A^T A$ (including multiplicities) are 21, 18, 0, and 0. Thus the (nonzero) singular values of A are $\sigma_1 = \sqrt{21}$ and $\sigma_2 = \sqrt{18}$. It can be shown that

$$\left\{ \frac{1}{\sqrt{7}}\begin{bmatrix}1\\2\\1\\1\end{bmatrix}, \frac{1}{\sqrt{3}}\begin{bmatrix}1\\-1\\0\\1\end{bmatrix}, \frac{1}{\sqrt{11}}\begin{bmatrix}1\\1\\-3\\0\end{bmatrix}, \frac{1}{\sqrt{2}}\begin{bmatrix}1\\0\\0\\-1\end{bmatrix} \right\}$$

is an orthonormal basis for $\mathcal{R}^4$ of eigenvectors of $A^T A$ corresponding to the eigenvalues 21, 18, 0, and 0, respectively. Let $\mathbf{v}_1$, $\mathbf{v}_2$, $\mathbf{v}_3$, and $\mathbf{v}_4$ denote the vectors in this basis. These vectors form a set of right singular vectors of A. To obtain left singular vectors, let

$$\mathbf{u}_1 = \frac{1}{\sigma_1}A\mathbf{v}_1 = \frac{1}{\sqrt{21}}\begin{bmatrix}3 & 0 & 1 & 3\\0 & 3 & 1 & 0\\0 & -3 & -1 & 0\end{bmatrix}\frac{1}{\sqrt{7}}\begin{bmatrix}1\\2\\1\\1\end{bmatrix}$$

$$= \frac{1}{\sqrt{3}}\begin{bmatrix}1\\1\\-1\end{bmatrix}$$

and

$$\mathbf{u}_2 = \frac{1}{\sigma_2}A\mathbf{v}_2 = \frac{1}{\sqrt{18}}\begin{bmatrix}3 & 0 & 1 & 3\\0 & 3 & 1 & 0\\0 & -3 & -1 & 0\end{bmatrix}\frac{1}{\sqrt{3}}\begin{bmatrix}1\\-1\\0\\1\end{bmatrix}$$

$$= \frac{1}{\sqrt{6}}\begin{bmatrix}2\\-1\\1\end{bmatrix}.$$

Since $\mathbf{u}_1$ and $\mathbf{u}_2$ are orthonormal, the set of these vectors can be extended to an orthonormal basis $\{\mathbf{u}_1, \mathbf{u}_2, \mathbf{u}_3\}$ for $\mathcal{R}^3$. So we must choose $\mathbf{u}_3$ to be a unit vector that is orthogonal to both $\mathbf{u}_1$ and $\mathbf{u}_2$. Proceeding as in Exercise 5, we obtain

$$\mathbf{u}_3 = \frac{1}{\sqrt{2}}\begin{bmatrix}0\\1\\1\end{bmatrix}.$$

Now let

$$U = [\mathbf{u}_1 \ \mathbf{u}_2 \ \mathbf{u}_3] = \begin{bmatrix} \frac{1}{\sqrt{3}} & \frac{2}{\sqrt{6}} & 0 \\ \frac{1}{\sqrt{3}} & \frac{-1}{\sqrt{6}} & \frac{1}{\sqrt{2}} \\ \frac{-1}{\sqrt{3}} & \frac{1}{\sqrt{6}} & \frac{1}{\sqrt{2}} \end{bmatrix},$$

$$\Sigma = \begin{bmatrix} \sigma_1 & 0 & 0 & 0 \\ 0 & \sigma_2 & 0 & 0 \\ 0 & 0 & 0 & 0 \end{bmatrix} = \begin{bmatrix} \sqrt{21} & 0 & 0 & 0 \\ 0 & \sqrt{18} & 0 & 0 \\ 0 & 0 & 0 & 0 \end{bmatrix}$$

and

$$V = [\mathbf{v}_1\ \mathbf{v}_2\ \mathbf{v}_3\ \mathbf{v}_4] = \begin{bmatrix} \frac{1}{\sqrt{7}} & \frac{1}{\sqrt{3}} & \frac{1}{\sqrt{11}} & \frac{1}{\sqrt{2}} \\ \frac{2}{\sqrt{7}} & \frac{-1}{\sqrt{3}} & \frac{1}{\sqrt{11}} & 0 \\ \frac{1}{\sqrt{7}} & 0 & \frac{-3}{\sqrt{11}} & 0 \\ \frac{1}{\sqrt{7}} & \frac{1}{\sqrt{3}} & 0 & \frac{-1}{\sqrt{2}} \end{bmatrix}$$

to obtain the singular value decomposition

$$\begin{bmatrix} \frac{1}{\sqrt{3}} & \frac{2}{\sqrt{6}} & 0 \\ \frac{1}{\sqrt{3}} & \frac{-1}{\sqrt{6}} & \frac{1}{\sqrt{2}} \\ \frac{-1}{\sqrt{3}} & \frac{1}{\sqrt{6}} & \frac{1}{\sqrt{2}} \end{bmatrix} \begin{bmatrix} \sqrt{21} & 0 & 0 & 0 \\ 0 & \sqrt{18} & 0 & 0 \\ 0 & 0 & 0 & 0 \end{bmatrix} \begin{bmatrix} \frac{1}{\sqrt{7}} & \frac{1}{\sqrt{3}} & \frac{1}{\sqrt{11}} & \frac{1}{\sqrt{2}} \\ \frac{2}{\sqrt{7}} & \frac{-1}{\sqrt{3}} & \frac{1}{\sqrt{11}} & 0 \\ \frac{1}{\sqrt{7}} & 0 & \frac{-3}{\sqrt{11}} & 0 \\ \frac{1}{\sqrt{7}} & \frac{1}{\sqrt{3}} & 0 & \frac{-1}{\sqrt{2}} \end{bmatrix}^T.$$

18. $\begin{bmatrix} \frac{-1}{3} & 0 & \frac{4}{\sqrt{18}} \\ \frac{2}{3} & \frac{1}{\sqrt{2}} & \frac{1}{\sqrt{18}} \\ \frac{2}{3} & \frac{-1}{\sqrt{2}} & \frac{1}{\sqrt{18}} \end{bmatrix} \begin{bmatrix} 36 & 0 & 0 & 0 \\ 0 & 18 & 0 & 0 \\ 0 & 0 & 0 & 0 \end{bmatrix} \begin{bmatrix} \frac{1}{3} & 0 & 0 & \frac{4}{\sqrt{18}} \\ \frac{-2}{3} & \frac{-1}{\sqrt{2}} & 0 & \frac{1}{\sqrt{18}} \\ 0 & 0 & 1 & 0 \\ \frac{2}{3} & \frac{-1}{\sqrt{2}} & 0 & \frac{-1}{\sqrt{18}} \end{bmatrix}^T$

19. In the following figure,

$$\mathbf{u}_1 = \frac{1}{\sqrt{2}} \begin{bmatrix} 1 \\ -1 \end{bmatrix}, \mathbf{u}_2 = \frac{1}{\sqrt{2}} \begin{bmatrix} 1 \\ 1 \end{bmatrix},$$

$OP = 2\sqrt{2}$, and $OQ = \sqrt{2}$.

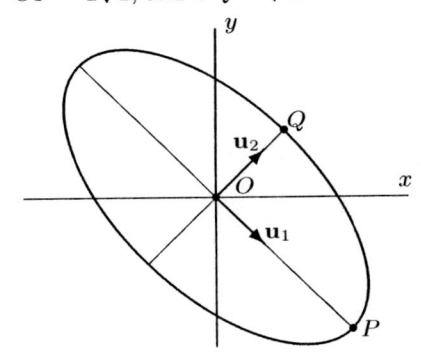

Figure for Exercise 19

20. In the following figure,

$$\mathbf{u}_1 = \frac{1}{\sqrt{2}} \begin{bmatrix} 1 \\ 1 \end{bmatrix}, \mathbf{u}_2 = \frac{1}{\sqrt{2}} \begin{bmatrix} 1 \\ -1 \end{bmatrix},$$

$OP = 3$, and $OQ = 1$.

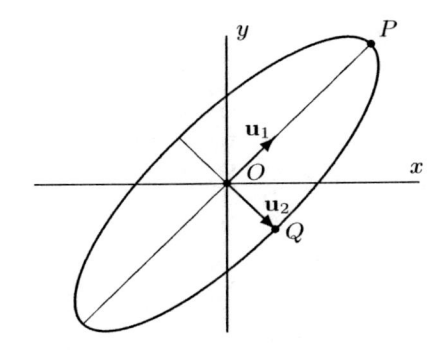

Figure for Exercise 20

21. Let $A = \begin{bmatrix} 1 & 1 \\ 2 & 2 \end{bmatrix}$, $\mathbf{b} = \begin{bmatrix} 2 \\ 4 \end{bmatrix}$, and $\mathbf{z}$ be the unique solution of $A\mathbf{x} = \mathbf{b}$ with least norm. Since

$$A^T A = \begin{bmatrix} 1 & 2 \\ 1 & 2 \end{bmatrix} \begin{bmatrix} 1 & 1 \\ 2 & 2 \end{bmatrix} = \begin{bmatrix} 5 & 5 \\ 5 & 5 \end{bmatrix},$$

its characteristic polynomial is

$$\det(A^T A - tI_2) = \det \begin{bmatrix} 5-t & 5 \\ 5 & 5-t \end{bmatrix}$$
$$= t(t-10).$$

So its eigenvalues 10 are and 0. Therefore $\sigma_1 = \sqrt{10}$ is the (nonzero) singular value of A. Let

$$\mathbf{v}_1 = \frac{1}{\sqrt{2}} \begin{bmatrix} 1 \\ 1 \end{bmatrix} \quad \text{and} \quad \mathbf{v}_2 = \frac{1}{\sqrt{2}} \begin{bmatrix} 1 \\ -1 \end{bmatrix}.$$

Then $\{\mathbf{v}_1, \mathbf{v}_2\}$ is an orthonormal basis for $\mathcal{R}^2$ consisting of eigenvectors of $A^T A$ corresponding to 10 and 0, respectively. Next, let

$$\mathbf{u}_1 = \frac{1}{\sigma_1} A \mathbf{v}_1 = \frac{1}{\sqrt{10}} \begin{bmatrix} 1 & 1 \\ 2 & 2 \end{bmatrix} \frac{1}{\sqrt{2}} \begin{bmatrix} 1 \\ 1 \end{bmatrix}$$
$$= \frac{1}{\sqrt{5}} \begin{bmatrix} 1 \\ 2 \end{bmatrix},$$

and choose a unit vector $\mathbf{u}_2$ orthogonal to $\mathbf{u}_1$ such as

$$\mathbf{u}_2 = \frac{1}{\sqrt{5}} \begin{bmatrix} 2 \\ -1 \end{bmatrix}.$$

Let

$$U = [\mathbf{u}_1\ \mathbf{u}_2] = \frac{1}{\sqrt{5}} \begin{bmatrix} 1 & 2 \\ 2 & -1 \end{bmatrix},$$

$$\Sigma = \begin{bmatrix} \sigma_1 & 0 \\ 0 & 0 \end{bmatrix} = \begin{bmatrix} \sqrt{10} & 0 \\ 0 & 0 \end{bmatrix},$$

and

$$V = [\mathbf{u}_1 \ \mathbf{u}_2] = \frac{1}{\sqrt{2}} \begin{bmatrix} 1 & 1 \\ 1 & -1 \end{bmatrix}.$$

Then $A = U\Sigma V^T$ is a singular value decomposition of A. So, by the boxed result on pages 444–449,

$$\mathbf{z} = V\Sigma^\dagger U^T \mathbf{b}$$

$$= \frac{1}{\sqrt{2}} \begin{bmatrix} 1 & 1 \\ 1 & -1 \end{bmatrix} \begin{bmatrix} \frac{1}{\sqrt{10}} & 0 \\ 0 & 0 \end{bmatrix} \frac{1}{\sqrt{5}} \begin{bmatrix} 1 & 2 \\ 2 & -1 \end{bmatrix}^T \begin{bmatrix} 2 \\ 4 \end{bmatrix}$$

$$= \begin{bmatrix} 1 \\ 1 \end{bmatrix}.$$

22. $\begin{bmatrix} 1 \\ -1 \end{bmatrix}$ 23. $\begin{bmatrix} 3 \\ -4 \\ 1 \end{bmatrix}$ 24. $\begin{bmatrix} 0 \\ 1 \\ 0 \end{bmatrix}$

25. $\frac{1}{35} \begin{bmatrix} 20 \\ -37 \\ 11 \end{bmatrix}$ 26. $\begin{bmatrix} 1.5 \\ 1.0 \\ 1.5 \end{bmatrix}$ 27. $\begin{bmatrix} 3 \\ 1 \\ 1 \end{bmatrix}$

28. $\frac{1}{3} \begin{bmatrix} 7 \\ 5 \\ -1 \end{bmatrix}$ 29. $\begin{bmatrix} 0.04 \\ 0.08 \end{bmatrix}$ 30. $\frac{1}{6} \begin{bmatrix} 1 \\ -1 \\ 2 \end{bmatrix}$

31. Let $\mathbf{z}$ be the unique solution of $A\mathbf{x} = \mathbf{b}$ with least norm, where

$$A = \begin{bmatrix} 1 & 1 \\ 1 & -1 \\ 1 & 2 \end{bmatrix} \quad \text{and} \quad \mathbf{b} = \begin{bmatrix} 3 \\ 1 \\ 2 \end{bmatrix}.$$

By Exercise 5, $A = U\Sigma V^T$ is a singular value decomposition of A, where

$$U = \begin{bmatrix} \frac{3}{\sqrt{35}} & \frac{1}{\sqrt{10}} & \frac{-3}{\sqrt{14}} \\ \frac{-1}{\sqrt{35}} & \frac{3}{\sqrt{10}} & \frac{1}{\sqrt{14}} \\ \frac{5}{\sqrt{35}} & 0 & \frac{2}{\sqrt{14}} \end{bmatrix}, \quad \Sigma = \begin{bmatrix} \sqrt{7} & 0 \\ 0 & \sqrt{2} \\ 0 & 0 \end{bmatrix},$$

and

$$V = \begin{bmatrix} \frac{1}{\sqrt{5}} & \frac{2}{\sqrt{5}} \\ \frac{2}{\sqrt{5}} & \frac{-1}{\sqrt{5}} \end{bmatrix}.$$

So, by the boxed result on pages 448–449,

$$\mathbf{z} = V\Sigma^\dagger U^T \mathbf{b}$$

$$= \begin{bmatrix} \frac{1}{\sqrt{5}} & \frac{2}{\sqrt{5}} \\ \frac{2}{\sqrt{5}} & \frac{-1}{\sqrt{5}} \end{bmatrix} \begin{bmatrix} \frac{1}{\sqrt{7}} & 0 & 0 \\ 0 & \frac{1}{\sqrt{2}} & 0 \end{bmatrix} \begin{bmatrix} \frac{3}{\sqrt{35}} & \frac{1}{\sqrt{10}} & \frac{-3}{\sqrt{14}} \\ \frac{-1}{\sqrt{35}} & \frac{3}{\sqrt{10}} & \frac{1}{\sqrt{14}} \\ \frac{5}{\sqrt{35}} & 0 & \frac{2}{\sqrt{14}} \end{bmatrix}^T \begin{bmatrix} 3 \\ 1 \\ 2 \end{bmatrix}$$

$$= \frac{1}{7} \begin{bmatrix} 12 \\ 3 \end{bmatrix}.$$

32. $\frac{1}{6} \begin{bmatrix} 10 \\ 3 \end{bmatrix}$ 33. $\frac{5}{6} \begin{bmatrix} 3 \\ 3 \\ 2 \end{bmatrix}$ 34. $\frac{1}{3} \begin{bmatrix} -3 \\ 4 \end{bmatrix}$

35. $\frac{1}{154} \begin{bmatrix} -11 \\ 30 \\ 27 \end{bmatrix}$ 36. $\frac{1}{6} \begin{bmatrix} -1 \\ 6 \\ -5 \end{bmatrix}$

37. Let $A = \begin{bmatrix} 1 \\ 2 \\ 2 \end{bmatrix}$. By Exercise 3, $A = U\Sigma V^T$ is a singular value decomposition of A, where

$$U = \begin{bmatrix} \frac{1}{3} & \frac{2}{\sqrt{5}} & \frac{2}{3\sqrt{5}} \\ \frac{2}{3} & \frac{-1}{\sqrt{5}} & \frac{4}{3\sqrt{5}} \\ \frac{2}{3} & 0 & \frac{-5}{3\sqrt{5}} \end{bmatrix}, \quad \Sigma = \begin{bmatrix} 3 \\ 0 \\ 0 \end{bmatrix},$$

and

$$V = [1].$$

Thus

$$A^\dagger = V\Sigma^\dagger U^T = [1] \begin{bmatrix} \frac{1}{3} & 0 & 0 \end{bmatrix} \begin{bmatrix} \frac{1}{3} & \frac{2}{3} & \frac{2}{3} \\ \frac{2}{\sqrt{5}} & \frac{-1}{\sqrt{5}} & 0 \\ \frac{2}{3\sqrt{5}} & \frac{4}{3\sqrt{5}} & \frac{-5}{3\sqrt{5}} \end{bmatrix}$$

$$= \frac{1}{9} [1 \ 2 \ 2].$$

38. $\frac{1}{10} \begin{bmatrix} 1 & -2 \\ -1 & 2 \end{bmatrix}$ 39. $\frac{1}{3} \begin{bmatrix} 2 & 1 \\ -1 & -2 \\ -1 & 1 \end{bmatrix}$

40. $\frac{1}{12} \begin{bmatrix} 1 & 3 & 1 & 1 \\ 4 & -2 & 0 & 2 \end{bmatrix}$ 41. $\frac{1}{14} \begin{bmatrix} 4 & 8 & 2 \\ 1 & -5 & 4 \end{bmatrix}$

42. $\begin{bmatrix} 1 & 0.0 \\ 0 & 0.5 \\ 0 & 0.0 \\ 0 & 0.0 \end{bmatrix}$

43. Let $A = \begin{bmatrix} 1 & 1 & 1 \\ 1 & -1 & -1 \end{bmatrix}$. By Exercise 7, $A = U\Sigma V^T$ is a singular value decomposition of A, where

$$U = \begin{bmatrix} \frac{1}{\sqrt{2}} & \frac{1}{\sqrt{2}} \\ \frac{-1}{\sqrt{2}} & \frac{1}{\sqrt{2}} \end{bmatrix}, \quad \Sigma = \begin{bmatrix} 2 & 0 & 0 \\ 0 & \sqrt{2} & 0 \end{bmatrix},$$

and

$$V = \begin{bmatrix} 0 & 1 & 0 \\ \frac{1}{\sqrt{2}} & 0 & \frac{1}{\sqrt{2}} \\ \frac{1}{\sqrt{2}} & 0 & \frac{-1}{\sqrt{2}} \end{bmatrix}.$$

Thus
$$A^\dagger = V\Sigma^\dagger U^T$$
$$= \begin{bmatrix} 0 & 1 & 0 \\ \frac{1}{\sqrt{2}} & 0 & \frac{1}{\sqrt{2}} \\ \frac{1}{\sqrt{2}} & 0 & \frac{-1}{\sqrt{2}} \end{bmatrix} \begin{bmatrix} \frac{1}{2} & 0 \\ 0 & \frac{1}{\sqrt{2}} \\ 0 & 0 \end{bmatrix} \begin{bmatrix} \frac{1}{\sqrt{2}} & \frac{-1}{\sqrt{2}} \\ \frac{1}{\sqrt{2}} & \frac{1}{\sqrt{2}} \end{bmatrix}$$
$$= \frac{1}{4} \begin{bmatrix} 2 & 2 \\ 1 & -1 \\ 1 & -1 \end{bmatrix}.$$

44. $\frac{1}{5} \begin{bmatrix} 2 & 1 \\ 5 & 5 \\ 4 & 2 \end{bmatrix}$
45. $\frac{1}{14} \begin{bmatrix} 4 & -1 \\ -2 & 4 \\ 8 & 5 \end{bmatrix}$

46. $\frac{1}{30} \begin{bmatrix} 0 & 5 & 0 \\ 4 & -3 & 0 \\ 2 & 1 & 0 \\ 2 & -4 & 0 \end{bmatrix}$
47. $\frac{1}{9} \begin{bmatrix} 1 & 2 & 2 \\ 2 & 4 & 4 \\ 2 & 4 & 4 \end{bmatrix}$

48. $\frac{1}{4} \begin{bmatrix} 1 & 1 & -1 & 1 \\ 1 & 1 & -1 & 1 \\ -1 & -1 & 1 & -1 \\ 1 & 1 & -1 & 1 \end{bmatrix}$

49. $\frac{1}{3} \begin{bmatrix} 2 & 1 & 1 \\ 1 & 2 & -1 \\ 1 & -1 & 2 \end{bmatrix}$
50. $\frac{1}{14} \begin{bmatrix} 13 & -3 & -2 \\ -3 & 5 & -6 \\ -2 & -6 & 10 \end{bmatrix}$

51. $\frac{1}{75} \begin{bmatrix} 74 & 7 & 5 \\ 7 & 26 & -35 \\ 5 & -35 & 50 \end{bmatrix}$

52. $\frac{1}{90} \begin{bmatrix} 65 & -40 & -5 \\ -40 & 26 & -8 \\ -5 & -8 & 89 \end{bmatrix}$

53. Let
$$A = \begin{bmatrix} 1 & 1 \\ 1 & -1 \\ 1 & 2 \end{bmatrix}.$$

By Exercise 5, $A = U\Sigma V^T$ is a singular value decomposition of A, where
$$U = \begin{bmatrix} \frac{3}{\sqrt{35}} & \frac{1}{\sqrt{10}} & \frac{-3}{\sqrt{14}} \\ \frac{-1}{\sqrt{35}} & \frac{3}{\sqrt{10}} & \frac{1}{\sqrt{14}} \\ \frac{5}{\sqrt{35}} & 0 & \frac{2}{\sqrt{14}} \end{bmatrix}.$$

Since rank $A = 2$, let
$$D = \begin{bmatrix} 1 & 0 & 0 \\ 0 & 1 & 0 \\ 0 & 0 & 0 \end{bmatrix}.$$

By equation (13), we have
$$P_W = UDU^T$$

$$= \begin{bmatrix} \frac{3}{\sqrt{35}} & \frac{1}{\sqrt{10}} & \frac{-3}{\sqrt{14}} \\ \frac{-1}{\sqrt{35}} & \frac{3}{\sqrt{10}} & \frac{1}{\sqrt{14}} \\ \frac{5}{\sqrt{35}} & 0 & \frac{2}{\sqrt{14}} \end{bmatrix} \begin{bmatrix} 1 & 0 & 0 \\ 0 & 1 & 0 \\ 0 & 0 & 0 \end{bmatrix} \begin{bmatrix} \frac{3}{\sqrt{35}} & \frac{-1}{\sqrt{35}} & \frac{5}{\sqrt{35}} \\ \frac{1}{\sqrt{10}} & \frac{3}{\sqrt{10}} & 0 \\ \frac{-3}{\sqrt{14}} & \frac{1}{\sqrt{14}} & \frac{2}{\sqrt{14}} \end{bmatrix}$$

$$= \frac{1}{14} \begin{bmatrix} 5 & 3 & 6 \\ 3 & 13 & -2 \\ 6 & -2 & 10 \end{bmatrix}.$$

54. $\frac{1}{12} \begin{bmatrix} 9 & -1 & 1 & 5 \\ -1 & 11 & 3 & 1 \\ 1 & 3 & 1 & 1 \\ 5 & 1 & 1 & 3 \end{bmatrix}$

55. False, σ^2 is an eigenvalue of $A^T A$.
56. True
57. False, see Example 7.
58. True
59. False, every matrix has a pseudoinverse.
60. False, $\mathcal{B}_1$ is an orthonormal basis of $A^T A$.
61. True   62. True
63. False, $\mathcal{B}_2$ is an orthonormal basis of AA^T.
64. True   65. True
66. False, if $A = U\Sigma V^T$ is a singular value decomposition of A, then $A = (-U)\Sigma(-V)^T$ is also a singular value decomposition of A.
67. False, only the nonzero diagonal entries are singular values.
68. True
69. False, $V\Sigma^T U^T$ is a singular value decomposition of A^T.
70. True   71. True
72. False, $\mathbf{u}$ is the unique vector of least norm that minimizes $\|A\mathbf{u} - \mathbf{b}\|$.
73. True
74. False, $A^\dagger = V\Sigma^\dagger U^T$.
75. True
76. (a) We apply equations (9) and (10) to obtain
$$A^T A \mathbf{v}_i = A^T \sigma_i \mathbf{u}_i = \sigma_i A^T \mathbf{u}_i = \sigma_i^2 \mathbf{v}_i$$
for $i \leq k$, and
$$A^T A \mathbf{v}_i = A^T \mathbf{0} = 0 \mathbf{v}_i$$
for $i > k$. The result now follows.
(b) The proof is similar to that in (a).

(c) By (a), $\sigma_1^2, \sigma_2^2, \ldots, \sigma_k^2, 0, \ldots, 0$ are the eigenvalues of $A^T A$. Thus the singular values of A are the positive square roots of the k largest eigenvalues of $A^T A$. Since the eigenvalues of $A^T A$ are unique, the singular values of A are unique.

Observe that $\{-\mathbf{v}_1, -\mathbf{v}_2, \ldots, -\mathbf{v}_n\}$ and $\{-\mathbf{u}_1, -\mathbf{u}_2, \ldots, -\mathbf{u}_m\}$ are orthonormal bases satisfying equations (9) and (10) that are different from the given bases $\mathcal{B}_1$ and $\mathcal{B}_2$, respectively.

77. (a) Let $\mathcal{B} = \{\mathbf{v}_1, \mathbf{v}_2, \ldots, \mathbf{v}_n\}$ be an orthonormal basis for $\mathcal{R}^n$ satisfying equation (9), and let $\mathbf{v}$ be a vector in $\mathcal{R}^n$. Then
$$\mathbf{v} = a_1 \mathbf{v}_1 + a_2 \mathbf{v}_2 + \cdots + a_n \mathbf{v}_n$$
for some scalars $a_1, a_2, \ldots, a_n$. Thus
$$\begin{aligned}
\|A\mathbf{v}\|^2 &= (A\mathbf{v}) \cdot (A\mathbf{v}) \\
&= (A\mathbf{v})^T (A\mathbf{v}) \\
&= \mathbf{v}^T A^T A \mathbf{v} \\
&= \mathbf{v} \cdot (A^T A \mathbf{v}) \\
&= (a_1 \mathbf{v}_1 + a_2 \mathbf{v}_2 + \cdots + a_n \mathbf{v}_n) \cdot \\
&\quad (a_1 \sigma_1^2 \mathbf{v}_1 + a_2 \sigma_2^2 \mathbf{v}_2 + \cdots + a_m \sigma_m^2 \mathbf{v}_m) \\
&= a_1^2 \sigma_1^2 + a_2^2 \sigma_2^2 + \cdots + a_m^2 \sigma_m^2 \\
&\leq a_1^2 \sigma_1^2 + a_2^2 \sigma_1^2 + \cdots + a_m^2 \sigma_1^2 \\
&\leq (a_1^2 + a_2^2 + \cdots + a_n^2)\sigma_1^2 \\
&= \|\mathbf{v}\|^2 \sigma_1^2,
\end{aligned}$$
and hence $\|A\mathbf{v}\| \leq \sigma_1 \|\mathbf{v}\|$. The proof that $\sigma_m \|\mathbf{v}\| \leq \|A\mathbf{v}\|$ is similar.

(b) Let $\mathbf{v} = \mathbf{v}_m$ and $\mathbf{w} = \mathbf{v}_1$, where $\mathbf{v}_1$ and $\mathbf{v}_m$ are as in (a).

78. (a) Let $\mathcal{B}_1 = \{\mathbf{v}_1, \mathbf{v}_2, \ldots, \mathbf{v}_n\}$ and $\mathcal{B}_2 = \{\mathbf{u}_1, \mathbf{u}_2, \ldots, \mathbf{u}_m\}$ be the sets of columns of V and U, respectively. Since V and U are orthogonal matrices, these sets are orthonormal bases for $\mathcal{R}^n$ and $\mathcal{R}^m$, respectively. Since $A = U\Sigma V^T$, we have that $AV = \Sigma U$. It follows that
$$[A\mathbf{v}_1 \ A\mathbf{v}_2 \ \cdots \ A\mathbf{v}_n] = AV = U\Sigma$$
$$= [\sigma_1 \mathbf{u}_1 \ \sigma_2 \mathbf{u}_2 \ \cdots \ \sigma_k \mathbf{u}_k \ \mathbf{0} \ \cdots \ \mathbf{0}],$$
and so $A\mathbf{v}_i = \sigma_i \mathbf{u}_i$ if $i \leq k$, and $A\mathbf{v}_i = \mathbf{0}$ if $i > k$.
To prove that $A^T \mathbf{u}_i = \sigma_i \mathbf{v}_i$ if $i \leq k$ and $A^T \mathbf{u}_i = \mathbf{0}$ if $i > k$, apply the method used in (a) to
$$A^T = (U\Sigma V^T)^T = V \Sigma^T U^T.$$
It follows that $\sigma_1, \sigma_2, \ldots, \sigma_k$ are the singular values of A and $\mathcal{B}_1$ and $\mathcal{B}_2$ are orthonormal bases for $\mathcal{R}^n$ and $\mathcal{R}^m$ satisfying equations (9) and (10), respectively.

(b) See (a).

79. Suppose that $A = U\Sigma V^T$ is a singular value decomposition of the $m \times n$ matrix A. Then $A^T = V\Sigma^T U^T$ is a factorization of A^T in which V and U are orthogonal matrices and Σ^T is an $n \times m$ matrix having the form of equation (11). It follows that $V\Sigma^T U^T$ is a singular value decomposition of A^T.

80. Applying the definition of *singular value*, we see that A and A^T have the same singular values. In solving Exercise 76, we saw that the nonzero eigenvalues of $A^T A$ are the squares of the (nonzero) singular values of A. Replacing A by A^T, we see that the nonzero eigenvalues of AA^T are the squares of the singular values of A^T. It follows that $A^T A$ and AA^T have the same nonzero eigenvalues.

81. Since $\Sigma = I_m \Sigma I_n^T$ is a singular value decomposition of Σ, it follows that the pseudoinverse of Σ is $I_n \Sigma^\dagger I_m^T = \Sigma^\dagger$.

82. Suppose that A is an $n \times n$ invertible matrix. For any $\mathbf{b}$ in $\mathcal{R}^n$, $A^{-1}\mathbf{b}$ is the unique solution of the system $A\mathbf{x} = \mathbf{b}$. It follows that $A^\dagger \mathbf{b}$ is also a solution of this system, and hence $A^\dagger \mathbf{b} = A^{-1}\mathbf{b}$. Therefore $A^\dagger = A^{-1}$.

83. Suppose that $A = U\Sigma V^T$ is a singular value decomposition of A. Observe that $(\Sigma^T)^\dagger = (\Sigma^\dagger)^T$ and $A^T = V\Sigma^T U^T$ is a singular value decomposition of A^T. Therefore
$$\begin{aligned}
(A^T)^\dagger &= U(\Sigma^T)^\dagger V^T \\
&= U(\Sigma^\dagger)^T V^T \\
&= (V\Sigma^\dagger U^T)^T \\
&= (A^\dagger)^T.
\end{aligned}$$

84. Let $\sigma_1, \sigma_2, \ldots, \sigma_k$ be the (nonzero) singular values of A. By Exercise 76, the squares of these singular values are the nonzero eigenvalues of $A^T A = A^2$. But the nonzero eigenvalues of A^2 are the squares of the nonzero eigenvalues of A. So if $\lambda_1, \lambda_2, \ldots, \lambda_k$ are the nonzero eigenvalues of A (listed in decreasing order of absolute value), then $\sigma_i^2 = \lambda_i^2$ for $1 \leq i \leq k$. Therefore $\sigma_i = |\lambda_i|$ for $1 \leq i \leq k$.

85. Suppose that A is a positive semidefinite matrix. Since A is symmetric, there exists an orthogonal matrix V and a diagonal matrix D such that $A = VDV^T$. Furthermore, the diagonal entries of D are the eigenvalues of A, and these are nonnegative by Exercise 60 of Section 6.6. Also, V and D can be chosen so that the diagonal

entries are listed in decreasing order of absolute value. Since D has the form given in equation (11), we see that $A = VDV^T$ is a singular value decomposition of A.

Now suppose that A is not positive semidefinite. Then A has a negative eigenvalue. Consider any factorization of the form $A = V\Sigma V^T$, where V is an orthogonal matrix and Σ is of the form given in equation (11). It follows that Σ is a diagonal matrix whose diagonal entries are the eigenvalues of A. Since A has a negative eigenvalue, at least one of the diagonal entries of Σ is negative. This entry cannot be a singular value of A, and it follows that $A = V\Sigma V^T$ is not a singular value decomposition of A.

86. (a) Since Q is invertible, it has rank n, and hence Q has n singular values. By Exercise 76, the square of each singular value is an eigenvalue of $Q^T Q = I_n$. But 1 is the only eigenvalue of I_n, and hence, each singular value of Q is $\sqrt{1} = 1$.
    (b) $Q = QI_n I_n^T$

87. If A is an orthogonal matrix, then 1 is the only singular value of A by Exercise 86. Conversely, suppose that 1 is the only singular value of A and $A = U\Sigma V^T$ is a singular value decomposition of A. Since A has rank n, it has n singular values, and the nonzero diagonal entries of Σ are the singular values of A by Exercise 78. Hence $\Sigma = I_n$. It follows that $A = UI_n V^T = UV^T$, which is an orthogonal matrix.

88. (a) Both UP and VQ are orthogonal matrices because they are products of orthogonal matrices; so $(UP)\Sigma(VQ)^T$ is a singular value decomposition of some $m \times n$ matrix. Furthermore,
$$(UP)\Sigma(VQ)^T = U(P\Sigma)Q^T V^T$$
$$= U(\Sigma Q)Q^T V^T = U\Sigma V^T$$
$$= A.$$

Therefore $(UP)\Sigma(VQ)^T$ is a singular value decomposition of A.
(b) Let $A = U = \Sigma = V = I_2$ and $P = Q = -I_2$.
(c) Suppose that $U_1 \Sigma V_1^T$ is another singular value decomposition of A. Then $U_1 \Sigma V_1^T = U\Sigma V^T$, and hence $U^T U_1 \Sigma = \Sigma V^T V_1$. Let $P = U^T U_1$ and $Q = V^T V_1$. Then $P\Sigma = \Sigma Q$, $U_1 = UP$, and $V_1 = VQ$.

89. If Σ is an $m \times n$ matrix of the form in equation (11) and $\Sigma^\dagger$ is an $n \times m$ matrix of the form in equation (14), their product is the $m \times m$ diagonal matrix whose first k diagonal entries are $\sigma_i \cdot \dfrac{1}{\sigma_i} = 1$ and whose last $n - k$ diagonal entries are zero.

90. Let $Z = \operatorname{Row} A$. Since $\operatorname{Row} A = \operatorname{Col} A^T$, we obtain $A^T(A^T)^\dagger = P_Z$ by applying the boxed result on page 451 to A^T. Observe that, by Exercise 83, $(A^T)^\dagger = (A^\dagger)^T$. Since orthogonal projection matrices are symmetric, we have
$$P_Z = P_Z^T = (A^T(A^T)^\dagger)^T = ((A^T)^\dagger)^T (A^T)^T$$
$$= ((A^\dagger)^T)^T A = A^\dagger A.$$

91. (a) For each i, let $\mathbf{e}_i$ denote the ith standard vector of $\mathcal{R}^m$. Then
$$A = U\Sigma V^T$$
$$= U[\sigma_1 \mathbf{e}_1 \; \sigma_2 \mathbf{e}_2 \; \cdots \; \sigma_k \mathbf{e}_k \; \mathbf{0} \; \cdots \; \mathbf{0}]V^T$$
$$= [U\sigma_1 \mathbf{e}_1 \; U\sigma_2 \mathbf{e}_2 \; \cdots \; U\sigma_k \mathbf{e}_k \; \mathbf{0} \; \cdots \; \mathbf{0}]V^T$$
$$= [\sigma_1 \mathbf{u}_1 \; \sigma_2 \mathbf{u}_2 \; \cdots \; \sigma_k \mathbf{u}_k \; \mathbf{0} \; \cdots \; \mathbf{0}] \begin{bmatrix} \mathbf{v}_1^T \\ \mathbf{v}_2^T \\ \vdots \\ \mathbf{v}_n^T \end{bmatrix}$$
$$= \sigma_1 \mathbf{u}_1 \mathbf{v}_1^T + \sigma_2 \mathbf{u}_2 \mathbf{v}_2^T + \cdots + \sigma_k \mathbf{u}_k \mathbf{v}_k^T$$
$$= \sigma_1 Q_1 + \sigma_2 Q_2 + \cdots \sigma_k Q_k.$$

(b) Exercise 52 of Section 2.5 implies that $\operatorname{rank} Q_i = 1$ for every i.
(c) First observe that, for each i,
$$Q_i Q_i^T = (\mathbf{u}_i \mathbf{v}_i^T)(\mathbf{u}_i \mathbf{v}_i^T)^T = \mathbf{u}_i \mathbf{v}_i^T \mathbf{v}_i \mathbf{u}_i^T$$
$$= \mathbf{u}_i(\mathbf{v}_i \cdot \mathbf{v}_i)\mathbf{u}_i^T = \mathbf{u}_i \mathbf{u}_i^T.$$

Thus
$$Q_i Q_i^T \mathbf{u}_i = (\mathbf{u}_i \mathbf{u}_i^T)\mathbf{u}_i = \mathbf{u}_i(\mathbf{u}_i^T \mathbf{u}_i)$$
$$= \mathbf{u}_i(\mathbf{u}_i \cdot \mathbf{u}_i) = \mathbf{u}_i \cdot 1 = \mathbf{u}_i.$$

Furthermore, for any vector $\mathbf{w}$ orthogonal to $\mathbf{u}_i$, we have
$$Q_i Q_i^T \mathbf{w} = (\mathbf{u}_i \mathbf{u}_i^T)\mathbf{w} = \mathbf{u}_i(\mathbf{u}_i^T \mathbf{w})$$
$$= \mathbf{u}_i(\mathbf{u}_i \cdot \mathbf{w}) = \mathbf{u}_i \cdot 0 = \mathbf{0}.$$

It follows that $Q_i Q_i^T$ is the orthogonal projection matrix for the span of $\{\mathbf{u}_i\}$.
(d) The proof is similar to the proof of (c).
(e) For $i \neq j$,
$$Q_i Q_j^T = (\mathbf{u}_i \mathbf{v}_i^T)(\mathbf{u}_j \mathbf{v}_j^T)^T = \mathbf{u}_i \mathbf{v}_i^T \mathbf{v}_j \mathbf{u}_j^T$$
$$= \mathbf{u}_i(\mathbf{v}_i \cdot \mathbf{v}_j)\mathbf{v}_j^T = \mathbf{u}_i \bullet 0 \bullet \mathbf{v}_j^T = O.$$

The proof of the second equation is similar.

92. (rounded to 4 places after the decimal)
$$U = \begin{bmatrix} 0.0845 & -0.9922 & 0.0919 \\ 0.9301 & 0.0454 & -0.3644 \\ 0.3574 & 0.1162 & 0.9267 \end{bmatrix},$$

$$\Sigma = \begin{bmatrix} 4.1246 & 0 & 0 & 0 \\ 0 & 2.4401 & 0 & 0 \\ 0 & 0 & 1.4261 & 0 \end{bmatrix},$$

$$V = \begin{bmatrix} 0.3531 & -0.7470 & 0.5231 & -0.2090 \\ 0.7632 & 0.1035 & -0.1168 & 0.6271 \\ 0.1593 & -0.4356 & -0.8409 & -0.2787 \\ 0.5172 & 0.4915 & 0.0743 & -0.6967 \end{bmatrix},$$

and

$$A^\dagger = \begin{bmatrix} 0.3447 & -0.0680 & 0.3350 \\ -0.0340 & 0.2039 & -0.0049 \\ 0.1262 & 0.2427 & -0.5534 \\ -0.1845 & 0.1068 & 0.1165 \end{bmatrix}.$$

93. (rounded to 4 places after the decimal)
$$U = \begin{bmatrix} 0.5836 & 0.7289 & -0.3579 \\ 0.7531 & -0.6507 & -0.0970 \\ 0.3036 & 0.2129 & 0.9287 \end{bmatrix},$$

$$\Sigma = \begin{bmatrix} 5.9073 & 0 & 0 & 0 \\ 0 & 2.2688 & 0 & 0 \\ 0 & 0 & 1.7194 & 0 \end{bmatrix},$$

$$V = \begin{bmatrix} 0.3024 & -0.3462 & -0.8612 & -0.2170 \\ 0.0701 & 0.9293 & -0.3599 & 0.0434 \\ 0.2777 & 0.1283 & 0.2755 & -0.9113 \\ 0.9091 & 0.0043 & 0.2300 & 0.3472 \end{bmatrix},$$

and

$$A^\dagger = \begin{bmatrix} 0.0979 & 0.1864 & -0.4821 \\ 0.3804 & -0.2373 & -0.1036 \\ 0.0113 & -0.0169 & 0.1751 \\ 0.0433 & 0.1017 & 0.1714 \end{bmatrix}.$$

6.8 PRINCIPAL COMPONENT ANALYSIS

1. $\overline{x} = \frac{1}{3}[2 - 3 + 4] = 1$
2. $\overline{y} = \frac{1}{3}[4 + 2 + 3] = 3$
3.
$$s_\mathbf{x}^2 = \frac{1}{2}[(2-1)^2 + (-3-1)^2 + (4-1)^2]$$
$$= \frac{1}{2}(1 + 16 + 9) = 13$$

4.
$$s_\mathbf{y}^2 = \frac{1}{2}[(4-3)^2 + (2-3)^2 + (3-3)^2]$$
$$= \frac{1}{2}(1 + 1 + 0) = 1$$

5.
$$\text{cov}(\mathbf{x}, \mathbf{y}) = \frac{1}{2}[(2-1)(4-3) + (-3-1)(2-3)$$
$$+ (4-1)(3-3)]$$
$$= \frac{1}{2}[1 + 4 + 0] = \frac{5}{2}$$

6. $\dfrac{5/2}{\sqrt{13}\sqrt{1}} \approx .6934$ 7. $\begin{bmatrix} 13 & \frac{5}{2} \\ \frac{5}{2} & 1 \end{bmatrix}$

8. $\begin{bmatrix} 1 & .6934 \\ .6934 & 1 \end{bmatrix}$

9. True
10. False, the sum should be divided by $m - 1$.
11. False, the covariance may be any real number.
12. False, the correlation may be any real number between -1 and 1, inclusively.
13. False, their correlation is either -1 or 1.
14. True
15. False, the covariance may be any real number.
16. True   17. True   18. True
19. True   20. True
21.
$$\text{cov}(\mathbf{x}, \mathbf{y}) = \frac{1}{m-1}(\mathbf{x} - \overline{\mathbf{x}}) \cdot (\mathbf{y} - \overline{\mathbf{y}})$$
$$= \frac{1}{m-1}(\mathbf{y} - \overline{\mathbf{y}}) \cdot (\mathbf{x} - \overline{\mathbf{x}}) = \text{cov}(\mathbf{y}, \mathbf{x})$$

22. (a) Using Exercise 28, we have
$$\text{cov}(c\mathbf{x}, \mathbf{y}) = \frac{1}{m-1}(c\mathbf{x} - \overline{c\mathbf{x}}) \cdot (\mathbf{y} - \overline{\mathbf{y}})$$
$$= \frac{1}{m-1}(c\mathbf{x} - c\overline{\mathbf{x}}) \cdot (\mathbf{y} - \overline{\mathbf{y}})$$
$$= \frac{1}{m-1}(c(\mathbf{x} - \overline{\mathbf{x}})) \cdot (\mathbf{y} - \overline{\mathbf{y}})$$
$$= c\frac{1}{m-1}(\mathbf{x} - \overline{\mathbf{x}}) \cdot (\mathbf{y} - \overline{\mathbf{y}})$$
$$= c \cdot \text{cov}(\mathbf{x}, \mathbf{y}).$$

(b) By Exercises 21 and 22(a),
$$\text{cov}(\mathbf{x}, c\mathbf{y}) = \text{cov}(c\mathbf{y}, \mathbf{x}) = c \cdot \text{cov}(\mathbf{y}, \mathbf{x})$$
$$= c \cdot \text{cov}(\mathbf{x}, \mathbf{y}).$$

23. (a) Using Exercise 28, we have
$$\text{cov}(\mathbf{x} + \mathbf{y}, \mathbf{z})$$
$$= \frac{1}{m-1}[(\mathbf{x} + \mathbf{y}) - \overline{(\mathbf{x} + \mathbf{y})}] \cdot (\mathbf{z} - \overline{\mathbf{z}})$$
$$= \frac{1}{m-1}[(\mathbf{x} + \mathbf{y}) - (\overline{\mathbf{x}} + \overline{\mathbf{y}})] \cdot (\mathbf{z} - \overline{\mathbf{z}})$$

$$= \frac{1}{m-1}[(\mathbf{x}-\overline{\mathbf{x}})+(\mathbf{y}-\overline{\mathbf{y}})]\cdot(\mathbf{z}-\overline{\mathbf{z}})$$
$$= \frac{1}{m-1}(\mathbf{x}-\overline{\mathbf{x}})\cdot(\mathbf{z}-\overline{\mathbf{z}})$$
$$\quad + \frac{1}{m-1}(\mathbf{y}-\overline{\mathbf{y}})\cdot(\mathbf{z}-\overline{\mathbf{z}})$$
$$= \operatorname{cov}(\mathbf{x},\mathbf{z}) + \operatorname{cov}(\mathbf{y},\mathbf{z}).$$

(b) By Exercises 21 and 23(a),
$$\operatorname{cov}(\mathbf{x},\mathbf{y}+\mathbf{z}) = \operatorname{cov}(\mathbf{y}+\mathbf{z},\mathbf{x})$$
$$= \operatorname{cov}(\mathbf{y},\mathbf{x}) + \operatorname{cov}(\mathbf{z},\mathbf{x})$$
$$= \operatorname{cov}(\mathbf{x},\mathbf{y}) + \operatorname{cov}(\mathbf{x},\mathbf{z}).$$

24. $\operatorname{cov}(\mathbf{x},\mathbf{x}) = \frac{1}{m-1}(\mathbf{x}-\overline{\mathbf{x}})\cdot(\mathbf{x}-\overline{\mathbf{x}}) = s_\mathbf{x}^2$

25. Suppose $\operatorname{cov}(\mathbf{x},\mathbf{x}) = 0$. Then $(\mathbf{x}-\overline{\mathbf{x}})\cdot(\mathbf{x}-\overline{\mathbf{x}}) = 0$. So $\mathbf{x}-\overline{\mathbf{x}} = \mathbf{0}$ or $\mathbf{x} = \overline{\mathbf{x}}$. It follows that $x_i = \overline{x}$ for all i. Now suppose all the components of $\mathbf{x}$ are equal. Then the mean of $\mathbf{x}$ equals this common value, and so $\mathbf{x} = \overline{\mathbf{x}}$. Thus $\operatorname{cov}(\mathbf{x},\mathbf{x}) = 0$.

26. (a) Suppose all the components of $\mathbf{w}$ equal the constant c. It follows that the mean of $\mathbf{w}$ is also c. So every component of $\overline{\mathbf{w}}$ equals c, that is, $\mathbf{w} = \overline{\mathbf{w}}$. Therefore
$$\operatorname{cov}(\mathbf{w},\mathbf{y}) = \frac{1}{m-1}(\mathbf{w}-\overline{\mathbf{w}})\cdot(\mathbf{y}-\overline{\mathbf{y}})$$
$$= \frac{1}{m-1}\mathbf{0}\cdot(\mathbf{y}-\overline{\mathbf{y}}) = 0.$$

(b) By Exercises 21 and 26(a),
$$\operatorname{cov}(\mathbf{x},\mathbf{u}) = \operatorname{cov}(\mathbf{u},\mathbf{x}) = 0.$$

27. (a) By Exercises 23 and 26, we have
$$\operatorname{cov}(\mathbf{x}+\mathbf{w},\mathbf{y}) = \operatorname{cov}(\mathbf{x},\mathbf{y}) + \operatorname{cov}(\mathbf{w},\mathbf{y})$$
$$= \operatorname{cov}(\mathbf{x},\mathbf{y}) + 0 = \operatorname{cov}(\mathbf{x},\mathbf{y}).$$

(b) Part (b) is a consequence of Exercises 21 and 27(a).

(c) By Exercises 23 and 26, we have
$$\operatorname{cov}(\mathbf{x}+\mathbf{w},\mathbf{y}+\mathbf{u})$$
$$= \operatorname{cov}(\mathbf{x},\mathbf{y}+\mathbf{u}) + \operatorname{cov}(\mathbf{w},\mathbf{y}+\mathbf{u})$$
$$= \operatorname{cov}(\mathbf{x},\mathbf{y}) + \operatorname{cov}(\mathbf{x},\mathbf{u}) + 0$$
$$= \operatorname{cov}(\mathbf{x},\mathbf{y}) + 0$$
$$= \operatorname{cov}(\mathbf{x},\mathbf{y}).$$

28. (a) The proof follows from (b) with $d = 0$.

(b) The sample mean of $c\mathbf{x} + d\mathbf{y}$ is
$$\frac{1}{m}\sum_{i=1}^{m}(cx_i + dx_i) = \frac{1}{m}\left[c\sum_{i=1}^{m}x_i + d\sum_{i=1}^{m}y_i\right]$$
$$= c\frac{1}{m}\sum_{i=1}^{m}x_i + d\frac{1}{m}\sum_{i=1}^{m}y_i$$
$$= c\overline{\mathbf{x}} + d\overline{\mathbf{y}}.$$

29. By Exercise 28(a), the variance of $c\mathbf{x}$ is
$$\frac{1}{m-1}(c\mathbf{x}-\overline{c\mathbf{x}})\cdot(c\mathbf{x}-\overline{c\mathbf{x}})$$
$$= \frac{1}{m-1}(c\mathbf{x}-c\overline{\mathbf{x}})\cdot(c\mathbf{x}-c\overline{\mathbf{x}})$$
$$= \frac{1}{m-1}(c(\mathbf{x}-\overline{\mathbf{x}}))\cdot(c(\mathbf{x}-\overline{\mathbf{x}}))$$
$$= c^2\frac{1}{m-1}(\mathbf{x}-\overline{\mathbf{x}})\cdot(\mathbf{x}-\overline{\mathbf{x}}) = c^2 s_\mathbf{x}^2.$$

30. By Exercises 21, 23, and 24, the variance of $\mathbf{x}+\mathbf{y}$ is
$$\operatorname{cov}(\mathbf{x}+\mathbf{y},\mathbf{x}+\mathbf{y})$$
$$= \operatorname{cov}(\mathbf{x},\mathbf{x}+\mathbf{y}) + \operatorname{cov}(\mathbf{y},\mathbf{x}+\mathbf{y})$$
$$= \operatorname{cov}(\mathbf{x},\mathbf{x}) + \operatorname{cov}(\mathbf{x},\mathbf{y})$$
$$\quad + \operatorname{cov}(\mathbf{y},\mathbf{x}) + \operatorname{cov}(\mathbf{y},\mathbf{y})$$
$$= s_\mathbf{x}^2 + \operatorname{cov}(\mathbf{x},\mathbf{y}) + \operatorname{cov}(\mathbf{x},\mathbf{y}) + s_\mathbf{y}^2$$
$$= s_\mathbf{x}^2 + s_\mathbf{y}^2 + 2\operatorname{cov}(\mathbf{x},\mathbf{y}).$$

The proof for $\mathbf{x} - \mathbf{y}$ is similar, but it also uses Exercise 22.

31. If variables $\mathbf{x}$ and $\mathbf{y}$ are measured in feet and pounds, respectively, then it is easy to show that $\overline{\mathbf{x}}$ and $\overline{\mathbf{y}}$ are also measured in feet and pounds, respectively. It follows that $\operatorname{cov}(\mathbf{x},\mathbf{y})$ is measured in feet $\times$ pounds. We know that $s_\mathbf{x}$ and $s_\mathbf{y}$ are measured in feet and pounds, respectively. So the "units" of correlation, $\frac{\operatorname{cov}(\mathbf{x},\mathbf{y})}{s_\mathbf{x} s_\mathbf{y}}$ has the form $\frac{feet \times pounds}{feet \times pounds}$, that is, the correlation is unit free.

32. For any i and j, we have
$$C_{ij} = \operatorname{cov}(\mathbf{x}_i, \mathbf{x}_j)$$
$$= \frac{1}{m-1}\sum_{k+1}^{m}(x_{ki}-\overline{x}_i)(x_{kj}-\overline{x}_j)$$

$$= \frac{1}{m-1}\sum_{k=1}^{m}(X-\overline{X})_{ki}(X-\overline{X})_{kj}$$

$$= \frac{1}{m-1}\sum_{k=1}^{m}(X-\overline{X})_{ik}^{T}(X-\overline{X})_{kj}$$

$$= \frac{1}{m-1}\left[(X-\overline{X})^{T}(X-\overline{X})\right]_{ij}$$

$$= \left[\frac{1}{m-1}(X-\overline{X})^{T}(X-\overline{X})\right]_{ij}.$$

So $C = \frac{1}{m-1}(X-\overline{X})^{T}(X-\overline{X})$.

33. The mean of $\frac{1}{s_{\mathbf{x}}}(\mathbf{x}-\overline{\mathbf{x}})$ is

$$\frac{1}{m}\sum_{i=1}^{m}\frac{1}{s_{\mathbf{x}}}(x_i - \overline{x}) = \frac{1}{ms_{\mathbf{x}}}\left[\sum_{i=1}^{m}x_i - \sum_{i=1}^{m}\overline{x}\right]$$

$$= \frac{1}{ms_{\mathbf{x}}}[m\overline{x} - m\overline{x}] = 0.$$

Using Exercise 29, we see that the variance of $\frac{1}{s_{\mathbf{x}}}(\mathbf{x}-\overline{\mathbf{x}})$ is $\frac{1}{s_{\mathbf{x}}^2}\times$ variance of $(\mathbf{x}-\overline{\mathbf{x}})$. By Exercises 27 and 24, the variance of $\mathbf{x}-\overline{\mathbf{x}}$ is $\mathrm{cov}(\mathbf{x}-\overline{\mathbf{x}},\mathbf{x}-\overline{\mathbf{x}}) = \mathrm{cov}(\mathbf{x},\mathbf{x}) = s_{\mathbf{x}}^2$. The result follows.

34. For any i and j, we have

$$(C_0)_{ij} = \frac{\mathrm{cov}(\mathbf{x}_i,\mathbf{x}_j)}{s_{\mathbf{x}_i}s_{\mathbf{x}_j}}$$

$$= \frac{1}{m-1}\frac{(\mathbf{x}_i-\overline{\mathbf{x}}_i)\cdot(\mathbf{x}_j-\overline{\mathbf{x}}_j)}{s_{\mathbf{x}_i}s_{\mathbf{x}_j}}$$

$$= \frac{1}{m-1}\mathbf{z}_i\cdot\mathbf{z}_j$$

$$= \frac{1}{m-1}(Z^T Z)_{ij}$$

So $C_0 = \frac{1}{m-1}Z^T Z$.

35. (a) Since $\overline{\mathbf{z}}_j = 0$ for all j, we have

$$\overline{Z\mathbf{w}} = \frac{1}{m}\sum_{i=1}^{m}(Z\mathbf{w})_i$$

$$= \frac{1}{m}\sum_{i=1}^{m}\sum_{j=1}^{m}z_{ij}w_j$$

$$= \sum_{j=1}^{m}\left(\sum_{i=1}^{m}\frac{1}{m}z_{ij}\right)w_j$$

$$= \sum_{j=1}^{m}\overline{\mathbf{z}}_j w_j$$

$$= 0.$$

(b) Any principal component $\mathbf{y}$ can be written as $\mathbf{y} = Z\mathbf{u}$, where $\mathbf{u}$ unit vector and an eigenvector of C_0 with eigenvalue λ. Thus $\mathbf{u}^T\mathbf{u} = \mathbf{u}\cdot\mathbf{u} = 1$ and $C_0\mathbf{u} = \lambda\mathbf{u}$. Because $Z\mathbf{u}$ has mean 0 from (a), we have

$$s_{\mathbf{y}}^2 = s_{Z\mathbf{u}}^2 = \frac{1}{m-1}(Z\mathbf{u})^T(Z\mathbf{u})$$

$$= \frac{1}{m-1}\mathbf{u}^T Z^T Z\mathbf{u} = \mathbf{u}^T C_0 \mathbf{u}$$

$$= \mathbf{u}^T\lambda\mathbf{u} = \lambda\mathbf{u}^T\mathbf{u} = \lambda.$$

36. Consider any two distinct principal components $\mathbf{y}_i$ and $\mathbf{y}_j$. Let $\mathbf{u}_i$ and $\mathbf{u}_j$ be the orthonormal vectors such that $\mathbf{y}_i = Z\mathbf{u}_i$ and $\mathbf{y}_j = Z\mathbf{u}_j$, and suppose that $C_0\mathbf{u}_j = \lambda_j\mathbf{u}_j$. By Exercises 35(a) and 34, we have

$$\mathrm{cov}(\mathbf{y}_i,\mathbf{y}_j) = \mathrm{cov}(Z\mathbf{u}_i, Z\mathbf{u}_j)$$

$$= \frac{1}{m-1}(Z\mathbf{u}_i)^T(Z\mathbf{u}_j)$$

$$= \frac{1}{m-1}\mathbf{u}_i^T Z^T Z\mathbf{u}_j = \mathbf{u}_i^T C_0 \mathbf{u}_j$$

$$= \mathbf{u}_i^T \lambda_j \mathbf{u}_j = \lambda_j \mathbf{u}_i\cdot\mathbf{u}_j$$

$$= 0.$$

37. (a) By Exercises 27 and 22, we have

$$r = \frac{\mathrm{cov}(\mathbf{x},\mathbf{y})}{s_{\mathbf{x}}s_{\mathbf{y}}} = \frac{\mathrm{cov}(\mathbf{x}-\overline{\mathbf{x}},\mathbf{y}-\overline{\mathbf{y}})}{s_{\mathbf{x}}s_{\mathbf{y}}}$$

$$= \mathrm{cov}\left(\frac{\mathbf{x}-\overline{\mathbf{x}}}{s_{\mathbf{x}}},\frac{\mathbf{y}-\overline{\mathbf{y}}}{s_{\mathbf{y}}}\right) = \mathrm{cov}(\mathbf{x}^*,\mathbf{y}^*).$$

(b) By Exercise 30 and (a) above, we have

$$0 \le s_{\mathbf{x}^*\pm\mathbf{y}^*}^2 = s_{\mathbf{x}^*}^2 + s_{\mathbf{y}^*}^2 \pm 2\mathrm{cov}(\mathbf{x}^*,\mathbf{y}^*)$$

$$= 2 \pm 2r.$$

So $\pm r \le 1$. But this implies that $|r| \le 1$.

38. (a) The last 4 columns of the expanded table are shown below.

PRE*	FE*	ACTE*	ACTM*
1.6348	1.6579	1.5107	1.4128
0.4639	1.0048	−0.7998	−0.7339
0.6395	0.8089	1.5107	0.2202
1.0493	1.0048	0.3555	1.1743
0.6981	−0.2361	−0.7998	0.9357
0.5224	−1.3465	−0.7998	−0.2569
0.2297	0.2211	−0.4147	0.9357
−0.5314	−0.2361	0.3555	−1.4495
−0.2387	0.2211	0.7405	−0.7339
−0.4143	−0.4974	0.3555	−1.4495
−0.6485	0.2864	−1.5700	0.6972
−1.8780	−1.6077	−1.1849	0.2202
−1.5267	−1.2811	0.7405	−0.9724

(b) Let Z be the 13×4 matrix whose entries are given in the last 13 rows of the table in (a). Then $C_0 = \frac{1}{12} Z^T Z =$

$$\begin{bmatrix} 1.0000 & 0.7499 & 0.3305 & 0.5201 \\ 0.7499 & 1.0000 & 0.4053 & 0.4406 \\ 0.3305 & 0.4053 & 1.0000 & -0.0924 \\ 0.5201 & 0.4406 & -0.0924 & 1.0000 \end{bmatrix}.$$

(c) The variables are PRE* and FE*. The correlation is .7499.

(d) The variables are ACTM* and ACTE*. The correlation is -0.0924.

(e) $P =$

$$\begin{bmatrix} 0.6025 & 0.0458 & 0.3294 & 0.7255 \\ 0.5974 & -0.0813 & 0.4168 & -0.6803 \\ 0.3131 & -0.7786 & -0.5427 & 0.0355 \\ 0.4267 & 0.6205 & -0.6506 & -0.0981 \end{bmatrix}$$

and

$$D = \begin{bmatrix} 2.2835 & 0 & 0 & 0 \\ 0 & 1.0966 & 0 & 0 \\ 0 & 0 & 0.3773 & 0 \\ 0 & 0 & 0 & 0.2427 \end{bmatrix}.$$

(f) $\mathbf{y}_1 = .6025\text{PRE}^* + .5974\text{FE}^*$
$\qquad + .3131\text{ACTE}^* + .4267\text{ACTM}^*$

$\mathbf{y}_2 = .0458\text{PRE}^* - .0813\text{FE}^*$
$\qquad - .7786\text{ACTE}^* + .6205\text{ACTM}^*$

(g) $2.2835/4 = .5709 = 57.09\%$

(h) $(2.2835 + 1.0966)/4 = .8450 = 84.50\%$

6.9 ROTATIONS OF $\mathcal{R}^3$ AND COMPUTER GRAPHICS

1. $\begin{bmatrix} 0 & 1 & 0 \\ 0 & 0 & -1 \\ -1 & 0 & 0 \end{bmatrix}$ 2. $\begin{bmatrix} 0 & 0 & 1 \\ 1 & 0 & 0 \\ 0 & 1 & 0 \end{bmatrix}$

3. We have

$$M = P_{90°} R_{45°} = \begin{bmatrix} 1 & 0 & 0 \\ 0 & 0 & -1 \\ 0 & 1 & 0 \end{bmatrix} \begin{bmatrix} \frac{1}{\sqrt{2}} & -\frac{1}{\sqrt{2}} & 0 \\ \frac{1}{\sqrt{2}} & \frac{1}{\sqrt{2}} & 0 \\ 0 & 0 & 1 \end{bmatrix}$$

$$= \frac{1}{\sqrt{2}} \begin{bmatrix} 1 & -1 & 0 \\ 0 & 0 & -\sqrt{2} \\ 1 & 1 & 0 \end{bmatrix}.$$

4. $\dfrac{1}{\sqrt{2}} \begin{bmatrix} 0 & 0 & \sqrt{2} \\ 1 & 1 & 0 \\ -1 & 1 & 0 \end{bmatrix}$

5. $\dfrac{1}{4} \begin{bmatrix} 2\sqrt{3} & 0 & 2 \\ 1 & 2\sqrt{3} & -\sqrt{3} \\ -\sqrt{3} & 2 & 3 \end{bmatrix}$

6. $\dfrac{1}{\sqrt{2}} \begin{bmatrix} 1 & 0 & 1 \\ 1 & 0 & -1 \\ 0 & \sqrt{2} & 0 \end{bmatrix}$

7. Let $\mathbf{v}_3 = \dfrac{1}{\sqrt{2}} \begin{bmatrix} 1 \\ 0 \\ 1 \end{bmatrix}$. We must select nonzero vectors $\mathbf{w}_1$ and $\mathbf{w}_2$ so that $\mathbf{w}_1$, $\mathbf{w}_2$, and $\mathbf{v}_3$ form an orthogonal set and $\mathbf{w}_2$ lies in the direction of the counterclockwise rotation of $\mathbf{w}_1$ by $90°$ with respect to the orientation defined by $\mathbf{v}_3$. First choose $\mathbf{w}_1$ to be any nonzero vector orthogonal to $\mathbf{v}_3$, say $\mathbf{w}_1 = \begin{bmatrix} 0 \\ 1 \\ 0 \end{bmatrix}$. Then choose $\mathbf{w}_2$ to be a nonzero vector orthogonal to $\mathbf{w}_1$ and $\mathbf{v}_3$. Two possibilities are $\begin{bmatrix} 1 \\ 0 \\ -1 \end{bmatrix}$ and $\begin{bmatrix} -1 \\ 0 \\ 1 \end{bmatrix}$. Since

$$\det \begin{bmatrix} 0 & 1 & 1 \\ 1 & 0 & 0 \\ 0 & -1 & 1 \end{bmatrix} < 0$$

and

$$\det \begin{bmatrix} 0 & -1 & 1 \\ 1 & 0 & 0 \\ 0 & 1 & 1 \end{bmatrix} > 0,$$

we choose $\mathbf{w}_2 = \begin{bmatrix} -1 \\ 0 \\ 1 \end{bmatrix}$ so that the determinant of the matrix $[\mathbf{w}_1 \ \mathbf{w}_2 \ \mathbf{w}_3]$ is positive. (Once we replace $\mathbf{w}_2$ by a unit vector in the same direction, we can apply Theorem 6.20.) Now let $\mathbf{v}_1 = \mathbf{w}_1$,

$$\mathbf{v}_2 = \frac{1}{\|\mathbf{w}_2\|} \mathbf{w}_2 = \frac{1}{\sqrt{2}} \begin{bmatrix} -1 \\ 0 \\ 1 \end{bmatrix},$$

and

$$V = [\mathbf{v}_1 \ \mathbf{v}_2 \ \mathbf{v}_3] = \begin{bmatrix} 0 & -\frac{1}{\sqrt{2}} & \frac{1}{\sqrt{2}} \\ 1 & 0 & 0 \\ 0 & \frac{1}{\sqrt{2}} & \frac{1}{\sqrt{2}} \end{bmatrix}.$$

Then

$P = V R_{180°} V^T$

$= \begin{bmatrix} 0 & -\frac{1}{\sqrt{2}} & \frac{1}{\sqrt{2}} \\ 1 & 0 & 0 \\ 0 & \frac{1}{\sqrt{2}} & \frac{1}{\sqrt{2}} \end{bmatrix} \begin{bmatrix} -1 & 0 & 0 \\ 0 & -1 & 0 \\ 0 & 0 & 1 \end{bmatrix} \begin{bmatrix} 0 & 1 & 0 \\ -\frac{1}{\sqrt{2}} & 0 & \frac{1}{\sqrt{2}} \\ \frac{1}{\sqrt{2}} & 0 & \frac{1}{\sqrt{2}} \end{bmatrix}$

180   Chapter 6   Orthogonality

$$= \begin{bmatrix} 0 & 0 & 1 \\ 0 & -1 & 0 \\ 1 & 0 & 0 \end{bmatrix}.$$

8. $\dfrac{1}{3}\begin{bmatrix} 1 & -\sqrt{3}-1 & -\sqrt{3}+1 \\ \sqrt{3}-1 & 1 & -\sqrt{3}-1 \\ \sqrt{3}+1 & \sqrt{3}-1 & 1 \end{bmatrix}$

9. $\dfrac{1}{2\sqrt{2}}\begin{bmatrix} \sqrt{2}+1 & \sqrt{2}-1 & -\sqrt{2} \\ \sqrt{2}-1 & \sqrt{2}+1 & \sqrt{2} \\ \sqrt{2} & -\sqrt{2} & 2 \end{bmatrix}$

10. $\dfrac{1}{4}\begin{bmatrix} \sqrt{2}+2 & -\sqrt{2}+2 & 2 \\ -\sqrt{2}+2 & \sqrt{2}+2 & -2 \\ -2 & 2 & 2\sqrt{2} \end{bmatrix}$

11. Let $\mathbf{v}_3 = \dfrac{1}{\sqrt{2}}\begin{bmatrix} 1 \\ -1 \\ 0 \end{bmatrix}$. As in Exercise 7, we select nonzero vectors $\mathbf{w}_1$ and $\mathbf{w}_2$ that are orthogonal to $\mathbf{v}_3$ and to each other so that

$$\det [\mathbf{w}_1\ \mathbf{w}_2\ \mathbf{v}_3] > 0.$$

We choose

$$\mathbf{w}_1 = \begin{bmatrix} 1 \\ 1 \\ 0 \end{bmatrix} \quad \text{and} \quad \mathbf{w}_2 = \begin{bmatrix} 0 \\ 0 \\ 1 \end{bmatrix}.$$

Next, set

$$\mathbf{v}_1 = \dfrac{1}{\|\mathbf{w}_1\|}\mathbf{w}_1 = \dfrac{1}{\sqrt{2}}\begin{bmatrix} 1 \\ 1 \\ 0 \end{bmatrix},$$

$\mathbf{v}_2 = \mathbf{w}_2$, and

$$V = [\mathbf{v}_1\ \mathbf{v}_2\ \mathbf{v}_3] = \begin{bmatrix} \frac{1}{\sqrt{2}} & 0 & \frac{1}{\sqrt{2}} \\ \frac{1}{\sqrt{2}} & 0 & -\frac{1}{\sqrt{2}} \\ 0 & 1 & 0 \end{bmatrix}.$$

Then

$P = VR_{30°}V^T$

$$= \begin{bmatrix} \frac{1}{\sqrt{2}} & 0 & \frac{1}{\sqrt{2}} \\ \frac{1}{\sqrt{2}} & 0 & -\frac{1}{\sqrt{2}} \\ 0 & 1 & 0 \end{bmatrix} \begin{bmatrix} \frac{\sqrt{3}}{2} & -\frac{1}{2} & 0 \\ \frac{1}{2} & \frac{\sqrt{3}}{2} & 0 \\ 0 & 0 & 1 \end{bmatrix} \begin{bmatrix} \frac{1}{\sqrt{2}} & \frac{1}{\sqrt{2}} & 0 \\ 0 & 0 & 1 \\ \frac{1}{\sqrt{2}} & -\frac{1}{\sqrt{2}} & 0 \end{bmatrix}$$

$$= \dfrac{1}{4}\begin{bmatrix} \sqrt{3}+2 & \sqrt{3}-2 & -\sqrt{2} \\ \sqrt{3}-2 & \sqrt{3}+2 & -\sqrt{2} \\ \sqrt{2} & \sqrt{2} & 2\sqrt{3} \end{bmatrix}.$$

12. $\dfrac{1}{4}\begin{bmatrix} \sqrt{3}+2 & \sqrt{3}-2 & \sqrt{2} \\ \sqrt{3}-2 & \sqrt{3}+2 & \sqrt{2} \\ -\sqrt{2} & -\sqrt{2} & 2\sqrt{3} \end{bmatrix}$

13. $\dfrac{1}{3}\begin{bmatrix} 2 & -2 & -1 \\ 1 & 2 & -2 \\ 2 & 1 & 2 \end{bmatrix}$

14. $\dfrac{1}{3\sqrt{2}}\begin{bmatrix} 2+\sqrt{2} & 1-\sqrt{2}-\sqrt{3} & -1+\sqrt{2}-\sqrt{3} \\ 1-\sqrt{2}+\sqrt{3} & 2+\sqrt{2} & 1-\sqrt{2}-\sqrt{3} \\ -1+\sqrt{2}+\sqrt{3} & 1-\sqrt{2}+\sqrt{3} & 2+\sqrt{2} \end{bmatrix}$

15. (a) $\begin{bmatrix} -1 \\ -1 \\ 1 \end{bmatrix}$ (b) $-\dfrac{1}{2}$ 16. (a) $\begin{bmatrix} 1 \\ 1 \\ 1 \end{bmatrix}$ (b) $-\dfrac{1}{2}$

17. Let

$$M = \dfrac{1}{\sqrt{2}}\begin{bmatrix} 1 & -1 & 0 \\ 0 & 0 & -\sqrt{2} \\ 1 & 1 & 0 \end{bmatrix},$$

the rotation matrix in Exercise 3.

(a) The axis of rotation is the span of an eigenvector of M corresponding to the eigenvalue 1, and hence we seek a nonzero solution of $(M - I_3)\mathbf{x} = \mathbf{0}$. The reduced row echelon form of the augmented matrix of the system of equations given in matrix form above is

$$\begin{bmatrix} 1 & 0 & -1-\sqrt{2} & 0 \\ 0 & 1 & 1 & 0 \end{bmatrix},$$

and so we the general solution is

$$\begin{bmatrix} x_1 \\ x_2 \\ x_3 \end{bmatrix} = x_3 \begin{bmatrix} 1+\sqrt{2} \\ -1 \\ 1 \end{bmatrix}.$$

Thus the span of $\begin{bmatrix} 1+\sqrt{2} \\ -1 \\ 1 \end{bmatrix}$ is the axis of rotation.

(b) Choose any nonzero vector that is orthogonal to the vector in (a), for example,

$$\mathbf{w} = \begin{bmatrix} 0 \\ 1 \\ 1 \end{bmatrix},$$

and let α be the angle between $\mathbf{w}$ and $M\mathbf{w}$. Notice that $\|M\mathbf{w}\| = \|\mathbf{w}\|$ because M is an orthogonal matrix. Therefore by Exercise 98 of Section 6.1,

$$\cos \alpha = \dfrac{M\mathbf{w} \cdot \mathbf{w}}{\|M\mathbf{w}\|\|\mathbf{w}\|}$$

$$= \dfrac{\dfrac{1}{\sqrt{2}}\begin{bmatrix} 1 & -1 & 0 \\ 0 & 0 & -\sqrt{2} \\ 1 & 1 & 0 \end{bmatrix}\begin{bmatrix} 0 \\ 1 \\ 1 \end{bmatrix} \cdot \begin{bmatrix} 0 \\ 1 \\ 1 \end{bmatrix}}{\left\|\begin{bmatrix} 0 \\ 1 \\ 1 \end{bmatrix}\right\|^2}$$

$$= \frac{1}{2\sqrt{2}} \begin{bmatrix} -1 \\ -\sqrt{2} \\ 1 \end{bmatrix} \cdot \begin{bmatrix} 0 \\ 1 \\ 1 \end{bmatrix} = \frac{1-\sqrt{2}}{2\sqrt{2}}.$$

18. (a) $\begin{bmatrix} 1 \\ 1+\sqrt{2} \\ 1 \end{bmatrix}$ (b) $\dfrac{\sqrt{2}-2}{4}$

19. (a) $\begin{bmatrix} 1 \\ 1 \\ 2-\sqrt{3} \end{bmatrix}$ (b) $\dfrac{4\sqrt{3}-1}{8}$

20. (a) $\begin{bmatrix} 1+\sqrt{2} \\ 1 \\ 1 \end{bmatrix}$ (b) $\dfrac{\sqrt{2}-2}{4}$

21. (a) $\begin{bmatrix} \sqrt{3} \\ \sqrt{2}+1 \\ 1 \end{bmatrix}$ (b) $\dfrac{3\sqrt{2}-2}{8}$

22. (a) $\begin{bmatrix} 2+\sqrt{3} \\ 1 \\ 1+\sqrt{2} \end{bmatrix}$ (b) $\dfrac{\sqrt{3}(\sqrt{2}+1)^2-2}{2[(\sqrt{2}+1)^2+1]}$

23. Let T be the reflection and A be the standard matrix of T. Choose a nonzero vector orthogonal to $\begin{bmatrix} 1 \\ 2 \\ 3 \end{bmatrix}$ and $\begin{bmatrix} 1 \\ 0 \\ -1 \end{bmatrix}$, for example, $\begin{bmatrix} 1 \\ -2 \\ 1 \end{bmatrix}$, and let

$$\mathcal{B} = \left\{ \begin{bmatrix} 1 \\ 2 \\ 3 \end{bmatrix}, \begin{bmatrix} 1 \\ 0 \\ -1 \end{bmatrix}, \begin{bmatrix} 1 \\ -2 \\ 1 \end{bmatrix} \right\}.$$

Then $\mathcal{B}$ is a basis for $\mathcal{R}^3$, and

$$T\left(\begin{bmatrix} 1 \\ 2 \\ 3 \end{bmatrix}\right) = \begin{bmatrix} 1 \\ 2 \\ 3 \end{bmatrix}, \quad T\left(\begin{bmatrix} 1 \\ 0 \\ -1 \end{bmatrix}\right) = \begin{bmatrix} 1 \\ 0 \\ -1 \end{bmatrix},$$

and

$$T\left(\begin{bmatrix} 1 \\ -2 \\ 1 \end{bmatrix}\right) = -\begin{bmatrix} 1 \\ -2 \\ 1 \end{bmatrix}.$$

Let B be the matrix whose columns are the vectors in $\mathcal{B}$. Then

$$[T]_\mathcal{B} = B^{-1}AB = \begin{bmatrix} 1 & 0 & 0 \\ 0 & 1 & 0 \\ 0 & 0 & -1 \end{bmatrix} = D,$$

and therefore

$$A = BDB^{-1}$$

$$= \begin{bmatrix} 1 & 1 & 1 \\ 2 & 0 & -2 \\ 3 & -1 & 1 \end{bmatrix} \begin{bmatrix} 1 & 0 & 0 \\ 0 & 1 & 0 \\ 0 & 0 & -1 \end{bmatrix} \begin{bmatrix} \frac{1}{6} & \frac{1}{6} & \frac{1}{6} \\ \frac{2}{3} & \frac{1}{6} & -\frac{1}{3} \\ \frac{1}{6} & -\frac{1}{3} & \frac{1}{6} \end{bmatrix}$$

$$= \frac{1}{3}\begin{bmatrix} 2 & 2 & -1 \\ 2 & -1 & 2 \\ -1 & 2 & 2 \end{bmatrix}.$$

24. $\begin{bmatrix} 0 & 1 & 0 \\ 1 & 0 & 0 \\ 0 & 0 & 1 \end{bmatrix}$ 25. $\dfrac{1}{3}\begin{bmatrix} 1 & -2 & -2 \\ -2 & 1 & -2 \\ -2 & -2 & 1 \end{bmatrix}$

26. Let T be the reflection and A be the standard matrix of T. Find a basis for the subspace of $\mathcal{R}^3$ consisting of the vectors orthogonal to $\mathbf{v}$. This subspace is the solution space of the system of linear equations

$$x_2 + 2x_2 - x_3 = 0,$$

whose general solution has the vector form

$$\begin{bmatrix} x_1 \\ x_2 \\ x_3 \end{bmatrix} = x_2 \begin{bmatrix} -2 \\ 1 \\ 0 \end{bmatrix} + x_3 \begin{bmatrix} 1 \\ 0 \\ 1 \end{bmatrix}.$$

It follows that

$$\left\{ \begin{bmatrix} -2 \\ 1 \\ 0 \end{bmatrix}, \begin{bmatrix} 1 \\ 0 \\ 1 \end{bmatrix} \right\}$$

is a basis for this subspace. We adjoin $\mathbf{v}$ to this set to obtain a basis

$$\mathcal{B} = \left\{ \begin{bmatrix} -2 \\ 1 \\ 0 \end{bmatrix}, \begin{bmatrix} 1 \\ 0 \\ 1 \end{bmatrix}, \begin{bmatrix} 1 \\ 2 \\ -1 \end{bmatrix} \right\}$$

for $\mathcal{R}^3$. Let B be the matrix whose columns are the vectors in $\mathcal{B}$. As in Exercise 23,

$$[T]_\mathcal{B} = B^{-1}AB = \begin{bmatrix} 1 & 0 & 0 \\ 0 & 1 & 0 \\ 0 & 0 & -1 \end{bmatrix} = D,$$

and therefore

$$A = BDB^{-1}$$

$$= \begin{bmatrix} -2 & 1 & 1 \\ 0 & 1 & 2 \\ 0 & 1 & -1 \end{bmatrix} \begin{bmatrix} 1 & 0 & 0 \\ 0 & 1 & 0 \\ 0 & 0 & -1 \end{bmatrix} \begin{bmatrix} -\frac{1}{3} & \frac{1}{3} & \frac{1}{3} \\ \frac{1}{6} & \frac{1}{3} & \frac{5}{6} \\ \frac{1}{6} & \frac{1}{3} & -\frac{1}{6} \end{bmatrix}$$

$$= \frac{1}{3}\begin{bmatrix} 2 & -2 & 1 \\ -2 & -1 & 2 \\ 1 & 2 & 2 \end{bmatrix}.$$

27. $\dfrac{1}{9}\begin{bmatrix} 7 & -4 & -4 \\ -4 & 1 & 8 \\ 4 & 8 & 1 \end{bmatrix}$ 28. $\dfrac{1}{9}\begin{bmatrix} 7 & 4 & 4 \\ 4 & 1 & -8 \\ 4 & -8 & 1 \end{bmatrix}$

29. $\dfrac{1}{25}\begin{bmatrix} 16 & 12 & -15 \\ 12 & 9 & 20 \\ -15 & 20 & 0 \end{bmatrix}$

30. $\dfrac{1}{45}\begin{bmatrix} 44 & -8 & 5 \\ -8 & -19 & 40 \\ 5 & 40 & 16 \end{bmatrix}$

31. First, obtain a basis $\{\mathbf{w}_1, \mathbf{w}_2\}$ for W by selecting two linearly independent vectors that are orthogonal to $\mathbf{v}$ such as

$$\mathbf{v}_1 = \begin{bmatrix} 1 \\ 0 \\ 1 \end{bmatrix} \quad \text{and} \quad \mathbf{v}_2 = \begin{bmatrix} 2 \\ -1 \\ 0 \end{bmatrix}.$$

Although we could proceed as in previous exercises, here is an alternate approach.

Let A be the standard matrix of T_W. Then

$$A\begin{bmatrix} 1 & 2 & 1 \\ 0 & -1 & 2 \\ 1 & 0 & -1 \end{bmatrix}$$

$$= \begin{bmatrix} A\begin{bmatrix}1\\0\\1\end{bmatrix} & A\begin{bmatrix}2\\-1\\0\end{bmatrix} & A\begin{bmatrix}1\\2\\-1\end{bmatrix} \end{bmatrix}$$

$$= \begin{bmatrix} 1 & 2 & -1 \\ 0 & -1 & -2 \\ 1 & 0 & 1 \end{bmatrix},$$

and therefore

$$A = \begin{bmatrix} 1 & 2 & -1 \\ 0 & -1 & -2 \\ 1 & 0 & 1 \end{bmatrix} \begin{bmatrix} 1 & 2 & 1 \\ 0 & -1 & 2 \\ 1 & 0 & -1 \end{bmatrix}^{-1}$$

$$= \dfrac{1}{3}\begin{bmatrix} 2 & -2 & 1 \\ -2 & -1 & 2 \\ 1 & 2 & 2 \end{bmatrix}.$$

32. $\dfrac{1}{3}\begin{bmatrix} 1 & 2 & 2 \\ 2 & 1 & -2 \\ 2 & -2 & 1 \end{bmatrix}$

33. $\dfrac{1}{5}\begin{bmatrix} 3 & 0 & -4 \\ 0 & 5 & 0 \\ -4 & 0 & 3 \end{bmatrix}$

34. $\dfrac{1}{7}\begin{bmatrix} 6 & 2 & 3 \\ 2 & 3 & -6 \\ 3 & -6 & 2 \end{bmatrix}$

35. $\dfrac{1}{25}\begin{bmatrix} 16 & -12 & -15 \\ -12 & 9 & -20 \\ -15 & -20 & 0 \end{bmatrix}$

36. $\dfrac{1}{5}\begin{bmatrix} -4 & 0 & 3 \\ 0 & 1 & 0 \\ 3 & 0 & 4 \end{bmatrix}$

37. $\dfrac{1}{9}\begin{bmatrix} 1 & 4 & -8 \\ 4 & 7 & 4 \\ -8 & 4 & 1 \end{bmatrix}$

38. $\dfrac{1}{9}\begin{bmatrix} 1 & 4 & -8 \\ 4 & 7 & 4 \\ -8 & 4 & 1 \end{bmatrix}$

39. (a) The given matrix does not have 1 as an eigenvalue. Therefore it is neither a rotation matrix nor the standard matrix of a reflection operator, both of which have 1 as an eigenvalue.

40. (a) a reflection (b) $\left\{ \begin{bmatrix}0\\1\\0\end{bmatrix}, \begin{bmatrix}1\\0\\1\end{bmatrix} \right\}$

41. (a) Since

$$\det \begin{bmatrix} 1 & 0 & 0 \\ 0 & -1 & 0 \\ 0 & 0 & -1 \end{bmatrix} = 1,$$

the matrix is a rotation matrix by Theorem 6.20.

(b) Observe that $\begin{bmatrix}1\\0\\0\end{bmatrix}$ is an eigenvector of the matrix corresponding to eigenvalue 1, and therefore this vector forms a generating set for the axis of rotation.

42. (a) a rotation (b) $\begin{bmatrix}1\\0\\-1\end{bmatrix}$

43. (a) a rotation (b) $\begin{bmatrix}2\\1\\-2\end{bmatrix}$

44. (a) a reflection (b) $\left\{ \begin{bmatrix}1\\0\\1\end{bmatrix}, \begin{bmatrix}1\\-2\\0\end{bmatrix} \right\}$

45. Let M denote the given matrix.

(a) Since

$$\det M = \det \begin{bmatrix} \tfrac{1}{\sqrt{2}} & 0 & \tfrac{1}{\sqrt{2}} \\ 0 & 1 & 0 \\ \tfrac{1}{\sqrt{2}} & 0 & \tfrac{-1}{\sqrt{2}} \end{bmatrix} = -1,$$

M is not a rotation matrix by Theorem 6.20. We can establish that M is the standard matrix of a reflection by showing that M has a 2-dimensional eigenspace corresponding to eigenvalue 1. For, in this case, it must have a third eigenvector corresponding to eigenvalue -1 because its determinant equals the product of its eigenvalues. The reduced row echelon form of $M - I_3$ is

$$\begin{bmatrix} 1 & 0 & -(1+\sqrt{2}) \\ 0 & 0 & 0 \\ 0 & 0 & 0 \end{bmatrix},$$

and hence the eigenspace corresponding to eigenvalue 1 is 2-dimensional. Therefore M is the standard matrix of a reflection.

(b) The matrix equation $(M - I_2)\mathbf{x} = \mathbf{0}$ is the system
$$x_1 - (1 + \sqrt{2})x_3 = 0,$$
and hence the vector form of its general solution is
$$\begin{bmatrix} x_1 \\ x_2 \\ x_3 \end{bmatrix} = x_2 \begin{bmatrix} 0 \\ 1 \\ 0 \end{bmatrix} + x_3 \begin{bmatrix} 1 + \sqrt{2} \\ 0 \\ 1 \end{bmatrix}.$$

It follows that
$$\left\{ \begin{bmatrix} 0 \\ 1 \\ 0 \end{bmatrix}, \begin{bmatrix} 1 + \sqrt{2} \\ 0 \\ 1 \end{bmatrix} \right\}$$
is a basis for the 2-dimensional subspace about which $\mathcal{R}^3$ is reflected.

46. (a) a rotation (b) $\begin{bmatrix} 0 \\ 1 \\ 0 \end{bmatrix}$

47. False, consider $P = \begin{bmatrix} -1 & 0 & 0 \\ 0 & 1 & 0 \\ 0 & 0 & 1 \end{bmatrix}$.

48. False, let P be the matrix in the solution to Exercise 47.

49. False, let $Q = \begin{bmatrix} -1 & 0 & 0 \\ 0 & 0 & -1 \\ 0 & 1 & 0 \end{bmatrix}$.

50. False, consider I_3.

51. True

52. False, consider the matrix Q in the solution to Exercise 49.

53. True   54. True   55. True

56. False, for example, if $\phi = \theta = 90°$, then
$$Q_\phi R_\theta = \begin{bmatrix} 0 & 0 & 1 \\ 1 & 0 & 0 \\ 0 & 1 & 0 \end{bmatrix},$$
but
$$R_\theta Q_\phi = \begin{bmatrix} 0 & -1 & 0 \\ 0 & 0 & 1 \\ -1 & 0 & 0 \end{bmatrix}.$$

57. True   58. True

59. False, the matrix is $Q_\theta = \begin{bmatrix} \cos\theta & 0 & \sin\theta \\ 0 & 1 & 0 \\ -\sin\theta & 0 & \cos\theta \end{bmatrix}$.

60. True   61. True   62. True

63. False, the rotation, as viewed from $\mathbf{v}_3$, is counterclockwise.

64. False, the determinant is equal to -1.

65. False, the eigenvector corresponds to the eigenvalue 1.

66. False, any nonzero solution of $(P_\theta - R_\phi^T)\mathbf{x} = \mathbf{0}$ forms a basis for the axis of rotation.

67. True

68. Let $\mathbf{v}$ be a vector in $\mathcal{R}^3$ that is not in L. Then there are vectors $\mathbf{w}$ in L and $\mathbf{z}$ in $L^\perp$ such that $\mathbf{z} \neq \mathbf{0}$ and $\mathbf{v} = \mathbf{w} + \mathbf{z}$. Thus
$$P(\mathbf{v}) = P(\mathbf{w} + \mathbf{z}) = P\mathbf{w} + P\mathbf{z} = \mathbf{w} + P\mathbf{z}.$$
To show that $P\mathbf{v} \neq \mathbf{v}$, it suffices to show that $P\mathbf{z} \neq \mathbf{z}$. Let $\{\mathbf{v}_1, \mathbf{v}_2\}$ be an orthonormal basis for $L^\perp$ as described on page 472, and suppose that $\mathbf{z} = a\mathbf{v}_1 + b\mathbf{v}_2$. Then, by equations (19) and (20),
$$P\mathbf{z} = aP\mathbf{v}_1 + bP\mathbf{v}_2$$
$$= a(\cos\theta\mathbf{v}_1 + \sin\theta\mathbf{v}_2) + b(-\sin\theta\mathbf{v}_1 + \cos\theta\mathbf{v}_2)$$
$$= (a\cos\theta - b\sin\theta)\mathbf{v}_1 + (a\sin\theta + b\cos\theta)\mathbf{v}_2.$$
Suppose that $P\mathbf{z} = \mathbf{z}$. Then
$$a\cos\theta - b\sin\theta = a \quad \text{and} \quad a\sin\theta + b\cos\theta = b.$$
Thus
$$A_\theta = \begin{bmatrix} \cos\theta & -\sin\theta \\ \sin\theta & \cos\theta \end{bmatrix} \begin{bmatrix} a \\ b \end{bmatrix} = \begin{bmatrix} a \\ b \end{bmatrix}.$$
Since $\mathbf{z} \neq \mathbf{0}$, it follows that $\begin{bmatrix} a \\ b \end{bmatrix} \neq \begin{bmatrix} 0 \\ 0 \end{bmatrix}$, and hence $\begin{bmatrix} a \\ b \end{bmatrix}$ is an eigenvector of the 2×2 rotation matrix A_θ corresponding to eigenvalue 1. It follows that $A_\theta = I_2$, and hence $P = I_3$, contrary to what is given.

69. We have
$$\det P_\theta = \det \begin{bmatrix} 1 & 0 & 0 \\ 0 & \cos\theta & -\sin\theta \\ 0 & \sin\theta & \cos\theta \end{bmatrix}$$
$$= \det \begin{bmatrix} \cos\theta & -\sin\theta \\ \sin\theta & \cos\theta \end{bmatrix}$$
$$= \cos^2\theta + \sin^2\theta = 1.$$
The other determinants are computed in a similar manner.

184   Chapter 6   Orthogonality

70. Since P is an orthogonal matrix, so is P^2. Furthermore, $\det P = \pm 1$ by Theorem 6.10. Thus $\det P^2 = (\det P)^2 = 1$, and hence P^2 is a rotation matrix by Theorem 6.20.

71. By the boxed result on page 472, $P = V R_\theta V^T$ for some orthogonal matrix V and angle θ. Thus $P^T = V(R_\theta)^T V^T = V R_{-\theta} V^T$, and hence P^T rotates $\mathcal{R}^3$ by the angle $-\theta$ about L.

72. Choose an orthonormal basis $\{\mathbf{v}_1, \mathbf{v}_2\}$ for W and a unit vector $\mathbf{v}_3$ in $W^\perp$. Then $\mathcal{B} = \{\mathbf{v}_1, \mathbf{v}_2, \mathbf{v}_3\}$ is an orthonormal basis for $\mathcal{R}^3$. Furthermore,
$$T_W(\mathbf{v}_1) = \mathbf{v}_1, \quad T_W(\mathbf{v}_2) = \mathbf{v}_2,$$
and
$$T_W(\mathbf{v}_3) = -\mathbf{v}_3,$$
and hence
$$[T_W]_\mathcal{B} = \begin{bmatrix} 1 & 0 & 0 \\ 0 & 1 & 0 \\ 0 & 0 & -1 \end{bmatrix},$$
an orthogonal matrix. Let $Q = \begin{bmatrix} \mathbf{v}_1 & \mathbf{v}_2 & \mathbf{v}_3 \end{bmatrix}$, the 3×3 matrix whose columns are the vectors in $\mathcal{B}$. Then Q is an orthogonal matrix because its columns form an orthonormal basis for $\mathcal{R}^3$. Therefore $B_W = Q[T_W]_\mathcal{B} Q^{-1}$ is an orthogonal matrix by Theorem 6.10.

73. (a) Clearly, $T_W(\mathbf{w}) = \mathbf{w}$ for all $\mathbf{w}$ in W, and hence 1 is an eigenvalue of T_W. Let Z denote the eigenspace corresponding to eigenvalue 1. Then W is contained in Z. Observe that $\dim Z < 3$, for otherwise T_W would be the identity transformation, which it is not. Since $\dim W = 2$, it follows that $W = Z$.

    (b) Since $T_W(\mathbf{z}) = -\mathbf{z}$ for all $\mathbf{z}$ in $W^\perp$, -1 is an eigenvalue and $W^\perp$ is contained in the corresponding eigenspace. Because the eigenspace corresponding to 1 has dimension 2, the eigenspace corresponding to -1 has dimension 1. But $\dim W^\perp = 1$, and hence these two subspaces are equal.

74. Let W and Z denote the eigenspaces of T corresponding to the eigenvalues 1 and -1, respectively. We begin by showing that W is contained in $Z^\perp$. Let $\mathbf{w}$ be in W and $\mathbf{z}$ in Z. Then $\mathbf{w} = T(\mathbf{w})$ and $\mathbf{z} = T(-\mathbf{z})$, and hence
$$\mathbf{w} \cdot \mathbf{z} = T(\mathbf{w}) \cdot T(-\mathbf{z}) = \mathbf{w} \cdot (-\mathbf{z}),$$
from which it follows that $\mathbf{w} \cdot \mathbf{z} = 0$. So W is contained in $Z^\perp$.

Since $\dim Z = 1$, we have $\dim Z^\perp = 2$. So if we show that $\dim W = 2$, then $W = Z^\perp$. Choose a unit vector $\mathbf{v}_1$ in W, and extend $\{\mathbf{v}_1\}$ to an orthonormal basis $\{\mathbf{v}_1, \mathbf{v}_2\}$ for $Z^\perp$. Now select a unit vector $\mathbf{v}_3$ in Z. Then $\mathcal{B} = \{\mathbf{v}_1, \mathbf{v}_2, \mathbf{v}_3\}$ is an orthonormal basis for $\mathcal{R}^3$. Observe that $T(\mathbf{v}_2)$ is in $Z^\perp$, and consider any $\mathbf{z}$ in Z. Then
$$T(\mathbf{v}_2) \cdot \mathbf{z} = T(\mathbf{v}_2) \cdot T(\mathbf{z}) = \mathbf{v}_2 \cdot \mathbf{z} = 0,$$
and hence $\mathbf{v}_2$ is in $Z^\perp$. It follows that $T(\mathbf{v}_2) = a\mathbf{v}_1 + b\mathbf{v}_2$ for some scalars a and b. Thus
$$[T]_\mathcal{B} = \begin{bmatrix} 1 & a & 0 \\ 0 & b & 0 \\ 0 & 0 & -1 \end{bmatrix},$$
and the characteristic polynomial of $[T]_\mathcal{B}$ is $-(t-1)(t-b)(t+1)$. Since the eigenvalue 1 has multiplicity 2, it follows that $b = 1$. Thus
$$1 = \|\mathbf{v}_2\|^2 = \|T(\mathbf{v}_2)\|^2 = \|a\mathbf{v}_1 + \mathbf{v}_2\|^2 = a^2 + 1,$$
and hence $a = 0$. It follows that $[T]_\mathcal{B}$ is a diagonal matrix, and so T is a diagonalizable linear operator. As a consequence, $\dim W = 2$ because this is the multiplicity of the eigenvalue 1. Thus $W = Z^\perp$, and therefore $T = T_W$.

75. Let $\mathcal{B} = \{\mathbf{v}_1, \mathbf{v}_2, \mathbf{v}_3\}$ be a basis for $\mathcal{R}^3$ such that $\mathbf{v}_1$ and $\mathbf{v}_2$ are in W and $\mathbf{v}_3$ is in $W^\perp$. Then
$$[T_W]_\mathcal{B} = \begin{bmatrix} 1 & 0 & 0 \\ 0 & 1 & 0 \\ 0 & 0 & -1 \end{bmatrix},$$
and hence $\det [T_W]_\mathcal{B} = -1$. Since B is similar to $[T_W]_\mathcal{B}$, we have $\det B = -1$.

76. (a) Let $\mathcal{B}$ and $[T_W]_\mathcal{B}$ be as in Exercise 75. Because B is similar to $[T_W]_\mathcal{B}$, there is an invertible matrix Q such that $B = Q[T_W]_\mathcal{B} Q^{-1}$. Hence
$$B^2 = Q[T_W]_\mathcal{B} Q^{-1} Q[T_W]_\mathcal{B} Q^{-1}$$
$$= Q \begin{bmatrix} 1 & 0 & 0 \\ 0 & 1 & 0 \\ 0 & 0 & -1 \end{bmatrix}^2 Q^{-1}$$
$$= Q I_3 Q^{-1} = I_3.$$

    (b) Since the basis $\mathcal{B}$ in Exercise 75 is an orthonormal basis, $B = Q[T_W]_\mathcal{B} Q^T$, where Q is the orthogonal matrix whose columns are the vectors in $\mathcal{B}$. Furthermore, $[T_W]_\mathcal{B}$ is symmetric since it is a diagonal matrix. Therefore
$$B^T = (Q[T_W]_\mathcal{B} Q^T)^T = Q[T_W]_\mathcal{B}^T Q^T$$

$$= Q[T_W]_\mathcal{B} Q^T = B,$$

and hence B is symmetric.

77. By Exercise 72, B and C are orthogonal matrices, and hence BC is an orthogonal matrix by Theorem 6.10. In addition, $\det B = \det C = -1$ by Exercise 75. So

$$\det BC = (\det B)(\det C) = (-1)(-1) = 1,$$

and hence BC is a rotation matrix by Theorem 6.20.

78. (a) Let C denote the axis of rotation of $B_2 B_1$ and let W denote the intersection W_1 and W_2. Observe that both C and W are 1-dimensional subspaces of $\mathcal{R}^3$. Let $\mathbf{v}$ be a vector in W. Then $B_1 \mathbf{v} = B_2 \mathbf{v} = \mathbf{v}$, and hence $B_2 B_1 \mathbf{v} = \mathbf{v}$. It follows that $\mathbf{v}$ is in C, and therefore W is a subspace of C. Since these subspaces are of equal dimension, $W = C$.

(b) First observe that $B_2^{-1} = B_2$ by Exercise 76(a). A vector $\mathbf{v}$ in $\mathcal{R}^3$ is in the axis of rotation if and only if $B_2 B_1 \mathbf{v} = \mathbf{v}$ if and only if $B_2 \mathbf{v} = B_1^{-1} \mathbf{v} = B_1 \mathbf{v}$ if and only if $B_2 \mathbf{v} - B_1 \mathbf{v} = (B_2 - B_1)\mathbf{v} = \mathbf{0}$ if and only if $\mathbf{v}$ is a solution of the system of linear equations $(B_2 - B_1)\mathbf{x} = \mathbf{0}$.

79. (a) Let C and W be as in the solution to Exercise 78. Since $\mathbf{n}_1$ is orthogonal to W_1 and $\mathbf{n}_2$ is orthogonal to W_2, both $\mathbf{n}_1$ and $\mathbf{n}_2$ are orthogonal to W, which is equal to C by Exercise 78.

(b) Since $B_2 B_1$ is an orthogonal matrix, it preserves lengths, and so $\|B_2 B_1 \mathbf{n}_1\| = \|\mathbf{n}_1\| = 1$. Furthermore, $B_2^T = B_2^{-1} = B_2$ by Exercise 76. Thus, by Exercise 98 of Section 6.1, we have

$$\cos \theta = \frac{(B_2 B_1 \mathbf{n}_1) \cdot \mathbf{n}_1}{\|B_2 B_1 \mathbf{n}_1\| \|\mathbf{n}_1\|} = (B_2 B_1 \mathbf{n}_1) \cdot \mathbf{n}_1$$

$$= (B_1 \mathbf{n}_1) \cdot (B_2^T \mathbf{n}_1) = -\mathbf{n}_1 \cdot (B_2 \mathbf{n}_1).$$

The second equation is proved similarly.

80. Let
$$D = \begin{bmatrix} 1 & 0 & 0 \\ 0 & 1 & 0 \\ 0 & 0 & -1 \end{bmatrix},$$

which is the standard matrix of the reflection about the z-axis. Recall that $R_{90°}$ is the 90°-rotation matrix about the z-axis. Let

$$C = DR_{90°} = \begin{bmatrix} 1 & 0 & 0 \\ 0 & 1 & 0 \\ 0 & 0 & -1 \end{bmatrix} \begin{bmatrix} 0 & -1 & 0 \\ 1 & 0 & 0 \\ 0 & 0 & 1 \end{bmatrix}$$

$$= \begin{bmatrix} 0 & -1 & 0 \\ 1 & 0 & 0 \\ 0 & 0 & -1 \end{bmatrix}.$$

Then C is not the standard matrix of a reflection because 1 is not an eigenvalue of C. However, C is an orthogonal matrix because it is a product of D and $R_{90°}$, which are orthogonal matrices. Finally,

$$\det C = \det(DR_{90°}) = (\det D)(\det R_{90°})$$
$$= (-1)(1) = -1.$$

81. Let $Q = CB^{-1}$. Then

$$[Q\mathbf{v}_1 \; Q\mathbf{v}_2 \; Q\mathbf{v}_3] = QB = CB^{-1}B$$
$$= C = [\mathbf{v}_1 \; \mathbf{v}_2 \; -\mathbf{v}_3],$$

and hence $Q\mathbf{v}_1 = \mathbf{v}_1$, $Q\mathbf{v}_2 = \mathbf{v}_2$, and $Q\mathbf{v}_3 = -\mathbf{v}_3$. Since $\{\mathbf{v}_1, \mathbf{v}_2\}$ and $\{\mathbf{v}_3\}$ are bases for W and $W^\perp$, respectively, we have that $Q\mathbf{v} = \mathbf{v}$ for every vector $\mathbf{v}$ in W, and $Q\mathbf{v} = -\mathbf{v}$ for every vector $\mathbf{v}$ in $W^\perp$. Therefore $Q = CB^{-1}$ is the standard matrix of the reflection of $\mathcal{R}^3$ about W.

82. (rounded to 4 places after the decimal)
$$\text{Span} \left\{ \begin{bmatrix} .8052 \\ .5822 \\ .1132 \end{bmatrix} \right\}, \quad 27°$$

83. (rounded to 4 places after the decimal)
$$\text{Span} \left\{ \begin{bmatrix} .4609 \\ .1769 \\ .8696 \end{bmatrix} \right\}, \quad 48°$$

CHAPTER 6 REVIEW EXERCISES

1. True   2. True
3. False, the vectors must belong to $\mathcal{R}^n$ for some n.
4. True   5. True   6. True   7. True
8. True
9. False, if W is a 1-dimensional subspace of $\mathcal{R}^3$, then $\dim W^\perp = 2$.
10. False, I_n is an invertible orthogonal projection matrix.
11. True
12. False, let W be the x-axis in $\mathcal{R}^2$, and let $\mathbf{v} = \begin{bmatrix} 1 \\ 2 \end{bmatrix}$. Then $\mathbf{w} = \begin{bmatrix} 1 \\ 0 \end{bmatrix}$, which is not orthogonal to $\mathbf{v}$.

186   Chapter 6   Orthogonality

13. False, the least-squares line minimizes the sum of the squares of the *vertical distances* from the data points to the line.

14. False, in addition, each column must have length equal to 1.

15. False, consider $\begin{bmatrix} 1 & 1 \\ 1 & 2 \end{bmatrix}$, which has determinant 1 but is not an orthogonal matrix.

16. True   17. True

18. False, only symmetric matrices have spectral decompositions.

19. True

20. (a) $\sqrt{13}$, $\sqrt{17}$ (b) $2\sqrt{5}$ (c) 5
    (d) not orthogonal

21. (a) $\|\mathbf{u}\| = \sqrt{45}$, $\|\mathbf{v}\| = \sqrt{20}$ (b) $d = \sqrt{65}$
    (c) $\mathbf{u} \cdot \mathbf{v} = 0$ (d) orthogonal.

22. (a) $\sqrt{14}$, $2\sqrt{5}$ (b) $\sqrt{30}$ (c) 2
    (d) not orthogonal

23. (a)
$$\|\mathbf{u}\| = \sqrt{1^2 + (-1)^2 + 2^2} = \sqrt{6} \quad \text{and}$$
$$\|\mathbf{v}\| = \sqrt{2^2 + 4^2 + 1^2} = \sqrt{21}$$

(b)
$$d = \|\mathbf{u} - \mathbf{v}\|$$
$$= \sqrt{(1-2)^2 + (-1-4)^2 + (2-1)^2}$$
$$= \sqrt{27}$$

(c) $\mathbf{u} \cdot \mathbf{v} = (1)(2) + (-1)(4) + (2)(1) = 0$
(d) $\mathbf{u}$ and $\mathbf{v}$ are orthogonal.

24. $\dfrac{23}{17} \begin{bmatrix} 1 \\ 4 \end{bmatrix}$, $\dfrac{7}{\sqrt{17}}$

25. $\mathbf{w} = \dfrac{1}{5} \begin{bmatrix} -1 \\ 2 \end{bmatrix}$, $d = 3.5777$ 26. 6

27.
$$(2\mathbf{u} + 3\mathbf{v}) \cdot \mathbf{w} = 2(\mathbf{u} \cdot \mathbf{w}) + 3(\mathbf{v} \cdot \mathbf{w})$$
$$= 2(5) + 3(-3) = 1$$

28. 37   29. 113

30. not orthogonal, $\left\{ \begin{bmatrix} 1 \\ 1 \\ 0 \end{bmatrix}, \begin{bmatrix} 1 \\ -1 \\ 1 \end{bmatrix}, \dfrac{1}{3}\begin{bmatrix} -1 \\ 1 \\ 2 \end{bmatrix} \right\}$

31. Let A denote the matrix whose columns are the vectors in S. Since the reduced row echelon form of A is
$$\begin{bmatrix} 1 & 0 & 0 \\ 0 & 1 & 0 \\ 0 & 0 & 1 \\ 0 & 0 & 0 \end{bmatrix},$$

which has rank 3, the columns of S are linearly independent. Furthermore, S is not orthogonal. Let $\mathbf{u}_1$, $\mathbf{u}_2$, and $\mathbf{u}_3$ denote the vectors in S, listed in the same order. Set $\mathbf{v}_1 = \mathbf{u}_1$,

$$\mathbf{v}_2 = \mathbf{u}_2 - \dfrac{\mathbf{u}_2 \cdot \mathbf{v}_1}{\|\mathbf{v}_1\|^2}\mathbf{v}_1 = \begin{bmatrix} 0 \\ 0 \\ 1 \\ 1 \end{bmatrix} - \dfrac{-1}{3}\begin{bmatrix} 1 \\ 1 \\ -1 \\ 0 \end{bmatrix}$$

$$= \dfrac{1}{3}\begin{bmatrix} 1 \\ 1 \\ 2 \\ 3 \end{bmatrix},$$

and

$$\mathbf{v}_3 = \mathbf{u}_3 - \dfrac{\mathbf{u}_3 \cdot \mathbf{v}_1}{\|\mathbf{v}_1\|^2}\mathbf{v}_1 - \dfrac{\mathbf{u}_3 \cdot \mathbf{v}_2}{\|\mathbf{v}_2\|^2}\mathbf{v}_2$$

$$= \begin{bmatrix} 1 \\ 2 \\ 0 \\ 1 \end{bmatrix} - \dfrac{3}{3}\begin{bmatrix} 1 \\ 1 \\ -1 \\ 0 \end{bmatrix} - \left(\dfrac{2}{\frac{15}{9}}\right)\left(\dfrac{1}{3}\right)\begin{bmatrix} 1 \\ 1 \\ 2 \\ 3 \end{bmatrix}$$

$$= \dfrac{1}{5}\begin{bmatrix} -2 \\ 3 \\ 1 \\ -1 \end{bmatrix}.$$

Therefore the orthogonal basis is

$$\left\{ \begin{bmatrix} 1 \\ 1 \\ -1 \\ 0 \end{bmatrix}, \dfrac{1}{3}\begin{bmatrix} 1 \\ 1 \\ 2 \\ 3 \end{bmatrix}, \dfrac{1}{5}\begin{bmatrix} -2 \\ 3 \\ 1 \\ -1 \end{bmatrix} \right\}.$$

32. $\left\{ \begin{bmatrix} 1 \\ 2 \\ 0 \end{bmatrix}, \begin{bmatrix} 0 \\ 3 \\ 1 \end{bmatrix} \right\}$

33. A vector $\mathbf{v}$ is in $S^\perp$ if and only if $\mathbf{v}$ is orthogonal to both vectors in S, which occurs if and only if $\mathbf{v}$ is a solution of the system

$$2x_1 + x_2 - x_3 \quad\quad = 0$$
$$3x_1 + 4x_2 - 2x_3 - 2x_4 = 0.$$

The augmented matrix of this system has reduced row echelon form

$$\begin{bmatrix} 1 & 0 & -\frac{6}{5} & \frac{2}{5} & 0 \\ 0 & 1 & \frac{7}{5} & -\frac{4}{5} & 0 \end{bmatrix},$$

from which we obtain the vector form of the general solution

$$\begin{bmatrix} x_1 \\ x_2 \\ x_3 \\ x_4 \end{bmatrix} = x_3 \begin{bmatrix} \frac{6}{5} \\ -\frac{7}{5} \\ 1 \\ 0 \end{bmatrix} + x_4 \begin{bmatrix} -\frac{2}{5} \\ \frac{4}{5} \\ 0 \\ 1 \end{bmatrix}.$$

The two vectors in this representation form a basis for $S^\perp$. We can also multiply each of these vectors by 5 to obtain the basis

$$\left\{ \begin{bmatrix} 6 \\ -7 \\ 5 \\ 0 \end{bmatrix}, \begin{bmatrix} -2 \\ 4 \\ 0 \\ 5 \end{bmatrix} \right\}.$$

34. $\mathbf{w} = \frac{1}{5} \begin{bmatrix} 14 \\ 7 \end{bmatrix}$, $\mathbf{z} = \frac{1}{5} \begin{bmatrix} -4 \\ 8 \end{bmatrix}$

35. We have

$$\mathbf{w} = (\mathbf{v} \cdot \mathbf{v}_1)\mathbf{v}_1 + (\mathbf{v} \cdot \mathbf{v}_2)\mathbf{v}_2$$

$$= \frac{5}{\sqrt{5}} \frac{1}{\sqrt{5}} \begin{bmatrix} 1 \\ 2 \\ 0 \end{bmatrix} + \frac{-9}{\sqrt{14}} \frac{1}{\sqrt{14}} \begin{bmatrix} -2 \\ 1 \\ 3 \end{bmatrix}$$

$$= \frac{1}{14} \begin{bmatrix} 32 \\ 19 \\ -27 \end{bmatrix}$$

and

$$\mathbf{z} = \mathbf{v} - \mathbf{w} = \begin{bmatrix} 1 \\ 2 \\ -3 \end{bmatrix} - \frac{1}{14} \begin{bmatrix} 32 \\ 19 \\ -27 \end{bmatrix} = \frac{1}{14} \begin{bmatrix} -18 \\ 9 \\ -15 \end{bmatrix}.$$

The distance from $\mathbf{v}$ to W is

$$\|\mathbf{z}\| = \frac{3}{14}\sqrt{70}.$$

36. $P_W = \frac{1}{3} \begin{bmatrix} 2 & 1 & 1 \\ 1 & 2 & -1 \\ 1 & -1 & 2 \end{bmatrix}$ and $\mathbf{w} = \begin{bmatrix} 3 \\ -2 \\ 5 \end{bmatrix}$

37. $P_W = \frac{1}{6} \begin{bmatrix} 1 & 2 & 0 & -1 \\ 2 & 4 & 0 & -2 \\ 0 & 0 & 0 & 0 \\ -1 & -2 & 0 & 1 \end{bmatrix}$ and $\mathbf{w} = \begin{bmatrix} 2 \\ 4 \\ 0 \\ -2 \end{bmatrix}$

38. $P_W = \frac{1}{2} \begin{bmatrix} 1 & 0 & 1 \\ 0 & 0 & 0 \\ 1 & 0 & 1 \end{bmatrix}$ and $\mathbf{w} = \begin{bmatrix} 3 \\ 0 \\ 3 \end{bmatrix}$

39. A vector is in W if and only if it is orthogonal to both of the vectors in the given set, that is, if and only if it is a solution of the system

$$\begin{aligned} x_1 - x_2 &= 0 \\ x_1 + x_3 &= 0. \end{aligned}$$

Thus a basis for W is

$$\left\{ \begin{bmatrix} -1 \\ -1 \\ 1 \\ 0 \end{bmatrix}, \begin{bmatrix} 0 \\ 0 \\ 0 \\ 1 \end{bmatrix} \right\}.$$

Let C be the matrix whose columns are the vectors in this basis. Then

$$P_W = C(C^T C)^{-1} C^T = \frac{1}{3} \begin{bmatrix} 1 & 1 & -1 & 0 \\ 1 & 1 & -1 & 0 \\ -1 & -1 & 1 & 0 \\ 0 & 0 & 0 & 3 \end{bmatrix},$$

and the vector $\mathbf{w}$ in W closest to $\mathbf{v}$ is

$$\mathbf{w} = P_W \mathbf{v} = \frac{1}{3} \begin{bmatrix} 1 & 1 & -1 & 0 \\ 1 & 1 & -1 & 0 \\ -1 & -1 & 1 & 0 \\ 0 & 0 & 0 & 3 \end{bmatrix} \begin{bmatrix} 2 \\ -1 \\ 1 \\ 2 \end{bmatrix}$$

$$= \begin{bmatrix} 0 \\ 0 \\ 0 \\ 2 \end{bmatrix}.$$

40. $1.8 + 2.4x$

41. Let

$$C = \begin{bmatrix} 1 & 1 \\ 1 & 2 \\ 1 & 3 \\ 1 & 4 \\ 1 & 5 \end{bmatrix} \quad \text{and} \quad \mathbf{y} = \begin{bmatrix} 3.2 \\ 5.1 \\ 7.1 \\ 9.2 \\ 11.4 \end{bmatrix}.$$

Then

$$C^T C = \begin{bmatrix} 1 & 1 & 1 & 1 & 1 \\ 1 & 2 & 3 & 4 & 5 \end{bmatrix} \begin{bmatrix} 1 & 1 \\ 1 & 2 \\ 1 & 3 \\ 1 & 4 \\ 1 & 5 \end{bmatrix}$$

$$= \begin{bmatrix} 5 & 15 \\ 15 & 55 \end{bmatrix},$$

and hence

$$\begin{bmatrix} c \\ v \end{bmatrix} = (C^T C)^{-1} C^T \mathbf{y}$$

$$= \begin{bmatrix} 5 & 15 \\ 15 & 55 \end{bmatrix}^{-1} \begin{bmatrix} 1 & 1 & 1 & 1 & 1 \\ 1 & 2 & 3 & 4 & 5 \end{bmatrix} \begin{bmatrix} 3.2 \\ 5.1 \\ 7.1 \\ 9.2 \\ 11.4 \end{bmatrix}$$

$$= \frac{1}{10} \begin{bmatrix} 11 & -3 \\ -3 & 1 \end{bmatrix} \begin{bmatrix} 36.0 \\ 128.5 \end{bmatrix}$$

$$= \begin{bmatrix} 1.05 \\ 2.05 \end{bmatrix}.$$

Therefore $v = 2.05$ and $c = 1.05$.

42. $y = \frac{1}{70}(238 - 199x + 95x^2)$

188 Chapter 6 Orthogonality

43. Let A denote the given matrix. Then
$$A^T A = \begin{bmatrix} 0.58 & 0.00 \\ 0.00 & 0.58 \end{bmatrix} \neq \begin{bmatrix} 1 & 0 \\ 0 & 1 \end{bmatrix}.$$
Therefore the matrix is not orthogonal by Theorem 6.9.

44. yes   45. yes   46. no

47. Since $\det \frac{1}{2} \begin{bmatrix} 1 & \sqrt{3} \\ -\sqrt{3} & 1 \end{bmatrix} = \frac{1}{4}(1+3) = 1$, the matrix is a rotation matrix. Comparing the first column of this matrix with the first column of the rotation matrix A_θ, we see that
$$A_\theta = \begin{bmatrix} \cos\theta & -\sin\theta \\ \sin\theta & \cos\theta \end{bmatrix} = \begin{bmatrix} \frac{1}{2} & \frac{\sqrt{3}}{2} \\ -\frac{\sqrt{3}}{2} & \frac{1}{2} \end{bmatrix},$$
and hence $\cos\theta = \frac{1}{2}$ and $\sin\theta = -\frac{\sqrt{3}}{2}$. Therefore $\theta = -60°$.

48. a rotation, $\theta = 60°$

49. Since $\det \frac{1}{2} \begin{bmatrix} 1 & \sqrt{3} \\ \sqrt{3} & -1 \end{bmatrix} = \frac{1}{4}(-1-3) = -1$, the matrix is the standard matrix of a reflection. To find the line of reflection, we find an eigenvector of the matrix corresponding to eigenvalue 1 by solving the equation
$$\begin{bmatrix} \frac{1}{2} - 1 & \frac{\sqrt{3}}{2} \\ \frac{\sqrt{3}}{2} & -\frac{1}{2} - 1 \end{bmatrix} \begin{bmatrix} x_1 \\ x_2 \end{bmatrix} = \begin{bmatrix} 0 \\ 0 \end{bmatrix}.$$
The vector form of the general solution of this equation is
$$\begin{bmatrix} x_1 \\ x_2 \end{bmatrix} = x_2 \begin{bmatrix} \sqrt{3} \\ 1 \end{bmatrix},$$
and hence the equation of the line is $y = \frac{1}{\sqrt{3}}x$.

50. a reflection, $y = 2x$

51. The standard matrix of T is
$$Q = \begin{bmatrix} 0 & -1 & 0 \\ 0 & 0 & 1 \\ 1 & 0 & 0 \end{bmatrix}.$$
Since $Q^T Q = I_3$, we see that Q is an orthogonal matrix. Thus T is an orthogonal operator.

52. Observe that the standard matrix of U is given by
$$Q = \frac{1}{\sqrt{2}} \begin{bmatrix} 1 & 1 \\ -1 & 1 \end{bmatrix}.$$

(a) Since
$$QQ^T = \frac{1}{\sqrt{2}} \begin{bmatrix} 1 & 1 \\ -1 & 1 \end{bmatrix} \frac{1}{\sqrt{2}} \begin{bmatrix} 1 & -1 \\ 1 & 1 \end{bmatrix}$$
$$= \begin{bmatrix} 1 & 0 \\ 0 & 1 \end{bmatrix},$$
it follows that Q is an orthogonal matrix. Therefore U is an orthogonal operator.

(b) Observe that
$$\det Q = \det \frac{1}{\sqrt{2}} \begin{bmatrix} 1 & 1 \\ -1 & 1 \end{bmatrix} = \frac{1}{2}(1+1) = 1,$$
and hence U is a rotation by Theorem 6.11. Let P be the standard matrix of T. Then $\det P = 1$ by Theorem 6.11. Also, PQ is the standard matrix of TU. Thus
$$\det PU = (\det P)(\det U) = 1 \cdot 1 = 1,$$
and hence TU is a rotation by Theorem 6.11.

(c) Let P be the standard matrix of T. Then $\det P = -1$ by Theorem 6.11. Using the method of (b), we see that $\det PQ = -1$, and hence TU is a reflection by Theorem 6.11.

53. The characteristic polynomial of A is
$$\det(A - tI_2) = \det \begin{bmatrix} 2-t & 3 \\ 3 & 2-t \end{bmatrix}$$
$$= (t+1)(t-5),$$
and hence A has the eigenvalues $\lambda_1 = -1$ and $\lambda_2 = 5$. The vectors $\begin{bmatrix} -1 \\ 1 \end{bmatrix}$ and $\begin{bmatrix} 1 \\ 1 \end{bmatrix}$ are corresponding eigenvectors. Normalizing these vectors, we obtain the orthonormal basis
$$\{\mathbf{u}_1, \mathbf{u}_2\} = \left\{ \frac{1}{\sqrt{2}} \begin{bmatrix} -1 \\ 1 \end{bmatrix}, \frac{1}{\sqrt{2}} \begin{bmatrix} 1 \\ 1 \end{bmatrix} \right\}$$
of eigenvectors of A. We use these eigenvectors and eigenvalues to obtain the spectral decomposition as follows:
$$A = \lambda_1 \mathbf{u}_1 \mathbf{u}_1^T + \lambda_2 \mathbf{u}_2 \mathbf{u}_2^T$$
$$= (-1)\frac{1}{\sqrt{2}} \begin{bmatrix} -1 \\ 1 \end{bmatrix} \frac{1}{\sqrt{2}}[-1 \ 1] + 5\frac{1}{\sqrt{2}} \begin{bmatrix} 1 \\ 1 \end{bmatrix} \frac{1}{\sqrt{2}}[1 \ 1]$$
$$= (-1) \begin{bmatrix} 0.5 & -0.5 \\ -0.5 & 0.5 \end{bmatrix} + 5 \begin{bmatrix} 0.5 & 0.5 \\ 0.5 & 0.5 \end{bmatrix}.$$

54. $\left\{ \dfrac{1}{\sqrt{5}} \begin{bmatrix} 2 \\ -1 \\ 0 \end{bmatrix}, \dfrac{1}{\sqrt{5}} \begin{bmatrix} 1 \\ 2 \\ 0 \end{bmatrix}, \begin{bmatrix} 0 \\ 0 \\ 1 \end{bmatrix} \right\}$, 5, 10, -9

$$A = 5 \begin{bmatrix} 0.8 & -0.4 & 0 \\ -0.4 & 0.2 & 0 \\ 0 & 0 & 0 \end{bmatrix} + 10 \begin{bmatrix} 0.2 & 0.4 & 0 \\ 0.4 & 0.8 & 0 \\ 0 & 0 & 0 \end{bmatrix}$$
$$+ (-9) \begin{bmatrix} 0 & 0 & 0 \\ 0 & 0 & 0 \\ 0 & 0 & 1 \end{bmatrix}$$

55. Let $a_{11} = 1$, $a_{22} = 1$, and $a_{12} = a_{21} = \frac{1}{2}(6) = 3$, so that
$$A = \begin{bmatrix} 1 & 3 \\ 3 & 1 \end{bmatrix}.$$
The eigenvalues of A are 4 and -2, and
$$\left\{ \dfrac{1}{\sqrt{2}} \begin{bmatrix} 1 \\ 1 \end{bmatrix}, \dfrac{1}{\sqrt{2}} \begin{bmatrix} -1 \\ 1 \end{bmatrix} \right\}$$
is an orthonormal basis of $\mathcal{R}^2$ consisting of corresponding eigenvectors. The first of these basis vectors has both positive components, and so we construct the rotation matrix with it as its first column
$$P = \dfrac{1}{\sqrt{2}} \begin{bmatrix} 1 & -1 \\ 1 & 1 \end{bmatrix}.$$
This matrix is the rotation matrix corresponding to an angle of
$$\cos^{-1}\left(\dfrac{1}{\sqrt{2}}\right) = 45°.$$
Thus if we rotate the x- and y-axes by 45°, the original equation becomes $4(x')^2 - 2(y')^2 = 16$. (The coefficients of $(x')^2$ and $(y')^2$ are the eigenvalues corresponding to the first and second columns of P.) The new equation can be written as
$$\dfrac{(x')^2}{4} - \dfrac{(y')^2}{8} = 1,$$
and so the conic section is a hyperbola.

56. $45°$, $\dfrac{(x')^2}{9} + \dfrac{5(y')^2}{9} = 1$, an ellipse

57. It suffices to show that $Q^T P_W Q\mathbf{z} = P_Z \mathbf{z}$ for all $\mathbf{z}$ in Z and $Q^T P_W Q\mathbf{y} = P_Z \mathbf{y}$ for all $\mathbf{y}$ in $Z^\perp$. (See the proof of Exercise 74 of Section 6.3.) If $\mathbf{z}$ in Z, then $\mathbf{z} = Q^T \mathbf{w}$ for some $\mathbf{w}$ in W. Hence
$$Q^T P_W Q\mathbf{z} = Q^T P_W Q Q^T \mathbf{w} = Q^T P_W \mathbf{w}$$
$$= Q^T \mathbf{w} = \mathbf{z} = P_Z \mathbf{z}.$$
Now suppose that $\mathbf{y}$ is in $Z^\perp$. Because Q^T, an orthogonal matrix, preserves dot products, it also preserves orthogonality of vectors. So $Z^\perp = \{Q^T \mathbf{y} : \mathbf{y} \text{ is in } W^\perp\}$. Thus $\mathbf{y} = Q^T \mathbf{u}$ for some $\mathbf{u}$ in $W^\perp$. Therefore
$$Q^T P_W Q\mathbf{y} = Q^T P_W Q Q^T \mathbf{u} = Q^T P_W \mathbf{u}$$
$$= Q^T \mathbf{0} = \mathbf{0} = P_Z \mathbf{y}.$$

58. We first show that $\operatorname{Col} P_W = W$. Suppose that $\mathbf{u}$ is in $\operatorname{Col} P_W$. Then $\mathbf{u} = P_W \mathbf{v}$ for some vector $\mathbf{v}$ in $\mathcal{R}^n$. Therefore
$$P_W \mathbf{u} = P_W^2 \mathbf{v} = P_W \mathbf{v} = \mathbf{u},$$
and hence $\mathbf{u}$ is in W by Exercise 68 of Section 6.3.

Now suppose that $\mathbf{w}$ is in W. Then $\mathbf{w} = P_W \mathbf{w}$ by Exercise 68 of Section 6.3, and hence $\mathbf{w}$ is in $\operatorname{Col} P_W$. It follows that $\operatorname{Col} P_W = W$. Therefore
$$\operatorname{rank} P_W = \dim \operatorname{Col} P_W = \dim W.$$

CHAPTER 6 MATLAB EXERCISES

1. (a) $\mathbf{u}_1 \cdot \mathbf{u}_2 = -2$, $\|\mathbf{u}_1\| = 4$, $\|\mathbf{u}_2\| = \sqrt{23} \approx 4.7958$

   (b) $\mathbf{u}_3 \cdot \mathbf{u}_4 = -56$, $\|\mathbf{u}_3\| = \sqrt{28} \approx 5.2915$, $\|\mathbf{u}_4\| = \sqrt{112}$

   (c) $|\mathbf{u}_1 \cdot \mathbf{u}_2| = 2 \leq 4\sqrt{23} = \|\mathbf{u}_1\| \cdot \|\mathbf{u}_2\|$

   (d) $|\mathbf{u}_3 \cdot \mathbf{u}_4| = 56 = \sqrt{28} \cdot \sqrt{112} = \|\mathbf{u}_3\| \cdot \|\mathbf{u}_4\|$

   (e) For vectors $\mathbf{u}$ and $\mathbf{v}$ in $\mathcal{R}^n$, the Cauchy-Schwarz inequality is an equality if and only if $\mathbf{u}$ is a multiple of $\mathbf{v}$ or $\mathbf{v}$ is a multiple of $\mathbf{u}$.

2. (a) First observe that for any vector $\mathbf{x}$ in $\mathcal{R}^n$, the entries of $A^T \mathbf{x}$ are the dot products of the corresponding columns of A and $\mathbf{x}$. It follows that $A^T \mathbf{x} = \mathbf{0}$ if and only if $\mathbf{x}$ is orthogonal to all of the columns of A, which is the case if and only if $\mathbf{x}$ is orthogonal to all of the vectors in W, that is, $\mathbf{x}$ is in $W^\perp$.

   (b) $\left\{ \begin{bmatrix} 0.7273 \\ 0.5402 \\ -0.3772 \\ 0.1891 \\ 0.0329 \end{bmatrix}, \begin{bmatrix} -0.6222 \\ 0.6071 \\ -0.0844 \\ 0.4660 \\ 0.1411 \end{bmatrix}, \begin{bmatrix} 0.0079 \\ -0.3191 \\ -0.3140 \\ 0.1003 \\ 0.8885 \end{bmatrix} \right\}$

3. Answers are given correct to 4 places after the decimal point.

(a) $\left\{ \begin{bmatrix} -0.1994 \\ 0.1481 \\ -0.1361 \\ -0.6282 \\ -0.5316 \\ 0.4924 \end{bmatrix}, \begin{bmatrix} 0.1153 \\ 0.0919 \\ -0.5766 \\ 0.6366 \\ -0.4565 \\ 0.1790 \end{bmatrix}, \begin{bmatrix} 0.3639 \\ -0.5693 \\ 0.5469 \\ 0.1493 \\ -0.4271 \\ 0.1992 \end{bmatrix} \right\}$

(b) (i) $\begin{bmatrix} 1.3980 \\ -1.5378 \\ 1.4692 \\ 2.7504 \\ 1.4490 \\ -1.6574 \end{bmatrix}$ (ii) $\begin{bmatrix} 1 \\ -2 \\ 2 \\ -1 \\ -3 \\ 2 \end{bmatrix}$ (iii) $\begin{bmatrix} 0 \\ 0 \\ 0 \\ 0 \\ 0 \\ 0 \end{bmatrix}$

(c) They are the same.

(d) If M is a matrix whose columns form an orthonormal basis for a subspace W of $\mathcal{R}^n$, then $P_W = MM^T$, that is, MM^T is the orthogonal projection matrix for W.

4. Answers are given correct to 4 places after the decimal point.

(a) $V =$
$\begin{bmatrix} 1.1000 & 2.7581 & -2.6745 & -0.3438 \\ 2.3000 & 5.8488 & 1.4345 & -1.0069 \\ 3.1000 & 2.3093 & -0.2578 & 3.1109 \\ 7.2000 & -1.9558 & 0.4004 & 1.5733 \\ 8.0000 & -1.1954 & -0.3051 & -2.2847 \end{bmatrix}$

(b) $D =$
$\begin{bmatrix} 131.9500 & 0.0000 & 0.0000 & 0.0000 \\ 0.0000 & 52.4032 & 0.0000 & 0.0000 \\ 0.0000 & 0.0000 & 9.5306 & 0.0000 \\ 0.0000 & 0.0000 & 0.0000 & 18.5046 \end{bmatrix}$

(c) $Q =$
$\begin{bmatrix} 0.0958 & 0.3810 & -0.8663 & -0.0799 \\ 0.2002 & 0.8080 & 0.4647 & -0.2341 \\ 0.2699 & 0.3190 & -0.0835 & 0.7232 \\ 0.6268 & -0.2702 & 0.1297 & 0.3657 \\ 0.6964 & -0.1651 & -0.0988 & -0.5311 \end{bmatrix}$

(d) $R =$
$\begin{bmatrix} 11.4869 & -3.7399 & 1.0804 & 13.1166 \\ 0.0000 & 7.2390 & -6.3751 & 2.6668 \\ 0.0000 & 0.0000 & 3.0872 & -5.9697 \\ 0.0000 & 0.0000 & 0.0000 & 4.3017 \end{bmatrix}$

(e) In this case, we have

$Q =$
$\begin{bmatrix} -0.0958 & -0.3810 & 0.8663 & -0.0799 \\ -0.2002 & -0.8080 & -0.4647 & -0.2341 \\ -0.2699 & -0.3190 & 0.0835 & 0.7232 \\ -0.6268 & 0.2702 & -0.1297 & 0.3657 \\ -0.6964 & 0.1651 & 0.0988 & -0.5311 \end{bmatrix}$

$R =$
$\begin{bmatrix} -11.4869 & 3.7399 & -1.0804 & -13.1166 \\ 0.0000 & -7.2390 & 6.3751 & -2.6668 \\ 0.0000 & 0.0000 & -3.0872 & 5.9697 \\ 0.0000 & 0.0000 & 0.0000 & 4.3017 \end{bmatrix}$

5. Answers are given correct to 4 places after the decimal point.

$Q =$
$\begin{bmatrix} 0.2041 & 0.4308 & 0.3072 & 0.3579 \\ 0.8165 & -0.1231 & 0.2861 & 0.2566 \\ -0.2041 & 0.3077 & 0.6264 & -0.1235 \\ 0.4082 & -0.2462 & -0.3222 & -0.2728 \\ 0.2041 & 0.8001 & -0.4849 & -0.1253 \\ 0.2041 & 0.0615 & 0.3042 & -0.8371 \end{bmatrix}$

$R =$
$\begin{bmatrix} 4.8990 & 3.2660 & -1.4289 & 1.8371 \\ 0.0000 & 5.4160 & -0.0615 & 3.5081 \\ 0.0000 & 0.0000 & 7.5468 & -2.2737 \\ 0.0000 & 0.0000 & 0.0000 & 2.2690 \end{bmatrix}$

6. Answers are given correct to 4 places after the decimal point.

(a) $\mathcal{B}_1 =$
$\left\{ \begin{bmatrix} -0.1994 \\ 0.1481 \\ -0.1361 \\ -0.6282 \\ -0.5316 \\ 0.4924 \end{bmatrix}, \begin{bmatrix} 0.1153 \\ 0.0919 \\ -0.5766 \\ 0.6366 \\ -0.4565 \\ 0.1790 \end{bmatrix}, \begin{bmatrix} 0.3639 \\ -0.5693 \\ 0.5469 \\ 0.1493 \\ -0.4271 \\ 0.1992 \end{bmatrix} \right\}$

(b) $\mathcal{B}_2 =$
$\left\{ \begin{bmatrix} 0.8986 \\ 0.3169 \\ -0.1250 \\ -0.2518 \\ 0.1096 \\ 0.0311 \end{bmatrix}, \begin{bmatrix} -0.0808 \\ 0.6205 \\ 0.5183 \\ 0.3372 \\ 0.1022 \\ 0.4644 \end{bmatrix}, \begin{bmatrix} 0.0214 \\ -0.4000 \\ -0.2562 \\ 0.0246 \\ 0.5514 \\ 0.6850 \end{bmatrix} \right\}$

(c) $PP^T = P^TP = I_6$

7. Answers are given correct to 4 places after the decimal point.

(a) $P_W = [B \ C]$, where

$$B = \begin{bmatrix} 0.3913 & 0.0730 & -0.1763 \\ 0.0730 & 0.7180 & -0.1688 \\ -0.1763 & -0.1688 & 0.8170 \\ -0.2716 & -0.1481 & -0.2042 \\ 0.2056 & 0.1328 & 0.1690 \\ -0.2929 & 0.3593 & 0.1405 \end{bmatrix}$$

and

$$C = \begin{bmatrix} -0.2716 & 0.2056 & -0.2929 \\ -0.1481 & 0.1328 & 0.3593 \\ -0.2042 & 0.1690 & 0.1405 \\ 0.7594 & 0.1958 & 0.0836 \\ 0.1958 & 0.8398 & -0.0879 \\ 0.0836 & -0.0879 & 0.4744 \end{bmatrix}$$

(b) same as (a)

(c) $P_W \mathbf{v} = \mathbf{v}$ for all $\mathbf{v}$ in $\mathcal{S}$.

(d) $\left\{ \begin{bmatrix} -1.75 \\ -0.50 \\ -1.00 \\ -1.25 \\ 1.00 \\ 0.00 \end{bmatrix}, \begin{bmatrix} 0.85 \\ -0.60 \\ -0.10 \\ 0.05 \\ 0.00 \\ 1.00 \end{bmatrix} \right\}$

In each case, $P_W \mathbf{v} = \mathbf{0}$.

8. Answers are given correct to 4 places after the decimal point.

   (b) $y = 0.5404 + 0.4091x$

   (c) $y = 0.2981 + 0.7279x - 0.0797x^2$

9. In the case of the least squares line, the ith entry of $C * a$ is $a_0 + a_1 x_i$, where x_i is the second entry of the ith row of C. Similarly, for the best quadratic fit, the ith entry of $C * a$ is

   $$a_0 + a_1 x_i + a_2 x_i^2.$$

10. Answers are given correct to 4 places after the decimal point.

    The standard matrix of T is

    $$A = \begin{bmatrix} \cos 35° & -\sin 35° \\ \sin 35° & \cos 35° \end{bmatrix}$$

    $$= \begin{bmatrix} 0.8192 & -0.5736 \\ 0.5736 & 0.8192 \end{bmatrix}.$$

    Let B be the standard matrix of U. Observe that $\left\{ \begin{bmatrix} 1.0 \\ 2.3 \end{bmatrix}, \begin{bmatrix} 2.3 \\ -1.0 \end{bmatrix} \right\}$ is a basis for $\mathcal{R}^2$ consisting of eigenvectors of U with corresponding eigenvalues 1 and -1, respectively. Let

    $$P = \begin{bmatrix} 1.0 & 2.3 \\ 2.3 & -1.0 \end{bmatrix} \text{ and } D = \begin{bmatrix} 1 & 0 \\ 0 & -1 \end{bmatrix}.$$

    Then

    $$B = PDP^{-1} = \begin{bmatrix} -0.6820 & 0.7313 \\ 0.7313 & 0.6820 \end{bmatrix}.$$

    (a) Applying the MATLAB function `eig` to BA (see Appendix D) shows that $\begin{bmatrix} 0.6560 \\ 0.7547 \end{bmatrix}$ is an eigenvector of BA, and hence of UT, with corresponding eigenvalue 1. This vector lies on the line through $\mathbf{0}$ about which UT reflects $\mathcal{R}^2$. The slope of this line is $0.7547/0.6560 = 1.1504$, and hence the equation of the line is $y = 1.1504x$.

    (b) The vector $\begin{bmatrix} -0.1045 \\ -0.9945 \end{bmatrix}$ is an eigenvector of AB, and hence of TU, with corresponding eigenvalue 1. The slope of the line containing this vector is $(-0.9945)/(-0.1045) = 9.5167$, and hence the equation of the line is $y = 9.5167x$.

11. Answers are given correct to 4 places after the decimal point.

    (a) $P =$

    $$\begin{bmatrix} -0.5 & -0.5477 & -0.5 & -0.4472 & 0.0000 \\ 0.5 & -0.5477 & 0.5 & -0.4472 & 0.0000 \\ -0.5 & 0.3651 & 0.5 & -0.4472 & 0.4082 \\ 0.0 & 0.3651 & 0.0 & -0.4472 & -0.8165 \\ 0.5 & 0.3651 & -0.5 & -0.4472 & 0.4082 \end{bmatrix}$$

    $$D = \begin{bmatrix} -4 & 0 & 0 & 0 & 0 \\ 0 & 0 & 0 & 0 & 0 \\ 0 & 0 & -8 & 0 & 0 \\ 0 & 0 & 0 & 5 & 0 \\ 0 & 0 & 0 & 0 & 12 \end{bmatrix}$$

    (b) The columns of P form an orthonormal basis and the diagonal entries of D (in the same order) are the corresponding eigenvalues.

    (c) $A =$

    $$-4 \begin{bmatrix} 0.25 & -0.25 & 0.25 & 0 & -0.25 \\ -0.25 & 0.25 & -0.25 & 0 & 0.25 \\ 0.25 & -0.25 & 0.25 & 0 & -0.25 \\ 0.00 & 0.00 & 0.00 & 0 & 0.00 \\ -0.25 & 0.25 & -0.25 & 0 & 0.25 \end{bmatrix}$$

    $$+ 0 \begin{bmatrix} .3 & .3 & -.2000 & -.2000 & -.2000 \\ .3 & .3 & -.2000 & -.2000 & -.2000 \\ -.2 & -.2 & .1333 & .1333 & .1333 \\ -.2 & -.2 & .1333 & .1333 & .1333 \\ -.2 & -.2 & .1333 & .1333 & .1333 \end{bmatrix}$$

192 Chapter 6 Orthogonality

$$-8\begin{bmatrix} 0.25 & -0.25 & -0.25 & 0 & 0.25 \\ -0.25 & 0.25 & 0.25 & 0 & -0.25 \\ -0.25 & 0.25 & 0.25 & 0 & -0.25 \\ 0.00 & 0.00 & 0.00 & 0 & 0.00 \\ 0.25 & -0.25 & -0.25 & 0 & 0.25 \end{bmatrix}$$

$$+5\begin{bmatrix} 0.2 & 0.2 & 0.2 & 0.2 & 0.2 \\ 0.2 & 0.2 & 0.2 & 0.2 & 0.2 \\ 0.2 & 0.2 & 0.2 & 0.2 & 0.2 \\ 0.2 & 0.2 & 0.2 & 0.2 & 0.2 \\ 0.2 & 0.2 & 0.2 & 0.2 & 0.2 \end{bmatrix}$$

$$+12\begin{bmatrix} 0 & 0 & 0.0000 & 0.0000 & 0.0000 \\ 0 & 0 & 0.0000 & 0.0000 & 0.0000 \\ 0 & 0 & 0.1667 & -0.3333 & 0.1667 \\ 0 & 0 & -0.3333 & 0.6667 & -0.3333 \\ 0 & 0 & 0.1667 & -0.3333 & 0.1667 \end{bmatrix}$$

(d) $A_2 = \begin{bmatrix} -2 & 2 & 2 & 0 & -2 \\ 2 & -2 & -2 & 0 & 2 \\ 2 & -2 & 0 & -4 & 4 \\ 0 & 0 & -4 & 8 & -4 \\ -2 & 2 & 4 & -4 & 0 \end{bmatrix}$

(e) $\|E_2\| = 6.4031 \qquad \|A\| = 15.7797$

(f) 40.58%

12. Answers are given correct to 4 places after the decimal point.

(a) $U =$

$$\begin{bmatrix} -0.5404 & 0.6121 & 0.2941 & 0.4968 \\ -0.6121 & -0.5404 & -0.4968 & 0.2941 \\ -0.5762 & 0.0359 & 0.2028 & -0.7909 \\ 0.0359 & 0.5762 & -0.7909 & -0.2028 \end{bmatrix}$$

$$S = \begin{bmatrix} 7.5622 & 0 & 0 & 0 & 0 & 0 \\ 0 & 2.9687 & 0 & 0 & 0 & 0 \\ 0 & 0 & 0 & 0 & 0 & 0 \\ 0 & 0 & 0 & 0 & 0 & 0 \end{bmatrix}$$

$V = [B \ C]$, where

$$B = \begin{bmatrix} -.2286 & .0363 & -.8821 \\ .0142 & .5823 & .3581 \\ -.7000 & -.4735 & .2658 \\ -.2286 & .0363 & .0538 \\ -.4430 & .6548 & -.0922 \\ -.4572 & .0725 & .1076 \end{bmatrix}$$

and

$$C = \begin{bmatrix} -.4036 & .0507 & -.0528 \\ -.7024 & .1371 & -.1429 \\ -.1473 & .3045 & -.3172 \\ -.0529 & -.9050 & -.3488 \\ .5550 & .1674 & -.1744 \\ -.1058 & -.1970 & .8510 \end{bmatrix}$$

(b) The last 4 columns of V are the columns of Null A.

(c) The first 2 columns of U are the columns of $\mathtt{orth}(A)$.

(d) Let $A = USV^T$ be a singular value decomposition of an $m \times n$ matrix A with k (not necessarily distinct) singular values. Then the first k columns of U form an orthonormal basis for Col A, and the last $n - k$ columns of V form an orthonormal basis for Null A.

(e) Let $\sigma_1, \sigma_2, \ldots, \sigma_k$ be the (nonzero) singular values of A. Then $s_{ii} = \sigma_i$, for $1 \le i \le k$, and $s_{ij} = 0$ otherwise.

Proof that the first k columns of U form an orthonormal basis for Col A.
Since the first k columns of U are orthonormal, it suffices to show that these columns form a generating set for Col A. Suppose that $\mathbf{x}$ is in Col A. Then $\mathbf{x} = A\mathbf{y}$ for some vector $\mathbf{y}$ in $\mathcal{R}^n$. Let $\mathbf{z} = V^T\mathbf{y}$. Then

$$\mathbf{x} = USV^T\mathbf{y} = U(S\mathbf{z}) = U\begin{bmatrix} \sigma_1 z_1 \\ \sigma_2 z_2 \\ \vdots \\ \sigma_n z_n \end{bmatrix}$$

$$= \sigma_1 z_1 \mathbf{u}_1 + \sigma_2 z_2 \mathbf{u}_2 + \cdots + \sigma_n z_n \mathbf{u}_n,$$

which is in the span of the first k columns of U.

Conversely, suppose that $\mathbf{x}$ is in the span of the first k columns of U. Then

$$\mathbf{x} = c_1 \mathbf{u}_1 + c_2 \mathbf{u}_2 + \cdots + c_k \mathbf{u}_k.$$

Let

$$\mathbf{z} = \begin{bmatrix} \frac{c_1}{\sigma_1} \\ \frac{c_2}{\sigma_2} \\ \vdots \\ \frac{c_k}{\sigma_k} \\ 0 \\ 0 \\ \vdots \\ 0 \end{bmatrix} \qquad \text{and} \qquad \mathbf{y} = V\mathbf{z}.$$

Then

$$A\mathbf{y} = USV^T V\mathbf{z} = US\mathbf{z} = U\begin{bmatrix} c_1 \\ c_2 \\ \vdots \\ c_k \\ 0 \\ \vdots \\ 0 \end{bmatrix}$$

$$= c_1\mathbf{u}_1 + c_2\mathbf{u}_2 + \cdots + c_k\mathbf{u}_k = \mathbf{x},$$

and hence $\mathbf{x}$ is in Col A.

Proof that the last $n - k$ columns of V form an orthonormal basis for Null A.

Since the last $n - k$ columns of V are orthonormal, it suffices to show that these columns form a generating set for Null A. Consider any $\mathbf{x}$ in $\mathcal{R}^n$, and let $\mathbf{z} = V^T\mathbf{x}$. Then $A\mathbf{x} = \mathbf{0}$ if and only if $USV^T\mathbf{x} = US\mathbf{z} = \mathbf{0}$, which is true if and only if

$$\sigma_1 z_1 \mathbf{u}_1 + \sigma_2 z_2 \mathbf{u}_2 + \cdots + \sigma_1 z_k \mathbf{u}_k = \mathbf{0}.$$

But the preceding equation is true if and only if $z_1 = z_2 = \cdots = z_k = 0$, which is true if and only if

$$\mathbf{x} = V\mathbf{z} = z_{k+1}\mathbf{v}_{k+1} + \cdots + z_n\mathbf{v}_n,$$

that is, $\mathbf{v}$ is in the span of the last $n - k$ columns of V.

13. The answer is given correct to 4 places after the decimal point.

    $P_W = [B \ C]$, where

    $$B = \begin{bmatrix} 0.3913 & 0.0730 & -0.1763 \\ 0.0730 & 0.7180 & -0.1688 \\ -0.1763 & -0.1688 & 0.8170 \\ -0.2716 & -0.1481 & -0.2042 \\ 0.2056 & 0.1328 & 0.1690 \\ -0.2929 & 0.3593 & 0.1405 \end{bmatrix}$$

    and

    $$C = \begin{bmatrix} -0.2716 & 0.2056 & -0.2929 \\ -0.1481 & 0.1328 & 0.3593 \\ -0.2042 & 0.1690 & 0.1405 \\ 0.7594 & 0.1958 & 0.0836 \\ 0.1958 & 0.8398 & -0.0879 \\ 0.0836 & -0.0879 & 0.4744 \end{bmatrix}$$

14. The answer is given correct to 4 places after the decimal point.

    $$\begin{bmatrix} 0.7550 \\ -0.0861 \\ 0.6556 \\ 0.9205 \\ -0.0795 \end{bmatrix}$$

15. Answers are given correct to 4 places after the decimal point.

    Let $Q_1 =$

    $$\begin{bmatrix} 0.0618 & -0.1724 & 0.2088 & 0.0391 & -0.0597 \\ 0.1709 & -0.4769 & 0.5774 & 0.1082 & -0.1651 \\ -0.0867 & 0.2420 & -0.2930 & -0.0549 & 0.0838 \\ 0.0767 & -0.2141 & 0.2592 & 0.0486 & -0.0741 \end{bmatrix},$$

    $Q_2 =$

    $$\begin{bmatrix} 0.3163 & 0.1374 & -0.0541 & 0.5775 & 0.1199 \\ 0.0239 & 0.0104 & -0.0041 & 0.0436 & 0.0090 \\ 0.3291 & 0.1429 & -0.0563 & 0.6008 & 0.1247 \\ 0.0640 & 0.0278 & -0.0109 & 0.1168 & 0.0243 \end{bmatrix},$$

    $Q_3 =$

    $$\begin{bmatrix} 0.1316 & -0.2635 & -0.2041 & -0.0587 & 0.1452 \\ -0.2036 & 0.4076 & 0.3157 & 0.0908 & -0.2246 \\ -0.1470 & 0.2943 & 0.2280 & 0.0656 & -0.1622 \\ 0.1813 & -0.3630 & -0.2811 & -0.0808 & 0.2000 \end{bmatrix},$$

    and $Q_4 =$

    $$\begin{bmatrix} -0.3747 & -0.2063 & -0.1482 & 0.2684 & -0.1345 \\ 0.0334 & 0.0184 & 0.0132 & -0.0239 & 0.0120 \\ 0.2571 & 0.1415 & 0.1016 & -0.1841 & 0.0922 \\ 0.5180 & 0.2852 & 0.2048 & -0.3710 & 0.1859 \end{bmatrix}.$$

    (a) $A = \sigma_1 Q_1 + \sigma_2 Q_2 + \sigma_3 Q_3 + \sigma_4 Q_4$, where

    $$\sigma_1 = 205.2916, \quad \sigma_2 = 123.3731,$$
    $$\sigma_3 = 50.3040, \text{ and } \sigma_4 = 6.2391.$$

    (b) $A_2^T =$

    $$\begin{bmatrix} 51.7157 & 38.0344 & 22.7926 & 23.6467 \\ -18.4559 & -96.6198 & 67.3103 & -40.5194 \\ 36.1913 & 118.0373 & -67.1013 & 51.8636 \\ 79.2783 & 27.5824 & 62.8508 & 24.3814 \\ 2.5334 & -32.7751 & 32.5841 & -12.2222 \end{bmatrix}$$

    $E_2^T =$

    $$\begin{bmatrix} 4.2843 & -10.0344 & -5.7926 & 12.3533 \\ -14.5441 & 20.6198 & 15.6897 & -16.4806 \\ -11.1913 & 15.9627 & 12.1013 & -12.8636 \\ -1.2783 & 4.4176 & 2.1492 & -6.3814 \\ 6.4666 & -11.2249 & -7.5841 & 11.2222 \end{bmatrix}$$

    (c) $\|E_2\| = 50.6894, \quad \|A\| = 244.8163$

    (d) $\dfrac{\|E_2\|}{\|A\|} = 0.2071$

16. (a) $\begin{bmatrix} 0.8298 \\ -0.1538 \\ 0.5364 \end{bmatrix}, \theta = 38°$

    (b) $\begin{bmatrix} 0.8298 \\ 0.1538 \\ 0.5364 \end{bmatrix}, \theta = 38°$

17. (a) We use the rational format in MATLAB to obtain the precise value

    $$A_W = \begin{bmatrix} 2/3 & -2/3 & 1/3 \\ -2/3 & -1/3 & 2/3 \\ 1/3 & 2/3 & 2/3 \end{bmatrix}.$$

    (b) Since $Q_{23°}$ is a rotation, its determinant is equal to 1. Observe that $\det A_W = -1$.

Since A_W and $Q_{23°}$ are orthogonal matrices, $A_W Q_{23°} A_W$ is an orthogonal matrix. Furthermore,

$$\det(A_W Q_{23°} A_W)$$
$$= (\det A_W)(\det Q_{23°})(\det A_W)$$
$$= (-1)(1)(-1) = 1.$$

So $A_W Q_{23°} A_W$ is a rotation by Theorem 6.20.

(c) $\mathbf{v} = \begin{bmatrix} -2 \\ -1 \\ 2 \end{bmatrix}$ is a vector that lies on the axis of rotation, and the angle of rotation is $23°$.

(d) Let R be a rotation matrix and A_W be the standard matrix of the reflection of $\mathcal{R}^3$ about a two-dimensional subspace W. A nonzero vector $\mathbf{w}$ lies on the axis of rotation of $A_W R A_W$ if and only if $\mathbf{w} = A_W \mathbf{v}$, where $\mathbf{v}$ is nonzero vector that lies on the axis of rotation of R. Furthermore, the angle of rotation of $A_W R A_W$ is equal to the angle of rotation of R.

18. (b) Let V and E be as in Exercise 18(a). Define $C = A_W V$, where A_W is the matrix in 17(a), and apply the MATLAB command `grfig(C, E)`.

Chapter 7

Vector Spaces

7.1 VECTOR SPACES AND THEIR SUBSPACES

1. Consider the matrix equation

$$\begin{bmatrix} 0 & 2 & 0 \\ 1 & 1 & 1 \end{bmatrix} = x_1 \begin{bmatrix} 1 & 2 & 1 \\ 0 & 0 & 0 \end{bmatrix} + x_2 \begin{bmatrix} 0 & 0 & 0 \\ 1 & 1 & 1 \end{bmatrix} + x_3 \begin{bmatrix} 1 & 0 & 1 \\ 1 & 2 & 3 \end{bmatrix}.$$

Comparing the corresponding entries of the right and left sides of this matrix equation yields the system of linear equations

$$\begin{aligned} x_1 + x_3 &= 0 \\ 2x_1 &= 2 \\ x_1 + x_3 &= 0 \\ x_2 + x_3 &= 1 \\ x_2 + 2x_3 &= 1 \\ x_2 + 3x_3 &= 1. \end{aligned}$$

The reduced row echelon form of the augmented matrix of this system is

$$R = \begin{bmatrix} 1 & 0 & 0 & 0 \\ 0 & 1 & 0 & 0 \\ 0 & 0 & 1 & 0 \\ 0 & 0 & 0 & 1 \\ 0 & 0 & 0 & 0 \\ 0 & 0 & 0 & 0 \end{bmatrix},$$

indicating that the system is inconsistent. Thus the matrix equation has no solution, and so the given matrix does not lie in the span of the given set.

2. Yes.

$$\begin{bmatrix} 1 & 2 & 1 \\ 1 & 1 & 1 \end{bmatrix} = 1 \begin{bmatrix} 1 & 2 & 1 \\ 0 & 0 & 0 \end{bmatrix} + 1 \begin{bmatrix} 0 & 0 & 0 \\ 1 & 1 & 1 \end{bmatrix} + 0 \begin{bmatrix} 1 & 0 & 1 \\ 1 & 2 & 3 \end{bmatrix}.$$

3. Yes.

$$\begin{bmatrix} 2 & 2 & 2 \\ 2 & 3 & 4 \end{bmatrix} = 1 \begin{bmatrix} 1 & 2 & 1 \\ 0 & 0 & 0 \end{bmatrix} + 1 \begin{bmatrix} 0 & 0 & 0 \\ 1 & 1 & 1 \end{bmatrix} + 1 \begin{bmatrix} 1 & 0 & 1 \\ 1 & 2 & 3 \end{bmatrix}.$$

4. No.

5. As in Exercise 1, the given matrix lies in the span of the given set if and only if the matrix equation

$$\begin{bmatrix} 2 & 2 & 2 \\ 1 & 1 & 1 \end{bmatrix} = x_1 \begin{bmatrix} 1 & 2 & 1 \\ 0 & 0 & 0 \end{bmatrix} + x_2 \begin{bmatrix} 0 & 0 & 0 \\ 1 & 1 & 1 \end{bmatrix} + x_3 \begin{bmatrix} 1 & 0 & 1 \\ 1 & 2 & 3 \end{bmatrix}$$

has a solution. Comparing the corresponding entries of the right and left sides of this matrix equation yields the system of linear equations

$$\begin{aligned} x_1 + x_3 &= 2 \\ 2x_1 &= 2 \\ x_1 + x_3 &= 2 \\ x_2 + x_3 &= 1 \\ x_2 + 2x_3 &= 1 \\ x_2 + 3x_3 &= 1. \end{aligned}$$

The reduced row echelon form of the augmented matrix of this system is the marix R in the solution to Exercise 1. Therefore the matrix equation has no solution, and so the given matrix does not lie in the span of the given set.

6. Yes. The given matrix is a linear combination if and only if the matrix equation

$$\begin{bmatrix} 1 & 10 & 1 \\ 3 & -1 & -5 \end{bmatrix} = x_1 \begin{bmatrix} 1 & 2 & 1 \\ 0 & 0 & 0 \end{bmatrix} + x_2 \begin{bmatrix} 0 & 0 & 0 \\ 1 & 1 & 1 \end{bmatrix} + x_3 \begin{bmatrix} 1 & 0 & 1 \\ 1 & 2 & 3 \end{bmatrix}$$

196   Chapter 7   Vector Spaces

has a solution. Comparing the corresponding entries of the right and left sides of the matrix equation yields the system of linear equations

$$\begin{aligned} x_1 \phantom{{}+{}} &+ x_3 = 1 \\ 2x_1 \phantom{{}+{}} &\phantom{{}+{}} = 10 \\ x_1 \phantom{{}+{}} &+ x_3 = 1 \\ x_2 &+ x_3 = 3 \\ x_2 &+ 2x_3 = -1 \\ x_2 &+ 3x_3 = -5. \end{aligned}$$

The reduced row echelon form of the augmented matrix of this system is

$$\begin{bmatrix} 1 & 0 & 0 & 5 \\ 0 & 1 & 0 & 7 \\ 0 & 0 & 1 & -4 \\ 0 & 0 & 0 & 0 \\ 0 & 0 & 0 & 0 \\ 0 & 0 & 0 & 0 \end{bmatrix},$$

indicating that the system has the unique solution $x_1 = 5$, $x_2 = 7$, and $x_3 = -4$. Therefore

$$\begin{bmatrix} 1 & 10 & 1 \\ 3 & -1 & -5 \end{bmatrix} = 5 \begin{bmatrix} 1 & 2 & 1 \\ 0 & 0 & 0 \end{bmatrix} + 7 \begin{bmatrix} 0 & 0 & 0 \\ 1 & 1 & 1 \end{bmatrix}$$
$$+ (-4) \begin{bmatrix} 1 & 0 & 1 \\ 1 & 2 & 3 \end{bmatrix}.$$

7. No.

8. No. Suppose that

$$\begin{bmatrix} 1 & 3 & 6 \\ 3 & 5 & 7 \end{bmatrix} = x_1 \begin{bmatrix} 1 & 2 & 1 \\ 0 & 0 & 0 \end{bmatrix} + x_2 \begin{bmatrix} 0 & 0 & 0 \\ 1 & 1 & 1 \end{bmatrix}$$
$$+ x_3 \begin{bmatrix} 1 & 0 & 1 \\ 1 & 2 & 3 \end{bmatrix}.$$

Comparing the $(1, 1)$- and the $(1, 3)$-entries of the right and left sides of the matrix equation yields the system of linear equations

$$\begin{aligned} x_1 + x_3 &= 1 \\ x_1 + x_3 &= 6, \end{aligned}$$

which is obviously inconsistent.

9. Yes.

$$\begin{bmatrix} -2 & -8 & -2 \\ 5 & 7 & 9 \end{bmatrix} = (-4) \begin{bmatrix} 1 & 2 & 1 \\ 0 & 0 & 0 \end{bmatrix} + 3 \begin{bmatrix} 0 & 0 & 0 \\ 1 & 1 & 1 \end{bmatrix}$$
$$+ 2 \begin{bmatrix} 1 & 0 & 1 \\ 1 & 2 & 3 \end{bmatrix}.$$

10. Yes. Consider the equation

$$-3 - x^2 + x^3$$
$$= a(1-x) + b(1+x^2) + c(1+x-x^3)$$
$$= (a+b+c) + (-a+c)x + bx^2 + (-c)x^3.$$

Comparing corresponding coefficients, we obtain the system of linear equations

$$\begin{aligned} a + b + c &= -3 \\ -a \phantom{{}+b{}} + c &= 0 \\ b \phantom{{}+c{}} &= -1 \\ -c &= 1. \end{aligned}$$

The reduced row echelon form of the augmented matrix of this system is

$$\begin{bmatrix} 1 & 0 & 0 & -1 \\ 0 & 1 & 0 & -1 \\ 0 & 0 & 1 & -1 \\ 0 & 0 & 0 & 0 \end{bmatrix},$$

indicating that the system has the unique solution $a = b = c = -1$. Therefore

$$-3 - x^2 + x^3 =$$
$$(-1)(1-x) + (-1)(1+x^2) + (-1)(1+x-x^3).$$

11. No. Consider the equation

$$1 + x + x^2 + x^3$$
$$= a(1-x) + b(1+x^2) + c(1+x-x^3)$$
$$= (a+b+c) + (-a+c)x + bx^2 + (-c)x^3.$$

Comparing corresponding coefficients, we obtain the system of linear equations

$$\begin{aligned} a + b + c &= 1 \\ -a \phantom{{}+b{}} + c &= 1 \\ b \phantom{{}+c{}} &= 1 \\ -c &= 1. \end{aligned}$$

The reduced row echelon form of the augmented matrix of this system is

$$\begin{bmatrix} 1 & 0 & 0 & 0 \\ 0 & 1 & 0 & 0 \\ 0 & 0 & 1 & 0 \\ 0 & 0 & 0 & 1 \end{bmatrix},$$

indicating that the system is inconsistent.

12. No.

13. Yes.

$$-2 + x + x^2 + x^3 =$$
$$(-2)(1-x) + 1(1-x^2) + (-1)(1+x-x^3).$$

14. No.

15. Yes.
$$1-2x-x^2 = 2(1+x)+(-1)(1+x^2)+0(1+x-x^3).$$

16. No. The given matrix is in the span if and only if the matrix equation
$$\begin{bmatrix} 1 & 0 \\ 0 & 1 \end{bmatrix} = x_1 \begin{bmatrix} 1 & 0 \\ -1 & 0 \end{bmatrix} + x_2 \begin{bmatrix} 0 & 1 \\ 0 & 1 \end{bmatrix} + x_3 \begin{bmatrix} 1 & 1 \\ 0 & 0 \end{bmatrix}$$
has a solution. Comparing the right and left sides of the corresponding entries of the matrix equation yields the system of linear equations
$$\begin{aligned} x_1 + x_3 &= 1 \\ x_2 + x_3 &= 0 \\ -x_1 &= 0 \\ x_2 &= 1. \end{aligned}$$
The reduced row echelon form of the this system is
$$\begin{bmatrix} 1 & 0 & 0 & 0 \\ 0 & 1 & 0 & 0 \\ 0 & 0 & 1 & 0 \\ 0 & 0 & 0 & 1 \end{bmatrix},$$
indicating that the system is inconsistent.

17. Yes. The given matrix is in the span if and only if the matrix equation
$$\begin{bmatrix} 1 & 2 \\ -3 & 4 \end{bmatrix} = x_1 \begin{bmatrix} 1 & 0 \\ -1 & 0 \end{bmatrix} + x_2 \begin{bmatrix} 0 & 1 \\ 0 & 1 \end{bmatrix} + x_3 \begin{bmatrix} 1 & 1 \\ 0 & 0 \end{bmatrix}$$
has a solution. Comparing the right and left sides of the corresponding entries of the matrix equation yields the system of linear equations
$$\begin{aligned} x_1 + x_3 &= 1 \\ x_2 + x_3 &= 2 \\ -x_1 &= -3 \\ x_2 &= 4. \end{aligned}$$
The reduced row echelon form of the augmented matrix of this system is
$$\begin{bmatrix} 1 & 0 & 0 & 3 \\ 0 & 1 & 0 & 4 \\ 0 & 0 & 1 & -2 \\ 0 & 0 & 0 & 0 \end{bmatrix},$$
indicating that the system has the unique solution $x_1 = 3$, $x_2 = 4$, and $x_3 = -2$. Therefore
$$\begin{bmatrix} 1 & 2 \\ -3 & 4 \end{bmatrix} = 3\begin{bmatrix} 1 & 0 \\ -1 & 0 \end{bmatrix} + 4\begin{bmatrix} 0 & 1 \\ 0 & 1 \end{bmatrix} + (-2)\begin{bmatrix} 1 & 1 \\ 0 & 0 \end{bmatrix}.$$

18. Yes.
$$\begin{bmatrix} 2 & -1 \\ -1 & -2 \end{bmatrix} = 1\begin{bmatrix} 1 & 0 \\ -1 & 0 \end{bmatrix} + (-2)\begin{bmatrix} 0 & 1 \\ 0 & 1 \end{bmatrix} + 1\begin{bmatrix} 1 & 1 \\ 0 & 0 \end{bmatrix}.$$

19. No.   **20.** No.

21. Yes.
$$\begin{bmatrix} 1 & -2 \\ -3 & 0 \end{bmatrix} = 3\begin{bmatrix} 1 & 0 \\ -1 & 0 \end{bmatrix} + 0\begin{bmatrix} 0 & 1 \\ 0 & 1 \end{bmatrix} + (-2)\begin{bmatrix} 1 & 1 \\ 0 & 0 \end{bmatrix}.$$

For Exercises 22–27, observe that a polynomial is in the span of the given set if and only if it is of the form
$$a(1+x) + b(1+x+x^2) + c(1+x+x^2+x^3)$$
$$= (a+b+c) + (a+b+c)x + (b+c)x^2 + cx^3.$$

22. No. By the equation above, the constant term and the coefficient of the x-term must be equal.

23. Yes. First observe that the constant term and the coefficient of the x-term are equal. For this reason, applying the equation above, we need only compare the constant terms, and the coefficients of the x^2 and x^3 terms to obtain the system of linear equations
$$\begin{aligned} a+b+c &= 3 \\ b+c &= 2 \\ c &= -1. \end{aligned}$$
The reduced row echelon form of augmented matrix of this system is
$$\begin{bmatrix} 1 & 0 & 0 & 1 \\ 0 & 1 & 0 & 3 \\ 0 & 0 & 1 & -1 \end{bmatrix},$$
indicating that the system has the unique solution $a = 1$, $b = 3$, and $c = -1$. Therefore
$$3 + 3x + 2x^2 - x^3 =$$
$$1(1+x) + 3(1+x+x^2) + (-1)(1+x+x^2+x^3).$$

24. Yes.
$$4x^2 - 3x^3 = (-4)(1+x) + 7(1+x+x^2)$$
$$+ (-3)(1+x+x^2+x^3).$$

25. Yes, because any vector in a set lies in the span of that set.

26. No, because the constant term and the coefficient of the x-term are not equal.

27. No, because the constant term and the coefficient of the x-term are not equal.

For Exercises 28–30, observe that a polynomial is in the span of S if and only if it is of the form
$$(9a - 3b) + (4a - 5b)x + (5a - 2b)x^2 + (-3a + b)x^3.$$

28. Comparing the coefficients of the given polynomial with the form above, we obtain the system of linear equations

$$9a - 3b = -6$$
$$4a - 5b = 12$$
$$5a - 3b = -2$$
$$-3a + 3b = 2.$$

The reduced row echelon form of the augmented matrix of this system is

$$\begin{bmatrix} 1 & 0 & -2 \\ 0 & 1 & -4 \\ 0 & 0 & 0 \\ 0 & 0 & 0 \end{bmatrix},$$

indicating that the system has the unique solution $a = -2$ and $b = -4$. Therefore

$$-6 + 12x - 2x^2 + 2x^3 =$$
$$+ (-2)(9 + 4x + 5x^2 - 3x^3)$$
$$+ (-4)(-3 - 5x - 2x^2 + x^3).$$

29. Proceeding as in the solution to Exercise 28, we obtain

$$12 - 13x + 5x^2 - 4x^3 = 3(9 + 4x + 5x^2 - 3x^3)$$
$$+ 5(-3 - 5x - 2x^2 + x^3).$$

30. As above, the given polynomial is a linear combination of the polynomials in S if and only if the system of linear equations

$$9a - 3b = 8$$
$$4a - 5b = 7$$
$$5a - 3b = -2$$
$$-3a + 3b = 3$$

is consistent. The reduced row echelon form of the augmented matrix of this system is

$$\begin{bmatrix} 1 & 0 & 0 \\ 0 & 1 & 0 \\ 0 & 0 & 1 \\ 0 & 0 & 0 \end{bmatrix},$$

indicating that the system is inconsistent.

31. Let $W = \text{Span}\{1+x, 1-x, 1+x^2, 1-x^2\}$. Since $1 = .5(1+x) + .5(1-x)$, it follows that 1 is in W. Since $x = .5(1+x) + (-.5)(1-x)$, it follows that x is in W. Since $x^2 = .5(1+x^2) + (-.5)(1-x^2)$, it follows that x^2 is in W. Since $\{1, x, x^2\}$ is a generating set for $\mathcal{P}_2$ and is contained in W, it follows that $W = \mathcal{P}_2$.

32. Consider any matrix $\begin{bmatrix} a & b \\ c & d \end{bmatrix}$ in $\mathcal{M}_{2\times 2}$. This matrix lies in the span of the given set if and only if the matrix equation

$$\begin{bmatrix} a & b \\ c & d \end{bmatrix} = x_1 \begin{bmatrix} 0 & 1 \\ 1 & 1 \end{bmatrix} + x_2 \begin{bmatrix} 1 & 0 \\ 1 & 1 \end{bmatrix} + x_3 \begin{bmatrix} 1 & 1 \\ 0 & 1 \end{bmatrix}$$
$$+ x_4 \begin{bmatrix} 1 & 1 \\ 1 & 0 \end{bmatrix}$$

has a solution. Comparing the right and left sides of the corresponding entries of the matrix equation yields the system of linear equations

$$\begin{aligned} x_2 + x_3 + x_4 &= a \\ x_1 + x_3 + x_4 &= b \\ x_1 + x_2 + x_4 &= c \\ x_1 + x_2 + x_3 &= d. \end{aligned}$$

Let

$$A = \begin{bmatrix} 0 & 1 & 1 & 1 \\ 1 & 0 & 1 & 1 \\ 1 & 1 & 0 & 1 \\ 1 & 1 & 1 & 0 \end{bmatrix},$$

the coefficient matrix of the system. It can be shown that A is invertible, for example, by showing that its reduced row echelon form is I_4. It follows that the system has a solution for any choice of a, b, c, d, and we conclude that the span of the set is $\mathcal{M}_{2\times 2}$.

33. True

34. False, by Theorem 7.2, the zero vector of a vector space is unique.

35. False, if $a = 0$ and $\mathbf{v} \neq \mathbf{0}$.

36. True

37. False, any two polynomials can be added.

38. False. For example, if $n = 1$, $p(x) = 1 + x$, and $q(x) = 1 - x$, then $p(x)$ and $q(x)$ each have degree 1, but $p(x) + q(x)$ has degree 0.

39. True   40. True   41. True

42. True   43. True

44. False, the empty set contains no zero vector.

45. True   46. True   47. True   48. True

49. True   50. True   51. True

52. True   53. True   54. True

55. Let f, g, and h be in $\mathcal{F}(S)$. Then for any s in S, we have

$$((f + g) + h)(s) = (f + g)(s) + h(s)$$
$$= (f(s) + g(s)) + h(s)$$

$$= f(s) + (g(s) + h(s))$$
$$= f(s) + (g+h)(s)$$
$$= (f + (g+h))(s),$$

and hence $(f+g) + h = f + (g+h)$.

56. Let f be in $\mathcal{F}(S)$, and let $-f$ be the function in $\mathcal{F}(S)$ defined by $(-f)(s) = -f(s)$ for all s in S. Then for any s in S, we have

$$(f + (-f))(s) = f(s) + (-f)(s)$$
$$= f(s) - f(s) = 0 = \mathbf{0}(s),$$

and hence $f + (-f) = \mathbf{0}$.

57. Let f be in $\mathcal{F}(S)$. Then for any s in S, we have $(1f)(s) = 1(f(s)) = f(s)$, and hence $1f = f$.

58. Let f be in $\mathcal{F}(S)$ and a and b be scalars. Then for any s in S, we have

$$[(ab)f](s) = (ab)f(s) = a(bf(s)) = a(bf)(s)$$
$$= [a(bf)](s),$$

and hence $(ab)f = a(bf)$.

59. Let f be in $\mathcal{F}(S)$ and a and b be scalars. Then for any s in S, we have

$$((a+b)f)(s) = (a+b)f(s) = af(s) + bf(s)$$
$$= (af)(s) + (bf)(s)$$
$$= (af + bf)(s),$$

and hence $(a+b)f = af + bf$.

60. Yes. Since every entry of O is 0, the zero matrix is symmetric and hence is in V. Suppose that A and B are in V. The (i,j)-entry of $A+B$ is $a_{ij} + b_{ij} = a_{ji} + b_{ji}$, which is the (j,i)-entry of $A+B$. Hence $A+B$ is in V. So V is closed under addition. Now let c be any scalar. Then the (i,j)-entry of cA is $ca_{ij} = ca_{ji}$, which is the (j,i)-entry of cA. So V is closed under scalar multiplication.

61. No, V is not closed under addition. For example, let $n = 2$, $A = \begin{bmatrix} 1 & 0 \\ 0 & 0 \end{bmatrix}$, and $B = \begin{bmatrix} 0 & 0 \\ 0 & 1 \end{bmatrix}$. Then $\det A = \det B = 0$, and hence A and B are in V, but $\det(A+B) = 1$, and so $A+B$ is not in V.

62. No, V is not closed under scalar multiplication. For example, let $A = I_n$. Then $A^2 = I_n^2 = I_n = A$, and so A is in V. However $(2A)^2 = 4A^2 = 4A \neq 2A$, and hence $2A$ is not in V.

63. Yes. Since $OB = BO = O$, the zero matrix is in V. Now suppose that A and C are in V. Then

$$(A+C)B = AB + CB = BA + BC$$
$$= B(A+C),$$

and hence $A+C$ is in V. So V is closed under addition. Now let c be any scalar. Then

$$(cA)B = c(AB) = c(BA) = B(cA),$$

and hence cA is in V. Therefore V is closed under scalar multiplication.

64. Yes. Setting $a = b = 0$, we see that O is in V. Now suppose that A_1 and A_2 are in V, where

$$A_1 = \begin{bmatrix} a_1 & 2a_1 \\ 0 & b_1 \end{bmatrix} \quad \text{and} \quad A_2 = \begin{bmatrix} a_2 & 2a_2 \\ 0 & b_2 \end{bmatrix}$$

for some scalars a_1, a_2, b_1, and b_2. Then

$$A_1 + A_2 = \begin{bmatrix} a_1 + a_2 & 2(a_1 + a_2) \\ 0 & b_1 + b_2 \end{bmatrix} \text{ is in } V,$$

which is therefore closed under addition. Let $A = \begin{bmatrix} a & 2a \\ 0 & b \end{bmatrix}$ be in V and c be any scalar. Then

$$cA = \begin{bmatrix} ca & 2(ca) \\ 0 & cb \end{bmatrix} \text{ is in } V,$$

and therefore V is closed under scalar multiplication.

65. Yes. Since $O^T = O = -O$, the zero matrix is in V. Suppose that A and B are in V. Then $(A+B)^T = A^T + B^T = -A + (-B) = -(A+B)$, and hence $A+B$ is in V. It follows that V is closed under addition. Let c be any scalar. Then $(cA)^T = cA^T = c(-A) = -(cA)$, and hence cA is in V. Therefore V is closed under scalar multiplication.

66. Yes. Since the zero polynomial is in V, it suffices to show that V is closed under addition and scalar multiplication. Suppose

$$p(x) = a_0 + a_1 x + \cdots + a_m x^m$$
$$q(x) = b_0 + b_1 x + \cdots + b_m x^m$$

are in V, and let $0 \leq k \leq m$ be an odd integer. Since

$$p(x) + q(x) = (a_0 + b_0) + (a_1 + b_1)x + \cdots$$
$$+ (a_m + b_m)x^m$$

and $a_k + b_k = 0 + 0 = 0$, it follows that $p(x) + q(x)$ is in V. Furthermore, for any scalar c,

$$cp(x) = ca_0 + (ca_1)x + \cdots + (ca_m)x^m,$$

and hence $ca_k = c \cdot 0 = 0$. It follows that $cp(x)$ is in V. Therefore V is closed under addition and scalar multiplication.

67. No, V is not closed under addition. Let $m = 4$, $p(x) = 1 + x^2 + x^4$ and $q(x) = 1 - x^2 + x^4$. Both $p(x)$ and $q(x)$ are in V, but $p(x) + q(x) = 2 + 2x^4$ is not in V.

68. No, V is not closed under scalar multiplication. Let $p(x) = 1 + x$. Then $p(x)$ is in V, but $(-1)p(x) = -1 + (-1)x$ is not in V.

69. Yes. Since the zero polynomial is in V, it suffices to show that V is closed under addition and scalar multiplication. Suppose
$$p(x) = a_0 + a_1 x + \cdots + a_m x^m$$
and
$$q(x) = b_0 + b_1 x + \cdots + b_m x^m$$
are in V. Then $a_0 + a_1 = b_0 + b_1 = 0$. Also
$$p(x) + q(x) = (a_0 + b_0) + (a_1 + b_1)x + \cdots + (a_m + b_m)x^m,$$
and hence
$$(a_0 + b_0) + (a_1 + b_1) = (a_0 + a_1) + (b_0 + b_1)$$
$$= 0 + 0 = 0.$$
Therefore $p(x) + q(x)$ is in V, and V is closed under addition. Furthermore, for any scalar c,
$$cp(x) = ca_0 + (ca_1)x + \cdots + (ca_m)x^m,$$
and hence $ca_0 + ca_1 = c(a_0 + a_1) = c \cdot 0 = 0$. So $cp(x)$ is in V, proving that V is closed under scalar multiplication.

70. Yes. The zero function is in V because the images of all the elements of S, including those in S', are 0. Let f and g be in V. Then, for any s in S',
$$(f + g)(s) = f(s) + g(s) = 0 + 0 = 0,$$
and hence $f + g$ is in V. Thus V is closed under addition. Let c be any scalar. Then, for any s in S', $(cf)(s) = c \cdot f(s) = c \cdot 0 = 0$, and hence cf is in V. Thus V is closed under scalar multiplication.

71. Yes. Let $\mathbf{0}$ denote the zero function. Then
$$\mathbf{0}(s_1) + \mathbf{0}(s_2) + \cdots + \mathbf{0}(s_n) = 0 + 0 + \cdots + 0 = 0,$$
and hence the zero function is in V. Suppose that f and g are in V. Then
$$(f + g)(s_1) + (f + g)(s_2) + \cdots + (f + g)(s_n)$$
$$= f(s_1) + f(s_2) + \cdots + f(s_n)$$
$$+ g(s_1) + g(s_2) + \cdots + g(s_n)$$
$$= 0 + 0 = 0,$$
and hence $f + g$ is in V. Thus V is closed under addition. Let c be any scalar. Then
$$(cf)(s_1) + (cf)(s_2) + \cdots + (cf)(s_n)$$
$$= c(f(s_1) + f(s_2) + \cdots + f(s_n))$$
$$= c(0 + 0 + \cdots + 0) = 0,$$
and hence cf is in V. Thus V is closed under scalar multiplication.

72. No, V is not closed under addition. Suppose that $S = \{s_1, s_2\}$ and f and g are defined by
$$f(s_1) = 0, \quad f(s_2) = 1, \quad g(s_1) = 1, \quad \text{and}$$
$$g(s_2) = 0.$$
Then $f(s_1) \cdot f(s_2) = g(s_1) \cdot g(s_2) = 0$, and hence both f and g are in V. However,
$$(f + g)(s_1) \cdot (f + g)(s_2)$$
$$= [f(s_1) + g(s_1)] \cdot [f(s_2) + g(s_2)]$$
$$= 1 \cdot 1 = 1 \neq 0,$$
and hence $f + g$ is not in V.

73. Let f and g be in S. Then
$$(f + g)(1) = f(1) + g(1) = 2 + 2 \neq 2,$$
and hence $f + g$ is not in S.

74. We verify some of the axioms of a vector space.

    Axiom 1 Let T and U be in $\mathcal{L}(\mathcal{R}^n, \mathcal{R}^m)$. Then, for any $\mathbf{v}$ in $\mathcal{R}^n$, we have
    $$(T + U)(\mathbf{v}) = T(\mathbf{v}) + U(\mathbf{v}) = U(\mathbf{v}) + T(\mathbf{v})$$
    $$= (U + T)(\mathbf{v}),$$
    and hence $T + U = U + T$.

    Axiom 3 Let T be in $\mathcal{L}(\mathcal{R}^n)$. Then, for any $\mathbf{v}$ in $\mathcal{R}^n$, we have
    $$(T + T_0)(\mathbf{v}) = T(\mathbf{v}) + T_0(\mathbf{v}) = T(\mathbf{v}) + \mathbf{0}$$
    $$= T(\mathbf{v}),$$
    and hence $T + T_0 = T$.

    Axiom 7 Let T and U be in $\mathcal{L}(\mathcal{R}^n, \mathcal{R}^m)$ and a be a scalar. Then, for any $\mathbf{v}$ in $\mathcal{R}^n$, we have
    $$(a(T + U))(\mathbf{v}) = a((T + U)(\mathbf{v}))$$
    $$= a(T(\mathbf{v}) + U(\mathbf{v}))$$

$$= a(T(\mathbf{v})) + a(U(\mathbf{v}))$$
$$= (aT)(\mathbf{v}) + (aU)(\mathbf{v})$$
$$= (aT + aU)(\mathbf{v}),$$

and hence $a(T + U) = aT + aU$.

75. Let
$$p(x) = a_0 + a_1 x + \cdots + a_n x^n$$
and
$$q(x) = b_0 + b_1 x + \cdots + b_n x^n,$$
be polynomials. With this notation, we verify some of the axioms of a vector space.

Axiom 1 We have

$p(x) + q(x)$
$= (a_0 + b_0) + (a_1 + b_1)x + \cdots + (a_n + b_n)x^n$
$= (b_0 + a_0) + (b_1 + a_1)x + \cdots + (b_n + a_n)x^n$
$= q(x) + p(x).$

Axiom 7 Let c be any scalar. Then

$c(p(x) + q(x))$
$= c(a_0 + b_0) + c(a_1 + b_1)x + \cdots$
$\quad + c(a_n + b_n)x^n$
$= (ca_0 + cb_0) + (ca_1 + cb_1)x + \cdots$
$\quad + (ca_n + cb_n)x^n$
$= (ca_0 + ca_1 x + \cdots + ca_n x^n)$
$\quad + (cb_0 + cb_1 x + \cdots + cb_n x^n)$
$= cp(x) + cq(x).$

76. We verify some of the axioms of a vector space.

    Axiom 1 Let f and g be in V. Then, for any s in S,
    $$(f + g)(s) = f(s) + g(s) = g(s) + f(s)$$
    $$= (g + f)(s),$$
    and hence $f + g = g + f$.

    Axiom 3 Let O be the function defined by $O(s) = \mathbf{0}$ for all s in S. Then, for any f in V and s in S, we have
    $$(f+O)(s) = f(s) + O(s) = f(s) + \mathbf{0} = f(s),$$
    and hence $f + O = f$.

    Axiom 8 Let f be in V and a and b be scalars. Then, for any s in S, we have
    $$[(a+b)f](s) = (a+b) \cdot f(s)$$

$$= af(s) + bf(s)$$
$$= (af)(s) + (bf)(s)$$
$$= (af + bf)(s),$$
and hence $(a + b)f = af + bf$.

77. Since $\mathcal{M}_{2 \times 2}$ is a vector space, it suffices to show that V is a subspace of $\mathcal{M}_{2 \times 2}$. Clearly, the 2×2 zero matrix O is in V. Suppose that A_1 and A_2 are in V, where
$$A_1 = \begin{bmatrix} a_1 & 2a_1 \\ b_1 & -b_1 \end{bmatrix} \quad \text{and} \quad A_2 = \begin{bmatrix} a_2 & 2a_2 \\ b_2 & -b_2 \end{bmatrix}$$
for scalars a_1, a_2, b_1, and b_2. Then
$$A_1 + A_2 = \begin{bmatrix} a_1 + a_2 & 2(a_1 + a_2) \\ b_1 + b_2 & -(b_1 + b_2) \end{bmatrix},$$
which is in V. Similarly, cA is in V for any A in V and any scalar c. Hence V is a subspace of $\mathcal{M}_{2 \times 2}$, and therefore it is a vector space.

78. Since $\mathcal{F}(\mathcal{R})$ is a vector space, it suffices to show that V is a subspace. Clearly, the zero function is in V. Suppose that f and g are in V. For any $t < 0$,
$$(f + g)(t) = f(t) + g(t) = 0 + 0 = 0,$$
and hence $f + g$ is in V. Similarly, cf is in V for any f in V and any scalar c. Hence V is a subspace of $\mathcal{F}(\mathcal{R})$, and therefore it is a vector space.

79. Suppose that $\mathbf{u} + \mathbf{v} = \mathbf{u} + \mathbf{w}$. Then $\mathbf{v} + \mathbf{u} = \mathbf{w} + \mathbf{u}$ by axiom 1, and hence $\mathbf{v} = \mathbf{w}$ by Theorem 7.2(a).

80. By axiom 4, each vector has at least one additive inverse. Suppose that $\mathbf{v}$ and $\mathbf{w}$ are additive inverses of the vector $\mathbf{u}$. Then $\mathbf{u} + \mathbf{v} = \mathbf{u} + \mathbf{w} = \mathbf{0}$, and hence $\mathbf{v} = \mathbf{w}$ by Theorem 7.2(b).

81. By Theorem 7.2(e), $0\mathbf{0} = \mathbf{0}$, and hence, by axiom 6, we have
$$a\mathbf{0} = a(0\mathbf{0}) = (a \cdot 0)\mathbf{0} = 0\mathbf{0} = \mathbf{0}.$$

82. We apply Theorem 7.2(g) and axiom 6 to obtain
$$(-a)\mathbf{v} = ((-1)a)\mathbf{v} = (-1)(a\mathbf{v}) = -(a\mathbf{v})$$
and
$$(-a)\mathbf{v} = (a(-1))\mathbf{v} = a((-1)\mathbf{v}) = a(-\mathbf{v}).$$

83. It follows from axioms 8 and 7, respectively, that
$$(a+b)(\mathbf{u} + \mathbf{v}) = a(\mathbf{u} + \mathbf{v}) + b(\mathbf{u} + \mathbf{v})$$
$$= a\mathbf{u} + a\mathbf{v} + b\mathbf{u} + b\mathbf{v}.$$

84. By Theorem 7.2(g) and axioms 6 and 5, we have
$$-(-\mathbf{v}) = (-1)((-1)\mathbf{v}) = ((-1)(-1))\mathbf{v}$$
$$= 1\mathbf{v} = \mathbf{v}.$$

85. By Theorem 7.2(g) and axiom 7, we have
$$-(\mathbf{u}+\mathbf{v}) = (-1)(\mathbf{u}+\mathbf{v}) = (-1)\mathbf{u} + (-1)\mathbf{v}$$
$$= -\mathbf{u} + (-\mathbf{v}).$$

86. Since $c \neq 0$, it has a multiplicative inverse c^{-1} such that $c^{-1}c = 1$. Thus
$$c^{-1}(c\mathbf{u}) = (c^{-1}c)\mathbf{u} = 1\mathbf{u} = \mathbf{u}.$$
Similarly, $c^{-1}(c\mathbf{v}) = \mathbf{v}$. So if $c\mathbf{u} = c\mathbf{v}$, then $c^{-1}c\mathbf{u} = c^{-1}c\mathbf{v}$, and hence $\mathbf{u} = \mathbf{v}$.

87. $(-c)(-\mathbf{v}) = (-c)((-1)\mathbf{v}) = ((-c)(-1))\mathbf{v} = c\mathbf{v}$

88. Since $O(\mathbf{v}) = \mathbf{0}$, the zero transformation is in V. Suppose that T and U are in V and c is a scalar. Then
$$(T+U)(\mathbf{v}) = T(\mathbf{v}) + U(\mathbf{v}) = \mathbf{0} + \mathbf{0} = \mathbf{0},$$
and hence $T+U$ is in V. Hence V is closed under addition. Furthermore, $(cT)(\mathbf{v}) = c(T(\mathbf{v})) = c\mathbf{0} = \mathbf{0}$, and so cT lies in V. Thus V is closed under scalar multiplication, and we conclude that V is a subspace.

89. The zero function is differentiable and hence lies in W. Since the sum of two differentiable functions is differentiable and any scalar multiple of a differentiable function is differentiable, W is closed under addition and scalar multiplication. Therefore W is a subspace.

90. Since $0' = 0$, the zero function is in S. Suppose that f and g are in S. Then
$$(f+g)' = f' + g' = f + g,$$
and hence $f+g$ is in S. Thus S is closed under addition. Let c be any scalar. Then $(cf)' = cf' = cf$, and hence cf is in S. It follows that S is closed under scalar multiplication, and therefore S is a subspace.

91. (a) Since $\mathbf{0}(t) = \mathbf{0}(-t) = 0$, the zero function is even. Suppose that f and g are even functions. Then
$$(f+g)(t) = f(t) + g(t) = f(-t) + g(-t)$$
$$= (f+g)(-t),$$
and hence $f+g$ is even. Furthermore, for any scalar a,
$$(af)(t) = a(f(t)) = a(f(-t)) = (af)(-t),$$
and hence af is even. Thus the subset of even functions is closed under addition and scalar multiplication. Therefore this set is a subspace.

(b) Since $\mathbf{0}(-t) = -\mathbf{0}(t) = 0$, the zero function is odd. Suppose that f and g are odd functions. Then
$$(f+g)(-t) = f(-t) + g(-t)$$
$$= -f(t) - g(t) = (-f-g)(t)$$
$$= (-(f+g))(t),$$
and hence $f+g$ is odd. Furthermore, for any scalar a,
$$(af)(-t) = a(f(-t)) = a(-f(t))$$
$$= -(af)(t),$$
and hence af is odd. Thus the subset of odd functions is closed under addition and scalar multiplication. Therefore this set is a subspace.

92. (a) The zero function is continuous, the sum of continuous functions is continuous, and every scalar multiple of a continuous function is continuous. Thus V contains the zero function and is closed under both addition and scalar multiplication. Therefore V is a subspace.

(b) Observe that $\int_0^1 \mathbf{0}(t)\,dt = \int_0^1 0\,dt = 0$, and hence the zero function is in W. Suppose that f and g are functions in W. Then
$$\int_0^1 (f+g)(t)\,dt = \int_0^1 f(t)\,dt + \int_0^1 g(t)\,dt$$
$$= 0 + 0 = 0,$$
and hence $f+g$ is in W. Thus W is closed under addition. Furthermore, for any scalar a,
$$\int_0^1 (af)(t)\,dt = a\int_0^1 f(t)\,dt = a \cdot 0 = 0,$$
and hence af is in W. Thus W is closed under scalar multiplication. Therefore W is a subspace.

93. Since $\mathbf{0}$ is in both W_1 and W_2, the zero vector lies in $W_1 \cap W_2$. Suppose that $\mathbf{u}$ and $\mathbf{v}$ are in $W_1 \cap W_2$. Since $\mathbf{u}$ and $\mathbf{v}$ are in W_1 and W_1 is a subspace of V, the sum $\mathbf{u}+\mathbf{v}$ is in W_1. Similarly, $\mathbf{u}+\mathbf{v}$ is in W_2. Hence $\mathbf{u}+\mathbf{v}$ is in $W_1 \cap W_2$. Thus $W_1 \cap W_2$ is closed under addition. Furthermore, for any scalar a, the scalar multiple $a\mathbf{u}$ lies in

W_1, a subspace of V. Similarly, $a\mathbf{u}$ lies in W_2. Thus $a\mathbf{u}$ is in $W_1 \cap W_2$, and hence $W_1 \cap W_2$ is closed under scalar multiplication. Therefore $W_1 \cap W_2$ is a subspace of V.

94. Since $\mathbf{0}$ is in both W_1 and W_2 and $\mathbf{0} = \mathbf{0} + \mathbf{0}$, it follows that $\mathbf{0}$ is in W. Now suppose that $\mathbf{u}$ and $\mathbf{v}$ are in W. Then $\mathbf{u} = \mathbf{u}_1 + \mathbf{u}_2$ and $\mathbf{v} = \mathbf{v}_1 + \mathbf{v}_2$ for some vectors $\mathbf{u}_1$ and $\mathbf{v}_1$ in W_1 and $\mathbf{u}_2$ and $\mathbf{v}_2$ in W_2. Hence

$$\mathbf{u} + \mathbf{v} = (\mathbf{u}_1 + \mathbf{u}_2) + (\mathbf{v}_1 + \mathbf{v}_2)$$
$$= (\mathbf{u}_1 + \mathbf{v}_1) + (\mathbf{u}_2 + \mathbf{v}_2),$$

which lies in W because $\mathbf{u}_1 + \mathbf{v}_1$ lies in W_1 and $\mathbf{u}_2 + \mathbf{v}_2$ lies in W_2. Thus W is closed under addition. Furthermore, for any scalar a,

$$a\mathbf{u} = a(\mathbf{u}_1 + \mathbf{u}_2) = a\mathbf{u}_1 + a\mathbf{u}_2,$$

which lies in W because $a\mathbf{u}_1$ lies in W_1 and $a\mathbf{u}_2$ lies in W_2. Thus W is closed under scalar multiplication. Therefore W is a subspace of V.

95. Suppose that W is a subspace of V. Then (i) is satisfied. Let $\mathbf{w}_1$ and $\mathbf{w}_2$ be in W, and let a be a scalar. Then $a\mathbf{w}_1$ is in W because W is closed under scalar multiplication, and hence $a\mathbf{w}_1 + \mathbf{w}_2$ is in W because W is closed under addition. Therefore (ii) is satisfied.

    Conversely, suppose conditions (i) and (ii) are satisfied. By (i), the zero vector lies in V. Let $\mathbf{w}_1$ and $\mathbf{w}_2$ be in W. Then $\mathbf{w}_1 + \mathbf{w}_2 = 1 \cdot \mathbf{w}_1 + \mathbf{w}_2$, which is in W by (ii). Hence W is closed under addition. Furthermore, for any scalar a, $a\mathbf{w}_1 = a\mathbf{w}_1 + \mathbf{0}$, which is in W by (ii). Hence W is closed under scalar multiplication. Therefore W is a subspace of V.

96. The set W satisfies vector space axioms 1, 2, 5, 6, 7, and 8 because the vectors in W are also in V, a vector space which also satisfies these axioms. Also, W satisfies axiom 3, since it is required as part of the definition of *subspace*. So we need only verify axiom 4. Let $\mathbf{u}$ be any vector in W. Then $(-1)\mathbf{u}$ is in W because W is a subspace of V. Therefore W is a vector space.

97. Clearly W satisfies the second part of the definition of *subspace* since W is closed under the operations of addition and scalar multiplication. So we need only verify the first part of the definition. Since W is a vector space, it has a zero vector, $\mathbf{0}'$. Let $\mathbf{0}$ denote the zero vector of V. We show that $\mathbf{0}' = \mathbf{0}$. Since W is a vector space with zero vector equal to $\mathbf{0}'$, we have $0\mathbf{0}' = \mathbf{0}'$ by Theorem 7.2(e). Since V is a vector space with zero vector equal to $\mathbf{0}$, we have $0\mathbf{0}' = \mathbf{0}$ by Theorem 7.2(e). Therefore $\mathbf{0}' = \mathbf{0}$.

7.2 LINEAR TRANSFORMATIONS

1. Yes. First observe that the matrix $C = \begin{bmatrix} 1 & 2 \\ 3 & 4 \end{bmatrix}$ is invertible. Suppose that $T(A) = AC = O$. Then $A = OC^{-1} = O$. Therefore T is one-to-one by Theorem 7.5.

2. No. $U\left(\begin{bmatrix} 1 & 0 \\ 0 & -1 \end{bmatrix}\right) = 0$, and hence U is not one-to-one by Theorem 7.5.

3. No. Since $T\left(\begin{bmatrix} 0 & 1 \\ 0 & 2 \end{bmatrix}\right) = \begin{bmatrix} 0 & 1 \\ 0 & 2 \end{bmatrix}\begin{bmatrix} 1 \\ 0 \end{bmatrix} = \begin{bmatrix} 0 \\ 0 \end{bmatrix}$, T is not one-to-one by Theorem 7.5.

4. No. Let $f(x) = 1 - 2x + x^2$. Then $f(1) = f'(1) = 0$, and hence $U(f(x)) = \begin{bmatrix} 0 \\ 0 \end{bmatrix}$. Therefore U is not one-to-one by Theorem 7.5.

5. No. $T(1) = x \cdot 0 = 0$, and hence T is not one-to-one by Theorem 7.5.

6. Yes. Suppose that $f(x) = a + bx + cx^2$. Then

$$T(f(x)) = (a + bx + cx^2) + (b + 2cx)$$
$$= (a + b) + (b + 2c)x + cx^2 = 0.$$

Since the system of linear equations

$$\begin{aligned} a + b &= 0 \\ b + 2c &= 0 \\ c &= 0 \end{aligned}$$

has only the trivial solution $a = b = c = 0$, it follows that T is one-to-one by Theorem 7.5.

7. Yes. Suppose that

$$U\left(\begin{bmatrix} s \\ t \\ u \end{bmatrix}\right) = \begin{bmatrix} s & t \\ t & u \end{bmatrix} = \begin{bmatrix} 0 & 0 \\ 0 & 0 \end{bmatrix}.$$

Then $s = t = u = 0$, and hence U is one-to-one by Theorem 7.5.

8. No. Let $\mathbf{v} = \begin{bmatrix} 1 \\ 3 \end{bmatrix}$. Then

$$U(\mathbf{v}) = \det\begin{bmatrix} 1 & 1 \\ 3 & 3 \end{bmatrix} = 3 - 3 = 0,$$

and hence U is not one-to-one by Theorem 7.5.

9. Yes. Let $C = \begin{bmatrix} 1 & 2 \\ 3 & 4 \end{bmatrix}$. Then for any matrix A in $\mathcal{M}_{2\times 2}$, $T(AC^{-1}) = AC^{-1}C = A$, and hence A is in the range of T.

10. Yes. Let a be any real number and $A = \begin{bmatrix} a & 0 \\ 0 & 0 \end{bmatrix}$. Then $U(A) = a$, and hence a is in the range of U.

11. Yes. For any vector $\mathbf{v} = \begin{bmatrix} v_1 \\ v_2 \end{bmatrix}$ in $\mathcal{R}^2$, we have

$$T\left(\begin{bmatrix} v_1 & 0 \\ v_2 & 0 \end{bmatrix}\right) = \begin{bmatrix} v_1 & 0 \\ v_2 & 0 \end{bmatrix} \begin{bmatrix} 1 \\ 0 \end{bmatrix} = \begin{bmatrix} v_1 \\ v_2 \end{bmatrix} = \mathbf{v},$$

and hence $\mathbf{v}$ is in the range of T.

12. Yes. The image of an arbitrary polynomial $f(x) = a + bx + cx^2$ in $\mathcal{P}_2$ is

$$T(f(x)) = \begin{bmatrix} f(1) \\ f'(1) \end{bmatrix} = \begin{bmatrix} a+b+c \\ b+2c \end{bmatrix}$$

$$= a \begin{bmatrix} 1 \\ 0 \end{bmatrix} + b \begin{bmatrix} 1 \\ 1 \end{bmatrix} + c \begin{bmatrix} 1 \\ 2 \end{bmatrix}.$$

It follows that the range of T is the span of the set

$$\left\{ \begin{bmatrix} 1 \\ 0 \end{bmatrix}, \begin{bmatrix} 1 \\ 1 \end{bmatrix}, \begin{bmatrix} 1 \\ 2 \end{bmatrix} \right\},$$

which is equal to $\mathcal{P}_2$.

13. No. The constant polynomial 1 is not the product of x and any polynomial.

14. Yes. Consider an arbitrary polynomial $f(x) = a + bx + cx^2$. Then, as in Exercise 6, $T(f(x)) = (a + b) + (b + 2c)x + cx^2$. Now consider any polynomial $g(x) = p + qx + rx^2$ in $\mathcal{P}_2$. The requirement that $g(x)$ is in the range of T is that the system of linear equations

$$\begin{aligned} a + b & = p \\ b + 2c & = q \\ c & = r \end{aligned}$$

has a solution. The coefficient matrix

$$\begin{bmatrix} 1 & 1 & 0 \\ 0 & 1 & 2 \\ 0 & 0 & 1 \end{bmatrix}$$

of this system is invertible. So the system is consistent, and hence $g(x)$ is in the range of T.

15. No. A matrix A lies in the range of U if and only if $a_{12} = a_{21}$. So, for example, the matrix $\begin{bmatrix} 1 & 2 \\ 3 & 4 \end{bmatrix}$ is not in the range of U.

16. Yes. Let a be any real number and $\mathbf{v} = \begin{bmatrix} 0 \\ -a \end{bmatrix}$. Then $U(\mathbf{v}) = \det \begin{bmatrix} 0 & 1 \\ -a & 3 \end{bmatrix} = a$, and hence a is in the range of U.

17. Let $C = \begin{bmatrix} 1 & 2 \\ 3 & 4 \end{bmatrix}$. For any A and B in $\mathcal{M}_{2 \times 2}$, and for any scalar s,

$$T(A + B) = (A + B)C = AC + BC$$

$$= T(A) + T(B)$$

and

$$T(sA) = (sA)C = s(AC) = sT(A).$$

Therefore T is linear.

18. This is Exercise 82 of Section 1.1.

19. For any A and B in $\mathcal{M}_{2 \times 2}$,

$$T(A + B) = (A + B)\mathbf{e}_1 = A\mathbf{e}_1 + B\mathbf{e}_1$$
$$= T(A) + T(B).$$

Similarly, $T(cA) = cT(A)$ for any scalar c, and therefore T is linear.

20. For any $f(x)$ and $g(x)$ in $\mathcal{P}_2$,

$$U(f(x) + g(x)) = \begin{bmatrix} (f+g)(1) \\ (f+g)'(1) \end{bmatrix}$$
$$= \begin{bmatrix} f(1) + g(1) \\ f'(1) + g'(1) \end{bmatrix}$$
$$= \begin{bmatrix} f(1) \\ f'(1) \end{bmatrix} + \begin{bmatrix} g(1) \\ g'(1) \end{bmatrix}$$
$$= U(f(x)) + U(g(x)).$$

Similarly, $U(cf(x)) = cU(f(x))$ for any scalar c, and therefore U is linear.

21. For any $f(x)$ and $g(x)$ in $\mathcal{P}_2$,

$$T(f(x) + g(x)) = x(f(x) + g(x))'$$
$$= x(f'(x) + g'(x))$$
$$= xf'(x) + xg'(x)$$
$$= T(f(x)) + T(g(x)).$$

Similarly, $T(cf(x)) = cT(f(x))$ for any scalar c, and therefore T is linear.

22. For any $f(x)$ and $g(x)$ in $\mathcal{P}_2$,

$$T(f(x) + g(x))$$
$$= (f(x) + g(x)) + (f(x) + g(x))'$$
$$= f(x) + g(x) + f'(x) + g'(x)$$
$$= (f(x) + f'(x)) + (g(x) + g'(x))$$
$$= T(f(x)) + T(g(x)).$$

Similarly, $T(cf(x)) = cT(f(x))$ for any scalar c, and therefore T is linear.

23. For any vectors $\mathbf{u}$ and $\mathbf{v}$ in $\mathcal{R}^3$,

$$U(\mathbf{u} + \mathbf{v}) = U\left(\begin{bmatrix} u_1 + v_1 \\ u_2 + v_2 \\ u_3 + v_3 \end{bmatrix}\right)$$

$$= \begin{bmatrix} u_1 + v_1 & u_2 + v_2 \\ u_2 + v_2 & u_3 + v_3 \end{bmatrix}$$

$$= \begin{bmatrix} u_1 & u_2 \\ u_2 & u_3 \end{bmatrix} + \begin{bmatrix} v_1 & v_2 \\ v_2 & v_3 \end{bmatrix}$$

$$= U(\mathbf{u}) + U(\mathbf{v}).$$

Similarly, $U(c\mathbf{u}) = cU(\mathbf{u})$ for any scalar c, and therefore U is linear.

24. For any vectors $\mathbf{u}$ and $\mathbf{v}$ in $\mathcal{R}^2$,

$$U(\mathbf{u} + \mathbf{v}) = \det \begin{bmatrix} u_1 + v_1 & 1 \\ u_2 + v_2 & 3 \end{bmatrix}$$

$$= 3(u_1 + v_1) - (u_2 + v_2)$$
$$= (3u_1 - u_2) + (3v_1 - v_2)$$
$$= \det \begin{bmatrix} u_1 & 1 \\ u_2 & 3 \end{bmatrix} + \det \begin{bmatrix} v_1 & 1 \\ v_2 & 3 \end{bmatrix}$$
$$= U(\mathbf{u}) + U(\mathbf{v}).$$

Similarly, $U(c\mathbf{u}) = cU(\mathbf{u})$ for any scalar c, and therefore U is linear.

25.
$$UT\left(\begin{bmatrix} a & b \\ c & d \end{bmatrix}\right) = U\left(\begin{bmatrix} a & b \\ c & d \end{bmatrix}\begin{bmatrix} 1 & 2 \\ 3 & 4 \end{bmatrix}\right)$$
$$= U\left(\begin{bmatrix} a+3b & 2a+4b \\ c+3d & 2c+4d \end{bmatrix}\right)$$
$$= \text{trace}\left(\begin{bmatrix} a+3b & 2a+4b \\ c+3d & 2c+4d \end{bmatrix}\right)$$
$$= a + 3b + 2c + 4d.$$

26.
$$UT(a + bx + cx^2) = U(x(b + 2cx))$$
$$= U(bx + 2cx^2) = \begin{bmatrix} b+2c \\ b+4c \end{bmatrix}.$$

27.
$$UT(a + bx + cx^2)$$
$$= U((a + bx + cx^2) + (b + 2cx))$$
$$= U((a + b) + (b + 2c)x + cx^2)$$
$$= \begin{bmatrix} a+b+b+2c+c \\ b+2c+2c \end{bmatrix}$$
$$= \begin{bmatrix} a+2b+3c \\ b+4c \end{bmatrix}.$$

28.
$$UT\left(\begin{bmatrix} a & b \\ c & d \end{bmatrix}\right) = U\left(\begin{bmatrix} a & b \\ c & d \end{bmatrix}\begin{bmatrix} 1 \\ 0 \end{bmatrix}\right) = U\left(\begin{bmatrix} a \\ b \end{bmatrix}\right)$$
$$= \det\begin{bmatrix} a & 1 \\ b & 3 \end{bmatrix} = 3a - b.$$

29.
$$TU\left(\begin{bmatrix} s \\ t \\ u \end{bmatrix}\right) = T\left(\begin{bmatrix} s & t \\ t & u \end{bmatrix}\right) = \begin{bmatrix} s & t \\ t & u \end{bmatrix}^T$$
$$= \begin{bmatrix} s & t \\ t & u \end{bmatrix}.$$

30. T is not linear if $n > 1$. Let A be any $n \times n$ invertible matrix. Then

$$T(2A) = \det(2A) = 2^n \det A = 2^n T(A)$$
$$\neq 2T(A)$$

because $T(A) = \det A \neq 0$.

31. T is linear but is not an isomorphism.
To prove that T is linear, let $f(x)$ and $g(x)$ be in $\mathcal{P}$. Then

$$T(f(x) + g(x)) = x(f(x) + g(x))$$
$$= xf(x) + xg(x)$$
$$= T(f(x)) + T(g(x)),$$

and hence T preserves addition. Also, for any scalar c,

$$T((cf)(x)) = x(cf)(x) = c(xf(x)) = cT(f(x)).$$

So T preserves scalar multiplication. Thus T is linear.

To show that T is not an isomorphism, we show that T is not onto. For any nonzero polynomial $f(x)$, $T(f(x)) = xf(x)$ has degree greater than zero and hence $T(f(x)) \neq 1$. Therefore the constant polynomial 1 is not in the range of T.

32. T is both linear and an isomorphism. To see that T is linear, let $f(x)$ and $g(x)$ be in $\mathcal{P}_2$. Then

$$T((f(x) + g(x))) = \begin{bmatrix} f(0) + g(0) \\ f(1) + g(1) \\ f(2) + g(2) \end{bmatrix}$$
$$= \begin{bmatrix} f(0) \\ f(1) \\ f(2) \end{bmatrix} + \begin{bmatrix} g(0) \\ g(1) \\ g(2) \end{bmatrix}$$
$$= T(f(x)) + T(g(x)).$$

Similarly $T(cf(x)) = cT(f(x))$ for any scalar c.
To show that T is an isomorphism, consider any polynomial $f(x) = a + bx + cx^2$ in $\mathcal{P}_2$ and suppose that $T(f(x)) = \mathbf{0}$. Then $f(0) = f(1) = f(2) = 0$, and hence

$$\begin{aligned} a &= 0 \\ a + b + c &= 0 \\ a + 2b + 4c &= 0. \end{aligned}$$

The coefficient matrix of this system of linear equations

$$A = \begin{bmatrix} 1 & 0 & 0 \\ 1 & 1 & 1 \\ 1 & 2 & 4 \end{bmatrix}$$

is invertible, and hence the only solution of this system is $a = b = c = 0$. It follows that $f(x)$ is the zero polynomial, and hence T is one-to-one by Theorem 7.5.

To show that T is onto, consider any polynomial $g(x) = p + qx + rx^2$ in $\mathcal{P}_2$. Then $g(x)$ is in the range of T if and only if the system of linear equations

$$\begin{aligned} a &= p \\ a + b + c &= q \\ a + 2b + 4c &= r \end{aligned}$$

has a solution. But this is guaranteed because the coefficient matrix A is invertible. We conclude that T is onto, and hence T is an isomorphism.

33. T is not linear. Let $f(x)$ be any nonzero polynomial. Then

$$T(2f(x)) = (2f(x))^2 = 4(f(x))^2 \neq 2T(f(x)).$$

34. T is linear but is not an isomorphism. Let A and B be in $\mathcal{M}_{2\times 2}$. Then

$$T(A + B) = \begin{bmatrix} 1 & 1 \\ 1 & 1 \end{bmatrix}(A+B)$$
$$= \begin{bmatrix} 1 & 1 \\ 1 & 1 \end{bmatrix}A + \begin{bmatrix} 1 & 1 \\ 1 & 1 \end{bmatrix}B$$
$$= T(A) + T(B).$$

Similarly, $T(cA) = cT(A)$ for any scalar c, and hence T is linear. However, T is not one-to-one because

$$T\left(\begin{bmatrix} 1 & 1 \\ -1 & -1 \end{bmatrix}\right) = \begin{bmatrix} 1 & 1 \\ 1 & 1 \end{bmatrix}\begin{bmatrix} 1 & 1 \\ -1 & -1 \end{bmatrix}$$
$$= \begin{bmatrix} 0 & 0 \\ 0 & 0 \end{bmatrix}.$$

35. T is both linear and an isomorphism. Let f and g be in $\mathcal{F}(\mathcal{R})$. Then, for any x in $\mathcal{R}$,

$$T(f + g)(x) = (f + g)(x + 1)$$
$$= f(x+1) + g(x+1)$$
$$= T(f)(x) + T(g)(x),$$

and hence $T(f + g) = T(f) + T(g)$. Similarly, $T(cf) = cT(f)$ for any scalar c, and hence T is linear. To show that T is one-to-one, suppose that $T(f)$ is the zero function, that is, $T(f)(x) = f(x+1) = 0$ for all x. Then for any x in $\mathcal{R}$, we have $f(x) = f((x-1)+1) = 0$, and hence f is the zero function. Thus T is one-to-one. To show T is onto, let f be any function in $\mathcal{F}(\mathcal{R})$ and g be the function in $\mathcal{F}(\mathcal{R})$ defined by $g(x) = f(x-1)$ for all x in $\mathcal{R}$. Then for all x in $\mathcal{R}$, $T(g)(x) = g(x+1) = f(x-1+1) = f(x)$. Hence $T(g) = f$, and we conclude that T is onto.

36. T is linear, but not an isomorphism. Let f and g be in $\mathcal{D}(\mathcal{R})$. Then

$$T(f+g) = (f+g)' = f' + g' = T(f) + T(g).$$

Similarly, $T(cf) = cT(f)$ for any scalar c, and hence T is linear. However, T is not one-to-one. Let f be the constant function 1. Then $T(f) = f' = 0$. Thus T is not one-to-one by Theorem 7.5. We conclude that T is not an isomorphism.

37. T is linear, but not an isomorphism. Let f and g be in $\mathcal{D}(\mathcal{R})$. Then

$$T(f+g) = \int_0^1 (f(t) + g(t))\, dt$$
$$= \int_0^1 f(t)\, dt + \int_0^1 g(t)\, dt$$
$$= T(f) + T(g).$$

Similarly, $T(cf) = cT(f)$ for any scalar c, and hence T is linear. However, T is not one-to-one. Let $f(t) = 2t - 1$. Then

$$T(f) = \int_0^1 (2t-1)\, dt = t^2 - t\Big|_0^1 = 0 - 0 = 0,$$

and hence T is not one-to-one by Theorem 7.5.

38. Let f and g be in $\mathcal{F}(S)$. Then

$$T(f+g) = \begin{bmatrix} (f+g)(s_1) \\ (f+g)(s_2) \\ \vdots \\ (f+g)(s_n) \end{bmatrix} = \begin{bmatrix} f(s_1) + g(s_1) \\ f(s_2) + g(s_2) \\ \vdots \\ f(s_n) + g(s_n) \end{bmatrix}$$

$$= \begin{bmatrix} f(s_1) \\ f(s_2) \\ \vdots \\ f(s_n) \end{bmatrix} + \begin{bmatrix} g(s_1) \\ g(s_2) \\ \vdots \\ g(s_n) \end{bmatrix} = T(f) + T(g).$$

Similarly, $T(cf) = cT(f)$ for any scalar c. Therefore T is linear.

To show that T is an isomorphism, we prove that T is one-to-one and onto. Suppose that $T(f) = \mathbf{0}$. Then $f(s_i) = 0$ for all i, and hence f is the zero function. Thus T is one-to-one. Consider any vector $\mathbf{v}$ in $\mathcal{R}^n$. Let $f\colon S \to \mathcal{R}^n$ be defined by $f(s_i) = v_i$ for all i. Then $T(f) = \mathbf{v}$, and we conclude that T is onto.

39. True
40. False, it may fail to be onto.
41. True   42. True   43. True
44. False, all polynomials are in C^∞.
45. True
46. False, the definite integral of a function in $\mathsf{C}([a,b])$ is a real number.
47. True
48. False, the zero function is not in the solution set.
49. *Proof of (a)*: The zero vector of $\mathcal{F}(\mathcal{N})$ is the function for which the image of every nonnegative integer is zero. Clearly this function is in V. Let f and g be in V. Then f is nonzero at only finitely many nonnegative integers $a_1, a_2, \ldots, a_r$, and g is nonzero at only finitely many nonnegative integers $b_1, b_2, \ldots, b_s$. Then $f+g$ is zero except possibly at the finitely many nonnegative integers $a_1, a_2, \ldots, a_r, b_1, b_2, \ldots, b_s$. So $f+g$ is in V. Finally, for any scalar c, the function cf can be nonzero at only $a_1, a_2, \ldots, a_r$; so cf is in V. It follows that V is a subspace of $\mathcal{F}(\mathcal{N})$.

Proof of (b): Let f and g be in V, and let n be a positive integer such that $f(k) = g(k) = 0$ for $k > n$. Then

$$T(f+g) = (f+g)(0) + (f+g)(1)x + \cdots + (f+g)(n)x^n$$
$$= [f(0) + f(1)x + \cdots + f(n)x^n]$$
$$\quad + [g(0) + g(1)x + \cdots + g(n)x^n]$$
$$= T(f) + T(g).$$

Similarly, $T(cf) = cT(f)$ for any scalar c, and hence T is linear.

We now show that T is an isomorphism. Suppose that $T(f) = 0$, the zero polynomial. Then $f(k) = 0$ for all k, and hence f is the zero function. So T is one-to-one. Now consider any polynomial $p(x) = a_0 + a_1 x + \cdots + a_n x^n$. Let $f\colon V \to \mathcal{R}$ be defined by

$$f(k) = \begin{cases} a_k & \text{if } k \leq n \\ 0 & \text{if } k > n. \end{cases}$$

Then $T(f) = p(x)$, and hence T is onto. Therefore T is an isomorphism.

50. Let $T\colon V \to \mathcal{P}_2$ be defined by

$$T\left(\begin{bmatrix} a & b \\ c & -a \end{bmatrix}\right) = a + bx + cx^2.$$

Consider the matrices

$$A_1 = \begin{bmatrix} a_1 & b_1 \\ c_1 & -a_1 \end{bmatrix} \quad \text{and} \quad A_2 = \begin{bmatrix} a_2 & b_2 \\ c_2 & -a_2 \end{bmatrix}$$

in V. Then

$$T(A_1 + A_2) = T\left(\begin{bmatrix} a_1 + a_2 & b_1 + b_2 \\ c_1 + c_2 & -(a_1 + a_2) \end{bmatrix}\right)$$
$$= (a_1 + a_2) + (b_1 + b_2)x$$
$$\quad + (c_1 + c_2)x^2$$
$$= (a_1 + b_1 x + c_1 x^2)$$
$$\quad + (a_2 + b_2 x + c_2 x^2)$$
$$= T(A_1) + T(A_2).$$

Similarly, $T(cA) = cT(A)$ for all A in V, and hence T is linear.

We now show that T is an isomorphism. Suppose that

$$T(A) = T\left(\begin{bmatrix} a & b \\ c & -a \end{bmatrix}\right) = a + bx + cx^2 = 0,$$

the zero polynomial. Then $a = b = c = 0$, and hence $A = O$. Thus T is one-to-one. Now consider any polynomial $a + bx + cx^2$ in $\mathcal{P}_2$. Then

$$T\left(\begin{bmatrix} a & b \\ c & -a \end{bmatrix}\right) = a + bx + cx^2,$$

and hence T is onto. Therefore T is an isomorphism.

51. (a) Since the zero polynomial can be written as $0x^4 + 0x^2 + 0$, it is in V. Let $p(x) = a_1 x^4 + b_1 x^2 + c_1$ and $q(x) = a_2 x^4 + b_2 x^2 + c_2$ be in V. Then

$$p(x) + q(x) = (a_1 + a_2)x^4 + (b_1 + b_2)x^2$$

$$+ (c_1 + c_2),$$

which is in V. Thus V is closed under addition. Similarly, V is closed under scalar multiplication, and hence V is a subspace of $\mathcal{P}_4$.

(b) Let $T\colon V \to \mathcal{P}_2$ be defined by

$$T(ax^4 + bx^2 + c) = ax^2 + bx + c.$$

We show that T is linear. Let $p(x) = a_1 x^4 + b_1 x^2 + c_1$ and $q(x) = a_2 x^4 + b_2 x^2 + c_2$ be in V. Then

$$\begin{aligned}T(p(x) + q(x))\\ &= T((a_1 + a_2)x^4 + (b_1 + b_2)x^2 + (c_1 + c_2))\\ &= (a_1 + a_2)x^2 + (b_1 + b_2)x + (c_1 + c_2)\\ &= (a_1 x^2 + b_1 x + c_1) + (a_2 x^2 + b_2 x + c_2)\\ &= T(p(x)) + T(q(x)).\end{aligned}$$

Similarly, $T(cp(x)) = cT(p(x))$ for any $p(x)$ in V and any scalar c. Therefore T is linear. The proof that T is one-to-one and onto is similar to the proof given in Exercise 50.

52. Consider any vector $\mathbf{v}$ in V, and suppose that $T(\mathbf{v}) = \mathbf{w}$. Then $\mathbf{v} = T^{-1}(\mathbf{w})$. Hence $\mathbf{w} = (T^{-1})^{-1}(\mathbf{v})$. Thus $(T^{-1})^{-1}(\mathbf{v}) = T(\mathbf{v})$. Since this equation is valid for every vector $\mathbf{v}$ is V, we conclude that $(T^{-1})^{-1} = T$.

53. Suppose that V, W, and Z be vector spaces and $T\colon V \to W$ and $U\colon W \to Z$ be isomorphisms. Since T and U are both one-to-one and onto, UT is one-to-one and onto. Consider any vectors $\mathbf{u}$ and $\mathbf{v}$ in V. Then

$$\begin{aligned}UT(\mathbf{u} + \mathbf{v}) &= U(T(\mathbf{u} + \mathbf{v}))\\ &= U(T(\mathbf{u}) + T(\mathbf{v}))\\ &= UT(\mathbf{u}) + UT(\mathbf{v}).\end{aligned}$$

Similarly, $UT(c\mathbf{u}) = cUT(\mathbf{u})$ for any scalar c. Therefore UT is linear, and hence it is an isomorphism.

Let $\mathbf{z}$ be in Z. Then

$$\begin{aligned}UT(T^{-1}U^{-1}(\mathbf{z})) &= U(TT^{-1})U^{-1}(\mathbf{z})\\ &= UU^{-1}(\mathbf{z})\\ &= \mathbf{z},\end{aligned}$$

and hence $T^{-1}U^{-1}(\mathbf{z}) = (UT)^{-1}(\mathbf{z})$. We conclude that $(UT)^{-1} = T^{-1}U^{-1}$.

54. First observe that

$$\begin{aligned}(T_1 T_2 \cdots T_k)^{-1} &= [T_1(T_2 \cdots T_k)]^{-1}\\ &= (T_2 \cdots T_k)^{-1} T_1^{-1}\end{aligned}$$

by Exercise 53. Now apply Exercise 53 to $(T_2 \cdots T_k)^{-1}$ in a similar way, and continue this process to obtain the final result.

(If the students are familiar with mathematical induction, then this proof can be reformulated using an induction argument.)

55. (a) For any $\mathbf{u}$ and $\mathbf{v}$ in V, we have

$$\begin{aligned}(T + U)(\mathbf{u} + \mathbf{v}) &= T(\mathbf{u} + \mathbf{v})\\ &\quad + U(\mathbf{u} + \mathbf{v})\\ &= T(\mathbf{u}) + T(\mathbf{v}) +\\ &\quad + U(\mathbf{u}) + U(\mathbf{v})\\ &= [T(\mathbf{u}) + U(\mathbf{u})]\\ &\quad + [T(\mathbf{v}) + U(\mathbf{v})]\\ &= (T + U)(\mathbf{u})\\ &\quad + (T + U)(\mathbf{v}).\end{aligned}$$

Similarly, $(T + U)(c(\mathbf{u})) = c(T + U)(\mathbf{u})$ for any scalar c, and hence $T + U$ is linear.

(b) The proof is similar to the proof of (a).

(c) The proof is similar to the proof of Exercise 74 in Section 7.1.

(d) The zero transformation T_0 is the zero vector of this vector space.

56. Since $T(\mathbf{0}) = \mathbf{0}$, it follows that $\mathbf{0}$ is in the null space of T. Suppose that $\mathbf{u}$ and $\mathbf{v}$ are in the null space of T and c is a scalar. Then

$$T(\mathbf{u} + \mathbf{v}) = T(\mathbf{u}) + T(\mathbf{v}) = \mathbf{0} + \mathbf{0} = \mathbf{0}$$

and

$$T(c\mathbf{u}) = cT(\mathbf{u}) = c\mathbf{0} = \mathbf{0}.$$

Hence the null space of T is closed under vector addition and scalar multiplication. We conclude that the null space of T is a subspace of V.

57. Since $T(\mathbf{0}) = \mathbf{0}$, it follows that $\mathbf{0}$ is in the range of T. Suppose that $\mathbf{w}_1$ and $\mathbf{w}_2$ are in the range of T and c is a scalar. Then there exist vectors $\mathbf{v}_1$ and $\mathbf{v}_2$ in V such that $T(\mathbf{v}_1) = \mathbf{w}_1$ and $T(\mathbf{v}_2) = \mathbf{w}_2$. Thus

$$T(\mathbf{v}_1 + \mathbf{v}_2) = T(\mathbf{v}_1) + T(\mathbf{v}_2) = \mathbf{w}_1 + \mathbf{w}_2,$$

and

$$T(c\mathbf{v}_1) = cT(\mathbf{v}_1) = c\mathbf{w}_1.$$

Hence the range of T is closed under vector addition and scalar multiplication. We conclude that the range of T is a subspace of V.

58. (a) The zero function has derivatives of all orders, the sum of two functions that have derivatives of all orders also has derivatives of all orders, and any scalar multiple of a function that has derivatives of all orders also has derivatives of all orders. Hence C^∞ contains the zero function and is closed under addition and scalar multiplication. Therefore it is a subspace.

(b) Let f and g be in C^∞. Then for all t in $\mathcal{R}$,

$$T(f+g)(t) = e^t(f+g)''(t)$$
$$= e^t f''(t) + e^t g''(t)$$
$$= (T(f) + T(g))(t),$$

and hence $T(f+g) = T(f) + T(g)$. Similarly, $T(cf) = cT(f)$ for any scalar c, and therefore T is linear.

59. (a) The zero function is continuous, the sum of two continuous functions is continuous, and any scalar multiple of a continuous function is continuous. Hence $C([a,b])$ contains the zero function and is closed under addition and scalar multiplication. Therefore it is a subspace.

(b) To show that T is linear, let f and g be in $C([a,b])$. Then for any x in $[a,b]$ we have

$$T(f+g)(x) = \int_a^x (f+g)(t)\,dt$$
$$= \int_a^x f(t)\,dt + \int_a^x g(t)\,dt$$
$$= T(f)(x) + T(g)(x)$$
$$= (T(f) + T(g))(x),$$

and hence $T(f+g) = T(f) + T(g)$. Similarly, $T(cf) = cT(f)$ for any scalar c, and hence T is linear.

To prove that T is one-to-one, suppose that $T(f) = 0$ for some f in $C([a,b])$. Then for any x in $[a,b]$,

$$0 = \frac{d}{dx}T(f)(x) = \frac{d}{dx}\int_a^x f(t)\,dt = f(x),$$

and hence f is the zero function. Therefore T is one-to-one.

60. For Example 5, let $T\colon \mathcal{F}(S) \to \mathcal{R}$ be the function defined by $T(f) = f(s_0)$. Observe that T is a linear transformation and that W is the null space of T. The result is now a consequence of Exercise 56.

For Example 6, observe that the trace defines a linear transformation with W as its null space. The result is now a consequence of Exercise 56.

7.3 BASIS AND DIMENSION

1. Consider the matrix equation

$$x_1 \begin{bmatrix} 1 & 2 \\ 3 & 1 \end{bmatrix} + x_2 \begin{bmatrix} 1 & -5 \\ -4 & 0 \end{bmatrix} + x_3 \begin{bmatrix} 3 & -1 \\ 2 & 2 \end{bmatrix}$$
$$= \begin{bmatrix} 0 & 0 \\ 0 & 0 \end{bmatrix}.$$

Equating corresponding entries, we obtain the system of linear equations

$$x_1 + x_2 + 3x_3 = 0$$
$$2x_1 - 5x_2 - x_3 = 0$$
$$3x_1 - 4x_2 + 2x_3 = 0$$
$$x_1 + 2x_3 = 0.$$

The reduced row echelon form of the augmented matrix of this system is

$$\begin{bmatrix} 1 & 0 & 2 & 0 \\ 0 & 1 & 1 & 0 \\ 0 & 0 & 0 & 0 \\ 0 & 0 & 0 & 0 \end{bmatrix}.$$

Thus the preceding system has nonzero solutions, for example, $x_1 = -2$, $x_2 = -1$, and $x_3 = 1$, and so the given set of matrices is linearly dependent.

2. linearly dependent

3. Notice that $\begin{bmatrix} 12 & 9 \\ -3 & 0 \end{bmatrix} = 3\begin{bmatrix} 4 & 3 \\ -1 & 0 \end{bmatrix}$, and hence the set is linearly dependent.

4. Consider the matrix equation

$$x_1 \begin{bmatrix} 1 & 2 \\ 2 & 1 \end{bmatrix} + x_2 \begin{bmatrix} 1 & 3 \\ 3 & 1 \end{bmatrix} + x_3 \begin{bmatrix} 1 & 2 \\ 3 & 1 \end{bmatrix} = \begin{bmatrix} 0 & 0 \\ 0 & 0 \end{bmatrix}.$$

Equating corresponding entries, we obtain the system of linear equations

$$x_1 + x_2 + x_3 = 0$$
$$2x_1 + 3x_2 + 2x_3 = 0$$
$$2x_1 + 3x_2 + 3x_3 = 0$$
$$x_1 + x_2 + x_3 = 0.$$

The reduced row echelon form of the augmented matrix of this system is

$$\begin{bmatrix} 1 & 0 & 0 \\ 0 & 1 & 0 \\ 0 & 0 & 1 \\ 0 & 0 & 0 \end{bmatrix}.$$

Consequently, the only solution of the system is the trivial one, $x_1 = x_2 = x_3 = 0$, and we conclude that the set is linearly independent.

5. linearly independent

6. The set is linearly dependent. For example, the linear combination using the coefficients $x_1 = 5$, $x_2 = 2$, and $x_3 = -1$ yields the zero matrix.

7. linearly independent

8. linearly independent

9. Consider the polynomial equation
$$a(1+x)+b(1-x)+c(1+x+x^2)+d(1+x-x^2) = 0.$$

Equating corresponding coefficients, we obtain the system of linear equations
$$\begin{aligned} a+b+c+d &= 0 \\ a-b+c+d &= 0 \\ c-d &= 0. \end{aligned}$$

The reduced row echelon form of the augmented matrix of this system is
$$\begin{bmatrix} 1 & 0 & 0 & 2 & 0 \\ 0 & 1 & 0 & 0 & 0 \\ 0 & 0 & 1 & -1 & 0 \end{bmatrix}.$$

Thus the preceding system has nonzero solutions, for example, $a = -2$, $b = 0$, $c = 1$, and $d = 1$, and so the given set of polynomials is linearly dependent.

10. Observe that the second polynomial in the set is a multiple of the first one, and hence the set is linearly dependent.

11. Consider the polynomial equation
$$a_1(x^2 - 2x + 5) + a_2(2x^2 - 5x + 10) + a_3 x^2 = 0.$$

Equating corresponding coefficients, we obtain the system of linear equations
$$\begin{aligned} a_1 + 2a_2 + a_3 &= 0 \\ -2a_1 - 5a_2 &= 0 \\ 5a_1 + 10a_2 &= 0. \end{aligned}$$

The reduced row echelon form of the augmented matrix of this system is
$$\begin{bmatrix} 1 & 0 & 0 \\ 0 & 1 & 0 \\ 0 & 0 & 0 \end{bmatrix}.$$

So the only solution of the preceding system is $a_1 = a_2 = a_3 = 0$, and hence the given set is linearly independent.

12. The set is linearly dependent. For example, the linear combination using the respective coefficients $-5, 2, 1$ yields the zero polynomial.

13. linearly independent

14. The set is linearly dependent. For example, the linear combination using the respective coefficients $4, -3, 2$ yields the zero polynomial.

15. linearly independent

16. linearly independent

17. Assume that $\{t, t\sin t\}$ is linearly dependent. Since these are both nonzero functions, there exists a nonzero scalar a such that $t \sin t = at$ for all t in $\mathcal{R}$. Setting $t = \frac{\pi}{2}$, we obtain $\frac{\pi}{2} \sin \frac{\pi}{2} = a \frac{\pi}{2}$, from which we see that $a = 1$. Setting $t = \frac{\pi}{4}$, we obtain $\frac{\pi}{4} \sin \frac{\pi}{4} = a \frac{\pi}{4}$, from which we see that $a = \frac{1}{\sqrt{2}}$. This is a contradiction, and it follows that the set is linearly independent.

18. linearly independent

19. Observe that
$$0 \sin t + 1 \sin^2 t + 1 \cos^2 t + (-1)1 = 0$$
for all t, and hence the set is linearly dependent.

20. linearly independent

21. We show that for any positive integer n, any subset consisting of n functions is linearly independent. This is certainly true for $n = 1$ because any set consisting of a single nonzero function is linearly independent.

Now suppose that we have established that any subset consisting of k functions is linearly independent, where k is a fixed positive integer. Consider any subset consisting of $k+1$ functions
$$\{e^{n_1 t}, e^{n_2 t}, \ldots, e^{n_{k+1} t}\}.$$

Let $a_1, a_2, \ldots, a_{k+1}$ be scalars such that
$$a_1 e^{n_1 t} + a_2 e^{n_2 t} + \cdots + a_k e^{n_k t} + a_{k+1} e^{n_{k+1} t} = 0$$

for all t. We form two equations from the equation above. The first equation is obtained by taking the derivative of both sides with respect to t, and the second equation is obtained by multiplying both sides of the equation by n_{k+1}. The results are
$$n_1 a_1 e^{n_1 t} + n_2 a_2 e^{n_2 t} + \cdots$$
$$+ n_k a_k e^{n_k t} + n_{k+1} a_{k+1} e^{n_{k+1} t} = 0$$

and

$$n_{k+1} a_1 e^{n_1 t} + n_{k+1} a_2 e^{n_2 t} + \cdots$$

$$+ n_{k+1}a_k e^{n_k t} + n_{k+1}a_{k+1}e^{n_{k+1} t} = 0.$$

Now subtract both sides of the second equation from both sides of the first equation to obtain

$$(n_1 - n_{k+1})a_1 e^{n_1 t} + (n_2 - n_{k+1})a_2 e^{n_2 t} + \cdots$$
$$+ (n_k - n_{k+1})a_k e^{n_k t} = 0.$$

Since this last equation involves a linear combination of a set of k functions, which is assumed to be linearly independent, each coefficient $(n_i - n_{k+1})a_i$ is zero. But $n_i \neq n_{k+1}$ for each i, $1 \leq i \leq k$, and hence each $a_i = 0$. Thus the original equation reduces to $a_{k+1}e^{n_{k+1} t} = 0$, from which we conclude that $a_{k+1} = 0$. It follows that any subset consisting of $k+1$ functions is linearly independent.

Since a set of 1 function is linearly independent, the preceding paragraph implies that a set of 2 functions is linearly independent. Repeating this reasoning, we see that any set of 3 functions is linearly independent. Continuing this argument $n - 1$ times, we conclude that any set of n functions is linearly independent.

(If students are familiar with the principle of mathematical induction, this proof can be formulated using induction.)

22. Since the equation $\cos 2t = \cos^2 t - \sin^2 t$ is a trigonometric identity, it follows that the set is linearly dependent.

23. linearly independent

24. This set is linearly independent. In fact, it is a basis for $\mathcal{P}$.

25. Let
$$p_1(x) = \frac{(x-1)(x-2)}{(0-1)(1-2)},$$
$$p_2(x) = \frac{(x-0)(x-2)}{(1-0)(1-2)},$$
and
$$p_3(x) = \frac{(x-0)(x-1)}{(2-0)(2-1)}.$$
Then
$$p(x) = 1p_1(x) + 0p_2(x) + 3p_3(x)$$
$$= \frac{1}{2}(x-1)(x-2) + \frac{3}{2}x(x-1)$$
$$= 2x^2 - 3x + 1.$$

26. $-3x^2 + 6x + 5$ 27. $-2x^2 + 6x - 3$

28. $5x - 3$ 29. $x^3 - 4x + 2$

30. $-x^3 - 3x^2 + 5$

31. False. For example, the infinite set $\{1, x, x^2, \ldots\}$ is a linearly independent subset of $\mathcal{P}$.

32. False. Only finite-dimensional vector spaces have finite bases.

33. False. The dimension of $\mathcal{P}_n$ is equal to $n + 1$.

34. False. For example, $\mathcal{P}_2$ is a 3-dimensional subspace of the infinite-dimensional vector space $\mathcal{P}$.

35. False. Finite-dimensional vector spaces only have finite bases, and infinite-dimensional vector spaces only have infinite bases.

36. True   37. True   38. True

39. False. It is linearly independent.

40. True   41. True

42. False. The dimension is mn.

43. True   44. True

45. False. The dimension is mn.

46. True   47. True

48. False. For example $\{1, x, 1 + x\}$ is a finite linearly dependent subset of $\mathcal{P}$, but $\mathcal{P}$ is infinite-dimensional.

49. This set is linearly dependent. To show this, we find a nonzero solution of
$$(af + bg + ch)(n) = a(n+1) + b + c(2n-1)$$
$$= (a + 2c)n + (a + b - c) = 0$$
for all n. Thus we set the coefficients equal to zero to obtain the system
$$a + 2c = 0$$
$$a + b - c = 0,$$
which has the nonzero solution $a = -2$, $b = 3$, and $c = 1$.

50. Let $\mathcal{B} = \{f_1, f_2, \ldots, f_n, \ldots\}$. To show $\mathcal{B}$ is linearly independent, it suffices to show that each nonempty finite subset of $\mathcal{B}$ is linearly independent. Consider any finite subset $\mathcal{S}$ of $\mathcal{B}$. There is a positive integer n such that $\mathcal{S}$ is contained in $\mathcal{S}_n = \{f_1, f_2, \ldots, f_n\}$. We first show that $\mathcal{S}_n$ is linearly independent. Suppose that $a_1 f_1 + a_2 f_2 + \cdots + a_n f_n = O$, where O is the zero function. Then for any $1 \leq k \leq n$, we have
$$0 = O(k)$$
$$= a_1 f_1(k) + a_2 f_2(k) + \cdots + a_n f_n(k) = a_k,$$

and hence $\mathcal{S}_n$ is linearly independent. Since $\mathcal{S}$ is contained in $\mathcal{S}_n$, it follows that $\mathcal{S}$ is linearly independent. Therefore $\mathcal{B}$ is linearly independent. We now show that $\mathcal{B}$ is a generating set V. Consider any f in V. There is a positive integer m such that $f(n) = 0$ for $n > m$. Let $a_i = f(i)$ for each $1 \leq i \leq m$. By considering the cases $n \leq m$ and $n > m$ separately, we can show that

$$f(n) = a_1 f_1(n) + a_2 f_2(n) + \cdots + a_m f_m(n)$$

for every n in N. It follows that

$$f = a_1 f_1 + a_2 f_2 + \cdots + a_m f_m,$$

and we conclude that $\mathcal{B}$ is a generating set, and hence a basis, for V.

51. $\left\{ \begin{bmatrix} 1 & 0 & 0 \\ 0 & 0 & 0 \\ 0 & 0 & 0 \end{bmatrix}, \begin{bmatrix} 0 & 0 & 0 \\ 0 & 1 & 0 \\ 0 & 0 & 0 \end{bmatrix}, \begin{bmatrix} 0 & 0 & 0 \\ 0 & 0 & 0 \\ 0 & 0 & 1 \end{bmatrix}, \right.$
$\left. \begin{bmatrix} 0 & 1 & 0 \\ 1 & 0 & 0 \\ 0 & 0 & 0 \end{bmatrix}, \begin{bmatrix} 0 & 0 & 1 \\ 0 & 0 & 0 \\ 1 & 0 & 0 \end{bmatrix}, \begin{bmatrix} 0 & 0 & 0 \\ 0 & 0 & 1 \\ 0 & 1 & 0 \end{bmatrix} \right\}$

52. $\left\{ \begin{bmatrix} 0 & 1 & 0 \\ -1 & 0 & 0 \\ 0 & 0 & 0 \end{bmatrix}, \begin{bmatrix} 0 & 0 & 1 \\ 0 & 0 & 0 \\ -1 & 0 & 0 \end{bmatrix}, \begin{bmatrix} 0 & 0 & 0 \\ 0 & 0 & 1 \\ 0 & -1 & 0 \end{bmatrix} \right\}$

53. A matrix $A = \begin{bmatrix} x_1 & x_2 \\ x_3 & x_4 \end{bmatrix}$ is in W if and only if $x_1 + x_4 = 0$. For any such matrix A,

$$A = \begin{bmatrix} x_1 & x_2 \\ x_3 & -x_1 \end{bmatrix}$$

$$= x_1 \begin{bmatrix} 1 & 0 \\ 0 & -1 \end{bmatrix} + x_2 \begin{bmatrix} 0 & 1 \\ 0 & 0 \end{bmatrix} + x_3 \begin{bmatrix} 0 & 0 \\ 1 & 0 \end{bmatrix}.$$

It follows that

$$\left\{ \begin{bmatrix} 1 & 0 \\ 0 & -1 \end{bmatrix}, \begin{bmatrix} 0 & 1 \\ 0 & 0 \end{bmatrix}, \begin{bmatrix} 0 & 0 \\ 1 & 0 \end{bmatrix} \right\}$$

is a basis for W.

54. $\{x, x^2, \ldots, x^n\}$

55. Consider any polynomial

$$p(x) = a_0 + a_1 x + \cdots + a_n x^n$$

in W. Since $p(1) = 0$, we have

$$a_0 + a_1 + \cdots + a_n = 0,$$

and hence

$$p(x) = a_0 + a_1 x + \cdots$$
$$- (a_0 + a_1 + \cdots + a_{n-1}) x^n$$

$$= a_0(1 - x^n) + a_1(x - x^n) + \cdots$$
$$+ a_{n-1}(x^{n-1} - x^n).$$

It follows that $\{1-x^n, x-x^n, \ldots, x^{n-1}-x^n\}$ is a generating set for W. Proceeding as in Exercise 11, we can also show that this set is linearly independent, and hence it is a basis for W.

56. Observe that $f' = f$ if and only if $f(t) = ce^t$ for some scalar c. Therefore $\{e^t\}$ is a basis for W.

57. Observe that $W = \mathcal{P}_1$, and hence $\{1, x\}$ is a basis for W.

58. A polynomial $p(x) = a_0 + a_1 x + \cdots + a_n x^n$ is in W if and only if $p(-x) = -p(x)$, that is, $a_k(-x)^k = -a_k x^k$ for all k. This condition is equivalent to $a_k = 0$ if k is even. Therefore the set $\{x, x^3, \ldots, x^{2n+1}, \ldots, \}$ is a basis for W.

59. Suppose that $\mathcal{S} = \{p_0(x), p_1(x), \ldots, p_n(x)\}$ is a subset of $\mathcal{P}_n$ such that each $p_k(x)$ has degree k. We show that $\mathcal{S}$ is linearly independent. Suppose

$$c_0 p_0(x) + c_1 p_1(x) + \cdots + c_n p_n(x) = 0 =$$
$$0 + 0 \cdot x + \cdots + 0 \cdot x^n.$$

Since $p_n(x)$ is the only polynomial in $\mathcal{S}$ with a nonzero coefficient for x^n, it follows that $c_n = 0$. Thus

$$c_0 p_0(x) + c_1 p_1(x) + \cdots + c_{n-1} p_{n-1}(x) = \mathbf{0}.$$

As above, $p_{n-1}(x)$ is the only polynomial on the left side of this equation that has a nonzero coefficient for x^{n-1}, and hence $c_{n-1} = 0$. Continuing this pattern of reasoning, we deduce that $c_n = c_{n-1} = \cdots = c_0 = 0$, and hence $\mathcal{S}$ is linearly independent. Now apply item 2 in the box labelled **Properties of Finite-Dimensional Vector Spaces** on page 515 to conclude that $\mathcal{S}$ is a basis for $\mathcal{P}_n$.
(If the students are familiar with mathematical induction, then this proof can be reformulated as an induction argument.)

60. (a) Clearly the $n \times n$ zero matrix is a magic square with sum zero, and hence is in V_n. Suppose that A and B are in V_n with sums a and b, respectively. Then the ith row of $A + B$ sums to

$$(a_{i1} + b_{i1}) + \cdots + (a_{in} + b_{in}) = a + b.$$

Similarly, each column and each of the two diagonals sum to $a + b$, and hence $A + B$ is a magic square with sum $a + b$. Using a similar approach, we can show that for any scalar c, the product cA is a magic square with sum ca. Thus V_n is closed under addition and scalar multiplication.

(b) As in (a), the zero magic square is in W_n. Also as in (a), if A and B are in W_n, then $A + B$ is a magic square with sum $0 + 0 = 0$, and hence $A + B$ is in W_n. So W_n is closed under addition. Similarly, W_n is closed under scalar multiplication, and is therefore a subspace of V_n.

61. (a) Since every entry of C_n is $\frac{1}{n}$, the sum of any row, column, and diagonal is equal to 1. Therefore C_n is a magic square with sum 1.

(b) Suppose that A is an $n \times n$ magic square with sum s, and let $B = A - sC_n$. Since A and C_n are in V_n, it follows that B is in V_n, and B has sum $s - s \cdot 1 = 0$. So B is in W_n and $A = B + sC_n$. Since $B = A - sC_n$, it is necessarily unique.

62. The condition that a matrix $\begin{bmatrix} x_1 & x_2 & x_3 \\ x_4 & x_5 & x_6 \\ x_7 & x_8 & x_9 \end{bmatrix}$ is in W_3 can be written as a system of 8 homogeneous linear equations in 9 unknowns. For example, the condition that the first row sums to zero corresponds to the equation

$$x_1 + x_2 + x_3 = 0.$$

The resulting coefficient matrix has rank 7, and hence the solution space of this system is a 2-dimensional subspace of $\mathcal{R}^9$. Since the mapping $\Phi \colon \mathcal{R}^9 \to \mathcal{M}_{3 \times 3}$ defined by

$$\Phi\left(\begin{bmatrix} x_1 \\ x_2 \\ \vdots \\ x_9 \end{bmatrix}\right) = \begin{bmatrix} x_1 & x_2 & x_3 \\ x_4 & x_5 & x_6 \\ x_7 & x_8 & x_9 \end{bmatrix}$$

is an isomorphism that takes the solution space of the system onto W_3, it follows that $\dim W_3 = 2$.

63. Suppose that $\mathcal{B}$ is a basis for W_3. By Exercise 62, $\mathcal{B}$ consists of two matrices. Since C_3 is not in W_3, the set $\mathcal{B} \cup \{C_3\}$ is linearly independent. Furthermore, this set is a generating set for V_3 as a consequence of the equation in Exercise 61(b). Hence it is a basis for V_3. Since this set contains 3 matrices, it follows that $\dim V_3 = 3$.

64. The proof for this general case is similar to the proof of Exercise 63.

65. The proof for this general case is similar to the proof of Exercise 62. The coefficient matrix of the homogeneous system of linear equations contains $2n + 2$ rows (because there are $2n + 2$ equations) and n^2 columns (because there are n^2 unknowns). Add rows 1 through $n - 1$ to row n, creating a row with all ones. Now subtract rows $n + 1$ through $2n$ from this row of ones (the new nth row) to obtain a zero row. The other $2n + 1$ rows are linearly independent, and hence the coefficient matrix has rank $2n+1$. Therefore the dimension of the solution space is $n^2 - (2n + 1) = n^2 - 2n - 1$. As in Exercise 62, this is also the dimension of W_n.

66. (a) We give a proof of Theorem 7.8. Suppose that $\{\mathbf{v}_1, \mathbf{v}_2, \ldots, \mathbf{v}_n\}$ is a linearly independent subset of V, and that

$$a_1 T(\mathbf{v}_1) + a_2 T(\mathbf{v}_2) + \cdots + a_n T(\mathbf{v}_n) = \mathbf{0}$$

for some scalars $a_1, a_2, \ldots, a_n$. Then

$$T(a_1 \mathbf{v}_1 + a_2 \mathbf{v}_2 + \cdots + a_n \mathbf{v}_n)$$
$$= a_1 T(\mathbf{v}_1) + a_2 T(\mathbf{v}_2) + \cdots + a_n T(\mathbf{v}_n)$$
$$= \mathbf{0}.$$

Hence

$$a_1 \mathbf{v}_1 + a_2 \mathbf{v}_2 + \cdots + a_n \mathbf{v}_n = \mathbf{0}$$

because T is an isomorphism. Therefore $a_1 = a_2 = \cdots = a_n$, showing that the set of images is linearly independent.

(b) By (a) this set is linearly independent. We now show show that it is a generating set for W. Let $\mathbf{w}$ be any vector in W. Since T is onto, there is a vector $\mathbf{v}$ in V such that $T(\mathbf{v}) = \mathbf{w}$. Since $\{\mathbf{v}_1, \mathbf{v}_2, \ldots, \mathbf{v}_n\}$ is a basis for V, we have

$$\mathbf{v} = a_1 \mathbf{v}_1 + a_2 \mathbf{v}_2 + \cdots + a_n \mathbf{v}_n$$

for some scalars $a_1, a_2, \ldots, a_n$. Therefore

$$\mathbf{w} = T(a_1 \mathbf{v}_1 + a_2 \mathbf{v}_2 + \cdots + a_n \mathbf{v}_n)$$
$$= a_1 T(\mathbf{v}_1) + a_2 T(\mathbf{v}_2) + \cdots + a_n T(\mathbf{v}_n),$$

and we conclude that the set of images $\{T(\mathbf{v}_1), T(\mathbf{v}_2), \ldots, T(\mathbf{v}_n)\}$ is a generating set for V. It follows that it is a basis for W.

(c) Suppose V has a finite basis. By (b), the set of images is a basis for W. This set is finite and contains the same number of vectors as the given basis for V. We conclude that W is finite-dimensional and its dimension is equal to the dimension of V.

67. For each $1 \leq i \leq m$ and $1 \leq j \leq n$, let E_{ij} be the $m \times n$ matrix whose (i,j)-entry equals 1 and whose other entries equal 0. Since
$$\{E_{ij}: 1 \leq i \leq m \text{ and } 1 \leq j \leq n\}$$
is a basis for $\mathcal{M}_{m \times n}$, the set
$$\{U(E_{ij}): 1 \leq i \leq m \text{ and } 1 \leq j \leq n\},$$
where U is the isomorphism defined in Example 8, is a basis for $\mathcal{L}(\mathcal{R}^n, \mathcal{R}^m)$.

68. Let $\mathcal{B}$ be a basis for V and $\Phi_\mathcal{B}: V \to \mathcal{R}^n$ be the isomorphism defined on page 513. Also, let $\mathcal{S}$ be a linearly independent subset of V consisting of n vectors, and let $\mathcal{S}'$ be the set of images of the vectors in $\mathcal{S}$ under $\Phi_\mathcal{B}$. Then $\mathcal{S}'$ is linearly independent by Theorem 7.8. Since $\mathcal{S}'$ contains n vectors, it is a basis for $\mathcal{R}^n$. By Theorem 7.6, $\Phi_\mathcal{B}^{-1}: \mathcal{R}^n \to V$ is an isomorphism. Furthermore, $\mathcal{S}$ is the image of $\mathcal{S}'$ under $\Phi_\mathcal{B}^{-1}$, and hence $\mathcal{S}$ is a basis for V by Exercise 66.

69. Let $\mathcal{B}$ be a basis for V, and let $\Phi_\mathcal{B}: V \to \mathcal{R}^n$ be the isomorphism defined on page 513.
    (a) Let $\mathcal{S}$ be a subset of V containing more than n vectors, and let $\mathcal{S}'$ be the set of images of these vectors under $\Phi_\mathcal{B}$. Then $\mathcal{S}'$ is a subset of $\mathcal{R}^n$ consisting of more than n vectors, and hence is linearly dependent. As a consequence, $\mathcal{S}$ is linearly dependent. For otherwise, the images of vectors in a linearly independent set under an isomorphism would be linearly dependent, contrary to Theorem 7.8.
    (b) Part (b) follows from (a) and Exercise 68.

70. Let $\mathcal{B}$ be a basis for V and $\Phi_\mathcal{B}: V \to \mathcal{R}^n$ be the isomorphism defined on page 513. Also, let $\mathcal{S}$ be a subset of V and $\mathcal{S}'$ the set of images of these vectors under $\Phi_\mathcal{B}$.
    (a) Suppose that $\mathcal{S}$ is a generating set for V. Then $\mathcal{S}'$ is a generating set for $\mathcal{R}^n$, and hence contains at least n vectors. Therefore $\mathcal{S}$ contains at least n vectors.
    (b) Suppose that $\mathcal{S}$ contains exactly n vectors. Then $\mathcal{S}'$ is a subset of $\mathcal{R}^n$ consisting of n linearly independent vectors, and hence is a basis for $\mathcal{R}^n$. Since $\Phi_\mathcal{B}^{-1}$ is an isomorphism (by Theorem 7.6) and $\mathcal{S}$ is the image of the vectors in $\mathcal{S}'$ under $\Phi_\mathcal{B}^{-1}$, it follows that $\mathcal{S}$ is a basis for V by Exercise 66.

71. Let $\mathcal{B}$ be a basis for an n-dimensional vector space V, and let $\Phi_\mathcal{B}: V \to \mathcal{R}^n$ be the isomorphism defined on page 513.
    (a) Since the set of images of W, $\Phi_\mathcal{B}(W)$, is a subspace of $\mathcal{R}^n$, its dimension is at most n. If $\mathcal{S}$ is a basis for W, then the set of images of $\mathcal{S}$ under $\Phi_\mathcal{B}$ is a basis for $\Phi_\mathcal{B}(W)$ by Exercise 66. Hence
    $$\dim W = \dim \Phi_\mathcal{B}(W) \leq n = \dim V.$$
    (b) Let $\mathcal{S}$ be a basis for W, and suppose that $\dim W = \dim V = n$. By Exercise 69(b), $\mathcal{S}$ is a basis for V, and hence $V = W$.

72. Using the hint, we obtain an infinite set $\{\mathbf{v}_1, \mathbf{v}_2, \ldots, \mathbf{v}_n, \ldots\}$. It can be shown that, for every n, any finite subset of the form $\{\mathbf{v}_1, \mathbf{v}_2, \ldots, \mathbf{v}_n\}$ is linearly independent. We show that this infinite set is linearly independent. Consider any finite subset $\mathcal{S}$ of this infinite set, and suppose that n is the largest integer for which $\mathbf{v}_n$ is in $\mathcal{S}$. Let $\mathcal{S}' = \{\mathbf{v}_1, \mathbf{v}_2, \ldots, \mathbf{v}_n\}$. Since $\mathcal{S}'$ is linearly independent and $\mathcal{S}$ is a subset of $\mathcal{S}'$, it follows that $\mathcal{S}$ is linearly independent. Hence the original infinite set is linearly independent.

73. Suppose that $\mathcal{B} = \{\mathbf{v}_1, \mathbf{v}_2, \ldots, \mathbf{v}_n\}$ is a basis for a vector space V, and let $\mathbf{u}$ and $\mathbf{v}$ be in V. Then there exist unique scalars $a_1, a_2, \ldots, a_n$ and $b_1, b_2, \ldots, b_n$ such that
$$\mathbf{u} = a_1\mathbf{v}_1 + a_2\mathbf{v}_2 + \cdots + a_n\mathbf{v}_n$$
and
$$\mathbf{v} = b_1\mathbf{v}_1 + b_2\mathbf{v}_2 + \cdots + b_n\mathbf{v}_n.$$
Thus
$$\Phi_\mathcal{B}(\mathbf{u} + \mathbf{v})$$
$$= \Phi_\mathcal{B}((a_1 + b_1)\mathbf{v}_1 + (a_2 + b_2)\mathbf{v}_2 + \cdots + (a_n + b_n)\mathbf{v}_n)$$
$$= \begin{bmatrix} a_1 + b_1 \\ a_2 + b_2 \\ \vdots \\ a_n + b_n \end{bmatrix}$$
$$= \begin{bmatrix} a_1 \\ a_2 \\ \vdots \\ a_n \end{bmatrix} + \begin{bmatrix} b_1 \\ b_2 \\ \vdots \\ b_n \end{bmatrix}$$

$$= \Phi_B(\mathbf{u}) + \Phi_B(\mathbf{v}).$$

Similarly $\Phi_B(c\mathbf{u}) = c\Phi_B(\mathbf{u})$ for every $\mathbf{u}$ in V, and hence Φ_B is linear.

74. By Exercise 66(c), If W is finite-dimensional, then V is finite-dimensional. This is the contrapositive of the statement in Exercise 74.

75. Suppose that $\dim V = n$ and $\dim W = m$. By Exercise 66(c), if V_1 and V_2 are isomorphic vector spaces and V_1 is finite-dimensional, then V_2 is finite-dimensional and $\dim V_1 = \dim V_2$. Also, $\dim \mathcal{L}(\mathcal{R}^n, \mathcal{R}^n) = \dim \mathcal{M}_{m \times n} = m \cdot n$ by Example 8. So to establish the desired result, it suffices to show that $\mathcal{L}(\mathcal{R}^n, \mathcal{R}^n)$ is isomorphic to $\mathcal{L}(V, W)$.

Let $\mathcal{A}$ and $\mathcal{B}$ be bases for V and W, respectively, and let $\Phi_\mathcal{A}$ and $\Phi_\mathcal{B}$ be the isomorphisms from V to $\mathcal{R}^n$ and from W to $\mathcal{R}^m$ defined on page 513. Then $\Phi_\mathcal{A}^{-1}$ and $\Phi_\mathcal{B}^{-1}$ are isomorphisms by Theorem 7.6. For all T in $\mathcal{L}(\mathcal{R}^n, \mathcal{R}^m)$, define $\Phi \colon \mathcal{L}(\mathcal{R}^n, \mathcal{R}^m) \to \mathcal{L}(V, W)$ by

$$\Phi(T) = \Phi_\mathcal{B}^{-1} T \Phi_\mathcal{A}.$$

This composition is defined because if T is a linear transformation from $\mathcal{R}^n$ to $\mathcal{R}^m$, then $\Phi(T) = \Phi_\mathcal{B}^{-1} T \Phi_\mathcal{A}$ is a linear transformation from V to W. We show that Φ is an isomorphism. First observe that Φ is linear. Let T_1 and T_2 be in $\mathcal{L}(\mathcal{R}^n, \mathcal{R}^m)$. Then for any $\mathbf{v}$ in V,

$$\begin{aligned}\Phi(T_1 + T_2)(\mathbf{v}) &= \Phi_\mathcal{B}^{-1}(T_1 + T_2)\Phi_\mathcal{A}(\mathbf{v}) \\ &= \Phi_\mathcal{B}^{-1}(T_1\Phi_\mathcal{A}(\mathbf{v}) + T_2\Phi_\mathcal{A}(\mathbf{v})) \\ &= \Phi_\mathcal{B}^{-1}T_1\Phi_\mathcal{A}(\mathbf{v}) + \Phi_\mathcal{B}^{-1}T_2\Phi_\mathcal{A}(\mathbf{v}) \\ &= \Phi(T_1)(\mathbf{v}) + \Phi(T_2)(\mathbf{v}) \\ &= [\Phi(T_1) + \Phi(T_2)](\mathbf{v}).\end{aligned}$$

Hence $\Phi(T_1 + T_2) = \Phi(T_1) + \Phi(T_2)$. Similarly, $\Phi(cT) = c\Phi(T)$ for any T in $\mathcal{L}(\mathcal{R}^n, \mathcal{R}^m)$ and any scalar c. Thus Φ is linear. To show that Φ is one-to-one and onto, we construct its inverse directly. Let $\Psi \colon \mathcal{L}(V, W) \to \mathcal{L}(\mathcal{R}^n, \mathcal{R}^m)$ be defined by

$$\Psi(T) = \Phi_\mathcal{B} T \Phi_\mathcal{A}^{-1}$$

for all T in $\mathcal{L}(V, W)$. Then, for any T in $\mathcal{L}(\mathcal{R}^n, \mathcal{R}^m)$,

$$\Psi\Phi(T) = \Psi(\Phi_\mathcal{B}^{-1} T \Phi_\mathcal{A}) = \Phi_\mathcal{B}(\Phi_\mathcal{B}^{-1} T \Phi_\mathcal{A})\Phi_\mathcal{A}^{-1}$$
$$= T.$$

Similarly, $\Phi\Psi(T) = T$ for all T in $\mathcal{L}(V, W)$. It follows that Φ is an isomorphism.

76. We first show that each T_i is linear. Let $f(x)$ and $g(x)$ be in $\mathcal{P}_n$, and consider any i such that $0 \leq i \leq n$. Then

$$\begin{aligned}T_i(f(x) + g(x)) &= f(i) + g(i) \\ &= T_i(f(x)) + T_i(g(x)).\end{aligned}$$

Similarly, $T_i(cf(x)) = cT_i(f(x))$ for any scalar c. Therefore T_i is linear.

We now show that $\{T_0, T_1, \ldots, T_n\}$ is linearly independent. For each i, let $p_i(x)$ be the ith Lagrange interpolating polynomial associated with $0, 1, \ldots, n$. Then, for each $0 \leq i, j \leq n$,

$$T_i(p_j(x)) = p_j(i) = \begin{cases} 0 & \text{if } i \neq j \\ 1 & \text{if } i = j.\end{cases}$$

Suppose that $a_0 T_0 + a_1 T_1 + \cdots + a_n T_n$ is the zero operator. Then, for any $1 \leq j \leq n$,

$$\begin{aligned}0 &= (a_0 T_0 + a_1 T_1 + \cdots + a_n T_n)(p_j(x)) \\ &= a_0 T_0(p_j(x)) + a_1 T_1(p_j(x)) + \cdots \\ &\quad + a_n T_n(p_j(x)) \\ &= a_0 \cdot 0 + \cdots + a_j \cdot 1 + \cdots + a_n \cdot 0 \\ &= a_j.\end{aligned}$$

Since each coefficient in the linear combination is zero, $\{T_0, T_1, \ldots, T_n\}$ is linearly independent. By Exercise 75,

$$\begin{aligned}\dim \mathcal{L}(\mathcal{P}_n, \mathcal{R}) &= \dim \mathcal{P}_n \cdot \dim \mathcal{R} \\ &= (n+1) \cdot 1 = n+1.\end{aligned}$$

It follows from this equality and Exercise 69(b) that $\{T_0, T_1, \ldots, T_n\}$ is a basis for $\mathcal{L}(\mathcal{P}_n, \mathcal{R})$.

77. Let $T \colon \mathcal{P}_n \to \mathcal{R}$ be defined by

$$T(f(x)) = \int_a^b f(x)\, dx.$$

Then T is linear because the integral of a sum of polynomials is the sum of the integrals, and the integral of a scalar multiple of a polynomial is the same multiple of the integral. Therefore T is in $\mathcal{L}(\mathcal{P}_n, \mathcal{R})$. By Exercise 76, $\{T_0, T_1, \ldots, T_n\}$ is a basis for $\mathcal{L}(\mathcal{P}_n, \mathcal{R})$, and hence there exist unique scalars $c_1, c_1, \ldots, c_n$ such that $T = c_0 T_0 + c_1 T_1 + \cdots + c_n T_n$. Therefore, for any polynomial $f(x)$ in $\mathcal{P}_n$,

$$\begin{aligned}\int_a^b f(x)\, dx &= T(f(x)) \\ &= (c_0 T_0 + c_1 T_1 + \cdots + c_n T_n)(f(x))\end{aligned}$$

$$= c_0 T_0(f(x)) + c_1 T_1(f(x)) + \cdots$$
$$+ c_n T_n(f(x))$$
$$= c_0 f(0) + c_1 f(1) + \cdots + c_n f(n).$$

78. (a) By Exercise 77, there exist scalars c_0, c_1, and c_2 such that
$$\int_0^2 g(t)\, dt = c_0 g(a) + c_1 g(1) + c_2 g(2)$$
for every polynomial $g(t)$ in $\mathcal{P}_2$. To evaluate these scalars, we substitute $g(t) = 1$, $g(t) = t$, and $g(t) = t^2$ into the equation above to generate a system of linear equations with the coefficients as the unknowns. For example, substituting $g(t) = 1$, we obtain
$$c_0 + c_1 + c_2 = 2.$$
Continuing with $g(t) = t$ and $g(t) = t^2$, we obtain the system
$$\begin{aligned} c_0 + c_1 + c_2 &= 2 \\ c_1 + c_2 &= 2 \\ c_1 + 4c_2 &= \tfrac{8}{3}, \end{aligned}$$
which has the solution $c_0 = \tfrac{1}{3}$, $c_1 = \tfrac{4}{3}$, and $c_2 = \tfrac{1}{3}$. Now consider any a and b with $a < b$, and let $f(x)$ be any polynomial in $\mathcal{P}_2$. If $g(t)$ is the polynomial defined by
$$g(t) = f\left(a + \left(\frac{b-a}{2}\right)t\right),$$
then
$$\int_0^2 g(t)\, dt = \tfrac{1}{3}g(0) + \tfrac{4}{3}g(1) + \tfrac{1}{3}g(2)$$
$$= \tfrac{1}{3}f(a) + \tfrac{4}{3}f\left(\frac{b+a}{2}\right) + \tfrac{1}{3}f(b).$$
On the other hand, we can evaluate this integral using the substitution
$$x = a + \left(\frac{b-a}{2}\right)t,$$
and hence, $dt = \frac{2}{b-a}dx$. Thus
$$\int_0^2 g(t)\, dt = \int_0^2 f\left(a + \left(\frac{b-a}{2}\right)t\right) dt$$
$$= \left(\frac{2}{b-a}\right)\int_a^b f(x)\, dx.$$

Combining these equations, we obtain
$$\left(\frac{2}{b-a}\right)\int_a^b f(x)\, dx$$
$$= \tfrac{1}{3}f(a) + \tfrac{4}{3}f\left(\frac{b+a}{2}\right) + \tfrac{1}{3}f(b),$$
which simplifies to
$$\int_a^b f(x)\, dx$$
$$= \frac{b-a}{6}\left[f(a) + 4f\left(\frac{a+b}{2}\right) + f(b)\right].$$

(b) Let $f(x) = x^3$. Then
$$\int_a^b x^3\, dx$$
$$= \frac{b-a}{6}\left[a^3 + 4\left(\frac{a+b}{2}\right)^3 + b^3\right].$$
Let $T\colon \mathcal{P}_3 \to \mathcal{R}$ and $U\colon \mathcal{P}_3 \to \mathcal{R}$ be defined by
$$T(f(x)) = \int_a^b f(x)\, dx$$
and
$$U(f(x))$$
$$= \frac{b-a}{6}\left[f(a) + 4f\left(\frac{a+b}{2}\right) + f(b)\right].$$
As in Exercise 77, T is linear, and we can also verify that U is linear. We have already shown that $T(p(x)) = U(p(x))$ for every polynomial $p(x)$ in $\mathcal{P}_2$ and that $T(x^3) = U(x^3)$. Our goal is to show that $T = U$ on $\mathcal{P}_3$. Let $f(x)$ be any polynomial in $\mathcal{P}_3$. Then there exists a polynomial $p(x)$ in $\mathcal{P}_2$ and a scalar c such that $f(x) = p(x) + cx^3$. Thus
$$T(f(x)) = T(p(x) + cx^3)$$
$$= T(p(x)) + cT(x^3)$$
$$= U(p(x)) + cU(x^3)$$
$$= U(p(x) + cx^3) = U(f(x)),$$
and hence T and U agree on $\mathcal{P}_3$.

79. The set is linearly independent.

80. The set is linearly dependent. Let $f(x)$, $g(x)$, and $h(x)$ denote the first, second, and third polynomials in the set, respectively. Then $h(x) = (-3)f(x) + 4g(x)$.

81. The set is linearly dependent, and
$$M_3 = (-3)M_1 + 2M_2,$$
where M_j is the jth matrix in the set.

82. The set is linearly independent.

83. (rounded to 4 places after the decimal) $c_0 = 0.3486$, $c_1 = 0.8972$, $c_2 = -0.3667$, $c_3 = 0.1472$, $c_4 = -0.0264$

7.4 MATRIX REPRESENTATIONS OF LINEAR OPERATORS

1. Since
$$\begin{bmatrix} 1 & 2 \\ 3 & 4 \end{bmatrix}$$
$$= 1\begin{bmatrix} 1 & 0 \\ 0 & 0 \end{bmatrix} + 3\begin{bmatrix} 0 & 0 \\ 1 & 0 \end{bmatrix} + 4\begin{bmatrix} 0 & 0 \\ 0 & 1 \end{bmatrix} + 2\begin{bmatrix} 0 & 1 \\ 0 & 0 \end{bmatrix},$$
it follows that
$$[A]_\mathcal{B} = \begin{bmatrix} 1 \\ 3 \\ 4 \\ 2 \end{bmatrix}.$$

2. $\begin{bmatrix} -3 \\ 1 \\ 2 \end{bmatrix}$

3. $[\sin 2t - \cos 2t]_\mathcal{B} = [2\sin t \cos t - \cos^2 t + \sin^2 t]_\mathcal{B}$
$$= \begin{bmatrix} -1 \\ 1 \\ 2 \end{bmatrix}$$

4. A vector $\mathbf{u}$ in $\mathcal{R}^3$ is in V if and only if $u_1 = u_2 - 2u_3$, that is,
$$\mathbf{u} = \begin{bmatrix} u_1 \\ u_2 \\ u_3 \end{bmatrix} = \begin{bmatrix} u_2 - 2u_3 \\ u_2 \\ u_3 \end{bmatrix} = u_2\begin{bmatrix} 1 \\ 1 \\ 0 \end{bmatrix} + u_3\begin{bmatrix} -2 \\ 0 \\ 1 \end{bmatrix}.$$
It follows, in general, that $[\mathbf{u}]_\mathcal{B} = \begin{bmatrix} u_2 \\ u_3 \end{bmatrix}$. So for the particular vector given in this exercise, we have $[\mathbf{u}]_\mathcal{B} = \begin{bmatrix} -1 \\ -3 \end{bmatrix}$.

5. $[\mathbf{u}]_\mathcal{B} = \begin{bmatrix} -3 \\ 2 \\ 1 \end{bmatrix}$

6. Suppose that
$$a(x^3 - x^2) + b(x^2 - x) + c(x-1) + d(x^3+1)$$
$$= 2x^3 - 5x^2 + 3x - 2.$$

Collecting like terms, we obtain
$$(a+d)x^3 + (-a+b)x^2 + (-b+c)x$$
$$+ (-c+d)$$
$$= 2x^3 - 5x^2 + 3x - 2.$$

Equating corresponding coefficients, we obtain the system of linear equations
$$\begin{array}{rcr} a \quad\quad\quad + d & = & 2 \\ -a + b \quad\quad & = & -5 \\ -b + c \quad & = & 3 \\ -c + d & = & -2, \end{array}$$
which has the unique solution $a = 3$, $b = -2$, $c = 1$, and $d = -1$. Therefore
$$[\mathbf{u}]_\mathcal{B} = \begin{bmatrix} 3 \\ -2 \\ 1 \\ -1 \end{bmatrix}.$$

7. $[\mathbf{u}]_\mathcal{B} = \begin{bmatrix} -3 \\ -2 \\ 1 \end{bmatrix}$ **8.** $[\mathbf{u}]_\mathcal{B} = \begin{bmatrix} 2 \\ 3 \\ -1 \end{bmatrix}$

9. Since
$$D(e^t) = 1e^t, \quad D(e^{2t}) = 2e^{2t}, \quad \text{and}$$
$$D(e^{3t}) = 3e^{3t},$$
we see that
$$[T]_\mathcal{B} = [D]_\mathcal{B} = \begin{bmatrix} 1 & 0 & 0 \\ 0 & 2 & 0 \\ 0 & 0 & 3 \end{bmatrix}.$$

10. $\begin{bmatrix} 1 & 1 & 0 \\ 0 & 1 & 2 \\ 0 & 0 & 1 \end{bmatrix}$ **11.** $\begin{bmatrix} 1 & 0 & 0 \\ 3 & 3 & 3 \\ 1 & 2 & 4 \end{bmatrix}$

12. $\begin{bmatrix} 1 & 0 & 2 & 0 \\ 0 & 1 & 0 & 2 \\ 3 & 0 & 2 & 0 \\ 0 & 3 & 0 & 2 \end{bmatrix}$ **13.** $\begin{bmatrix} 0 & 1 & -2 & 0 \\ 0 & 0 & 2 & -6 \\ 0 & 0 & 0 & 3 \\ 0 & 0 & 0 & 0 \end{bmatrix}$

14. $\begin{bmatrix} 0 & 0 & 0 & 1 \\ 0 & 0 & 1 & 0 \\ 0 & 1 & 0 & 0 \\ 1 & 0 & 0 & 0 \end{bmatrix}$ **15.** $\begin{bmatrix} 1 & 0 & 0 & 0 \\ 0 & 0 & 1 & 0 \\ 0 & 1 & 0 & 0 \\ 0 & 0 & 0 & 1 \end{bmatrix}$

16. Denote the matrices in $\mathcal{B}$ by M_1, M_2 and M_3, respectively. Then
$$T(M_1) = \begin{bmatrix} 1 & 2 \\ -1 & -2 \end{bmatrix}\begin{bmatrix} 1 & 0 \\ 0 & 0 \end{bmatrix}\begin{bmatrix} 1 & -1 \\ 2 & -2 \end{bmatrix}$$
$$= \begin{bmatrix} 1 & -1 \\ -1 & 1 \end{bmatrix} = M_1 + M_2 - M_3,$$

$$T(M_2) = \begin{bmatrix} 1 & 2 \\ -1 & -2 \end{bmatrix} \begin{bmatrix} 0 & 0 \\ 0 & 1 \end{bmatrix} \begin{bmatrix} 1 & -1 \\ 2 & -2 \end{bmatrix}$$

$$= \begin{bmatrix} 4 & -4 \\ -4 & 4 \end{bmatrix} = 4M_1 + 4M_2 - 4M_3,$$

and

$$T(M_3) = \begin{bmatrix} 1 & 2 \\ -1 & -2 \end{bmatrix} \begin{bmatrix} 0 & 1 \\ 1 & 0 \end{bmatrix} \begin{bmatrix} 1 & -1 \\ 2 & -2 \end{bmatrix}$$

$$= \begin{bmatrix} 4 & -4 \\ -4 & 4 \end{bmatrix} = 4M_1 + 4M_2 - 4M_3.$$

Therefore

$$[T]_\mathcal{B} = \begin{bmatrix} 1 & 4 & 4 \\ 1 & 4 & 4 \\ -1 & -4 & -4 \end{bmatrix}.$$

17. (a) As in Example 3, let $\mathcal{B} = \{1, x, x^2\}$. Then

$$[D]_\mathcal{B} = \begin{bmatrix} 0 & 1 & 0 \\ 0 & 0 & 2 \\ 0 & 0 & 0 \end{bmatrix}.$$

So

$$[p'(x)]_\mathcal{B} = [D(p(x))]_\mathcal{B} = [D]_\mathcal{B} [p(x)]_\mathcal{B}$$

$$= \begin{bmatrix} 0 & 1 & 0 \\ 0 & 0 & 2 \\ 0 & 0 & 0 \end{bmatrix} \begin{bmatrix} 6 \\ 0 \\ -4 \end{bmatrix} = \begin{bmatrix} 0 \\ -8 \\ 0 \end{bmatrix},$$

and hence $p'(x) = -8x$.

(b) Let $\mathcal{B}$ be as in (a). Then

$$[p'(x)]_\mathcal{B} = [D(p(x))]_\mathcal{B} = [D]_\mathcal{B} [p(x)]_\mathcal{B}$$

$$= \begin{bmatrix} 0 & 1 & 0 \\ 0 & 0 & 2 \\ 0 & 0 & 0 \end{bmatrix} \begin{bmatrix} 2 \\ 3 \\ 5 \end{bmatrix} = \begin{bmatrix} 3 \\ 10 \\ 0 \end{bmatrix},$$

and hence $p'(x) = 3 + 10x$.

(c) Let $\mathcal{B} = \{1, x, x^2, x^3\}$. Then

$$[D]_\mathcal{B} = \begin{bmatrix} 0 & 1 & 0 & 0 \\ 0 & 0 & 2 & 0 \\ 0 & 0 & 0 & 3 \\ 0 & 0 & 0 & 0 \end{bmatrix}.$$

So

$$[p'(x)]_\mathcal{B} = [D(p(x))]_\mathcal{B} = [D]_\mathcal{B} [p(x)]_\mathcal{B}$$

$$= \begin{bmatrix} 0 & 1 & 0 & 0 \\ 0 & 0 & 2 & 0 \\ 0 & 0 & 0 & 3 \\ 0 & 0 & 0 & 0 \end{bmatrix} \begin{bmatrix} 0 \\ 0 \\ 0 \\ 1 \end{bmatrix} = \begin{bmatrix} 0 \\ 0 \\ 3 \\ 0 \end{bmatrix},$$

and hence $p'(x) = 3x^2$.

18. $\frac{1}{2}(e^t \cos t + e^t \sin t)$

19. (a) $-e^t + te^t$ (b) $2e^t - 2te^t + t^2 e^t$
    (c) $11e^t - 8te^t + 2t^2 e^t$

20. $1, \{e^t\}$

21. Let $T = D$, the differential operator on $V = \text{Span}\{e^t, t^{2t}, e^{3t}\}$. By Exercise 9,

$$[T]_\mathcal{B} = [D]_\mathcal{B} = \begin{bmatrix} 1 & 0 & 0 \\ 0 & 2 & 0 \\ 0 & 0 & 3 \end{bmatrix},$$

and hence 1, 2, and 3 are the eigenvalues of D with corresponding bases $\{e^t\}$, $\{e^{2t}\}$, $\{e^{3t}\}$.

22. The matrix $[T]_\mathcal{B}$, obtained in Exercise 12, has the eigenvalues -1 and 4 with bases for the corresponding eigenspaces

$$\left\{ \begin{bmatrix} 1 \\ 0 \\ -1 \\ 0 \end{bmatrix}, \begin{bmatrix} 0 \\ 1 \\ 0 \\ -1 \end{bmatrix} \right\} \text{ and } \left\{ \begin{bmatrix} 2 \\ 0 \\ 3 \\ 0 \end{bmatrix}, \begin{bmatrix} 0 \\ 2 \\ 0 \\ 3 \end{bmatrix} \right\},$$

respectively. It follows that -1 and 4 are the eigenvalues of T with bases for the corresponding eigenspaces of $\left\{ \begin{bmatrix} 1 & 0 \\ -1 & 0 \end{bmatrix}, \begin{bmatrix} 0 & 1 \\ 0 & -1 \end{bmatrix} \right\}$ and $\left\{ \begin{bmatrix} 2 & 0 \\ 3 & 0 \end{bmatrix}, \begin{bmatrix} 0 & 2 \\ 0 & 3 \end{bmatrix} \right\}$, respectively.

23. $1, 6, \{3x - 2x^2\}, \{x + x^2\}$

24. $1, -1, \{x + x^2, 1 + x^3\}, \{-x + x^2, -1 + x^3\}$

25. $0, \{1\}$

26. $0, 1, \left\{ \begin{bmatrix} -4 & 0 \\ 0 & 1 \end{bmatrix}, \begin{bmatrix} -4 & 1 \\ 1 & 0 \end{bmatrix} \right\}, \left\{ \begin{bmatrix} 1 & -1 \\ -1 & 1 \end{bmatrix} \right\}$

27. $1, -1,$

$$\left\{ \begin{bmatrix} 1 & 0 \\ 0 & 0 \end{bmatrix}, \begin{bmatrix} 0 & 1 \\ 1 & 0 \end{bmatrix}, \begin{bmatrix} 0 & 0 \\ 0 & 1 \end{bmatrix} \right\}, \left\{ \begin{bmatrix} 0 & 1 \\ -1 & 0 \end{bmatrix} \right\}$$

28. False, let T be the 90^c-rotation operator on $\mathcal{R}^2$.

29. False, the vector space on which the operator is defined must be finite-dimensional.

30. True   31. True   32. True   33. True

34. False. Let D be the linear operator on $\mathbf{C}^\infty$ defined in Example 2 of Section 7.2. Then, as is shown in Example 5 of this section, every real number is an eigenvalue of D.

35. False, the eigenspace is the set of *symmetric* matrices.

36. False,

$$[T]_\mathcal{B} = [\, [T(\mathbf{v}_1)]_\mathcal{B} \;\; [T(\mathbf{v}_2)]_\mathcal{B} \;\; \ldots \;\; [T(\mathbf{v}_n)]_\mathcal{B} \,].$$

37. False. As written, this expression may make no sense. What is true is that $[T(\mathbf{v})]_\mathcal{B} = [T]_\mathcal{B}[\mathbf{v}]_\mathcal{B}$.
38. True   39. True
40. (a) $T(1) = (D^2 + D)(1) = 1'' + 1' = 0$ and
$$T(e^{-t}) = (D^2 + D)(e^{-t}) = (e^{-t})'' + (e^{-t})'$$
$$= e^{-t} - e^{-t} = 0$$

(b)
$$T(e^{at}) = (D^2 + D)(e^{at})$$
$$= (e^{at})'' + (e^{at})' = a^2 e^{at} + a e^{at}$$
$$= (a^2 + a) e^{at}$$

41. For $\mathcal{B} = \{1, x, x^2\}$, which is a basis for $\mathcal{P}_2$, we see that
$$[D]_\mathcal{B} = \begin{bmatrix} 0 & 1 & 0 \\ 0 & 0 & 2 \\ 0 & 0 & 0 \end{bmatrix}.$$

(a) Since the characteristic polynomial of $[D]_\mathcal{B}$ is $-t^3$, D has only one eigenvalue, $\lambda = 0$.
(b) Since
$$\{[1]_\mathcal{B}\} = \left\{ \begin{bmatrix} 1 \\ 0 \\ 0 \end{bmatrix} \right\}$$
is a basis for the eigenspace of $[D]_\mathcal{B}$ corresponding to the eigenvalue 0, it follows that $\{1\}$ is a basis for the eigenspace of D corresponding to the eigenvalue 0.

42. Both 1 and -1 are eigenvalues of T because $T(A) = A$ for every symmetric matrix and $T(A) = -A$ for every skew-symmetric matrix. Suppose that A is an eigenvector of T with corresponding eigenvalue λ. Then $A^T = \lambda A$, and hence
$$A = (A^T)^T = (\lambda A)^T = \lambda A^T = \lambda^2 A.$$
Since $A \ne O$, it follows that $\lambda^2 = 1$, and therefore $\lambda = \pm 1$.

43. (a) For any f and g in $\mathcal{F}(N)$ and any n in N,
$$E(f+g)(n) = (f+g)(n+1)$$
$$= f(n+1) + g(n+1)$$
$$= (E(f) + E(g))(n),$$
and hence $E(f+g) = E(f) + E(g)$. Similarly, $E(cf) = cE(f)$ for any scalar c, and therefore E is linear.

(b) Observe that for any f in $\mathcal{F}(N)$ and any n in N,
$$(E^2 - E)(f)(n) = E(f)(n+1) - f(n+1)$$
$$= f(n+2) - f(n+1),$$
and hence a nonzero function f in $\mathcal{F}(N)$ is an eigenvector of $E^2 - E$ with corresponding eigenvalue 1 if and only if
$$f(n+2) - f(n+1) = f(n)$$
for all n in N. But this equation is equivalent to the defining condition for a Fibonacci sequence.

44. (a) Consider
$$A_1 = \begin{bmatrix} a_1 & b_1 \\ c_1 & d_1 \end{bmatrix} \quad \text{and} \quad A_2 = \begin{bmatrix} a_2 & b_2 \\ c_2 & d_2 \end{bmatrix}$$
in $\mathcal{M}_{2\times 2}$. Then
$$T(A_1 + A_2) = \begin{bmatrix} b_1 + b_2 & a_1 + a_2 + c_1 + c_2 \\ 0 & d_1 + d_2 \end{bmatrix}$$
$$= T(A_1) + T(A_2).$$
Similarly, $T(cA) = cT(A)$ for any A in $\mathcal{M}_{2\times 2}$, and therefore T is linear.

(b) $\begin{bmatrix} 0 & 1 & 0 & 0 \\ 1 & 0 & 1 & 0 \\ 0 & 0 & 0 & 0 \\ 0 & 0 & 0 & 1 \end{bmatrix}$ (c) $0, 1, -1$

(d) $\left\{ \begin{bmatrix} 1 & 0 \\ -1 & 0 \end{bmatrix} \right\}, \left\{ \begin{bmatrix} 1 & 1 \\ 0 & 0 \end{bmatrix}, \begin{bmatrix} 0 & 0 \\ 0 & 1 \end{bmatrix} \right\},$
$\left\{ \begin{bmatrix} 1 & -1 \\ 0 & 0 \end{bmatrix} \right\}$

45. (a) For any matrices A and C in $\mathcal{M}_{2\times 2}$,
$$T(A + C) = (\text{trace}(A+C))B$$
$$= (\text{trace}\,A + \text{trace}\,C)B$$
$$= (\text{trace}\,A)B + (\text{trace}\,C)B$$
$$= T(A) + T(C).$$
Similarly, $T(cA) = cT(A)$ for any scalar c, and therefore T is linear.

(b) Let
$$\mathcal{B} = \left\{ \begin{bmatrix} 1 & 0 \\ 0 & 0 \end{bmatrix}, \begin{bmatrix} 0 & 1 \\ 0 & 0 \end{bmatrix}, \begin{bmatrix} 0 & 0 \\ 1 & 0 \end{bmatrix}, \begin{bmatrix} 0 & 0 \\ 0 & 1 \end{bmatrix} \right\}.$$
Then
$$T\left(\begin{bmatrix} 1 & 0 \\ 0 & 0 \end{bmatrix} \right) = \left(\text{trace} \begin{bmatrix} 1 & 0 \\ 0 & 0 \end{bmatrix} \right) \begin{bmatrix} 1 & 2 \\ 3 & 4 \end{bmatrix}$$
$$= 1 \begin{bmatrix} 1 & 2 \\ 3 & 4 \end{bmatrix}$$

$$= 1\begin{bmatrix} 1 & 0 \\ 0 & 0 \end{bmatrix} + 2\begin{bmatrix} 0 & 1 \\ 0 & 0 \end{bmatrix}$$
$$+ 3\begin{bmatrix} 0 & 0 \\ 1 & 0 \end{bmatrix} + 4\begin{bmatrix} 0 & 0 \\ 0 & 1 \end{bmatrix},$$

$$T\left(\begin{bmatrix} 0 & 1 \\ 0 & 0 \end{bmatrix}\right) = \left(\text{trace}\begin{bmatrix} 0 & 1 \\ 0 & 0 \end{bmatrix}\right)\begin{bmatrix} 1 & 2 \\ 3 & 4 \end{bmatrix}$$
$$= 0\begin{bmatrix} 1 & 2 \\ 3 & 4 \end{bmatrix} = \begin{bmatrix} 0 & 0 \\ 0 & 0 \end{bmatrix},$$

$$T\left(\begin{bmatrix} 0 & 0 \\ 1 & 0 \end{bmatrix}\right) = \left(\text{trace}\begin{bmatrix} 0 & 0 \\ 1 & 0 \end{bmatrix}\right)\begin{bmatrix} 1 & 2 \\ 3 & 4 \end{bmatrix}$$
$$= 0\begin{bmatrix} 1 & 2 \\ 3 & 4 \end{bmatrix} = \begin{bmatrix} 0 & 0 \\ 0 & 0 \end{bmatrix},$$

and

$$T\left(\begin{bmatrix} 0 & 0 \\ 0 & 1 \end{bmatrix}\right) = \left(\text{trace}\begin{bmatrix} 0 & 0 \\ 0 & 1 \end{bmatrix}\right)\begin{bmatrix} 1 & 2 \\ 3 & 4 \end{bmatrix}$$
$$= 1\begin{bmatrix} 1 & 2 \\ 3 & 4 \end{bmatrix}$$
$$= 1\begin{bmatrix} 1 & 0 \\ 0 & 0 \end{bmatrix} + 2\begin{bmatrix} 0 & 1 \\ 0 & 0 \end{bmatrix}$$
$$+ 3\begin{bmatrix} 0 & 0 \\ 1 & 0 \end{bmatrix} + 4\begin{bmatrix} 0 & 0 \\ 0 & 1 \end{bmatrix}.$$

Hence

$$[T]_\mathcal{B} = \begin{bmatrix} 1 & 0 & 0 & 1 \\ 2 & 0 & 0 & 2 \\ 3 & 0 & 0 & 3 \\ 4 & 0 & 0 & 4 \end{bmatrix}.$$

(c) Suppose that A is a nonzero matrix with trace equal to zero. Then
$$T(A) = (\text{trace}\,A)B = 0B = O = 0A,$$
and hence A is an eigenvector of T with corresponding eigenvalue equal to 0.

(d) Suppose that A is an eigenvector of T with a corresponding nonzero eigenvalue λ. Then
$$\lambda A = T(A) = (\text{trace}\,A)B,$$
and hence $A = \left(\frac{\text{trace}\,A}{\lambda}\right)B$ because $\lambda \neq 0$.

46. (a) For any A and C in $\mathcal{M}_{n \times n}$,
$$T(A + C) = B(A + C) = BA + BC$$
$$= T(A) + T(C).$$
Similarly, $T(cA) = cT(A)$ for any scalar c, and therefore T is linear.

(b) Suppose that B is invertible. Let $U \colon \mathcal{M}_{n \times n} \to \mathcal{M}_{n \times n}$ be defined by $U(A) = B^{-1}A$. Then U is linear by (a). Also $UT(A) = B^{-1}BA = A$ for any A in $\mathcal{M}_{n \times n}$. Similarly, $TU(A) = A$ for any A in $\mathcal{M}_{n \times n}$. It follows that T is invertible, and, in fact, $U = T^{-1}$.

Now suppose that B is not invertible. Then there is a nonzero vector $\mathbf{v}$ such that $B\mathbf{v} = \mathbf{0}$. Let A be the $n \times n$ matrix with all of its columns equal to $\mathbf{v}$. Then $A \neq O$, but
$$T(A) = B[\mathbf{v}\ \mathbf{v}\ \cdots\ \mathbf{v}] = [B\mathbf{v}\ B\mathbf{v}\ \cdots\ B\mathbf{v}]$$
$$= O,$$
and hence T is not invertible.

(c) C is an eigenvector of T with corresponding eigenvalue λ if and only if $T(C) = BC = \lambda C$, and hence if and only if
$$[B\mathbf{c}_1\ B\mathbf{c}_2\ \cdots\ B\mathbf{c}_n] = [\lambda\mathbf{c}_1\ \lambda\mathbf{c}_2\ \cdots\ \lambda\mathbf{c}_n].$$
This condition is equivalent to $B\mathbf{c}_j = \lambda\mathbf{c}_j$ for all $1 \leq j \leq n$, which is true if and only if each column of C lies in the eigenspace of B corresponding to λ.

47. Since
$$\mathbf{v}_j = 0\mathbf{v}_1 + 0\mathbf{v}_2 + \cdots + 1\mathbf{v}_j + \cdots + 0\mathbf{v}_n,$$
it follows that $[\mathbf{v}_j]_\mathcal{B} = \mathbf{e}_j$.

48. Suppose that $\mathcal{B} = \{\mathbf{v}_1, \mathbf{v}_2, \ldots, \mathbf{v}_n\}$. Then for $1 \leq j \leq n$, we have $[UT(\mathbf{v}_j)]_\mathcal{B} = [U]_\mathcal{B}[T(\mathbf{v}_j)]_\mathcal{B}$ by Theorem 7.10. Hence
$$[UT]_\mathcal{B}$$
$$= [[UT(\mathbf{v}_1)]_\mathcal{B}\ [UT(\mathbf{v}_2)]_\mathcal{B}\ \cdots\ [UT(\mathbf{v}_n)]_\mathcal{B}]$$
$$= [[U]_\mathcal{B}[T(\mathbf{v}_1)]_\mathcal{B}\ [U]_\mathcal{B}[T(\mathbf{v}_2)]_\mathcal{B}\ \cdots\ [U]_\mathcal{B}[T(\mathbf{v}_n)]_\mathcal{B}]$$
$$= [U]_\mathcal{B}[[T(\mathbf{v}_1)]_\mathcal{B}\ [T(\mathbf{v}_2)]_\mathcal{B}\ \cdots\ [T(\mathbf{v}_n)]_\mathcal{B}]$$
$$= [U]_\mathcal{B}[T]_\mathcal{B}.$$

49. Let W be the eigenspace of T corresponding to λ. Since $T(\mathbf{0}) = \mathbf{0} = \lambda\mathbf{0}$, the zero vector is in W. For any $\mathbf{u}$ and $\mathbf{v}$ in W,
$$T(\mathbf{u} + \mathbf{v}) = T(\mathbf{u}) + T(\mathbf{v}) = \lambda\mathbf{u} + \lambda\mathbf{v} = \lambda(\mathbf{u} + \mathbf{v}),$$
and hence $\mathbf{u} + \mathbf{v}$ is in W. Similarly, any scalar multiple of $\mathbf{u}$ is in W, and hence W is closed under addition and scalar multiplication. Therefore W is a subspace of V. For any nonzero vector $\mathbf{u}$ in V, the equation $T(\mathbf{u}) = \lambda\mathbf{u}$ is satisfied if and only if $\mathbf{u}$ is an eigenvector corresponding to λ, and hence if and only if $\mathbf{u}$ is in W.

50. (a) Suppose that T is diagonalizable and $\mathcal{B} = \{\mathbf{v}_1, \mathbf{v}_2, \ldots, \mathbf{v}_n\}$ is a basis for V consisting of eigenvectors of T with corresponding eigenvalues $\lambda_1, \lambda_2, \ldots, \lambda_n$. For each j,

$$[T(\mathbf{v}_j)]_\mathcal{B} = [\lambda_j \mathbf{v}_j]_\mathcal{B} = \lambda_j [\mathbf{v}_j]_\mathcal{B} = \lambda_j \mathbf{e}_j,$$

by Exercise 47, and hence

$$[T]_\mathcal{B} = [\,[T(\mathbf{v}_1)]_\mathcal{B} \ \ [T(\mathbf{v}_2)]_\mathcal{B} \ \ \ldots \ \ [T(\mathbf{v}_n)]_\mathcal{B}\,]$$

$$= \begin{bmatrix} \lambda_1 & 0 & \cdots & 0 \\ 0 & \lambda_2 & \cdots & 0 \\ \vdots & \vdots & \vdots & \vdots \\ 0 & 0 & \cdots & \lambda_n \end{bmatrix},$$

which is a diagonal matrix.
Simply reverse the steps to prove the converse.

(b) Suppose that T is diagonalizable, and $\{\mathbf{v}_1, \mathbf{v}_2, \ldots, \mathbf{v}_n\}$ is a basis for V consisting of eigenvectors of T with corresponding eigenvalues $\lambda_1, \lambda_2, \ldots, \lambda_n$, respectively. Then for each j,

$$[T]_\mathcal{B} [\mathbf{v}_j]_\mathcal{B} = [T(\mathbf{v}_j)]_\mathcal{B} = [\lambda_j \mathbf{v}_j]_\mathcal{B} = \lambda_j [\mathbf{v}_j]_\mathcal{B},$$

and hence $\{[\mathbf{v}_1]_\mathcal{B}, [\mathbf{v}_2]_\mathcal{B}, \ldots, [\mathbf{v}_n]_\mathcal{B}\}$ is a basis for $\mathcal{R}^n$ of eigenvectors of $[T]_\mathcal{B}$. Therefore $[T]_\mathcal{B}$ is diagonalizable.
Conversely, suppose that $[T]_\mathcal{B}$ is diagonalizable and $\{\mathbf{w}_1, \mathbf{w}_2, \ldots, \mathbf{w}_n\}$ is a basis for $\mathcal{R}^n$ consisting of eigenvectors of $[T]_\mathcal{B}$ with corresponding eigenvalues $\lambda_1, \lambda_2, \ldots, \lambda_n$, respectively. For each j, let $\mathbf{v}_j$ be the unique vector in V such that $[\mathbf{v}_j]_\mathcal{B} = \mathbf{w}_j$. Then, as above, it can be shown that $T(\mathbf{v}_j) = \lambda_j \mathbf{v}_j$, for all j, and therefore T is diagonalizable.

(c) Observe that

$$\left\{ \begin{bmatrix} 1 & 0 \\ 0 & 0 \end{bmatrix}, \begin{bmatrix} 0 & 0 \\ 0 & 1 \end{bmatrix}, \begin{bmatrix} 0 & 1 \\ 1 & 0 \end{bmatrix} \right\}$$

and

$$\left\{ \begin{bmatrix} 0 & 1 \\ -1 & 0 \end{bmatrix} \right\}$$

are bases for the eigenspaces of U corresponding to the eigenvalues 1 and -1, respectively. Since the set consisting of all of these matrices is a basis for $\mathcal{M}_{2\times 2}$, we conclude that U is diagonalizable.

51. (a) For any matrices A and B in $\mathcal{M}_{2\times 2}$,

$$T(A+B) = (A+B)\mathbf{v} = A\mathbf{v} + B\mathbf{v}$$

$$= T(A) + T(B).$$

Similarly, $T(cA) = cT(A)$ for any scalar c, and hence T is linear.

(b) Observe that

$$\left[T\left(\begin{bmatrix} 1 & 0 \\ 0 & 0 \end{bmatrix}\right)\right]_\mathcal{C} = \left[\begin{bmatrix} 1 & 0 \\ 0 & 0 \end{bmatrix}\begin{bmatrix} 1 \\ 3 \end{bmatrix}\right]_\mathcal{C}$$

$$= \left[\begin{bmatrix} 1 \\ 0 \end{bmatrix}\right]_\mathcal{C} = \begin{bmatrix} 1 \\ 0 \end{bmatrix},$$

$$\left[T\left(\begin{bmatrix} 0 & 1 \\ 0 & 0 \end{bmatrix}\right)\right]_\mathcal{C} = \left[\begin{bmatrix} 0 & 1 \\ 0 & 0 \end{bmatrix}\begin{bmatrix} 1 \\ 3 \end{bmatrix}\right]_\mathcal{C}$$

$$= \left[\begin{bmatrix} 3 \\ 0 \end{bmatrix}\right]_\mathcal{C} = \begin{bmatrix} 3 \\ 0 \end{bmatrix},$$

$$\left[T\left(\begin{bmatrix} 0 & 0 \\ 1 & 0 \end{bmatrix}\right)\right]_\mathcal{C} = \left[\begin{bmatrix} 0 & 0 \\ 1 & 0 \end{bmatrix}\begin{bmatrix} 1 \\ 3 \end{bmatrix}\right]_\mathcal{C}$$

$$= \left[\begin{bmatrix} 0 \\ 1 \end{bmatrix}\right]_\mathcal{C} = \begin{bmatrix} 0 \\ 1 \end{bmatrix},$$

and

$$\left[T\left(\begin{bmatrix} 0 & 0 \\ 0 & 1 \end{bmatrix}\right)\right]_\mathcal{C} = \left[\begin{bmatrix} 0 & 0 \\ 0 & 1 \end{bmatrix}\begin{bmatrix} 1 \\ 3 \end{bmatrix}\right]_\mathcal{C}$$

$$= \left[\begin{bmatrix} 0 \\ 3 \end{bmatrix}\right]_\mathcal{C} = \begin{bmatrix} 0 \\ 3 \end{bmatrix}.$$

Therefore

$$[T]_\mathcal{B}^\mathcal{C} = \begin{bmatrix} 1 & 3 & 0 & 0 \\ 0 & 0 & 1 & 3 \end{bmatrix}.$$

(c) First observe that $[T(M)]_\mathcal{C} = T(M)$ for every matrix M in $\mathcal{B}$ because $\mathcal{C}$ is the standard basis for $\mathcal{R}^2$. Let $D = \begin{bmatrix} 1 & 1 \\ 1 & 2 \end{bmatrix}$. By Theorem 4.11, $[T(M)]_\mathcal{D} = D^{-1}[T(M)]$ for every matrix M in $\mathcal{B}$, and hence

$$[T]_\mathcal{B}^\mathcal{D}$$

$$= \left[D^{-1}\begin{bmatrix} 1 \\ 0 \end{bmatrix} \ \ D^{-1}\begin{bmatrix} 3 \\ 0 \end{bmatrix} \ \ D^{-1}\begin{bmatrix} 0 \\ 1 \end{bmatrix} \ \ D^{-1}\begin{bmatrix} 0 \\ 3 \end{bmatrix} \right]$$

$$= D^{-1}\begin{bmatrix} 1 & 3 & 0 & 0 \\ 0 & 0 & 1 & 3 \end{bmatrix}$$

$$= \begin{bmatrix} 2 & -1 \\ -1 & 1 \end{bmatrix}\begin{bmatrix} 1 & 3 & 0 & 0 \\ 0 & 0 & 1 & 3 \end{bmatrix}$$

$$= \begin{bmatrix} 2 & 6 & -1 & -3 \\ -1 & -3 & 1 & 3 \end{bmatrix}.$$

52. Let $\mathcal{B} = \{\mathbf{b}_1, \mathbf{b}_2, \ldots, \mathbf{b}_n\}$.

(a) Then
$$[sT]_\mathcal{B}^\mathcal{C} = [[sT(\mathbf{b}_1)]_\mathcal{C}\,[sT(\mathbf{b}_2)]_\mathcal{C}\cdots[sT(\mathbf{b}_n)]_\mathcal{C}]$$
$$= [s[T(\mathbf{b}_1)]_\mathcal{C}\,s[T(\mathbf{b}_2)]_\mathcal{C}\cdots s[T(\mathbf{b}_n)]_\mathcal{C}]$$
$$= s\left[[T(\mathbf{b}_1)]_\mathcal{C}\,[T(\mathbf{b}_2)]_\mathcal{C}\cdots[T(\mathbf{b}_n)]_\mathcal{C}\right]$$
$$= s[T]_\mathcal{B}^\mathcal{C}.$$

(b) For each j, the jth column of $[T+U]_\mathcal{B}^\mathcal{C}$ equals
$$(T+U)(\mathbf{b}_j) = [T(\mathbf{b}_j) + U(\mathbf{b}_j)]_\mathcal{C}$$
$$= [T(\mathbf{b}_j)]_\mathcal{C} + [U(\mathbf{b}_j)]_\mathcal{C},$$
which is the jth column of $[T]_\mathcal{B}^\mathcal{C} + [U]_\mathcal{B}^\mathcal{C}$. Thus $[T+U]_\mathcal{B}^\mathcal{C} = [T]_\mathcal{B}^\mathcal{C} + [U]_\mathcal{B}^\mathcal{C}$.

(c) Suppose that $\dim W = m$, and let L_1 and L_2 be the functions from V to $\mathcal{R}^m$ defined by
$$L_1(\mathbf{v}) = [T(\mathbf{v})]_\mathcal{C} \quad\text{and}\quad L_2(\mathbf{v}) = [T]_\mathcal{B}^\mathcal{C}[\mathbf{v}]_\mathcal{B}$$
for every vector $\mathbf{v}$ in V. It is easily shown that both L_1 and L_2 are linear. Since, for each j, $[T]_\mathcal{B}^\mathcal{C}\mathbf{e}_j$ is the jth column of $[T]_\mathcal{B}^\mathcal{C}$ and $[\mathbf{b}_j]_\mathcal{B} = \mathbf{e}_j$ by Exercise 47, we also have, for each j that
$$L_1(\mathbf{b}_j) = [T(\mathbf{b}_j)]_\mathcal{C} = [T]_\mathcal{B}^\mathcal{C}\mathbf{e}_j$$
$$= [T]_\mathcal{B}^\mathcal{C}[\mathbf{b}_j]_\mathcal{B} = L_2(\mathbf{b}_j).$$

Thus L_1 and L_2 agree on the basis $\mathcal{B}$ for V. Because L_1 and L_2 are linear, it follows that $L_1 = L_2$. So $L_1(\mathbf{v}) = L_2(\mathbf{v})$ for every vector $\mathbf{v}$ in V, which gives us the result.

(d) In the following sequence of equations, the conclusion of (c) is used several times. For any j,
$$[UT]_\mathcal{B}^\mathcal{C}\mathbf{e}_j = [UT]_\mathcal{B}^\mathcal{C}[\mathbf{b}_j]_\mathcal{B} = [UT(\mathbf{b}_j)]_\mathcal{D}$$
$$= [U]_\mathcal{C}^\mathcal{D}[T(\mathbf{b}_j)]_\mathcal{C} = [U]_\mathcal{C}^\mathcal{D}[T]_\mathcal{B}^\mathcal{C}[\mathbf{b}]_\mathcal{B}$$
$$= [U]_\mathcal{C}^\mathcal{D}[T]_\mathcal{B}^\mathcal{C}\mathbf{e}_j.$$

It follows that the jth column of $[UT]_\mathcal{B}^\mathcal{C}$ is equal to the jth column of $[U]_\mathcal{C}^\mathcal{D}[T]_\mathcal{B}^\mathcal{C}$ for all j, and we conclude that the two matrices are equal.

53. (a) For any polynomials $f(x)$ and $g(x)$ in $\mathcal{P}_2$,
$$T(f(x) + g(x)) = \begin{bmatrix} f(1) + g(1) \\ g(1) + g(2) \end{bmatrix}$$
$$= \begin{bmatrix} f(1) \\ f(2) \end{bmatrix} + \begin{bmatrix} g(1) \\ g(2) \end{bmatrix}$$
$$= T(f(x)) + T(g(x)).$$
Similarly $T(cf(x)) = cT(f(x))$. Hence T is linear.

(b) Since
$$T(1) = \begin{bmatrix} 1 \\ 1 \end{bmatrix}, \quad T(x) = \begin{bmatrix} 1 \\ 2 \end{bmatrix},$$
and
$$T(x^2) = \begin{bmatrix} 1 \\ 4 \end{bmatrix},$$
it follows that $[T]_\mathcal{B}^\mathcal{C} = \begin{bmatrix} 1 & 1 & 1 \\ 1 & 2 & 4 \end{bmatrix}$.

(c) (i) We have
$$T(f(x)) = \begin{bmatrix} a+b+c \\ a+2b+4b \end{bmatrix}$$
$$= [T(f(x))]_\mathcal{C}.$$

(ii) Clearly, $[f(x)]_\mathcal{B} = \begin{bmatrix} a \\ b \\ c \end{bmatrix}$ and
$$[T]_\mathcal{B}^\mathcal{C}[f(x)]_\mathcal{B} = \begin{bmatrix} 1 & 1 & 1 \\ 1 & 2 & 4 \end{bmatrix}\begin{bmatrix} a \\ b \\ c \end{bmatrix}$$
$$= \begin{bmatrix} a+b+c \\ a+2b+4b \end{bmatrix}$$
$$= [T(f(x))]_\mathcal{C}.$$

54. (rounded to 4 places after the decimal)
  (a) 6.3059, 1.4798, 0.2143, 1.0000
  (b) The vectors in the basis are
  $$-0.4589 - 0.3134x - 0.8314x^2,$$
  $$-0.9404 + 0.3306x + 0.0793x^2,$$
  $$0.1110 - 0.9250x + 0.3634x^2,$$
  and
  $$0.8807 - 0.4003x - 0.2402x^2 + 0.0801x^3.$$

  (c)
  $$T^{-1}(f(x)) = (0.5a_0 - 0.5a_1 + 3a_3)$$
  $$+ (a_0 + 4a_1 - 2a_2 - 2a_3)x$$
  $$+ (-0.5a_0 - 1.5a_1 + a_2 - 2a_3)x^2 + x^3$$

55. (rounded to 4 places after the decimal)
  (a) $-1.6533, 2.6277, 6.6533, 8.3723$
  (b) One possible basis of eigenvectors consists of the matrices
  $$\begin{bmatrix} -0.1827 & -0.7905 \\ 0.5164 & 0.2740 \end{bmatrix},$$

$$\begin{bmatrix} 0.6799 & -0.4655 \\ -0.4655 & 0.3201 \end{bmatrix},$$
$$\begin{bmatrix} 0.4454 & 0.0772 \\ 0.5909 & -0.6681 \end{bmatrix},$$
and
$$\begin{bmatrix} 0.1730 & 0.3783 \\ 0.3783 & 0.8270 \end{bmatrix}$$

(c) The entries in row 1 of the matrix $T^{-1}(A)$ are $0.2438a - 0.1736b + 0.0083c + 0.0496d$ and $-0.2603a - 0.2893b + 0.3471c + 0.0826d$, and the entries in row 2 of the matrix are $0.0124a + 0.3471b - 0.0165c - 0.0992d$ and $-0.1116a + 0.1240b - 0.1488c + 0.1074d$.

7.5 INNER PRODUCT SPACES

1. $\dfrac{15}{4}$ 2. $\dfrac{5}{3}$ 3. $\dfrac{21}{4}$ 4. $\dfrac{31}{5}$

5. $\dfrac{21}{2}$ 6. $\dfrac{3}{2}$

7. We have
$$\langle f, g \rangle = \int_1^2 f(t)g(t)\,dt = \int_1^2 te^t\,dt$$
$$= te^t - e^t \big|_1^2 = (2e^2 - e^2) - (e - e) = e^2.$$

8. $e^2 - 2$ 9. 25 10. 2

11.
$$\langle A, B \rangle = \operatorname{trace}\left(AB^T\right)$$
$$= \operatorname{trace}\left(\begin{bmatrix} 1 & -1 \\ 2 & 3 \end{bmatrix} \begin{bmatrix} 2 & 1 \\ 4 & 0 \end{bmatrix}\right)$$
$$= \operatorname{trace}\left(\begin{bmatrix} -2 & 1 \\ 16 & 2 \end{bmatrix}\right) = 0.$$

12. 11 13. -3 14. 5 15. -3

16. 15 17. 12 18. $\dfrac{4}{3}$

19. We have
$$\langle f(x), g(x) \rangle = \int_{-1}^{1} (x^2 - 2)(3x + 5)\,dx$$
$$= \int_{-1}^{1} (3x^3 + 5x^2 - 6x - 10)\,dx$$
$$= x^4 + \dfrac{5}{3}x^3 - 3x^2 - 10x \Big|_{-1}^{1}$$
$$= \dfrac{10}{3} - 20 = -\dfrac{50}{3}.$$

20. $-\dfrac{4}{3}$ 21. 0 22. $\dfrac{2}{3}$ 23. $-\dfrac{8}{3}$

24. $-\dfrac{44}{15}$

25. False, it is a scalar.

26. True

27. False, it has scalar values.

28. False, any positive scalar multiple of an inner product is an inner product.

29. True

30. False, if the set contains the zero vector, it is linearly dependent.

31. True 32. True 33. True 34. True

35. False, the indefinite integral of functions is not a scalar.

36. True 37. True

38. False, the norm of a vector equals $\sqrt{\langle \mathbf{v}, \mathbf{v} \rangle}$.

39. False, the equality must hold for *every* vector $\mathbf{u}$.

40. False, $\langle A, B \rangle = \operatorname{trace}(AB^T)$.

41. True 42. True

43. False, to obtain the normalized Legendre polynomials, these polynomials must be normalized.

44. False, $\mathcal{B}$ must be an orthonormal basis.

45. Let f, g, and h be in $\mathsf{C}([a,b])$.
   Axiom 3 We have
$$\langle f + g, h \rangle = \int_a^b (f + g)(t)h(t)\,dt$$
$$= \int_a^b f(t)h(t)\,dt + \int_a^b g(t)h(t)\,dt$$
$$= \langle f, h \rangle + \langle g, h \rangle.$$

   Axiom 4 For any scalar c, we have
$$\langle cf, g \rangle = \int_a^b (cf(t))g(t)\,dt$$
$$= c\int_a^b f(t)g(t)\,dt = c\langle f, g \rangle.$$

46. Let A, B, and C be in $\mathcal{M}_{n \times n}$.
   Axiom 3 We have
$$\langle A + B, C \rangle = \operatorname{trace}(A + B)C^T$$
$$= \operatorname{trace}(AC^T + BC^T)$$
$$= \operatorname{trace}(AC^T) + \operatorname{trace}(BC^T)$$
$$= \langle A, C \rangle + \langle B, C \rangle.$$

Axiom 4 For any scalar a, we have
$$\langle aA, B \rangle = \operatorname{trace}(aA)B^T = a\operatorname{trace}(AB^T)$$
$$= a\langle A, B \rangle.$$

47. Let $\mathbf{u}$, $\mathbf{v}$, and $\mathbf{w}$ be in V. If $\mathbf{u} \neq \mathbf{0}$, then $[\mathbf{u}]_\mathcal{B} \neq \mathbf{0}$, and hence $\langle \mathbf{u}, \mathbf{u} \rangle = [\mathbf{u}]_\mathcal{B} \cdot [\mathbf{u}]_\mathcal{B} > 0$, verifying axiom 1.
    Since $\langle \mathbf{u}, \mathbf{v} \rangle = [\mathbf{u}]_\mathcal{B} \cdot [\mathbf{v}]_\mathcal{B} = [\mathbf{v}]_\mathcal{B} \cdot [\mathbf{u}]_\mathcal{B} = \langle \mathbf{v}, \mathbf{u} \rangle$, axiom 2 is established.
    Notice that
$$\langle \mathbf{u} + \mathbf{v}, \mathbf{w} \rangle = [\mathbf{u} + \mathbf{v}]_\mathcal{B} \cdot [\mathbf{w}]_\mathcal{B}$$
$$= ([\mathbf{u}]_\mathcal{B} + [\mathbf{v}]_\mathcal{B}) \cdot [\mathbf{w}]_\mathcal{B}$$
$$= [\mathbf{u}]_\mathcal{B} \cdot [\mathbf{w}]_\mathcal{B} + [\mathbf{v}]_\mathcal{B} \cdot [\mathbf{w}]_\mathcal{B}$$
$$= \langle \mathbf{u}, \mathbf{w} \rangle + \langle \mathbf{v}, \mathbf{w} \rangle,$$
    establishing axiom 3.
    Finally, for any scalar a,
$$\langle a\mathbf{u}, \mathbf{v} \rangle = [a\mathbf{u}]_\mathcal{B} \cdot [\mathbf{v}]_\mathcal{B} = a[\mathbf{u}]_\mathcal{B} \cdot [\mathbf{v}]_\mathcal{B} = a\langle \mathbf{u}, \mathbf{v} \rangle,$$
    establishing axiom 4.

48. For any $\mathbf{u}$ and $\mathbf{v}$ in $\mathcal{R}^n$,
$$\langle \mathbf{u}, \mathbf{v} \rangle = (A\mathbf{u}) \cdot (A\mathbf{v}) = (A\mathbf{u})^T(A\mathbf{v})$$
$$= \mathbf{u}^T A^T A \mathbf{v} = (A^T A \mathbf{u})^T \mathbf{v}$$
$$= (A^T A \mathbf{u}) \cdot \mathbf{v}.$$
    But $A^T A$ is positive definite by Exercise 72 of Section 6.6, and hence we have an inner product by Exercise 47.

49. Let $\mathbf{u}$, $\mathbf{v}$, and $\mathbf{w}$ be in $\mathcal{R}^n$. If $\mathbf{u} \neq \mathbf{0}$, then
$$\langle \mathbf{u}, \mathbf{u} \rangle = A\mathbf{u} \cdot \mathbf{u} = (A\mathbf{u})^T \mathbf{u} = \mathbf{u}^T A \mathbf{u} > 0$$
    because A is positive definite, establishing axiom 1.
    We have
$$\langle \mathbf{u}, \mathbf{v} \rangle = (A\mathbf{u}) \cdot \mathbf{v} = (A\mathbf{u})^T \mathbf{v} = \mathbf{u}^T A \mathbf{v}$$
$$= \mathbf{u} \cdot (A\mathbf{v}) = (A\mathbf{v}) \cdot \mathbf{u} = \langle \mathbf{v}, \mathbf{u} \rangle,$$
    establishing axiom 2.
    Notice that
$$\langle \mathbf{u} + \mathbf{v}, \mathbf{w} \rangle = (A(\mathbf{u} + \mathbf{v})) \cdot \mathbf{w} = (A\mathbf{u} + A\mathbf{v}) \cdot \mathbf{w}$$
$$= (A\mathbf{u}) \cdot \mathbf{w} + (A\mathbf{v}) \cdot \mathbf{w}$$
$$= \langle \mathbf{u}, \mathbf{w} \rangle + \langle \mathbf{u}, \mathbf{w} \rangle,$$
    establishing axiom 3.
    Finally, for any scalar a,
$$\langle a\mathbf{u}, \mathbf{v} \rangle = (A(a\mathbf{u})) \cdot \mathbf{v} = a((A\mathbf{u}) \cdot \mathbf{v}) = a\langle \mathbf{u}, \mathbf{v} \rangle,$$
    establishing axiom 4.

50. No. We show that axiom 4 is not satisfied. Choose any vectors $\mathbf{u}$ and $\mathbf{v}$ in $\mathcal{R}^n$ such that $\mathbf{u} \cdot \mathbf{v} \neq 0$. Then
$$\langle (-1)\mathbf{u}, \mathbf{v} \rangle = |(-1)\mathbf{u} \cdot \mathbf{v}| = |\mathbf{u} \cdot \mathbf{v}| \neq (-1)\langle \mathbf{u}, \mathbf{v} \rangle.$$

51. Yes. We verify the axioms of an inner product.
    Axiom 1 Let $\mathbf{u}$ be a nonzero vector in V. Then $\langle \mathbf{u}, \mathbf{u} \rangle = 2(\mathbf{u} \cdot \mathbf{u}) \neq 0$ because $\mathbf{u} \cdot \mathbf{u} \neq 0$.
    Axiom 2 Let $\mathbf{u}$ and $\mathbf{v}$ be in V. Then
$$\langle \mathbf{u}, \mathbf{v} \rangle = 2(\mathbf{u} \cdot \mathbf{v}) = 2(\mathbf{v} \cdot \mathbf{u}) = \langle \mathbf{v}, \mathbf{u} \rangle.$$
    Axiom 3 Let $\mathbf{u}$, $\mathbf{v}$, and $\mathbf{w}$ be in V. Then
$$\langle \mathbf{u} + \mathbf{v}, \mathbf{w} \rangle = 2(\mathbf{u} + \mathbf{v}) \cdot \mathbf{w} = 2\mathbf{u} \cdot \mathbf{w} + 2\mathbf{v} \cdot \mathbf{w}$$
$$= \langle \mathbf{u}, \mathbf{w} \rangle + \langle \mathbf{v}, \mathbf{w} \rangle.$$
    Axiom 4 Let $\mathbf{u}$ and $\mathbf{v}$ be in V, and let a be a scalar. Then
$$\langle a\mathbf{u}, \mathbf{v} \rangle = 2(a\mathbf{u} \cdot \mathbf{v}) = 2a(\mathbf{v} \cdot \mathbf{u})$$
$$= a(2(\mathbf{v} \cdot \mathbf{u})) = a\langle \mathbf{u}, \mathbf{v} \rangle.$$

52. Yes. By Exercise 49, the rule defines an inner product because D is a positive definite matrix.

53. No. We show that axiom 1 is not satisfied. Let $f: [0, 2] \to \mathcal{R}$ be defined by
$$f(t) = \begin{cases} 0 & \text{if } 0 \leq t \leq 1 \\ t - 1 & \text{if } 1 < t \leq 2. \end{cases}$$
    Since f is continuous, it is in V. Furthermore, f is not the zero function. However
$$\langle f, f \rangle = \int_0^1 f(t)^2 \, dt = \int_0^1 0 \, dt = 0.$$

54. No. We show that axiom 1 is not satisfied. Choose any nonzero vector $\mathbf{u}$ in $\mathcal{R}^n$. Then
$$\langle \mathbf{u}, \mathbf{u} \rangle = -2(\mathbf{u} \cdot \mathbf{u}) < 0.$$

55. Yes. We verify the axioms of an inner product.
    Axiom 1 Let $\mathbf{u}$ be a nonzero vector in V. Then
$$\langle \mathbf{u}, \mathbf{u} \rangle = \langle \mathbf{u}, \mathbf{u} \rangle_1 + \langle \mathbf{u}, \mathbf{u} \rangle_2 > 0$$
    since $\langle \mathbf{u}, \mathbf{u} \rangle_1 > 0$ and $\langle \mathbf{u}, \mathbf{u} \rangle_2 > 0$.
    Axiom 2 Let $\mathbf{u}$ and $\mathbf{v}$ be in V. Then
$$\langle \mathbf{u}, \mathbf{v} \rangle = \langle \mathbf{u}, \mathbf{v} \rangle_1 + \langle \mathbf{u}, \mathbf{v} \rangle_2$$
$$= \langle \mathbf{v}, \mathbf{u} \rangle_1 + \langle \mathbf{v}, \mathbf{u} \rangle_2 = \langle \mathbf{v}, \mathbf{u} \rangle.$$

7.5 Inner Product Spaces

Axiom 3 Let $\mathbf{u}$, $\mathbf{v}$, and $\mathbf{w}$ be in V. Then

$$\langle \mathbf{u}+\mathbf{v}, \mathbf{w}\rangle = \langle \mathbf{u}+\mathbf{v}, \mathbf{w}\rangle_1 + \langle \mathbf{u}+\mathbf{v}, \mathbf{w}\rangle_2$$
$$= \langle \mathbf{u}, \mathbf{w}\rangle_1 + \langle \mathbf{v}, \mathbf{w}\rangle_1 + \langle \mathbf{u}, \mathbf{w}\rangle_2$$
$$+ \langle \mathbf{v}, \mathbf{w}\rangle_2$$
$$= \langle \mathbf{u}, \mathbf{w}\rangle_1 + \langle \mathbf{u}, \mathbf{w}\rangle_2 + \langle \mathbf{v}, \mathbf{w}\rangle_1$$
$$+ \langle \mathbf{v}, \mathbf{w}\rangle_2$$
$$= \langle \mathbf{u}, \mathbf{w}\rangle + \langle \mathbf{v}, \mathbf{w}\rangle.$$

Axiom 4 Let $\mathbf{u}$ and $\mathbf{v}$ be in V, and let a be a scalar. Then

$$\langle a\mathbf{u}, \mathbf{v}\rangle = \langle a\mathbf{u}, \mathbf{v}\rangle_1 + \langle a\mathbf{u}, \mathbf{v}\rangle_2$$
$$= a\langle \mathbf{u}, \mathbf{v}\rangle_1 + a\langle \mathbf{u}, \mathbf{v}\rangle_2$$
$$= a(\langle \mathbf{u}, \mathbf{v}\rangle_1 + \langle \mathbf{u}, \mathbf{v}\rangle_2)$$
$$= a\langle \mathbf{u}, \mathbf{v}\rangle.$$

56. The result depends on the choice of the given inner products. Notice that this rule always satisfies axioms 2, 3, and 4. Axiom 1 is satisfied (and hence the rule defines an inner product) if and only if $\langle \mathbf{u}, \mathbf{u}\rangle_1 > \langle \mathbf{u}, \mathbf{u}\rangle_2$ for every nonzero vector $\mathbf{u}$ in V. As a simple example, let $V = \mathcal{R}^n$, and consider the inner products defined by $\langle \mathbf{u}, \mathbf{v}\rangle_1 = 2(\mathbf{u}\cdot\mathbf{v})$ and $\langle \mathbf{u}, \mathbf{v}\rangle_2 = \mathbf{u}\cdot\mathbf{v}$ for every $\mathbf{u}$ and $\mathbf{v}$ in $\mathcal{R}^n$. In this case, $\langle \mathbf{u}, \mathbf{v}\rangle = \mathbf{u}\cdot\mathbf{v}$ for every $\mathbf{u}$ and $\mathbf{v}$ in $\mathcal{R}^n$, and hence the rule defines an inner product. On the other hand, if the definitions are reversed, we obtain $\langle \mathbf{u}, \mathbf{v}\rangle = -(\mathbf{u}\cdot\mathbf{v})$ for every $\mathbf{u}$ and $\mathbf{v}$ in $\mathcal{R}^n$, and hence the rule does not define an inner product.

57. Yes. With some minor modifications, the proof is similar to that in Exercise 55, which is a special case.

58. Suppose that $m \neq n$. Then

$$\langle \sin mt, \sin nt\rangle = \int_0^{2\pi} \sin mt \sin nt\, dt$$
$$= \frac{1}{2}\int_0^{2\pi} \cos(m+n)t\, dt$$
$$- \frac{1}{2}\int_0^{2\pi} \cos(m-n)t\, dt$$
$$= \frac{1}{2(m+n)}\sin(m+n)t\Big|_0^{2\pi}$$
$$- \frac{1}{2(m-n)}\sin(m-n)t\Big|_0^{2\pi}$$
$$= 0+0 = 0.$$

59. This is similar to Exercise 58.

60. (a) $p_3(x) = \sqrt{\frac{7}{8}}(5x^3 - 3x)$ (b) $\frac{180}{91}x - \frac{15}{13}x^3$

61. Let $\mathbf{u}_1 = 1$, $\mathbf{u}_2 = e^t$, and $\mathbf{u}_3 = e^{-t}$. We apply the Gram-Schmidt process to $\{\mathbf{u}_1, \mathbf{u}_2, \mathbf{u}_3\}$ to obtain an orthogonal basis $\{\mathbf{v}_1, \mathbf{v}_2, \mathbf{v}_3\}$. Let $\mathbf{v}_1 = \mathbf{u}_1 = 1$,

$$\mathbf{v}_2 = \mathbf{u}_2 - \frac{\langle \mathbf{u}_2, \mathbf{v}_1\rangle}{\|\mathbf{v}_1\|^2}\mathbf{v}_1$$
$$= e^t - \frac{\int_0^1 e^t 1\, dt}{\int_0^1 1^2\, dt}1$$
$$= e^t - \frac{e-1}{1}1 = e^t - e + 1,$$

and

$$\mathbf{v}_3 = \mathbf{u}_3 - \frac{\langle \mathbf{u}_3, \mathbf{v}_1\rangle}{\|\mathbf{v}_1\|^2}\mathbf{v}_1 - \frac{\langle \mathbf{u}_3, \mathbf{v}_2\rangle}{\|\mathbf{v}_2\|^2}\mathbf{v}_2$$
$$= e^{-t} - \frac{\int_0^1 e^{-t}1\, dt}{\int_0^1 1^2\, dt}1 -$$
$$\frac{\int_0^1 e^{-t}(e^t - e + 1)\, dt}{\int_0^1 (e^t - e + 1)^2\, dt}(e^t - e + 1)$$
$$= e^{-t} - \frac{e-1}{e} - \frac{2(e^2 - 3e + 1)}{e(e-1)(e-3)}(e^t - e + 1)$$
$$= e^{-t} + \frac{e^2 - 2e - 1}{e(e-3)} - \frac{2(e^2 - 3e + 1)}{e(e-1)(e-3)}e^t.$$

Thus

$$\{\mathbf{v}_1, \mathbf{v}_2, \mathbf{v}_3\} = \Big\{1, e^t - e + 1,$$
$$e^{-t} + \frac{e^2 - 2e - 1}{e(e-3)} - \frac{2(e^2 - 3e + 1)}{e(e-1)(e-3)}e^t\Big\}.$$

62. (a) Let $\mathbf{u}$, $\mathbf{v}$, and $\mathbf{w}$ be in V. If $\mathbf{u} \neq \mathbf{0}$, then $\langle\langle \mathbf{u}, \mathbf{u}\rangle\rangle = r\langle \mathbf{u}, \mathbf{u}\rangle > 0$ since $\langle \mathbf{u}, \mathbf{u}\rangle > 0$. This establishes axiom 1.
Since
$$\langle\langle \mathbf{u}, \mathbf{v}\rangle\rangle = r\langle \mathbf{u}, \mathbf{v}\rangle = r\langle \mathbf{v}, \mathbf{u}\rangle = \langle\langle \mathbf{v}, \mathbf{u}\rangle\rangle,$$
axiom 2 follows.
Because
$$\langle\langle \mathbf{u}+\mathbf{v}, \mathbf{w}\rangle\rangle = r\langle \mathbf{u}+\mathbf{v}, \mathbf{w}\rangle$$
$$= r\langle \mathbf{u}, \mathbf{w}\rangle + r\langle \mathbf{v}, \mathbf{w}\rangle$$
$$= \langle\langle \mathbf{u}, \mathbf{w}\rangle\rangle + \langle\langle \mathbf{v}, \mathbf{w}\rangle\rangle,$$
axiom 3 follows.

Finally, for any scalar a,

$$\langle\langle a\mathbf{u}, \mathbf{v}\rangle\rangle = r\langle a\mathbf{u}, \mathbf{v}\rangle = a(r\langle \mathbf{u}, \mathbf{v}\rangle)$$
$$= a\langle\langle \mathbf{u}, \mathbf{v}\rangle\rangle,$$

and hence axiom 4 is satisfied.

(b) If $r \leq 0$ and $\mathbf{u}$ is a nonzero vector, then $\langle\langle \mathbf{u}, \mathbf{u}\rangle\rangle = r\langle \mathbf{u}, \mathbf{u}\rangle \leq 0$ because $\langle \mathbf{u}, \mathbf{u}\rangle > 0$.

63. Suppose that $\|\mathbf{v}\| = 0$. Then $0 = \|\mathbf{v}\|^2 = \langle \mathbf{v}, \mathbf{v}\rangle$, and hence $\mathbf{v} = \mathbf{0}$ by axiom 1. Conversely, suppose that $\mathbf{v} = \mathbf{0}$. Then

$$\|\mathbf{v}\|^2 = \langle \mathbf{0}, \mathbf{0}\rangle = \langle 0\mathbf{0}, \mathbf{0}\rangle = 0\langle \mathbf{0}, \mathbf{0}\rangle = 0,$$

and hence $\|\mathbf{v}\| = 0$.

64. Since

$$\|c\mathbf{v}\|^2 = \langle c\mathbf{v}, c\mathbf{v}\rangle = c\langle \mathbf{v}, c\mathbf{v}\rangle = c\langle c\mathbf{v}, \mathbf{v}\rangle$$
$$= c^2\|\mathbf{v}\|^2,$$

we have $\|c\mathbf{v}\| = \sqrt{c^2\|\mathbf{v}\|^2} = |c|\|\mathbf{v}\|$.

65. We have

$$\langle \mathbf{u}, \mathbf{0}\rangle = \langle \mathbf{0}, \mathbf{u}\rangle = \langle 0\mathbf{0}, \mathbf{u}\rangle = 0\langle \mathbf{0}, \mathbf{u}\rangle = 0.$$

66. We have

$$\langle \mathbf{u} - \mathbf{w}, \mathbf{v}\rangle = \langle \mathbf{u} + (-1)\mathbf{w}, \mathbf{v}\rangle$$
$$= \langle \mathbf{u}, \mathbf{v}\rangle + (-1)\langle \mathbf{w}, \mathbf{v}\rangle$$
$$= \langle \mathbf{u}, \mathbf{v}\rangle - \langle \mathbf{w}, \mathbf{v}\rangle.$$

67. By Exercise 66,

$$\langle \mathbf{v}, \mathbf{u} - \mathbf{w}\rangle = \langle \mathbf{u} - \mathbf{w}, \mathbf{v}\rangle = \langle \mathbf{u}, \mathbf{v}\rangle - \langle \mathbf{w}, \mathbf{v}\rangle$$
$$= \langle \mathbf{v}, \mathbf{u}\rangle - \langle \mathbf{v}, \mathbf{w}\rangle.$$

68. We have

$$\langle \mathbf{u}, c\mathbf{v}\rangle = \langle c\mathbf{v}, \mathbf{u}\rangle = c\langle \mathbf{v}, \mathbf{u}\rangle = c\langle \mathbf{u}, \mathbf{v}\rangle.$$

69. Suppose that $\langle \mathbf{u}, \mathbf{v}\rangle = 0$ for all $\mathbf{u}$ in V. Since $\mathbf{w}$ is in V, we have $\langle \mathbf{w}, \mathbf{w}\rangle = 0$, and hence $\mathbf{w} = \mathbf{0}$ by axiom 1.

70. Suppose that $\langle \mathbf{u}, \mathbf{v}\rangle = \langle \mathbf{u}, \mathbf{w}\rangle$ for all $\mathbf{u}$ in W. Then

$$\langle \mathbf{u}, \mathbf{v} - \mathbf{w}\rangle = \langle \mathbf{u}, \mathbf{v}\rangle - \langle \mathbf{u}, \mathbf{w}\rangle = 0,$$

for all $\mathbf{u}$ in V, and hence $\mathbf{v} - \mathbf{w} = \mathbf{0}$ by Exercise 69. Therefore $\mathbf{v} = \mathbf{w}$.

71. Suppose that $\mathcal{B} = \{\mathbf{v}_1, \mathbf{v}_2, \ldots, \mathbf{v}_n\}$,

$$[\mathbf{u}]_\mathcal{B} = \begin{bmatrix} a_1 \\ a_2 \\ \vdots \\ a_n \end{bmatrix}, \quad \text{and} \quad [\mathbf{v}]_\mathcal{B} = \begin{bmatrix} b_1 \\ b_2 \\ \vdots \\ b_n \end{bmatrix}.$$

Then

$$\mathbf{u} = a_1\mathbf{v}_1 + a_2\mathbf{v}_2 + \cdots + a_n\mathbf{v}_n \quad \text{and}$$
$$\mathbf{v} = b_1\mathbf{v}_1 + b_2\mathbf{v}_2 + \cdots + b_n\mathbf{v}_n.$$

For any i and j,

$$\langle a_i\mathbf{v}_i, b_j\mathbf{v}_j\rangle = a_ib_j\langle \mathbf{v}_i, \mathbf{v}_j\rangle = \begin{cases} a_ib_j & \text{if } i = j \\ 0 & \text{if } i \neq j, \end{cases}$$

and hence

$$\langle \mathbf{u}, \mathbf{v}\rangle = \langle a_1\mathbf{v}_1 + a_2\mathbf{v}_2 + \cdots + a_n\mathbf{v}_n,$$
$$b_1\mathbf{v}_1 + b_2\mathbf{v}_2 + \cdots + b_n\mathbf{v}_n\rangle$$
$$= a_1b_1 + a_2b_2 + \cdots + a_nb_n = [\mathbf{u}]_\mathcal{B} \bullet [\mathbf{v}]_\mathcal{B}.$$

72. Observe that

$$\langle A, B\rangle = \text{trace}\,(AB^T) = \text{trace}\,(A(-B))$$
$$= -\text{trace}\,(AB)$$

and

$$\langle A, B\rangle = \langle B, A\rangle = \text{trace}\,(BA^T) = \text{trace}\,(BA).$$

By Exercise 65 of Section 2.1, trace $(AB) = $ trace (BA), and hence $\langle A, B\rangle = -\langle A, B\rangle$. It follows, therefore, that $\langle A, B\rangle = 0$.

73. Observe that

$$AB^T = \begin{bmatrix} a_{11} & a_{12} \\ a_{21} & a_{22} \end{bmatrix} \begin{bmatrix} b_{11} & b_{21} \\ b_{12} & b_{22} \end{bmatrix}$$
$$= \begin{bmatrix} a_{11}b_{11} + a_{12}b_{12} & a_{11}b_{21} + a_{12}b_{22} \\ a_{21}b_{11} + a_{22}b_{12} & a_{21}b_{21} + a_{22}b_{22} \end{bmatrix},$$

and hence

$$\langle A, B\rangle = \text{trace}\,(AB^T)$$
$$= a_{11}b_{11} + a_{12}b_{12} + a_{21}b_{21} + a_{22}b_{22}.$$

74. Let $C = AB^T$. Then

$$c_{ii} = a_{i1}b_{i1} + a_{i2}b_{i2} + \cdots + a_{in}b_{in},$$

and hence

$$\langle A, B\rangle = \text{trace}\,C = c_{11} + c_{22} + \cdots + c_{nn}$$
$$= a_{11}b_{11} + a_{12}b_{12} + \cdots + a_{nn}b_{nn}.$$

75. (a) $\left\{ \begin{bmatrix} 1 & 0 \\ 0 & 0 \end{bmatrix}, \frac{1}{\sqrt{2}} \begin{bmatrix} 0 & 1 \\ 1 & 0 \end{bmatrix}, \begin{bmatrix} 0 & 0 \\ 0 & 1 \end{bmatrix} \right\}$

(b) $\begin{bmatrix} 1 & 3 \\ 3 & 8 \end{bmatrix}$

76. Let $\mathcal{S}$ be any finite subset of $\mathcal{B}$. The proof that $\mathcal{S}$ is linearly independent is identical to the proof of Theorem 6.5 except that the inner product $\langle \mathbf{u}, \mathbf{v} \rangle$ is used in place of the dot product $\mathbf{u} \cdot \mathbf{v}$. It follows that every finite nonempty subset of $\mathcal{B}$ is linearly independent, and hence $\mathcal{B}$ is linearly independent.

77. If $\mathbf{u}$ or $\mathbf{v}$ is the zero vector, then both sides of the equality have the value zero. So suppose that $\mathbf{u} \neq \mathbf{0}$ and $\mathbf{v} \neq \mathbf{0}$. Then there exists a scalar c such that $\mathbf{v} = c\mathbf{u}$. Hence $\langle \mathbf{u}, \mathbf{v} \rangle^2 = \langle \mathbf{u}, c\mathbf{u} \rangle^2 = c^2 \langle \mathbf{u}, \mathbf{u} \rangle^2$, and

$$\langle \mathbf{u}, \mathbf{u} \rangle \langle \mathbf{v}, \mathbf{v} \rangle = \langle \mathbf{u}, \mathbf{u} \rangle \langle c\mathbf{u}, c\mathbf{u} \rangle = \langle \mathbf{u}, \mathbf{u} \rangle c^2 \langle \mathbf{u}, \mathbf{u} \rangle$$
$$= c^2 \langle \mathbf{u}, \mathbf{u} \rangle^2.$$

Therefore $\langle \mathbf{u}, \mathbf{v} \rangle^2 = \langle \mathbf{u}, \mathbf{u} \rangle \langle \mathbf{v}, \mathbf{v} \rangle$.

78. If $\mathbf{u}$ or $\mathbf{v}$ is the zero vector, then they are linearly dependent. So suppose that $\mathbf{u} \neq \mathbf{0}$, $\mathbf{v} \neq \mathbf{0}$, and $c = \frac{\langle \mathbf{u}, \mathbf{v} \rangle}{\langle \mathbf{u}, \mathbf{u} \rangle}$. Then

$$\|\mathbf{v} - c\mathbf{u}\|^2 = \langle \mathbf{v}, \mathbf{v} \rangle - 2c\langle \mathbf{v}, \mathbf{u} \rangle + c^2 \langle \mathbf{u}, \mathbf{u} \rangle$$
$$= \langle \mathbf{v}, \mathbf{v} \rangle - 2\frac{\langle \mathbf{u}, \mathbf{v} \rangle^2}{\langle \mathbf{u}, \mathbf{u} \rangle} + \frac{\langle \mathbf{u}, \mathbf{v} \rangle^2}{\langle \mathbf{u}, \mathbf{u} \rangle^2} \langle \mathbf{u}, \mathbf{u} \rangle$$
$$= \langle \mathbf{v}, \mathbf{v} \rangle - 2\frac{\langle \mathbf{u}, \mathbf{u} \rangle \langle \mathbf{v}, \mathbf{v} \rangle}{\langle \mathbf{u}, \mathbf{u} \rangle}$$
$$\quad + \frac{\langle \mathbf{u}, \mathbf{u} \rangle \langle \mathbf{v}, \mathbf{v} \rangle \langle \mathbf{u}, \mathbf{u} \rangle}{\langle \mathbf{u}, \mathbf{u} \rangle^2}$$
$$= \langle \mathbf{v}, \mathbf{v} \rangle - 2\langle \mathbf{v}, \mathbf{v} \rangle + \langle \mathbf{v}, \mathbf{v} \rangle$$
$$= 0.$$

Therefore $\mathbf{v} = c\mathbf{u}$.

79. For any $\mathbf{v}_1$ and $\mathbf{v}_2$ in V,

$$F_{\mathbf{u}}(\mathbf{v}_1 + \mathbf{v}_2) = \langle \mathbf{v}_1 + \mathbf{v}_2, \mathbf{u} \rangle = \langle \mathbf{v}_1, \mathbf{u} \rangle + \langle \mathbf{v}_2, \mathbf{u} \rangle$$
$$= F_{\mathbf{u}}(\mathbf{v}_1) + F_{\mathbf{u}}(\mathbf{v}_2).$$

Similarly, $F_{\mathbf{u}}(c\mathbf{v}) = cF_{\mathbf{u}}(\mathbf{v})$ for any vector $\mathbf{v}$ in V and any scalar c, and therefore $F_{\mathbf{u}}$ is linear.

80. Let $\{\mathbf{v}_1, \mathbf{v}_2, \ldots, \mathbf{v}_n\}$ be an orthonormal basis for V, and let

$$\mathbf{u} = T(\mathbf{v}_1)\mathbf{v}_1 + \cdots + T(\mathbf{v}_n)\mathbf{v}_n.$$

For each i,

$$F_{\mathbf{u}}(\mathbf{v}_i) = \langle \mathbf{v}_i, T(\mathbf{v}_1)\mathbf{v}_1 + \cdots + T(\mathbf{v}_n)\mathbf{v}_n \rangle$$
$$= \langle \mathbf{v}_i, T(\mathbf{v}_1)\mathbf{v}_1 \rangle + \cdots + \langle \mathbf{v}_i, T(\mathbf{v}_i)\mathbf{v}_i \rangle$$
$$\quad + \cdots + \langle \mathbf{v}_i, T(\mathbf{v}_n)\mathbf{v}_n \rangle$$
$$= \langle \mathbf{v}_i, \mathbf{v}_1 \rangle T(\mathbf{v}_1) + \cdots + \langle \mathbf{v}_i, \mathbf{v}_i \rangle T(\mathbf{v}_i)$$
$$\quad + \cdots + \langle \mathbf{v}_i, \mathbf{v}_n \rangle T(\mathbf{v}_n)$$
$$= T(\mathbf{v}_i).$$

Now consider any vector $\mathbf{v}$ in V, and suppose that

$$\mathbf{v} = a_1 \mathbf{v}_1 + a_2 \mathbf{v}_2 + \cdots + a_n \mathbf{v}_n.$$

Then

$$F_{\mathbf{u}}(\mathbf{v}) = a_1 F_{\mathbf{u}}(\mathbf{v}_1) + a_2 F_{\mathbf{u}}(\mathbf{v}_2) + \cdots + a_n F_{\mathbf{u}}(\mathbf{v}_n)$$
$$= a_1 T(\mathbf{v}_1) + a_2 T(\mathbf{v}_2) + \cdots + a_n T(\mathbf{v}_n)$$
$$= T(a_1 \mathbf{v}_1 + a_2 \mathbf{v}_2 + \cdots + a_n \mathbf{v}_n)$$
$$= T(\mathbf{v}).$$

Therefore $T = F_{\mathbf{u}}$.

81. (a) This is identical to Exercise 72 in Section 6.6.

(b) Let $\mathcal{B}$ be a basis for $\mathcal{R}^n$ that is orthonormal with respect to the given inner product, let B be the $n \times n$ matrix whose columns are the vectors in $\mathcal{B}$, and let $A = (B^{-1})^T B^{-1}$. (Although $\mathcal{B}$ is orthonormal with respect to the given inner product, it need not be orthonormal with respect to the usual dot product on $\mathcal{R}^n$.) Then A is positive definite by Exercise 72 in Section 6.6. Furthermore, by Theorem 4.11, $[\mathbf{u}]_\mathcal{B} = B^{-1}\mathbf{u}$ for any vector $\mathbf{u}$ in $\mathcal{R}^n$. So, for any vectors $\mathbf{u}$ and $\mathbf{v}$ in $\mathcal{R}^n$, we may apply Exercise 71 to obtain

$$\langle \mathbf{u}, \mathbf{v} \rangle = [\mathbf{u}]_\mathcal{B} \cdot [\mathbf{v}]_\mathcal{B} = (B^{-1}\mathbf{u}) \cdot (B^{-1}\mathbf{v})$$
$$= (B^{-1}\mathbf{u})^T (B^{-1}\mathbf{v})$$
$$= \mathbf{u}^T (B^{-1})^T (B^{-1}\mathbf{v})$$
$$= \mathbf{u}^T A \mathbf{v} = (A\mathbf{u})^T \mathbf{v}$$
$$= (A\mathbf{u}) \cdot \mathbf{v}.$$

82. (a) Let $T = I$, the identity operator on V. Then T is a linear isometry, and hence V is isometric to itself.

(b) Suppose that $T: V \to W$ is a linear isometry. Then T^{-1} is an isomorphism by Theorem 7.6. Let $\mathbf{u}$ and $\mathbf{v}$ be in W, and suppose that $\mathbf{u}' = T^{-1}(\mathbf{u})$ and $\mathbf{v}' = T^{-1}(\mathbf{v})$. Then $T(\mathbf{u}') = \mathbf{u}$ and $T(\mathbf{v}') = \mathbf{v}$. Thus

$$\langle \mathbf{u}, \mathbf{v} \rangle = \langle T(\mathbf{u}'), T(\mathbf{v}') \rangle = \langle \mathbf{u}', \mathbf{v}' \rangle$$

$$= \langle T^{-1}(\mathbf{u}), T^{-1}(\mathbf{v})\rangle,$$

and hence T^{-1} is a linear isometry. Therefore W is isometric to V.

(c) Let $T\colon V \to W$ and $U\colon W \to Z$ be linear isometries. Then UT is an isomorphism by Exercise 53 of Section 7.2. Furthermore, for any $\mathbf{u}$ and $\mathbf{v}$ in V,

$$\langle UT(\mathbf{u}), UT(\mathbf{v})\rangle = \langle U(T(\mathbf{u})), U(T(\mathbf{v}))\rangle$$
$$= \langle T(\mathbf{u}), T(\mathbf{v})\rangle$$
$$= \langle \mathbf{u}, \mathbf{v}\rangle,$$

and hence UT is a linear isometry. Therefore V is isomorphic to Z.

83. First, note that Φ_B is an isomorphism by the boxed result on page 514. Furthermore, by Exercise 71, for all $\mathbf{u}$ and $\mathbf{v}$ in V we have

$$\Phi_B(\mathbf{u}) \cdot \Phi_B(\mathbf{v}) = [\mathbf{u}]_B \cdot [\mathbf{u}]_B = \langle \mathbf{u}, \mathbf{v}\rangle,$$

and hence Φ_B is a linear isometry.

84. Let $\mathbf{v}$ be a vector in V. Then

$$\mathbf{v} = a_1\mathbf{w}_1 + a_2\mathbf{w}_2 + \cdots + a_n\mathbf{w}_n$$

for some scalars $a_1, a_2, \ldots, a_n$. Hence, for any i,

$$\langle \mathbf{v}, \mathbf{w}_i\rangle = \langle a_1\mathbf{w}_1 + \cdots + a_i\mathbf{w}_i + \cdots + a_n\mathbf{w}_n, \mathbf{w}_i\rangle$$
$$= a_1\langle \mathbf{w}_1, \mathbf{w}_i\rangle + \cdots + a_i\langle \mathbf{w}_i, \mathbf{w}_i\rangle + \cdots$$
$$\quad + a_n\langle \mathbf{w}_n, \mathbf{w}_i\rangle$$
$$= a_i.$$

The result now follows.

85. Let $\mathbf{u}$ and $\mathbf{v}$ be in W. Then by Exercise 84,

$$\mathbf{u} + \mathbf{v} = \langle \mathbf{u}+\mathbf{v}, \mathbf{w}_1\rangle \mathbf{w}_1 + \langle \mathbf{u}+\mathbf{v}, \mathbf{w}_2\rangle \mathbf{w}_2$$
$$+ \cdots + \langle \mathbf{u}+\mathbf{v}, \mathbf{w}_n\rangle \mathbf{w}_n$$
$$= (\langle \mathbf{u}, \mathbf{w}_1\rangle + \langle \mathbf{v}, \mathbf{w}_1\rangle)\mathbf{w}_1 + \cdots$$
$$+ (\langle \mathbf{u}, \mathbf{w}_n\rangle + \langle \mathbf{v}, \mathbf{w}_n\rangle)\mathbf{w}_n.$$

86. Suppose that $\mathbf{u} = a_1\mathbf{w}_1 + a_2\mathbf{w}_2 + \cdots + a_n\mathbf{w}_n$ and $\mathbf{v} = b_1\mathbf{w}_1 + b_2\mathbf{w}_2 + \cdots + b_n\mathbf{w}_n$. Then

$$\langle \mathbf{u}, \mathbf{v}\rangle = \langle a_1\mathbf{w}_1 + a_2\mathbf{w}_2 + \cdots + a_n\mathbf{w}_n, b_1\mathbf{w}_1$$
$$+ b_2\mathbf{w}_2 + \cdots + b_n\mathbf{w}_n\rangle$$
$$= a_1b_1 + a_2b_2 + \cdots + a_nb_n$$
$$= \langle \mathbf{u}, \mathbf{w}_1\rangle \langle \mathbf{v}, \mathbf{w}_1\rangle + \langle \mathbf{u}, \mathbf{w}_2\rangle \langle \mathbf{v}, \mathbf{w}_2\rangle$$
$$+ \cdots + \langle \mathbf{u}, \mathbf{w}_n\rangle \langle \mathbf{v}, \mathbf{w}_n\rangle$$

by Exercise 84.

87. Since $\left\{\frac{1}{\sqrt{n}}I_n\right\}$ is an orthonormal basis for W, the orthogonal projection of A on W is given by

$$\left\langle A, \frac{1}{\sqrt{n}}I_n\right\rangle\left(\frac{1}{\sqrt{n}}I_n\right) = \frac{1}{n}\langle A, I_n\rangle I_n$$
$$= \frac{1}{n}\operatorname{trace}(AI_n^T)I_n$$
$$= \left(\frac{\operatorname{trace} A}{n}\right)I_n.$$

88. (rounded to 4 places after the decimal)

$$\begin{bmatrix} 13.2353 & 12.7059 & 12.1765 & 11.6471 & 11.1176 \\ 10.5882 & 10.0588 & 9.5294 & 9.0000 & 8.4706 \\ 7.9412 & 7.4118 & 6.8824 & 6.3529 & 5.8235 \\ 5.2941 & 4.7647 & 4.2353 & 3.7059 & 3.1765 \\ 2.6471 & 2.1176 & 1.5882 & 1.0588 & 0.5294 \end{bmatrix}$$

CHAPTER 7 REVIEW EXERCISES

1. False, for example, $\mathbf{C}^\infty$ is not a subset of $\mathcal{R}^n$ for any n.
2. True
3. False, the dimension is mn.
4. False, it is an $mn \times mn$ matrix.
5. True
6. False, for example, let $\mathbf{u}$ and $\mathbf{w}$ be any vectors in an inner product space that are not orthogonal, and let $\mathbf{v} = \mathbf{0}$.
7. True
8. Yes. We verify a few of the axioms of a vector space.

    Axiom 1 For any sequences $\{a_n\}$ and $\{b_n\}$,
    $$\{a_n\} + \{b_n\} = \{a_n + b_n\} = \{b_n + a_n\}$$
    $$= \{b_n\} + \{a_n\}.$$

    Axiom 8 For any sequence $\{a_n\}$ and any scalars b and c,
    $$(b+c)\{a_n\} = \{(b+c)a_n\} = \{ba_n + ca_n\}$$
    $$= \{ba_n\} + \{ca_n\}$$
    $$= b\{a_n\} + c\{a_n\}.$$

9. No. First observe that $a \oplus 0 = a + 0 + a \cdot 0 = a$ for all a in V, and hence 0 is the (necessarily unique) additive identity for V. However, for any v in V,

    $$(-1) \oplus v = (-1) + v + (-1)v = -1 \neq 0,$$

    and hence -1 has no additive inverse, so that axiom 4 fails.

10. No. $0 \odot \begin{bmatrix} 0 & 0 \\ 1 & 1 \end{bmatrix} = \begin{bmatrix} 0 & 0 \\ 1 & 1 \end{bmatrix} \neq \begin{bmatrix} 0 & 0 \\ 0 & 0 \end{bmatrix}$, contrary to Theorem 7.2(e).

11. Yes. We verify a few of the axioms of a vector space.

    Axiom 2 Let f, g, and h be in V. Then for any x in $\mathcal{R}$,
    $$\begin{aligned}((f \oplus g) \oplus h)(x) &= (f \oplus g)(x)h(x) \\ &= (f(x)g(x))h(x) \\ &= f(x)(g(x)h(x)) \\ &= f(x)(g \oplus h)(x) \\ &= (f \oplus (g \oplus h))(x),\end{aligned}$$
    and therefore $(f \oplus g) \oplus h = f \oplus (g \oplus h)$.

    Axiom 7 Let f and g be in V, and let a be a scalar. Then for any x in $\mathcal{R}$,
    $$\begin{aligned}(a \odot (f \oplus g))(x) &= ((f \oplus g)(x))^a \\ &= (f(x)g(x))^a \\ &= f(x)^a g(x)^a \\ &= (a \odot f)(x)(a \odot g)(x) \\ &= ((a \odot f) \oplus (a \odot g))(x),\end{aligned}$$
    and hence $a \odot (f \oplus g) = (a \odot f) \oplus (a \odot g)$.

12. No. Let f be defined by $f(x) = 1$ for all x. Then f is in V, but $(-1)f$ is not in V because $((-1)f)(x) = (-1)f(x) = (-1)(1) = -1 < 0$ for all x.

13. No. V is not closed under addition. For example, although $x - x^2$ and $x + x^2$ are in V, the sum $(x - x^2) + (x + x^2) = 2x$ is not in V.

14. Yes. Observe that O is in W because $O\mathbf{v} = \mathbf{0} = 0\mathbf{v}$, and hence $\mathbf{v}$ is an eigenvector of O with corresponding eigenvalue 0. Now suppose that A and B are in W. Then there exist scalars λ and μ such that $A\mathbf{v} = \lambda\mathbf{v}$ and $B\mathbf{v} = \mu\mathbf{v}$. Hence
    $$(A + B)\mathbf{v} = A\mathbf{v} + B\mathbf{v} = \lambda\mathbf{v} + \mu\mathbf{v} = (\lambda + \mu)\mathbf{v},$$
    and so $\mathbf{v}$ is an eigenvector of $A + B$ with corresponding eigenvalue $\lambda + \mu$. Thus $A + B$ is in W, and so W is closed under addition. Similarly, W is closed under scalar multiplication, and therefore W is a subspace of V.

15. No. Since $\lambda \neq 0$, it follows that λ is not an eigenvalue of O, and hence O is not in W. Therefore W is not a subspace of V.

16. Yes.
    $$\begin{bmatrix} 1 & 10 \\ 9 & -1 \end{bmatrix} = 2\begin{bmatrix} 1 & 2 \\ 1 & -1 \end{bmatrix} + 3\begin{bmatrix} 0 & 1 \\ 2 & 0 \end{bmatrix} + 1\begin{bmatrix} -1 & 3 \\ 1 & 1 \end{bmatrix}.$$

17. No. Consider the matrix equation
    $$x_1\begin{bmatrix} 1 & 2 \\ 1 & -1 \end{bmatrix} + x_2\begin{bmatrix} 0 & 1 \\ 2 & 0 \end{bmatrix} + x_3\begin{bmatrix} -1 & 3 \\ 1 & 1 \end{bmatrix} = \begin{bmatrix} 2 & 8 \\ 1 & -5 \end{bmatrix}.$$
    Comparing the $(1,1)$-entries of both sides of this equation, we obtain $x_1 - x_3 = 2$. Comparing the $(2,2)$-entries of both sides of this equation, we obtain $-x_1 + x_3 = 5$, which is equivalent to $x_1 - x_3 = -5$. Since the two equations are inconsistent, the matrix equation has no solution, and hence the given matrix is not a linear combination of the matrices in the given set.

18. no

19. Yes. Consider the matrix equation
    $$x_1\begin{bmatrix} 1 & 2 \\ 1 & -1 \end{bmatrix} + x_2\begin{bmatrix} 0 & 1 \\ 2 & 0 \end{bmatrix} + x_3\begin{bmatrix} -1 & 3 \\ 1 & 1 \end{bmatrix} = \begin{bmatrix} 4 & 1 \\ -2 & -4 \end{bmatrix}.$$
    Comparing the corresponding entries on both sides of this equation, we obtain the system
    $$\begin{aligned}x_1 \quad\quad\quad - x_3 &= 4 \\ 2x_1 + x_2 + 3x_3 &= 1 \\ x_1 + 2x_2 + x_3 &= -2 \\ -x_1 \quad\quad\quad + x_3 &= -4,\end{aligned}$$
    whose coefficient matrix has the reduced row echelon form
    $$\begin{bmatrix} 1 & 0 & 0 & 3 \\ 0 & 1 & 0 & -2 \\ 0 & 0 & 1 & -1 \\ 0 & 0 & 0 & 0 \end{bmatrix}.$$
    Therefore the system has the solution $x_1 = 3$, $x_2 = -2$, and $x_3 = -1$. These are the coefficients of the linear combination that produces the given matrix.

20. Observe that the polynomial equation
    $$a(x^3 - x^2 + x + 1) + b(3x^2 + x + 2) + c(x - 1) = x^3 + 2x^2 + 5$$
    can be rewritten as
    $$ax^3 + (-a + 3b)x^2 + (a + b + c)x + (a + 2b - c) = x^3 + 2x^2 + 5,$$
    which is equivalent to the system of linear equations
    $$\begin{aligned}a \quad\quad\quad\quad &= 1 \\ -a + 3b \quad\quad &= 2 \\ a + b + c &= 0 \\ a + 2b - c &= 5.\end{aligned}$$
    Since this system has the solution $a = 1$, $b = 1$, $c = -2$, it follows that
    $$x^3 + 2x^2 + 5 = 1(x^3 - x^2 + x + 1) + 1(3x^2 + x + 2) + (-2)(x - 1).$$

21. $c = 5$

22. A matrix $A = \begin{bmatrix} x_1 & x_2 \\ x_3 & x_4 \end{bmatrix}$ is in W if and only if

$$\begin{bmatrix} 1 & 2 \\ 1 & 2 \end{bmatrix} \begin{bmatrix} x_1 & x_2 \\ x_3 & x_4 \end{bmatrix} = \begin{bmatrix} x_1 + 2x_3 & x_2 + 2x_4 \\ x_1 + 2x_3 & x_2 + 2x_4 \end{bmatrix}$$
$$= \begin{bmatrix} 0 & 0 \\ 0 & 0 \end{bmatrix},$$

and hence if and only if $x_1 = -2x_3$ and $x_2 = -2x_4$. Thus A is in W if and only if it has the form

$$A = \begin{bmatrix} -2x_3 & -2x_4 \\ x_3 & x_4 \end{bmatrix} = x_3 \begin{bmatrix} -2 & 0 \\ 1 & 0 \end{bmatrix} + x_4 \begin{bmatrix} 0 & -2 \\ 0 & 1 \end{bmatrix}.$$

It follows that

$$\left\{ \begin{bmatrix} -2 & 0 \\ 1 & 0 \end{bmatrix}, \begin{bmatrix} 0 & -2 \\ 0 & 1 \end{bmatrix} \right\}$$

is a generating set for W. Clearly, this set is linearly independent and hence is a basis for W. Thus $\dim W = 2$.

23. A polynomial $f(x) = a + bx + cx^2 + dx^3$ is in W if and only if

$$f(0) + f'(0) + f''(0) = a + b + 2c = 0,$$

that is,

$$a = -b - 2c.$$

So $f(x)$ is in W if and only if

$$f(x) = (-b - 2c) + bx + cx^2 + dx^3$$
$$= b(-1 - x) + c(-2 + x^2) + dx^3.$$

It follows that W is the span of

$$\{-1 + x, -2 + x^2, x^3\}.$$

Since this set is linearly independent, it is a basis for W. Therefore $\dim W = 3$.

24. T is both linear and an isomorphism. Let $\mathbf{u}_1 = \begin{bmatrix} a_1 \\ b_1 \\ c_1 \end{bmatrix}$ and $\mathbf{u}_2 = \begin{bmatrix} a_2 \\ b_2 \\ c_2 \end{bmatrix}$ be in $\mathcal{R}^3$. Then

$$T\left(\begin{bmatrix} a_1 \\ b_1 \\ c_1 \end{bmatrix} + \begin{bmatrix} a_2 \\ b_2 \\ c_2 \end{bmatrix}\right) = T\left(\begin{bmatrix} a_1 + a_2 \\ b_1 + b_2 \\ c_1 + c_2 \end{bmatrix}\right)$$

$$= ((a_1 + a_2) + (b_1 + b_2)) + ((a_1 + a_2) - (b_1 + b_2))x + (c_1 + c_2)x^2$$

$$= (a_1 + b_1) + (a_1 - b_1)x + c_1 x^2 + (a_2 + b_2) + (a_2 - b_2)x + c_2 x^2$$

$$= T\left(\begin{bmatrix} a_1 \\ b_1 \\ c_1 \end{bmatrix}\right) + T\left(\begin{bmatrix} a_2 \\ b_2 \\ c_2 \end{bmatrix}\right).$$

Similarly, $T(s\mathbf{u}) = sT(\mathbf{u})$ for every vector $\mathbf{u}$ in $\mathcal{R}^3$ and every scalar s. Hence T is linear.

To show that T is one-to-one, observe that if $T\left(\begin{bmatrix} a \\ b \\ c \end{bmatrix}\right)$ is the zero polynomial, then we have $a + b = 0$, $a - b = 0$, and $c = 0$, which implies that $\begin{bmatrix} a \\ b \\ c \end{bmatrix} = \mathbf{0}$.

To show that T is onto, consider any polynomial $p + qx + rx^2$. Set $a = \frac{1}{2}(p+q)$, $b = \frac{1}{2}(p-q)$, and $c = r$. Then $T\left(\begin{bmatrix} a \\ b \\ c \end{bmatrix}\right) = p + qx + rx^2$, and hence T is onto.

25. T is not linear. For example, let $A = I_2$. Then $T(2A) = \text{trace}(4I_2) = 8$, but $2T(A) = 2\,\text{trace}(I_2) = 4$.

26. T is linear, but T is not an isomorphism. If T were an isomorphism, then by the comment on page 515, T preserves dimension. However, by Example 7 of Section 7.3, we have $\dim \mathcal{M}_{2 \times 2} = 4$, which does not equal $\dim \mathcal{R}^3 = 3$.

27. T is both linear and an isomorphism. Let $f(x)$ and $g(x)$ be in $\mathcal{P}_2$. Then

$$T(f(x) + g(x)) = \begin{bmatrix} (f+g)(0) \\ (f+g)'(0) \\ \int_0^1 (f+g)(t)\,dt \end{bmatrix}$$

$$= \begin{bmatrix} f(0) + g(0) \\ f'(0) + g'(0) \\ \int_0^1 f(t)\,dt + \int_0^1 g(t)\,dt \end{bmatrix}$$

$$= \begin{bmatrix} f(0) \\ f'(0) \\ \int_0^1 f(t)\,dt \end{bmatrix} + \begin{bmatrix} g(0) \\ g'(0) \\ \int_0^1 g(t)\,dt \end{bmatrix}$$

$$= T(f(x)) + T(g(x)).$$

Thus T preserves addition. Furthermore, for

any scalar c,

$$T(cf(x)) = \begin{bmatrix} (cf)(0) \\ (cf)'(0) \\ \int_0^1 cf(t)\,dt \end{bmatrix} = \begin{bmatrix} cf(0) \\ cf'(0) \\ c\int_0^1 f(t)\,dt \end{bmatrix}$$

$$= c\begin{bmatrix} f(0) \\ f'(0) \\ \int_0^1 f(t)\,dt \end{bmatrix} = cT(f(x)),$$

and hence T preserves scalar multiplication. Therefore T is linear.

To show that T is an isomorphism, it suffices to show that T is one-to-one because the domain and the codomain of T are finite-dimensional vector spaces with the same dimension. Suppose $f(x) = a + bx + cx^2$ is a polynomial in $\mathcal{P}_2$ such that $T(f(x)) = 0$, the zero polynomial. Comparing components in this vector equation, we have

$$f(0) = 0, \ f'(0) = 0, \ \text{and} \ \int_0^1 f(t)\,dt = 0.$$

Since $f(0) = a + b0 + c0^2 = a$, we have $a = 0$. Similarly, we obtain $b = 0$ from the second equation and $c = 0$ from the third equation. Therefore f is the zero polynomial, and so the null space of T is the zero subspace. We conclude that T is one-to-one, and hence T is an isomorphism.

28. Since
$$T(1) = 1, \quad T(x) = 1 + 2x,$$
and
$$T(x^2) = 1 + 4x - 2x^2,$$
it follows that $[T]_\mathcal{B} = \begin{bmatrix} 1 & 1 & 1 \\ 0 & 2 & 4 \\ 0 & 0 & -2 \end{bmatrix}$.

29. $\begin{bmatrix} a & b \\ -b & a \end{bmatrix}$ 30. $\begin{bmatrix} 2 & 2 \\ -4 & -4 \end{bmatrix}$

31. We have

$$T\left(\begin{bmatrix} 1 & 0 \\ 0 & 0 \end{bmatrix}\right) = 2\begin{bmatrix} 1 & 0 \\ 0 & 0 \end{bmatrix} + \begin{bmatrix} 1 & 0 \\ 0 & 0 \end{bmatrix}^T = 3\begin{bmatrix} 1 & 0 \\ 0 & 0 \end{bmatrix},$$

$$T\left(\begin{bmatrix} 0 & 1 \\ 0 & 0 \end{bmatrix}\right) = 2\begin{bmatrix} 0 & 1 \\ 0 & 0 \end{bmatrix} + \begin{bmatrix} 0 & 1 \\ 0 & 0 \end{bmatrix}^T$$
$$= 2\begin{bmatrix} 0 & 1 \\ 0 & 0 \end{bmatrix} + \begin{bmatrix} 0 & 0 \\ 1 & 0 \end{bmatrix},$$

$$T\left(\begin{bmatrix} 0 & 0 \\ 1 & 0 \end{bmatrix}\right) = 2\begin{bmatrix} 0 & 0 \\ 1 & 0 \end{bmatrix} + \begin{bmatrix} 0 & 0 \\ 1 & 0 \end{bmatrix}^T$$
$$= 2\begin{bmatrix} 0 & 0 \\ 1 & 0 \end{bmatrix} + \begin{bmatrix} 0 & 1 \\ 0 & 0 \end{bmatrix},$$

and

$$T\left(\begin{bmatrix} 0 & 0 \\ 0 & 1 \end{bmatrix}\right) = 2\begin{bmatrix} 0 & 0 \\ 0 & 1 \end{bmatrix} + \begin{bmatrix} 0 & 0 \\ 0 & 1 \end{bmatrix}^T = 3\begin{bmatrix} 0 & 0 \\ 0 & 1 \end{bmatrix}.$$

Therefore

$$[T]_\mathcal{B} = \begin{bmatrix} 3 & 0 & 0 & 0 \\ 0 & 2 & 1 & 0 \\ 0 & 1 & 2 & 0 \\ 0 & 0 & 0 & 3 \end{bmatrix}.$$

32. Using the matrix computed in Exercise 28, we have

$$[T^{-1}]_\mathcal{B} = ([T]_\mathcal{B})^{-1} = \begin{bmatrix} 1 & 1 & 1 \\ 0 & 2 & 4 \\ 0 & 0 & -2 \end{bmatrix}^{-1}$$

$$= \begin{bmatrix} 1.0 & -0.5 & -0.5 \\ 0.0 & 0.5 & 1.0 \\ 0.0 & 0.0 & -0.5 \end{bmatrix}.$$

Hence for any polynomial $a + bx + cx^2$ in $\mathcal{P}_2$,

$$[T^{-1}(a+bx+cx^2)]_\mathcal{B} = [T^{-1}]_\mathcal{B}(a+bx+cx^2)_\mathcal{B}$$

$$= \begin{bmatrix} 1.0 & -0.5 & -0.5 \\ 0.0 & 0.5 & 1.0 \\ 0.0 & 0.0 & -0.5 \end{bmatrix} \begin{bmatrix} a \\ b \\ c \end{bmatrix}$$

$$= \begin{bmatrix} a - .5b - .5c \\ .5b + c \\ -.5c \end{bmatrix}.$$

Therefore
$$T^{-1}(a + bx + cx^2) =$$
$$(a - .5b - .5c) + (.5b + c)x - .5cx^2$$

33. $\dfrac{1}{a^2+b^2}(ac_1 - bc_2)e^{at}\cos bt$
$\quad + \dfrac{1}{a^2+b^2}(bc_1 + ac_2)e^{at}\sin bt$

34. $(\tfrac{1}{4}c_1 - \tfrac{1}{8}c_2)e^t \cos t + (\tfrac{1}{4}c_1 + \tfrac{1}{8}c_2)e^t \sin t$

35. Using the matrix computed in Exercise 31, we have

$$[T^{-1}]_\mathcal{B} = \begin{bmatrix} 3 & 0 & 0 & 0 \\ 0 & 2 & 1 & 0 \\ 0 & 1 & 2 & 0 \\ 0 & 0 & 0 & 3 \end{bmatrix}^{-1} = \frac{1}{3}\begin{bmatrix} 1 & 0 & 0 & 0 \\ 0 & 2 & -1 & 0 \\ 0 & -1 & 2 & 0 \\ 0 & 0 & 0 & 1 \end{bmatrix}.$$

Hence for any matrix $\begin{bmatrix} a & b \\ c & d \end{bmatrix}$ in $\mathcal{M}_{2\times 2}$,

$$\left[T^{-1}\left(\begin{bmatrix} a & b \\ c & d \end{bmatrix}\right)\right]_\mathcal{B} = [T^{-1}]_\mathcal{B} \left[\begin{bmatrix} a & b \\ c & d \end{bmatrix}\right]_\mathcal{B}$$

$$= \frac{1}{3}\begin{bmatrix} 1 & 0 & 0 & 0 \\ 0 & 2 & -1 & 0 \\ 0 & -1 & 2 & 0 \\ 0 & 0 & 0 & 1 \end{bmatrix} \begin{bmatrix} a \\ b \\ c \\ d \end{bmatrix}$$

$$= \frac{1}{3}\begin{bmatrix} a \\ 2b-c \\ -b+2c \\ d \end{bmatrix}.$$

Therefore

$$T^{-1}\left(\begin{bmatrix} a & b \\ c & d \end{bmatrix}\right) = \frac{1}{3}\begin{bmatrix} a & 2b-c \\ -b+2c & d \end{bmatrix}.$$

36. Let $A = [T]_\mathcal{B}$, where $\mathcal{B} = \{1, x, x^2\}$. By Exercise 28,

$$A = \begin{bmatrix} 1 & 1 & 1 \\ 0 & 2 & 4 \\ 0 & 0 & -2 \end{bmatrix}.$$

Since A is an upper triangular matrix, its eigenvalues are the diagonal entries, 1, 2, and -2.

The eigenspace of A corresponding to the eigenvalue 1 is the solution space of the system of linear equations

$$(A - I_3)\mathbf{x} = \begin{bmatrix} 0 & 0 & 0 \\ 0 & 1 & 4 \\ 0 & 0 & -3 \end{bmatrix}\begin{bmatrix} x_1 \\ x_2 \\ x_3 \end{bmatrix}$$

$$= \begin{bmatrix} 0 \\ x_2 + 4x_4 \\ -3x_3 \end{bmatrix} = \begin{bmatrix} 0 \\ 0 \\ 0 \end{bmatrix}.$$

Hence $x_2 = x_3 = 0$, and the vector form of the general solution is $\begin{bmatrix} x_1 \\ 0 \\ 0 \end{bmatrix} = x_1 \begin{bmatrix} 1 \\ 0 \\ 0 \end{bmatrix}$. It follows that $\{\mathbf{e}_1\}$ is a basis for the eigenspace of A corresponding to the eigenvalue 3. Since the constant polynomial 1 is the polynomial in $\mathcal{P}_2$ whose coordinate vector equals $\mathbf{e}_1$, it follows that $\{1\}$ is a basis for the eigenspace of T corresponding to the eigenvalue 1.

Continuing in this manner, it can be shown that $\{1+x\}$ is a basis for the eigenspace of T corresponding to the eigenvalue 2 and $\{x - x^2\}$ is a basis for the eigenspace of T corresponding to the eigenvalue -2.

37. T has no (real) eigenvalues.

38. T has no (real) eigenvalues.

39. Let $A = [T]_\mathcal{B}$. By Exercise 31,

$$A = \begin{bmatrix} 3 & 0 & 0 & 0 \\ 0 & 2 & 1 & 0 \\ 0 & 1 & 2 & 0 \\ 0 & 0 & 0 & 3 \end{bmatrix}.$$

We first find the eigenvalues of A. The characteristic polynomial of A is

$$\det(A - tI_4) = \det\begin{bmatrix} 3-t & 0 & 0 & 0 \\ 0 & 2-t & 1 & 0 \\ 0 & 1 & 2-t & 0 \\ 0 & 0 & 0 & 3-t \end{bmatrix}$$

$$= (t-3)^3(t-1),$$

and hence the eigenvalues of A are 3 and 1.

Next, we find a basis for the eigenspace of A corresponding to the eigenvalue 3. This eigenspace is the solution space of the system of equations

$$(A - 3I_4)\mathbf{x} = \begin{bmatrix} 0 & 0 & 0 & 0 \\ 0 & -1 & 1 & 0 \\ 0 & 1 & -1 & 0 \\ 0 & 0 & 0 & 0 \end{bmatrix}\begin{bmatrix} x_1 \\ x_2 \\ x_3 \\ x_4 \end{bmatrix} = \begin{bmatrix} 0 \\ 0 \\ 0 \\ 0 \end{bmatrix}.$$

Hence $x_2 = x_3$, and the vector form of the general solution is

$$\begin{bmatrix} x_1 \\ x_2 \\ x_3 \\ x_4 \end{bmatrix} = x_1\begin{bmatrix} 1 \\ 0 \\ 0 \\ 0 \end{bmatrix} + x_2\begin{bmatrix} 0 \\ 1 \\ 1 \\ 0 \end{bmatrix} + x_4\begin{bmatrix} 0 \\ 0 \\ 0 \\ 1 \end{bmatrix}.$$

Thus

$$\left\{\begin{bmatrix} 1 \\ 0 \\ 0 \\ 0 \end{bmatrix}, \begin{bmatrix} 0 \\ 1 \\ 1 \\ 0 \end{bmatrix}, \begin{bmatrix} 0 \\ 0 \\ 0 \\ 1 \end{bmatrix}\right\}$$

is a basis for the eigenspace of A corresponding to the eigenvalue 3. Let A_1, A_2, and A_3 be the matrices whose coordinate vectors relative to $\mathcal{B}$ are the vectors in the basis above. Then

$$\{A_1, A_2, A_3\} = \left\{\begin{bmatrix} 1 & 0 \\ 0 & 0 \end{bmatrix}, \begin{bmatrix} 0 & 1 \\ 1 & 0 \end{bmatrix}, \begin{bmatrix} 0 & 0 \\ 0 & 1 \end{bmatrix}\right\}$$

is a basis for the eigenspace of T corresponding to the eigenvalue 3.

Finally, we find a basis for the eigenspace of A corresponding to the eigenvalue 1. This eigenspace is the solution space of the system of equations

$$(A - I_4)\mathbf{x} = \begin{bmatrix} 2 & 0 & 0 & 0 \\ 0 & 1 & 1 & 0 \\ 0 & 1 & 1 & 0 \\ 0 & 0 & 0 & 2 \end{bmatrix} \begin{bmatrix} x_1 \\ x_2 \\ x_3 \\ x_4 \end{bmatrix} = \begin{bmatrix} 0 \\ 0 \\ 0 \\ 0 \end{bmatrix}.$$

Thus $x_1 = 0$, $x_3 = -x_2$, and $x_4 = 0$. So the vector form of the general solution is

$$\begin{bmatrix} x_1 \\ x_2 \\ x_3 \\ x_4 \end{bmatrix} = x_2 \begin{bmatrix} 0 \\ 1 \\ -1 \\ 0 \end{bmatrix}.$$

Since $[A_4] = \begin{bmatrix} 0 & 1 \\ -1 & 0 \end{bmatrix}$ is the matrix whose co-ordinate vector relative to $\mathcal{B}$ is $\begin{bmatrix} 0 \\ 1 \\ -1 \\ 0 \end{bmatrix}$, we see

that $\{A_4\}$ is a basis for the eigenspace of T corresponding to the eigenvalue 1.

40. 2

41. Let $M = \begin{bmatrix} 1 & 3 \\ 4 & 2 \end{bmatrix}$ and Z be the set of all matrices orthogonal to M. A matrix $A = \begin{bmatrix} x_1 & x_2 \\ x_3 & x_4 \end{bmatrix}$ is orthogonal to M if and only if

$$\langle M, A \rangle = \text{trace}(MA^T) = x_1 + 3x_2 + 4x_4 + 2x_4$$
$$= 0,$$

and hence, if and only if $x_1 = -(3x_2 + 4x_4 + 2x_4)$. So the matrix A is in Z if and only if

$$A = \begin{bmatrix} -3x_2 - 4x_4 - 2x_4 & x_2 \\ x_3 & x_4 \end{bmatrix}$$

$$= x_2 \begin{bmatrix} -3 & 1 \\ 0 & 0 \end{bmatrix} + x_3 \begin{bmatrix} -4 & 0 \\ 1 & 0 \end{bmatrix} x_4 \begin{bmatrix} -2 & 0 \\ 0 & 1 \end{bmatrix}.$$

It follows that the set

$$\left\{ \begin{bmatrix} -3 & 1 \\ 0 & 0 \end{bmatrix}, \begin{bmatrix} -4 & 0 \\ 1 & 0 \end{bmatrix}, \begin{bmatrix} -2 & 0 \\ 0 & 1 \end{bmatrix} \right\}$$

is a generating set for Z. In addition, this set is linearly independent, and so it is a basis for Z.

42. $\left\{ \dfrac{1}{\sqrt{2}} \begin{bmatrix} 1 & 0 \\ 0 & -1 \end{bmatrix}, \begin{bmatrix} 0 & 1 \\ 0 & 0 \end{bmatrix}, \begin{bmatrix} 0 & 0 \\ 1 & 0 \end{bmatrix} \right\}$

43. For any matrix $\begin{bmatrix} a & b \\ c & d \end{bmatrix}$ in $\mathcal{M}_{2\times 2}$,

$$\text{trace}\left(\begin{bmatrix} 0 & 1 \\ 1 & 0 \end{bmatrix} \begin{bmatrix} a & b \\ c & d \end{bmatrix} \right) = \text{trace} \begin{bmatrix} c & d \\ a & b \end{bmatrix} = c + b,$$

and so the matrix is in W if and only if $c = -b$. Thus the matrix is in W if and only if it is of the form

$$\begin{bmatrix} a & b \\ -b & d \end{bmatrix} = a \begin{bmatrix} 1 & 0 \\ 0 & 0 \end{bmatrix} + b \begin{bmatrix} 0 & 1 \\ -1 & 0 \end{bmatrix} + d \begin{bmatrix} 0 & 0 \\ 0 & 1 \end{bmatrix}.$$

Observe that the vectors in this linear combination form an orthogonal set. If we divide each vector by its length, we obtain the orthonormal basis

$$\{M_1, M_2, M_3\} = \left\{ \begin{bmatrix} 1 & 0 \\ 0 & 0 \end{bmatrix}, \frac{1}{\sqrt{2}} \begin{bmatrix} 0 & 1 \\ -1 & 0 \end{bmatrix}, \begin{bmatrix} 0 & 0 \\ 0 & 1 \end{bmatrix} \right\}$$

for W. Thus the orthogonal projection B of $A = \begin{bmatrix} 2 & 5 \\ 9 & -3 \end{bmatrix}$ is

$$B = \langle M_1, A \rangle M_1 + \langle M_2, A \rangle M_2 + \langle M_2, A \rangle M_2$$

$$= \left\langle \begin{bmatrix} 1 & 0 \\ 0 & 0 \end{bmatrix}, \begin{bmatrix} 2 & 5 \\ 9 & -3 \end{bmatrix} \right\rangle \begin{bmatrix} 1 & 0 \\ 0 & 0 \end{bmatrix}$$

$$+ \left\langle \frac{1}{\sqrt{2}} \begin{bmatrix} 0 & 1 \\ -1 & 0 \end{bmatrix}, \begin{bmatrix} 2 & 5 \\ 9 & -3 \end{bmatrix} \right\rangle \frac{1}{\sqrt{2}} \begin{bmatrix} 0 & 1 \\ -1 & 0 \end{bmatrix}$$

$$+ \left\langle \begin{bmatrix} 0 & 0 \\ 0 & 1 \end{bmatrix}, \begin{bmatrix} 2 & 5 \\ 9 & -3 \end{bmatrix} \right\rangle \begin{bmatrix} 0 & 0 \\ 0 & 1 \end{bmatrix}$$

$$= 2 \begin{bmatrix} 1 & 0 \\ 0 & 0 \end{bmatrix} + \frac{-4}{2} \begin{bmatrix} 0 & 1 \\ -1 & 0 \end{bmatrix} + (-3) \begin{bmatrix} 0 & 0 \\ 0 & 1 \end{bmatrix}$$

$$= \begin{bmatrix} 2 & -2 \\ 2 & -3 \end{bmatrix}.$$

44. We have

$$\langle f, g \rangle = \int_0^1 (\cos 2\pi t)(\sin 2\pi t)\, dt$$

$$= \frac{1}{4\pi} \sin^2(2\pi t) \Big|_0^{2\pi} = 0.$$

45. We apply the Gram-Schmidt process to the basis $\{\mathbf{u}_1, \mathbf{u}_2, \mathbf{u}_3\} = \{1, x, x^2\}$ to obtain an orthogonal basis for W. Let $\mathbf{v}_1 = \mathbf{u}_1 = 1$,

$$\mathbf{v}_2 = \mathbf{u}_2 - \frac{\langle \mathbf{u}_2, \mathbf{v}_1 \rangle}{\|\mathbf{v}_1\|^2} \mathbf{v}_1$$

$$= x - \frac{\int_0^1 x\, dx}{\int_0^1 1^2\, dx} 1 = x - \frac{1}{2},$$

and

$$\mathbf{v}_3 = \mathbf{u}_3 - \frac{\langle \mathbf{u}_3, \mathbf{v}_1 \rangle}{\|\mathbf{v}_1\|^2} \mathbf{v}_1 - \frac{\langle \mathbf{u}_3, \mathbf{v}_2 \rangle}{\|\mathbf{v}_2\|^2} \mathbf{v}_2$$

$$= x^2 - \frac{\int_0^1 x^2\, dx}{\int_1^0 1^2\, dx} 1$$

$$- \frac{\int_0^1 x^2(x - \frac{1}{2})\, dx}{\int_0^1 (x - \frac{1}{2})^2}\left(x - \frac{1}{2}\right)$$

$$= x^2 - \frac{1}{3} - \left(x - \frac{1}{2}\right) = x^2 - x + \frac{1}{6}.$$

Next, we divide each $\mathbf{v}_i$ by its norm to obtain the orthonormal basis

$$\{\mathbf{w}_1, \mathbf{w}_2, \mathbf{w}_3\} =$$
$$\left\{1, \sqrt{3}(2x - 1), \sqrt{5}(6x^2 - 6x + 1)\right\}$$

for W.

46. Since $f(t) = t$ is in W, it is equal to its orthogonal projection on W.

47. We use the orthonormal basis $\{\mathbf{w}_1, \mathbf{w}_2, \mathbf{w}_3\}$ from Exercise 45 to obtain the desired orthogonal projection. Let $\mathbf{w}$ denote the function $\mathbf{w}(x) = \sqrt{x}$. Then

$$\mathbf{w} = \langle \mathbf{w}, \mathbf{w}_1 \rangle \mathbf{w}_1 + \langle \mathbf{w}, \mathbf{w}_2 \rangle \mathbf{w}_2 + \langle \mathbf{w}, \mathbf{w}_3 \rangle \mathbf{w}_3$$

$$= \left(\int_0^1 1\sqrt{x}\, dx\right) 1$$

$$+ \left(\int_0^1 \sqrt{3}(2x-1)\sqrt{x}\, dx\right)\sqrt{3}(2x-1)$$

$$+ \left(\int_0^1 \sqrt{5}(6x^2 - 6x + 1)\sqrt{x}\, dx\right)\sqrt{5}(6x^2 - 6x + 1)$$

$$= \tfrac{2}{3} + \tfrac{2}{5}(2x-1) + \tfrac{(-2)}{21}(6x^2 - 6x + 1)$$

$$= \tfrac{6}{35} + \tfrac{48}{35}x - \tfrac{4}{7}x^2$$

48. We have

$$\langle A, I_n \rangle = \text{trace}\,(AI_n^T) = \text{trace}\,(AI_n) = \text{trace}\, A.$$

49. First observe that $T(\mathbf{0}) = \mathbf{0}$, and hence $\mathbf{0}$ is in $T(W)$. Suppose that $\mathbf{w}_1$ and $\mathbf{w}_2$ are in $T(W)$. Then there exist vectors $\mathbf{u}_1$ and $\mathbf{u}_2$ in W such that $T(\mathbf{u}_1) = \mathbf{w}_1$ and $T(\mathbf{u}_2) = \mathbf{w}_2$, and so

$$T(\mathbf{u}_1 + \mathbf{u}_2) = T(\mathbf{u}_1) + T(\mathbf{u}_2) = \mathbf{w}_1 = \mathbf{w}_2.$$

Thus $\mathbf{w}_1 + \mathbf{w}_2$ is in $T(W)$, and therefore $T(W)$ is closed under vector addition.

Next, suppose that $\mathbf{w}$ is in $T(W)$ and c is any scalar. Then there exists a vector $\mathbf{u}$ in W such that $T(\mathbf{u}) = \mathbf{w}$. It follows that

$$T(c\mathbf{u}) = cT(\mathbf{u}) = c\mathbf{w},$$

and hence $c\mathbf{w}$ is in $T(W)$. Thus $T(W)$ is closed under scalar multiplication, and therefore $T(W)$ is a subspace of V_2.

50. Since $\mathbf{p}$ is the orthogonal projection of $\mathbf{v}$ on W, it is the unique vector in W of minimum distance to $\mathbf{v}$. Consider any vector $\mathbf{z}$ in Z such that $\mathbf{z} \neq T(\mathbf{p})$. There exists a vector $\mathbf{w}$ in W such that $\mathbf{w} \neq \mathbf{p}$ and $T(\mathbf{w}) = \mathbf{z}$. Hence

$$\|\mathbf{z} - T(\mathbf{v})\| = \|T(\mathbf{w}) - T(\mathbf{v})\| = \|\mathbf{w} - \mathbf{v}\|$$
$$> \|\mathbf{p} - \mathbf{v}\| = \|T(\mathbf{p}) - T(\mathbf{v})\|.$$

It follows that $T(\mathbf{p})$ is the unique vector in Z of minimum distance to $T(\mathbf{v})$, and hence $T(\mathbf{p})$ is the orthogonal projection of $T(\mathbf{v})$ on Z. Thus $T(\mathbf{p}) = P_Z T(\mathbf{v})$, and so it follows that $\mathbf{p} = T^{-1}(P_Z T(\mathbf{v}))$.

CHAPTER 7 MATLAB EXERCISES

1. The set is linearly independent.
2. The set is linearly dependent.

$$\begin{bmatrix} 1 & -3 \\ 4 & 1 \end{bmatrix} = 2\begin{bmatrix} 1 & -1 \\ 3 & 1 \end{bmatrix} + (-1)\begin{bmatrix} 1 & 2 \\ 1 & 2 \end{bmatrix} + (1)\begin{bmatrix} 0 & 1 \\ -1 & 1 \end{bmatrix}$$

3. (a) Define $U: \mathcal{P}_n \to R$ by

$$U(f(x)) = f(-1) + f(-2)$$

for every polynomial $f(x)$ in $\mathcal{P}_n$. It is easily shown that U is linear. Hence, by Exercise 76 of Section 7.3, there exist unique scalars $c_0, c_0 \ldots, c_n$ such that

$$U = c_0 T_0 + c_1 T_1 + \cdots + c_n t_n,$$

where T_i is defined in Exercise 76. So, for every polynomial $f(x)$ in $\mathcal{P}_n$, we have

$$f(-1) + f(-2)$$
$$= U(f(x))$$
$$= (c_0 T_0 + c_1 T_1 + \cdots + c_n t_n)(f(x))$$
$$= c_0 T_0(f(x)) + c_1 T_1(f(x)) + \cdots$$
$$+ c_n T_n(f(x))$$
$$= c_0 f(0) + c_1 f(1) + \cdots + c_n f(n).$$

(b) $c_0 = 20$, $c_1 = -50$, $c_2 = 55$, $c_3 = -29$, $c_4 = 6$

4. Let A be the standard matrix of T. Then

$$A = \begin{bmatrix} 1 & -3 & -3 & 2 & 2 & 0 \\ 3 & 1 & -2 & -3 & 0 & 2 \\ 0 & 0 & 1 & -3 & -6 & -4 \\ 0 & 0 & 3 & 1 & -4 & -6 \\ 0 & 0 & 0 & 0 & 1 & -3 \\ 0 & 0 & 0 & 0 & 3 & 1 \end{bmatrix}.$$

(a) Using appropriate software, it can be shown that A has rank 4, and hence is invertible. Therefore, T is invertible.

(b) Observe that $[t^2 \sin t]_\mathcal{B} = \mathbf{e}_6$, and hence

$$[T^{-1}(t^2 \sin t)]_\mathcal{B} = A^{-1}\mathbf{e}_6 = \begin{bmatrix} 0.324 \\ -0.532 \\ 0.680 \\ -0.240 \\ 0.300 \\ 0.100 \end{bmatrix}.$$

It follows that

$$\begin{aligned} T^{-1}(t^2 \sin t) &= 0.324 \cos t - 0.532 \sin t \\ &+ 0.680 t \cos t - 0.240 t \sin t \\ &+ 0.300 t^2 \cos t + 0.100 t^3 \sin t. \end{aligned}$$

5. (a) $8, 4, -4, -8$

(b) $\left\{ \begin{bmatrix} -3 & 3 & 3 \\ -1 & 1 & 1 \end{bmatrix}, \begin{bmatrix} 3 & 0 & 3 \\ 0 & 1 & 0 \end{bmatrix}, \begin{bmatrix} 2 & 1 & 2 \\ 1 & 0 & 1 \end{bmatrix}, \right.$

$\left. \begin{bmatrix} -3 & 2 & -3 \\ 0 & 1 & 0 \end{bmatrix}, \begin{bmatrix} 0 & -1 & 0 \\ 1 & 0 & 1 \end{bmatrix}, \right.$

$\left. \begin{bmatrix} 1 & -1 & -1 \\ -1 & 1 & 1 \end{bmatrix} \right\}$

The corresponding eigenvalues for these matrices are 8, 4, 4, -4, -4, and -8, respectively.

6. (i) $B = \begin{bmatrix} 1 & 1 & 1 & 0 & 0 & 0 & 0 & 0 & 0 \\ 0 & 0 & 0 & 1 & 1 & 1 & 0 & 0 & 0 \\ 0 & 0 & 0 & 0 & 0 & 0 & 1 & 1 & 1 \\ 1 & 0 & 0 & 1 & 0 & 0 & 1 & 0 & 0 \\ 0 & 1 & 0 & 0 & 1 & 0 & 0 & 1 & 0 \\ 1 & 0 & 0 & 0 & 1 & 0 & 0 & 0 & 1 \\ 0 & 0 & 1 & 0 & 1 & 0 & 1 & 0 & 0 \end{bmatrix}$

(ii) $P_Z = I_9 - B^T(BB^T)^{-1}B = [S_1 \ S_2]$, where

$$S_1 = \frac{1}{6} \begin{bmatrix} 1 & -1 & 0 & -1 & 0 \\ -1 & 2 & -1 & 0 & 0 \\ 0 & -1 & 1 & 1 & 0 \\ -1 & 0 & 1 & 2 & 0 \\ 0 & 0 & 0 & 0 & 0 \\ 1 & 0 & -1 & -2 & 0 \\ 0 & 1 & -1 & -1 & 0 \\ 1 & -2 & 1 & 0 & 0 \\ -1 & 1 & 0 & 1 & 0 \end{bmatrix}$$

and

$$S_2 = \frac{1}{6} \begin{bmatrix} 1 & 0 & 1 & -1 \\ 0 & 1 & -2 & 1 \\ -1 & -1 & 1 & 0 \\ -2 & -1 & 0 & 1 \\ 0 & 0 & 0 & 0 \\ 2 & 1 & 0 & -1 \\ 1 & 1 & -1 & 0 \\ 0 & -1 & 2 & -1 \\ -1 & 0 & -1 & 1 \end{bmatrix}$$

(iii) Observe that

$$C_3 = \frac{1}{3} \begin{bmatrix} 1 & 1 & 1 \\ 1 & 1 & 1 \\ 1 & 1 & 1 \end{bmatrix},$$

and hence for any 3×3 matrix A, the Frobenius inner product of C_3 and A is $\frac{1}{3}$ times the sum of all of the entries of A. If A is in W_3, then each row of A sums to 0 and therefore the sum of all of the entries of A is equal to 0. It follows that C_3 is orthogonal to every vector in W_3.

We have $C_3 C_3^T = \frac{1}{9} \begin{bmatrix} 3 & 3 & 3 \\ 3 & 3 & 3 \\ 3 & 3 & 3 \end{bmatrix}$, and hence $\|C_3\|^2 = \text{trace}(C_3 C_3^T) = \frac{1}{9} \cdot 9 = 1$. Thus $\|C_3\| = 1$.

(iv) $P = \frac{1}{18}[P_1 \ P_2]$, where

$$P_1 = \begin{bmatrix} 5 & -1 & 2 & -1 & 2 \\ -1 & 8 & -1 & 2 & 2 \\ 2 & -1 & 5 & 5 & 2 \\ -1 & 2 & 5 & 8 & 2 \\ 2 & 2 & 2 & 2 & 2 \\ 5 & 2 & -1 & -4 & 2 \\ 2 & 5 & -1 & -1 & 2 \\ 5 & -4 & 5 & 2 & 2 \\ -1 & 5 & 2 & 5 & 2 \end{bmatrix}$$

and

$$P_2 = \begin{bmatrix} 5 & 2 & 5 & -1 \\ 2 & 5 & -4 & 5 \\ -1 & -1 & 5 & 2 \\ -4 & -1 & 2 & 5 \\ 2 & 2 & 2 & 2 \\ 8 & 5 & 2 & -1 \\ 5 & 5 & -1 & 2 \\ 2 & -1 & 8 & -1 \\ -1 & 2 & -1 & 5 \end{bmatrix}$$